DATE DUE

Demco, Inc. 38-293

VOLUME FIVE HUNDRED AND FOUR

Methods in ENZYMOLOGY

Imaging and Spectroscopic Analysis of Living Cells

Optical and Spectroscopic Techniques

METHODS IN ENZYMOLOGY

Editors-in-Chief

JOHN N. ABELSON AND MELVIN I. SIMON

Division of Biology
California Institute of Technology
Pasadena, California

Founding Editors

SIDNEY P. COLOWICK AND NATHAN O. KAPLAN

VOLUME FIVE HUNDRED AND FOUR

METHODS IN ENZYMOLOGY

Imaging and Spectroscopic Analysis of Living Cells

Optical and Spectroscopic Techniques

EDITED BY

P. MICHAEL CONN
*Divisions of Reproductive Sciences and Neuroscience (ONPRC)
Departments of Pharmacology and Physiology
Cell and Developmental Biology, and Obstetrics and Gynecology
(OHSU) Beaverton, OR, USA*

AMSTERDAM • BOSTON • HEIDELBERG • LONDON
NEW YORK • OXFORD • PARIS • SAN DIEGO
SAN FRANCISCO • SINGAPORE • SYDNEY • TOKYO
Academic Press is an imprint of Elsevier

Academic Press is an imprint of Elsevier
525 B Street, Suite 1900, San Diego, CA 92101-4495, USA
225 Wyman Street, Waltham, MA 02451, USA
32 Jamestown Road, London NW1 7BY, UK

First edition 2012

Copyright © 2012, Elsevier Inc. All Rights Reserved.

No part of this publication may be reproduced, stored in a retrieval system or transmitted in any form or by any means electronic, mechanical, photocopying, recording or otherwise without the prior written permission of the publisher

Permissions may be sought directly from Elsevier's Science & Technology Rights Department in Oxford, UK: phone (+44) (0) 1865 843830; fax (+44) (0) 1865 853333; email: permissions@elsevier.com. Alternatively you can submit your request online by visiting the Elsevier web site at http://elsevier.com/locate/permissions, and selecting *Obtaining permission to use Elsevier material*

Notice
No responsibility is assumed by the publisher for any injury and/or damage to persons or property as a matter of products liability, negligence or otherwise, or from any use or operation of any methods, products, instructions or ideas contained in the material herein. Because of rapid advances in the medical sciences, in particular, independent verification of diagnoses and drug dosages should be made

For information on all Academic Press publications
visit our website at elsevierdirect.com

ISBN: 978-0-12-391857-4
ISSN: 0076-6879

Printed and bound in United States of America
12 13 14 15 10 9 8 7 6 5 4 3 2 1

**Working together to grow
libraries in developing countries**

www.elsevier.com | www.bookaid.org | www.sabre.org

ELSEVIER   BOOK AID International   Sabre Foundation

Contents

Contributors xi
Preface xvii
Volumes in Series xix

Section I. Techniques 1

1. **Laser-Induced Radiation Microbeam Technology and Simultaneous Real-Time Fluorescence Imaging in Live Cells** 3

    Stanley W. Botchway, Pamela Reynolds, Anthony W. Parker, and Peter O'Neill

    1. Introduction 4
    2. Experimental Procedure 10
    3. Conclusion/Forward Look 25
    Acknowledgments 26
    References 26

2. **A Cell Biologist's Guide to High Resolution Imaging** 29

    Graeme Ball, Richard M. Parton, Russell S. Hamilton, and Ilan Davis

    1. Introduction 30
    2. Physical Limitations on the Resolution of Conventional Microscopy 31
    3. Preparations for High Resolution Fluorescence Imaging 35
    4. High Resolution Image Data Acquisition and Processing 39
    5. Processing 43
    6. Analysis of High Resolution Image Data 47
    7. Conclusions 51
    Acknowledgments 51
    References 51

3. **Applications of Fluorescence Lifetime Spectroscopy and Imaging to Lipid Domains *In Vivo*** 57

    André E. P. Bastos, Silvia Scolari, Martin Stöckl, and Rodrigo F. M. de Almeida

    1. The Challenge of Studying Lipid Domains *In Vivo* Defied by Fluorescence Lifetimes 58

2. Importance of Studies *In Vitro* to Design and Rationalize Studies *In Vivo*   61
3. Labeling Cell Membranes *In Vivo*: Lipophilic Probes Versus GFP-Tagged Membrane Proteins   67
4. Cuvette Lifetimes and FLIM: *In Vivo* Applications   70
Acknowledgments   78
References   79

4. Detecting and Tracking Nonfluorescent Nanoparticle Probes in Live Cells   83

Gufeng Wang and Ning Fang

1. Introduction   84
2. Techniques and Tools   84
3. Biological Applications   90
4. Cytotoxicity of Nanoparticle Probes   100
5. Conclusions and Future Perspective   101
Acknowledgments   102
References   102

5. Fluorescence Lifetime Microscopy of Tumor Cell Invasion, Drug Delivery, and Cytotoxicity   109

Gert-Jan Bakker, Volker Andresen, Robert M. Hoffman, and Peter Friedl

1. Introduction   110
2. Preparation of Mouse Mammary Tumor Cell Spheroids in a Collagen Matrix   112
3. Simultaneous Acquisition of Fluorescence Lifetime and Intensity to Monitor DOX Uptake   113
4. Monitoring Drug Uptake Kinetics by FLIM   119
5. Conclusions and Outlook   123
Acknowledgment   123
References   124

6. Measuring Membrane Protein Dynamics in Neurons Using Fluorescence Recovery after Photobleach   127

Inmaculada M. González-González, Frédéric Jaskolski, Yves Goldberg, Michael C. Ashby, and Jeremy M. Henley

1. Introduction   128
2. Experimental Parameters and Preparation   130
3. Equipment   133
4. Establishing FRAP Conditions   134
5. A Basic FRAP Protocol   136
6. Analysis of FRAP   137

Acknowledgments	145
References	145

7. Fluorescent Speckle Microscopy in Cultured Cells — 147
Marin Barisic, António J. Pereira, and Helder Maiato

1. Introduction	148
2. Technical Challenges of FSM	151
3. Detailed Methodology for FSM in Cultured Cells	153
4. Future Challenges	159
Acknowledgments	159
References	159

8. Green-to-Red Photoconvertible mEosFP-Aided Live Imaging in Plants — 163
Jaideep Mathur, Sarah Griffiths, Kiah Barton, and Martin H. Schattat

1. Introduction	164
2. Expression of mEosFP Fusion Proteins in Plants	167
3. Visualization of mEosFP Probes in Plants	170
4. Uses of mEosFP Probes in Plants	173
5. Post Acquisition Image Processing and Data Creation	177
Acknowledgments	178
References	179

9. Methods for Cell and Particle Tracking — 183
Erik Meijering, Oleh Dzyubachyk, and Ihor Smal

1. Introduction	184
2. Tracking Approaches	185
3. Tracking Tools	187
4. Tracking Measures	192
5. Tips and Tricks	195
Acknowledgments	197
References	197

10. Correlative Light-Electron Microscopy: A Potent Tool for the Imaging of Rare or Unique Cellular and Tissue Events and Structures — 201
Alexander A. Mironov and Galina V. Beznoussenko

1. Introduction	202
2. Observation of Living Cells and Fixation	203
3. Immunolabeling with NANOGOLD	207
4. Enhancement of Sample Contrast, Sample Locating, and Embedding	208

5.	Identification of the Cell of Interest on EPON Blocks	211
6.	Sample Orientation and EM Sectioning from the Very First Section	212
7.	Picking up Serial Sections with the Empty Slot Grid	214
8.	EM Analysis	216
	Acknowledgments	217
	References	218

11. Optical Techniques for Imaging Membrane Domains in Live Cells (Live-Cell Palm of Protein Clustering) — 221

Dylan M. Owen, David Williamson, Astrid Magenau, and Katharina Gaus

1.	Introduction	222
2.	Sample Preparation and Data Acquisition	224
3.	Data Analysis	228
	Acknowledgments	234
	References	234

12. Single Live Cell Topography and Activity Imaging with the Shear-Force-Based Constant-Distance Scanning Electrochemical Microscope — 237

Albert Schulte, Michaela Nebel, and Wolfgang Schuhmann

1.	Introduction	238
2.	Shear-Force-Based Constant-Distance Scanning Electrochemical Microscopy for Live Cell Studies: Apparatus, Probes, and Operation	240
3.	Selected Applications of SF-CD-SECM Live Cell Studies	248
	Acknowledgments	253
	References	254

13. Visualization of TGN-Endosome Trafficking in Mammalian and Drosophila Cells — 255

Satoshi Kametaka and Satoshi Waguri

1.	Introduction	256
2.	Molecular Tools	258
3.	Live-Cell Imaging in Mammalian Cells	260
4.	Live-Cell Imaging in Drosophila Cells	266
5.	Conclusion Remarks	269
	References	269

14. Live Cell Imaging with Chemical Specificity Using Dual Frequency CARS Microscopy — 273

Iestyn Pope, Wolfgang Langbein, Paola Borri, and Peter Watson

1.	Introduction	274
2.	"Noninvasive" Live Cell Imaging	275

3. Experimental Setup	279
4. Maximizing Collection Efficiency for Live Cell Imaging	287
Acknowledgments	289
References	290

15. Imaging Intracellular Protein Dynamics by Spinning Disk Confocal Microscopy — 293

Samantha Stehbens, Hayley Pemble, Lyndsay Murrow, and Torsten Wittmann

1. Introduction	294
2. Instrument Design	298
3. Combination with Other Imaging Techniques	304
4. Specimen Preparation	307
Acknowledgments	312
References	312

Section II. Tools — 315

16. Visualizing Dynamic Activities of Signaling Enzymes Using Genetically Encodable FRET-Based Biosensors: From Designs to Applications — 317

Xin Zhou, Katie J. Herbst-Robinson, and Jin Zhang

1. Introduction	318
2. Generalizable Modular Designs	319
3. FRET-Based Biosensors for Monitoring Signaling Enzymes	324
4. Example: A-kinase Activity Reporter (AKAR)	332
5. Summary and Perspectives	336
Acknowledgments	337
References	337

17. Live-Cell Imaging of Aquaporin-4 Supramolecular Assembly and Diffusion — 341

A. S. Verkman, Andrea Rossi, and Jonathan M. Crane

1. Aquaporin-4 (AQP4) and Orthogonal Arrays of Particles	342
2. Approaches to Image AQP4 and OAPs	343
3. AQP4 Diffusion and OAPs Studied by Quantum Dot Single Particle Tracking	345
4. OAP Dynamics and Structure Studied with GFP-AQP4 Chimeras	347
5. Single-Molecule Analysis Shows AQP4 Heterotetramers	349
6. Photobleaching Reveals Post-Golgi Assembly of OAPs	351
7. Super-Resolution Imaging of AQP4 OAPs	351
References	352

18. **Coiled-Coil Tag–Probe Labeling Methods for Live-Cell Imaging of Membrane Receptors**   355

Yoshiaki Yano, Kenichi Kawano, Kaoru Omae, and Katsumi Matsuzaki

1. Introduction	356
2. Various Principles Used for Tag–Probe Labeling	357
3. Coiled-Coil Tag–Probe Labeling	360
4. Applications	366
Acknowledgments	368
References	368

19. **Monitoring Protein Interactions in Living Cells with Fluorescence Lifetime Imaging Microscopy**   371

Yuansheng Sun, Nicole M. Hays, Ammasi Periasamy, Michael W. Davidson, and Richard N. Day

1. Introduction	372
2. FD FLIM Measurements	376
3. Measuring Protein–Protein Interactions in Living Cells Using FLIM–FRET	385
4. The Strengths and Limitations of FLIM	388
Acknowledgment	389
References	389

20. **Open Source Tools for Fluorescent Imaging**   393

Nicholas A. Hamilton

1. Why Open Source Software?	394
2. Open Source Software for Microscopy Imaging and Analysis	406
3. The Future	414
Acknowledgments	415
References	415

21. **Nanoparticle PEBBLE Sensors in Live Cells**   419

Yong-Eun Koo Lee and Raoul Kopelman

1. Introduction	420
2. PEBBLE Sensor Designs	424
3. Preparation/Characterization	430
4. Examples	431
5. Summary and Critical Issues	461
Acknowledgments	462
References	462

Author Index   *471*
Subject Index   *495*

Contributors

Volker Andresen
Rudolf-Virchow Center for Experimental Biomedicine and Department of Dermatology, Venerology and Allergology, University of Würzburg, Würzburg, Germany; and LaVision BioTec GmbH, Bielefeld, Germany

Michael C. Ashby
MRC Centre for Synaptic Plasticity, School of Biochemistry, Medical Sciences Building and School of Physiology and Pharmacology, University of Bristol, Bristol, United Kingdom

Gert-Jan Bakker
Microscopical Imaging of the Cell, Department of Cell Biology, Nijmegen Center for Molecular Life Science, Radboud University Nijmegen Medical Centre, Nijmegen, The Netherlands

Graeme Ball
Department of Biochemistry, The University of Oxford, Oxford, United Kingdom

Marin Barisic
Chromosome Instability & Dynamics Laboratory, Instituto de Biologia Molecular e Celular, Universidade do Porto, Porto, Portugal

Kiah Barton
Department of Molecular and Cellular Biology, Laboratory of Plant Development and Interactions, University of Guelph, Guelph, Canada

André E.P. Bastos
Centro de Química e Bioquímica, Faculdade de Ciências da Universidade de Lisboa, Lisboa, Portugal

Galina V. Beznoussenko
Istituto FIRC di Oncologia Molecolare, Milan, Italy

Paola Borri
School of Biosciences, Cardiff University, Cardiff, United Kingdom

Stanley W. Botchway
Research Complex at Harwell, Central Laser Facility, STFC, Rutherford Appleton Laboratory, Harwell-Oxford, Didcot, Oxford, Oxfordshire, United Kingdom

Jonathan M. Crane
Departments of Medicine and Physiology, University of California, San Francisco, San Francisco, California, USA

Michael W. Davidson
National High Magnetic Field Laboratory and Department of Biological Science, The Florida State University, Tallahassee, Florida, USA

Ilan Davis
Department of Biochemistry, The University of Oxford, Oxford, United Kingdom

Richard N. Day
Department of Cellular and Integrative Physiology, Indiana University School of Medicine, Indianapolis, Indiana, USA

Rodrigo F.M. de Almeida
Centro de Química e Bioquímica, Faculdade de Ciências da Universidade de Lisboa, Lisboa, Portugal

Oleh Dzyubachyk
Biomedical Imaging Group Rotterdam, Departments of Medical Informatics and Radiology, Erasmus MC—University Medical Center Rotterdam, Rotterdam, The Netherlands

Ning Fang
Ames Laboratory, U.S. Department of Energy and Department of Chemistry, Iowa State University, Ames, Iowa, USA

Peter Friedl
Microscopical Imaging of the Cell, Department of Cell Biology, Nijmegen Center for Molecular Life Science, Radboud University Nijmegen Medical Centre, Nijmegen, The Netherlands; Rudolf-Virchow Center for Experimental Biomedicine and Department of Dermatology, Venerology and Allergology, University of Würzburg, Würzburg, Germany

Katharina Gaus
Centre for Vascular Research, University of New South Wales, Sydney, Australia

Yves Goldberg
MRC Centre for Synaptic Plasticity, School of Biochemistry, Medical Sciences Building, University of Bristol, Bristol, United Kingdom

Inmaculada M. González-González
MRC Centre for Synaptic Plasticity, School of Biochemistry, Medical Sciences Building, University of Bristol, Bristol, United Kingdom

Sarah Griffiths
Department of Molecular and Cellular Biology, Laboratory of Plant Development and Interactions, University of Guelph, Guelph, Canada

Nicholas A. Hamilton
Division of Genomics Computational Biology, Institute for Molecular Bioscience, The University of Queensland, St. Lucia, Brisbane, Queensland, Australia

Russell S. Hamilton
Department of Biochemistry, The University of Oxford, Oxford, United Kingdom

Nicole M. Hays
Department of Cellular and Integrative Physiology, Indiana University School of Medicine, Indianapolis, Indiana, USA

Jeremy M. Henley
MRC Centre for Synaptic Plasticity, School of Biochemistry, Medical Sciences Building, University of Bristol, Bristol, United Kingdom

Katie J. Herbst-Robinson
Department of Pharmacology and Molecular Sciences, The Johns Hopkins University School of Medicine, Baltimore, Maryland, USA

Robert M. Hoffman
AntiCancer, Inc. and Department of Surgery, University of California San Diego, San Diego, California, USA

Frédéric Jaskolski
MRC Centre for Synaptic Plasticity, School of Biochemistry, Medical Sciences Building, University of Bristol, Bristol, United Kingdom

Satoshi Kametaka
Department of Anatomy and Histology, Fukushima Medical University School of Medicine, Fukushima, Japan

Kenichi Kawano
Graduate School of Pharmaceutical Sciences, Kyoto University, Sakyo-ku, Kyoto, Japan

Yong-Eun Koo Lee
Department of Chemistry, University of Michigan, Ann Arbor, Michigan, USA

Raoul Kopelman
Department of Chemistry, University of Michigan, Ann Arbor, Michigan, USA

Wolfgang Langbein
School of Physics and Astronomy, Cardiff University, Cardiff, United Kingdom

Astrid Magenau
Centre for Vascular Research, University of New South Wales, Sydney, Australia

Helder Maiato
Chromosome Instability & Dynamics Laboratory, Instituto de Biologia Molecular e Celular and Department of Experimental Biology, Faculdade de Medicina, Universidade do Porto, Porto, Portugal

Jaideep Mathur
Department of Molecular and Cellular Biology, Laboratory of Plant Development and Interactions, University of Guelph, Guelph, Canada

Katsumi Matsuzaki
Graduate School of Pharmaceutical Sciences, Kyoto University, Sakyo-ku, Kyoto, Japan

Erik Meijering
Biomedical Imaging Group Rotterdam, Departments of Medical Informatics and Radiology, Erasmus MC—University Medical Center Rotterdam, Rotterdam, The Netherlands

Alexander A. Mironov
Istituto FIRC di Oncologia Molecolare, Milan, Italy

Lyndsay Murrow
Department of Pathology, University of California, San Francisco, San Francisco, California, USA

Michaela Nebel
Analytische Chemie, Elektroanalytik & Sensorik, Ruhr-Universität Bochum, Bochum, Germany

Peter O'Neill
MRC/CRUK Gray Institute for Radiation Oncology and Biology, University of Oxford, Oxford, Oxfordshire, United Kingdom

Kaoru Omae
Graduate School of Pharmaceutical Sciences, Kyoto University, Sakyo-ku, Kyoto, Japan

Dylan M. Owen
Centre for Vascular Research, University of New South Wales, Sydney, Australia

Anthony W. Parker
Research Complex at Harwell, Central Laser Facility, STFC, Rutherford Appleton Laboratory, Harwell-Oxford, Didcot, Oxford, Oxfordshire, United Kingdom

Richard M. Parton
Department of Biochemistry, The University of Oxford, Oxford, United Kingdom

Hayley Pemble
Department of Cell & Tissue Biology, University of California, San Francisco, San Francisco, California, USA

António J. Pereira
Chromosome Instability & Dynamics Laboratory, Instituto de Biologia Molecular e Celular, Universidade do Porto, Porto, Portugal

Ammasi Periasamy
W.M. Keck Center for Cellular Imaging; Departments of Biology and Biomedical Engineering, University of Virginia, Charlottesville, Virginia, USA

Iestyn Pope
School of Biosciences, Cardiff University, Cardiff, United Kingdom

Pamela Reynolds
MRC/CRUK Gray Institute for Radiation Oncology and Biology, University of Oxford, Oxford, Oxfordshire, United Kingdom

Andrea Rossi
Departments of Medicine and Physiology, University of California, San Francisco, San Francisco, California, USA

Martin H. Schattat
Department of Molecular and Cellular Biology, Laboratory of Plant Development and Interactions, University of Guelph, Guelph, Canada

Wolfgang Schuhmann
Analytische Chemie, Elektroanalytik & Sensorik, Ruhr-Universität Bochum, Bochum, Germany

Albert Schulte
Biochemistry-Electrochemistry Research Unit, Schools of Chemistry and Biochemistry, Institute of Science, Suranaree University of Technology, Nakhon Ratchasima, Thailand

Silvia Scolari
Centro de Química e Bioquímica, Faculdade de Ciências da Universidade de Lisboa, Lisboa, Portugal

Ihor Smal
Biomedical Imaging Group Rotterdam, Departments of Medical Informatics and Radiology, Erasmus MC—University Medical Center Rotterdam, Rotterdam, The Netherlands

Martin Stöckl
Nanobiophysics, MESA+ Institute for Nanotechnology, University of Twente, Enschede, The Netherlands

Samantha Stehbens
Department of Cell & Tissue Biology, University of California, San Francisco, San Francisco, California, USA

Yuansheng Sun
W.M. Keck Center for Cellular Imaging; Departments of Biology and Biomedical Engineering, University of Virginia, Charlottesville, Virginia, USA

A.S. Verkman
Departments of Medicine and Physiology, University of California, San Francisco, San Francisco, California, USA

Satoshi Waguri
Department of Anatomy and Histology, Fukushima Medical University School of Medicine, Fukushima, Japan

Gufeng Wang
Department of Chemistry, North Carolina State University, Raleigh, North Carolina, USA

Peter Watson
School of Biosciences, Cardiff University, Cardiff, United Kingdom

David Williamson
Centre for Vascular Research, University of New South Wales, Sydney, Australia

Torsten Wittmann
Department of Cell & Tissue Biology, University of California, San Francisco, San Francisco, California, USA

Yoshiaki Yano
Graduate School of Pharmaceutical Sciences, Kyoto University, Sakyo-ku, Kyoto, Japan

Jin Zhang
Department of Pharmacology and Molecular Sciences and The Solomon H. Snyder Departments of Neuroscience and Oncology, The Johns Hopkins University School of Medicine, Baltimore, Maryland, USA

Xin Zhou
Department of Pharmacology and Molecular Sciences, The Johns Hopkins University School of Medicine, Baltimore, Maryland, USA

Preface

Going back to the dawn of light microscopy, imaging techniques have provided the opportunity for developing models of cellular function. Over the past 40 years, the availability of technology for high-resolution imaging and for evaluation of those images in live cells has extended our reach and, accordingly, our ability to understand cell function. Given the large number of choices in equipment and approaches, sorting out the best approach can be challenging, even to seasoned investigators.

The present volumes provide descriptions of methods used to image living cells, with particular reference to the technical approaches and reagents needed and approaches to selecting the best techniques. The authors explain how these methods are able to provide important biological insights in normal and pathological cells.

Authors were selected based on research contributions in the area about which they have written and based on their ability to describe their methodological contribution in a clear and reproducible way. They have been encouraged to make use of graphics, comparisons to other methods, and to provide tricks and approaches not revealed in prior publications that make it possible to adapt methods to other systems.

The editor wants to express appreciation to the contributors for providing their contributions in a timely fashion, to the senior editors for guidance, and to the staff at Academic Press for helpful input.

P. Michael Conn

Methods in Enzymology

VOLUME I. Preparation and Assay of Enzymes
Edited by SIDNEY P. COLOWICK AND NATHAN O. KAPLAN

VOLUME II. Preparation and Assay of Enzymes
Edited by SIDNEY P. COLOWICK AND NATHAN O. KAPLAN

VOLUME III. Preparation and Assay of Substrates
Edited by SIDNEY P. COLOWICK AND NATHAN O. KAPLAN

VOLUME IV. Special Techniques for the Enzymologist
Edited by SIDNEY P. COLOWICK AND NATHAN O. KAPLAN

VOLUME V. Preparation and Assay of Enzymes
Edited by SIDNEY P. COLOWICK AND NATHAN O. KAPLAN

VOLUME VI. Preparation and Assay of Enzymes *(Continued)*
Preparation and Assay of Substrates
Special Techniques
Edited by SIDNEY P. COLOWICK AND NATHAN O. KAPLAN

VOLUME VII. Cumulative Subject Index
Edited by SIDNEY P. COLOWICK AND NATHAN O. KAPLAN

VOLUME VIII. Complex Carbohydrates
Edited by ELIZABETH F. NEUFELD AND VICTOR GINSBURG

VOLUME IX. Carbohydrate Metabolism
Edited by WILLIS A. WOOD

VOLUME X. Oxidation and Phosphorylation
Edited by RONALD W. ESTABROOK AND MAYNARD E. PULLMAN

VOLUME XI. Enzyme Structure
Edited by C. H. W. HIRS

VOLUME XII. Nucleic Acids (Parts A and B)
Edited by LAWRENCE GROSSMAN AND KIVIE MOLDAVE

VOLUME XIII. Citric Acid Cycle
Edited by J. M. LOWENSTEIN

VOLUME XIV. Lipids
Edited by J. M. LOWENSTEIN

VOLUME XV. Steroids and Terpenoids
Edited by RAYMOND B. CLAYTON

VOLUME XVI. Fast Reactions
Edited by KENNETH KUSTIN

VOLUME XVII. Metabolism of Amino Acids and Amines (Parts A and B)
Edited by HERBERT TABOR AND CELIA WHITE TABOR

VOLUME XVIII. Vitamins and Coenzymes (Parts A, B, and C)
Edited by DONALD B. MCCORMICK AND LEMUEL D. WRIGHT

VOLUME XIX. Proteolytic Enzymes
Edited by GERTRUDE E. PERLMANN AND LASZLO LORAND

VOLUME XX. Nucleic Acids and Protein Synthesis (Part C)
Edited by KIVIE MOLDAVE AND LAWRENCE GROSSMAN

VOLUME XXI. Nucleic Acids (Part D)
Edited by LAWRENCE GROSSMAN AND KIVIE MOLDAVE

VOLUME XXII. Enzyme Purification and Related Techniques
Edited by WILLIAM B. JAKOBY

VOLUME XXIII. Photosynthesis (Part A)
Edited by ANTHONY SAN PIETRO

VOLUME XXIV. Photosynthesis and Nitrogen Fixation (Part B)
Edited by ANTHONY SAN PIETRO

VOLUME XXV. Enzyme Structure (Part B)
Edited by C. H. W. HIRS AND SERGE N. TIMASHEFF

VOLUME XXVI. Enzyme Structure (Part C)
Edited by C. H. W. HIRS AND SERGE N. TIMASHEFF

VOLUME XXVII. Enzyme Structure (Part D)
Edited by C. H. W. HIRS AND SERGE N. TIMASHEFF

VOLUME XXVIII. Complex Carbohydrates (Part B)
Edited by VICTOR GINSBURG

VOLUME XXIX. Nucleic Acids and Protein Synthesis (Part E)
Edited by LAWRENCE GROSSMAN AND KIVIE MOLDAVE

VOLUME XXX. Nucleic Acids and Protein Synthesis (Part F)
Edited by KIVIE MOLDAVE AND LAWRENCE GROSSMAN

VOLUME XXXI. Biomembranes (Part A)
Edited by SIDNEY FLEISCHER AND LESTER PACKER

VOLUME XXXII. Biomembranes (Part B)
Edited by SIDNEY FLEISCHER AND LESTER PACKER

VOLUME XXXIII. Cumulative Subject Index Volumes I-XXX
Edited by MARTHA G. DENNIS AND EDWARD A. DENNIS

VOLUME XXXIV. Affinity Techniques (Enzyme Purification: Part B)
Edited by WILLIAM B. JAKOBY AND MEIR WILCHEK

VOLUME XXXV. Lipids (Part B)
Edited by JOHN M. LOWENSTEIN

VOLUME XXXVI. Hormone Action (Part A: Steroid Hormones)
Edited by BERT W. O'MALLEY AND JOEL G. HARDMAN

VOLUME XXXVII. Hormone Action (Part B: Peptide Hormones)
Edited by BERT W. O'MALLEY AND JOEL G. HARDMAN

VOLUME XXXVIII. Hormone Action (Part C: Cyclic Nucleotides)
Edited by JOEL G. HARDMAN AND BERT W. O'MALLEY

VOLUME XXXIX. Hormone Action (Part D: Isolated Cells, Tissues, and Organ Systems)
Edited by JOEL G. HARDMAN AND BERT W. O'MALLEY

VOLUME XL. Hormone Action (Part E: Nuclear Structure and Function)
Edited by BERT W. O'MALLEY AND JOEL G. HARDMAN

VOLUME XLI. Carbohydrate Metabolism (Part B)
Edited by W. A. WOOD

VOLUME XLII. Carbohydrate Metabolism (Part C)
Edited by W. A. WOOD

VOLUME XLIII. Antibiotics
Edited by JOHN H. HASH

VOLUME XLIV. Immobilized Enzymes
Edited by KLAUS MOSBACH

VOLUME XLV. Proteolytic Enzymes (Part B)
Edited by LASZLO LORAND

VOLUME XLVI. Affinity Labeling
Edited by WILLIAM B. JAKOBY AND MEIR WILCHEK

VOLUME XLVII. Enzyme Structure (Part E)
Edited by C. H. W. HIRS AND SERGE N. TIMASHEFF

VOLUME XLVIII. Enzyme Structure (Part F)
Edited by C. H. W. HIRS AND SERGE N. TIMASHEFF

VOLUME XLIX. Enzyme Structure (Part G)
Edited by C. H. W. HIRS AND SERGE N. TIMASHEFF

VOLUME L. Complex Carbohydrates (Part C)
Edited by VICTOR GINSBURG

VOLUME LI. Purine and Pyrimidine Nucleotide Metabolism
Edited by PATRICIA A. HOFFEE AND MARY ELLEN JONES

VOLUME LII. Biomembranes (Part C: Biological Oxidations)
Edited by SIDNEY FLEISCHER AND LESTER PACKER

VOLUME LIII. Biomembranes (Part D: Biological Oxidations)
Edited by SIDNEY FLEISCHER AND LESTER PACKER

VOLUME LIV. Biomembranes (Part E: Biological Oxidations)
Edited by SIDNEY FLEISCHER AND LESTER PACKER

VOLUME LV. Biomembranes (Part F: Bioenergetics)
Edited by SIDNEY FLEISCHER AND LESTER PACKER

VOLUME LVI. Biomembranes (Part G: Bioenergetics)
Edited by SIDNEY FLEISCHER AND LESTER PACKER

VOLUME LVII. Bioluminescence and Chemiluminescence
Edited by MARLENE A. DELUCA

VOLUME LVIII. Cell Culture
Edited by WILLIAM B. JAKOBY AND IRA PASTAN

VOLUME LIX. Nucleic Acids and Protein Synthesis (Part G)
Edited by KIVIE MOLDAVE AND LAWRENCE GROSSMAN

VOLUME LX. Nucleic Acids and Protein Synthesis (Part H)
Edited by KIVIE MOLDAVE AND LAWRENCE GROSSMAN

VOLUME 61. Enzyme Structure (Part H)
Edited by C. H. W. HIRS AND SERGE N. TIMASHEFF

VOLUME 62. Vitamins and Coenzymes (Part D)
Edited by DONALD B. MCCORMICK AND LEMUEL D. WRIGHT

VOLUME 63. Enzyme Kinetics and Mechanism (Part A: Initial Rate and Inhibitor Methods)
Edited by DANIEL L. PURICH

VOLUME 64. Enzyme Kinetics and Mechanism
(Part B: Isotopic Probes and Complex Enzyme Systems)
Edited by DANIEL L. PURICH

VOLUME 65. Nucleic Acids (Part I)
Edited by LAWRENCE GROSSMAN AND KIVIE MOLDAVE

VOLUME 66. Vitamins and Coenzymes (Part E)
Edited by DONALD B. MCCORMICK AND LEMUEL D. WRIGHT

VOLUME 67. Vitamins and Coenzymes (Part F)
Edited by DONALD B. MCCORMICK AND LEMUEL D. WRIGHT

VOLUME 68. Recombinant DNA
Edited by RAY WU

VOLUME 69. Photosynthesis and Nitrogen Fixation (Part C)
Edited by ANTHONY SAN PIETRO

VOLUME 70. Immunochemical Techniques (Part A)
Edited by HELEN VAN VUNAKIS AND JOHN J. LANGONE

VOLUME 71. Lipids (Part C)
Edited by JOHN M. LOWENSTEIN

VOLUME 72. Lipids (Part D)
Edited by JOHN M. LOWENSTEIN

VOLUME 73. Immunochemical Techniques (Part B)
Edited by JOHN J. LANGONE AND HELEN VAN VUNAKIS

VOLUME 74. Immunochemical Techniques (Part C)
Edited by JOHN J. LANGONE AND HELEN VAN VUNAKIS

VOLUME 75. Cumulative Subject Index Volumes XXXI, XXXII, XXXIV–LX
Edited by EDWARD A. DENNIS AND MARTHA G. DENNIS

VOLUME 76. Hemoglobins
Edited by ERALDO ANTONINI, LUIGI ROSSI-BERNARDI, AND EMILIA CHIANCONE

VOLUME 77. Detoxication and Drug Metabolism
Edited by WILLIAM B. JAKOBY

VOLUME 78. Interferons (Part A)
Edited by SIDNEY PESTKA

VOLUME 79. Interferons (Part B)
Edited by SIDNEY PESTKA

VOLUME 80. Proteolytic Enzymes (Part C)
Edited by LASZLO LORAND

VOLUME 81. Biomembranes (Part H: Visual Pigments and Purple Membranes, I)
Edited by LESTER PACKER

VOLUME 82. Structural and Contractile Proteins (Part A: Extracellular Matrix)
Edited by LEON W. CUNNINGHAM AND DIXIE W. FREDERIKSEN

VOLUME 83. Complex Carbohydrates (Part D)
Edited by VICTOR GINSBURG

VOLUME 84. Immunochemical Techniques (Part D: Selected Immunoassays)
Edited by JOHN J. LANGONE AND HELEN VAN VUNAKIS

VOLUME 85. Structural and Contractile Proteins (Part B: The Contractile Apparatus and the Cytoskeleton)
Edited by DIXIE W. FREDERIKSEN AND LEON W. CUNNINGHAM

VOLUME 86. Prostaglandins and Arachidonate Metabolites
Edited by WILLIAM E. M. LANDS AND WILLIAM L. SMITH

VOLUME 87. Enzyme Kinetics and Mechanism (Part C: Intermediates, Stereo-chemistry, and Rate Studies)
Edited by DANIEL L. PURICH

VOLUME 88. Biomembranes (Part I: Visual Pigments and Purple Membranes, II)
Edited by LESTER PACKER

VOLUME 89. Carbohydrate Metabolism (Part D)
Edited by WILLIS A. WOOD

VOLUME 90. Carbohydrate Metabolism (Part E)
Edited by WILLIS A. WOOD

VOLUME 91. Enzyme Structure (Part I)
Edited by C. H. W. HIRS AND SERGE N. TIMASHEFF

VOLUME 92. Immunochemical Techniques (Part E: Monoclonal Antibodies and General Immunoassay Methods)
Edited by JOHN J. LANGONE AND HELEN VAN VUNAKIS

VOLUME 93. Immunochemical Techniques (Part F: Conventional Antibodies, Fc Receptors, and Cytotoxicity)
Edited by JOHN J. LANGONE AND HELEN VAN VUNAKIS

VOLUME 94. Polyamines
Edited by HERBERT TABOR AND CELIA WHITE TABOR

VOLUME 95. Cumulative Subject Index Volumes 61–74, 76–80
Edited by EDWARD A. DENNIS AND MARTHA G. DENNIS

VOLUME 96. Biomembranes [Part J: Membrane Biogenesis: Assembly and Targeting (General Methods; Eukaryotes)]
Edited by SIDNEY FLEISCHER AND BECCA FLEISCHER

VOLUME 97. Biomembranes [Part K: Membrane Biogenesis: Assembly and Targeting (Prokaryotes, Mitochondria, and Chloroplasts)]
Edited by SIDNEY FLEISCHER AND BECCA FLEISCHER

VOLUME 98. Biomembranes (Part L: Membrane Biogenesis: Processing and Recycling)
Edited by SIDNEY FLEISCHER AND BECCA FLEISCHER

VOLUME 99. Hormone Action (Part F: Protein Kinases)
Edited by JACKIE D. CORBIN AND JOEL G. HARDMAN

VOLUME 100. Recombinant DNA (Part B)
Edited by RAY WU, LAWRENCE GROSSMAN, AND KIVIE MOLDAVE

VOLUME 101. Recombinant DNA (Part C)
Edited by RAY WU, LAWRENCE GROSSMAN, AND KIVIE MOLDAVE

VOLUME 102. Hormone Action (Part G: Calmodulin and Calcium-Binding Proteins)
Edited by ANTHONY R. MEANS AND BERT W. O'MALLEY

VOLUME 103. Hormone Action (Part H: Neuroendocrine Peptides)
Edited by P. MICHAEL CONN

VOLUME 104. Enzyme Purification and Related Techniques (Part C)
Edited by WILLIAM B. JAKOBY

VOLUME 105. Oxygen Radicals in Biological Systems
Edited by LESTER PACKER

VOLUME 106. Posttranslational Modifications (Part A)
Edited by FINN WOLD AND KIVIE MOLDAVE

VOLUME 107. Posttranslational Modifications (Part B)
Edited by FINN WOLD AND KIVIE MOLDAVE

VOLUME 108. Immunochemical Techniques (Part G: Separation and Characterization of Lymphoid Cells)
Edited by GIOVANNI DI SABATO, JOHN J. LANGONE, AND HELEN VAN VUNAKIS

VOLUME 109. Hormone Action (Part I: Peptide Hormones)
Edited by LUTZ BIRNBAUMER AND BERT W. O'MALLEY

VOLUME 110. Steroids and Isoprenoids (Part A)
Edited by JOHN H. LAW AND HANS C. RILLING

VOLUME 111. Steroids and Isoprenoids (Part B)
Edited by JOHN H. LAW AND HANS C. RILLING

VOLUME 112. Drug and Enzyme Targeting (Part A)
Edited by KENNETH J. WIDDER AND RALPH GREEN

VOLUME 113. Glutamate, Glutamine, Glutathione, and Related Compounds
Edited by ALTON MEISTER

VOLUME 114. Diffraction Methods for Biological Macromolecules (Part A)
Edited by HAROLD W. WYCKOFF, C. H. W. HIRS, AND SERGE N. TIMASHEFF

VOLUME 115. Diffraction Methods for Biological Macromolecules (Part B)
Edited by HAROLD W. WYCKOFF, C. H. W. HIRS, AND SERGE N. TIMASHEFF

VOLUME 116. Immunochemical Techniques
(Part H: Effectors and Mediators of Lymphoid Cell Functions)
Edited by GIOVANNI DI SABATO, JOHN J. LANGONE, AND HELEN VAN VUNAKIS

VOLUME 117. Enzyme Structure (Part J)
Edited by C. H. W. HIRS AND SERGE N. TIMASHEFF

VOLUME 118. Plant Molecular Biology
Edited by ARTHUR WEISSBACH AND HERBERT WEISSBACH

VOLUME 119. Interferons (Part C)
Edited by SIDNEY PESTKA

VOLUME 120. Cumulative Subject Index Volumes 81–94, 96–101

VOLUME 121. Immunochemical Techniques (Part I: Hybridoma Technology and Monoclonal Antibodies)
Edited by JOHN J. LANGONE AND HELEN VAN VUNAKIS

VOLUME 122. Vitamins and Coenzymes (Part G)
Edited by FRANK CHYTIL AND DONALD B. MCCORMICK

VOLUME 123. Vitamins and Coenzymes (Part H)
Edited by FRANK CHYTIL AND DONALD B. MCCORMICK

VOLUME 124. Hormone Action (Part J: Neuroendocrine Peptides)
Edited by P. MICHAEL CONN

VOLUME 125. Biomembranes (Part M: Transport in Bacteria, Mitochondria, and Chloroplasts: General Approaches and Transport Systems)
Edited by SIDNEY FLEISCHER AND BECCA FLEISCHER

VOLUME 126. Biomembranes (Part N: Transport in Bacteria, Mitochondria, and Chloroplasts: Protonmotive Force)
Edited by SIDNEY FLEISCHER AND BECCA FLEISCHER

VOLUME 127. Biomembranes (Part O: Protons and Water: Structure and Translocation)
Edited by LESTER PACKER

VOLUME 128. Plasma Lipoproteins (Part A: Preparation, Structure, and Molecular Biology)
Edited by JERE P. SEGREST AND JOHN J. ALBERS

VOLUME 129. Plasma Lipoproteins (Part B: Characterization, Cell Biology, and Metabolism)
Edited by JOHN J. ALBERS AND JERE P. SEGREST

VOLUME 130. Enzyme Structure (Part K)
Edited by C. H. W. HIRS AND SERGE N. TIMASHEFF

VOLUME 131. Enzyme Structure (Part L)
Edited by C. H. W. HIRS AND SERGE N. TIMASHEFF

VOLUME 132. Immunochemical Techniques (Part J: Phagocytosis and Cell-Mediated Cytotoxicity)
Edited by GIOVANNI DI SABATO AND JOHANNES EVERSE

VOLUME 133. Bioluminescence and Chemiluminescence (Part B)
Edited by MARLENE DELUCA AND WILLIAM D. MCELROY

VOLUME 134. Structural and Contractile Proteins (Part C: The Contractile Apparatus and the Cytoskeleton)
Edited by RICHARD B. VALLEE

VOLUME 135. Immobilized Enzymes and Cells (Part B)
Edited by KLAUS MOSBACH

VOLUME 136. Immobilized Enzymes and Cells (Part C)
Edited by KLAUS MOSBACH

VOLUME 137. Immobilized Enzymes and Cells (Part D)
Edited by KLAUS MOSBACH

VOLUME 138. Complex Carbohydrates (Part E)
Edited by VICTOR GINSBURG

VOLUME 139. Cellular Regulators (Part A: Calcium- and Calmodulin-Binding Proteins)
Edited by ANTHONY R. MEANS AND P. MICHAEL CONN

VOLUME 140. Cumulative Subject Index Volumes 102–119, 121–134

VOLUME 141. Cellular Regulators (Part B: Calcium and Lipids)
Edited by P. MICHAEL CONN AND ANTHONY R. MEANS

VOLUME 142. Metabolism of Aromatic Amino Acids and Amines
Edited by SEYMOUR KAUFMAN

VOLUME 143. Sulfur and Sulfur Amino Acids
Edited by WILLIAM B. JAKOBY AND OWEN GRIFFITH

VOLUME 144. Structural and Contractile Proteins (Part D: Extracellular Matrix)
Edited by LEON W. CUNNINGHAM

VOLUME 145. Structural and Contractile Proteins (Part E: Extracellular Matrix)
Edited by LEON W. CUNNINGHAM

VOLUME 146. Peptide Growth Factors (Part A)
Edited by DAVID BARNES AND DAVID A. SIRBASKU

VOLUME 147. Peptide Growth Factors (Part B)
Edited by DAVID BARNES AND DAVID A. SIRBASKU

VOLUME 148. Plant Cell Membranes
Edited by LESTER PACKER AND ROLAND DOUCE

VOLUME 149. Drug and Enzyme Targeting (Part B)
Edited by RALPH GREEN AND KENNETH J. WIDDER

VOLUME 150. Immunochemical Techniques (Part K: *In Vitro* Models of B and T Cell Functions and Lymphoid Cell Receptors)
Edited by GIOVANNI DI SABATO

VOLUME 151. Molecular Genetics of Mammalian Cells
Edited by MICHAEL M. GOTTESMAN

VOLUME 152. Guide to Molecular Cloning Techniques
Edited by SHELBY L. BERGER AND ALAN R. KIMMEL

VOLUME 153. Recombinant DNA (Part D)
Edited by RAY WU AND LAWRENCE GROSSMAN

VOLUME 154. Recombinant DNA (Part E)
Edited by RAY WU AND LAWRENCE GROSSMAN

VOLUME 155. Recombinant DNA (Part F)
Edited by RAY WU

VOLUME 156. Biomembranes (Part P: ATP-Driven Pumps and Related Transport: The Na, K-Pump)
Edited by SIDNEY FLEISCHER AND BECCA FLEISCHER

VOLUME 157. Biomembranes (Part Q: ATP-Driven Pumps and Related Transport: Calcium, Proton, and Potassium Pumps)
Edited by SIDNEY FLEISCHER AND BECCA FLEISCHER

VOLUME 158. Metalloproteins (Part A)
Edited by JAMES F. RIORDAN AND BERT L. VALLEE

VOLUME 159. Initiation and Termination of Cyclic Nucleotide Action
Edited by JACKIE D. CORBIN AND ROGER A. JOHNSON

VOLUME 160. Biomass (Part A: Cellulose and Hemicellulose)
Edited by WILLIS A. WOOD AND SCOTT T. KELLOGG

VOLUME 161. Biomass (Part B: Lignin, Pectin, and Chitin)
Edited by WILLIS A. WOOD AND SCOTT T. KELLOGG

VOLUME 162. Immunochemical Techniques (Part L: Chemotaxis and Inflammation)
Edited by GIOVANNI DI SABATO

VOLUME 163. Immunochemical Techniques (Part M: Chemotaxis and Inflammation)
Edited by GIOVANNI DI SABATO

VOLUME 164. Ribosomes
Edited by HARRY F. NOLLER, JR., AND KIVIE MOLDAVE

VOLUME 165. Microbial Toxins: Tools for Enzymology
Edited by SIDNEY HARSHMAN

VOLUME 166. Branched-Chain Amino Acids
Edited by ROBERT HARRIS AND JOHN R. SOKATCH

VOLUME 167. Cyanobacteria
Edited by LESTER PACKER AND ALEXANDER N. GLAZER

VOLUME 168. Hormone Action (Part K: Neuroendocrine Peptides)
Edited by P. MICHAEL CONN

VOLUME 169. Platelets: Receptors, Adhesion, Secretion (Part A)
Edited by JACEK HAWIGER

VOLUME 170. Nucleosomes
Edited by PAUL M. WASSARMAN AND ROGER D. KORNBERG

VOLUME 171. Biomembranes (Part R: Transport Theory: Cells and Model Membranes)
Edited by SIDNEY FLEISCHER AND BECCA FLEISCHER

VOLUME 172. Biomembranes (Part S: Transport: Membrane Isolation and Characterization)
Edited by SIDNEY FLEISCHER AND BECCA FLEISCHER

VOLUME 173. Biomembranes [Part T: Cellular and Subcellular Transport: Eukaryotic (Nonepithelial) Cells]
Edited by SIDNEY FLEISCHER AND BECCA FLEISCHER

VOLUME 174. Biomembranes [Part U: Cellular and Subcellular Transport: Eukaryotic (Nonepithelial) Cells]
Edited by SIDNEY FLEISCHER AND BECCA FLEISCHER

VOLUME 175. Cumulative Subject Index Volumes 135–139, 141–167

VOLUME 176. Nuclear Magnetic Resonance (Part A: Spectral Techniques and Dynamics)
Edited by NORMAN J. OPPENHEIMER AND THOMAS L. JAMES

VOLUME 177. Nuclear Magnetic Resonance (Part B: Structure and Mechanism)
Edited by NORMAN J. OPPENHEIMER AND THOMAS L. JAMES

VOLUME 178. Antibodies, Antigens, and Molecular Mimicry
Edited by JOHN J. LANGONE

VOLUME 179. Complex Carbohydrates (Part F)
Edited by VICTOR GINSBURG

VOLUME 180. RNA Processing (Part A: General Methods)
Edited by JAMES E. DAHLBERG AND JOHN N. ABELSON

VOLUME 181. RNA Processing (Part B: Specific Methods)
Edited by JAMES E. DAHLBERG AND JOHN N. ABELSON

VOLUME 182. Guide to Protein Purification
Edited by MURRAY P. DEUTSCHER

VOLUME 183. Molecular Evolution: Computer Analysis of Protein and Nucleic Acid Sequences
Edited by RUSSELL F. DOOLITTLE

VOLUME 184. Avidin-Biotin Technology
Edited by MEIR WILCHEK AND EDWARD A. BAYER

VOLUME 185. Gene Expression Technology
Edited by DAVID V. GOEDDEL

VOLUME 186. Oxygen Radicals in Biological Systems (Part B: Oxygen Radicals and Antioxidants)
Edited by LESTER PACKER AND ALEXANDER N. GLAZER

VOLUME 187. Arachidonate Related Lipid Mediators
Edited by ROBERT C. MURPHY AND FRANK A. FITZPATRICK

VOLUME 188. Hydrocarbons and Methylotrophy
Edited by MARY E. LIDSTROM

VOLUME 189. Retinoids (Part A: Molecular and Metabolic Aspects)
Edited by LESTER PACKER

VOLUME 190. Retinoids (Part B: Cell Differentiation and Clinical Applications)
Edited by LESTER PACKER

VOLUME 191. Biomembranes (Part V: Cellular and Subcellular Transport: Epithelial Cells)
Edited by SIDNEY FLEISCHER AND BECCA FLEISCHER

VOLUME 192. Biomembranes (Part W: Cellular and Subcellular Transport: Epithelial Cells)
Edited by SIDNEY FLEISCHER AND BECCA FLEISCHER

VOLUME 193. Mass Spectrometry
Edited by JAMES A. MCCLOSKEY

VOLUME 194. Guide to Yeast Genetics and Molecular Biology
Edited by CHRISTINE GUTHRIE AND GERALD R. FINK

VOLUME 195. Adenylyl Cyclase, G Proteins, and Guanylyl Cyclase
Edited by ROGER A. JOHNSON AND JACKIE D. CORBIN

VOLUME 196. Molecular Motors and the Cytoskeleton
Edited by RICHARD B. VALLEE

VOLUME 197. Phospholipases
Edited by EDWARD A. DENNIS

VOLUME 198. Peptide Growth Factors (Part C)
Edited by DAVID BARNES, J. P. MATHER, AND GORDON H. SATO

VOLUME 199. Cumulative Subject Index Volumes 168–174, 176–194

VOLUME 200. Protein Phosphorylation (Part A: Protein Kinases: Assays, Purification, Antibodies, Functional Analysis, Cloning, and Expression)
Edited by TONY HUNTER AND BARTHOLOMEW M. SEFTON

VOLUME 201. Protein Phosphorylation (Part B: Analysis of Protein Phosphorylation, Protein Kinase Inhibitors, and Protein Phosphatases)
Edited by TONY HUNTER AND BARTHOLOMEW M. SEFTON

VOLUME 202. Molecular Design and Modeling: Concepts and Applications (Part A: Proteins, Peptides, and Enzymes)
Edited by JOHN J. LANGONE

VOLUME 203. Molecular Design and Modeling: Concepts and Applications (Part B: Antibodies and Antigens, Nucleic Acids, Polysaccharides, and Drugs)
Edited by JOHN J. LANGONE

VOLUME 204. Bacterial Genetic Systems
Edited by JEFFREY H. MILLER

VOLUME 205. Metallobiochemistry (Part B: Metallothionein and Related Molecules)
Edited by JAMES F. RIORDAN AND BERT L. VALLEE

VOLUME 206. Cytochrome P450
Edited by MICHAEL R. WATERMAN AND ERIC F. JOHNSON

VOLUME 207. Ion Channels
Edited by BERNARDO RUDY AND LINDA E. IVERSON

VOLUME 208. Protein–DNA Interactions
Edited by ROBERT T. SAUER

VOLUME 209. Phospholipid Biosynthesis
Edited by EDWARD A. DENNIS AND DENNIS E. VANCE

VOLUME 210. Numerical Computer Methods
Edited by LUDWIG BRAND AND MICHAEL L. JOHNSON

VOLUME 211. DNA Structures (Part A: Synthesis and Physical Analysis of DNA)
Edited by DAVID M. J. LILLEY AND JAMES E. DAHLBERG

VOLUME 212. DNA Structures (Part B: Chemical and Electrophoretic Analysis of DNA)
Edited by DAVID M. J. LILLEY AND JAMES E. DAHLBERG

VOLUME 213. Carotenoids (Part A: Chemistry, Separation, Quantitation, and Antioxidation)
Edited by LESTER PACKER

VOLUME 214. Carotenoids (Part B: Metabolism, Genetics, and Biosynthesis)
Edited by LESTER PACKER

VOLUME 215. Platelets: Receptors, Adhesion, Secretion (Part B)
Edited by JACEK J. HAWIGER

VOLUME 216. Recombinant DNA (Part G)
Edited by RAY WU

VOLUME 217. Recombinant DNA (Part H)
Edited by RAY WU

VOLUME 218. Recombinant DNA (Part I)
Edited by RAY WU

VOLUME 219. Reconstitution of Intracellular Transport
Edited by JAMES E. ROTHMAN

VOLUME 220. Membrane Fusion Techniques (Part A)
Edited by NEJAT DÜZGÜNEŞ

VOLUME 221. Membrane Fusion Techniques (Part B)
Edited by NEJAT DÜZGÜNEŞ

VOLUME 222. Proteolytic Enzymes in Coagulation, Fibrinolysis, and Complement Activation (Part A: Mammalian Blood Coagulation Factors and Inhibitors)
Edited by LASZLO LORAND AND KENNETH G. MANN

VOLUME 223. Proteolytic Enzymes in Coagulation, Fibrinolysis, and Complement Activation (Part B: Complement Activation, Fibrinolysis, and Nonmammalian Blood Coagulation Factors)
Edited by LASZLO LORAND AND KENNETH G. MANN

VOLUME 224. Molecular Evolution: Producing the Biochemical Data
Edited by ELIZABETH ANNE ZIMMER, THOMAS J. WHITE, REBECCA L. CANN, AND ALLAN C. WILSON

VOLUME 225. Guide to Techniques in Mouse Development
Edited by PAUL M. WASSARMAN AND MELVIN L. DEPAMPHILIS

VOLUME 226. Metallobiochemistry (Part C: Spectroscopic and Physical Methods for Probing Metal Ion Environments in Metalloenzymes and Metalloproteins)
Edited by JAMES F. RIORDAN AND BERT L. VALLEE

VOLUME 227. Metallobiochemistry (Part D: Physical and Spectroscopic Methods for Probing Metal Ion Environments in Metalloproteins)
Edited by JAMES F. RIORDAN AND BERT L. VALLEE

VOLUME 228. Aqueous Two-Phase Systems
Edited by HARRY WALTER AND GÖTE JOHANSSON

VOLUME 229. Cumulative Subject Index Volumes 195–198, 200–227

VOLUME 230. Guide to Techniques in Glycobiology
Edited by WILLIAM J. LENNARZ AND GERALD W. HART

VOLUME 231. Hemoglobins (Part B: Biochemical and Analytical Methods)
Edited by JOHANNES EVERSE, KIM D. VANDEGRIFF, AND ROBERT M. WINSLOW

VOLUME 232. Hemoglobins (Part C: Biophysical Methods)
Edited by JOHANNES EVERSE, KIM D. VANDEGRIFF, AND ROBERT M. WINSLOW

VOLUME 233. Oxygen Radicals in Biological Systems (Part C)
Edited by LESTER PACKER

VOLUME 234. Oxygen Radicals in Biological Systems (Part D)
Edited by LESTER PACKER

VOLUME 235. Bacterial Pathogenesis (Part A: Identification and Regulation of Virulence Factors)
Edited by VIRGINIA L. CLARK AND PATRIK M. BAVOIL

VOLUME 236. Bacterial Pathogenesis (Part B: Integration of Pathogenic Bacteria with Host Cells)
Edited by VIRGINIA L. CLARK AND PATRIK M. BAVOIL

VOLUME 237. Heterotrimeric G Proteins
Edited by RAVI IYENGAR

VOLUME 238. Heterotrimeric G-Protein Effectors
Edited by RAVI IYENGAR

VOLUME 239. Nuclear Magnetic Resonance (Part C)
Edited by THOMAS L. JAMES AND NORMAN J. OPPENHEIMER

VOLUME 240. Numerical Computer Methods (Part B)
Edited by MICHAEL L. JOHNSON AND LUDWIG BRAND

VOLUME 241. Retroviral Proteases
Edited by LAWRENCE C. KUO AND JULES A. SHAFER

VOLUME 242. Neoglycoconjugates (Part A)
Edited by Y. C. LEE AND REIKO T. LEE

VOLUME 243. Inorganic Microbial Sulfur Metabolism
Edited by HARRY D. PECK, JR., AND JEAN LEGALL

VOLUME 244. Proteolytic Enzymes: Serine and Cysteine Peptidases
Edited by ALAN J. BARRETT

VOLUME 245. Extracellular Matrix Components
Edited by E. RUOSLAHTI AND E. ENGVALL

VOLUME 246. Biochemical Spectroscopy
Edited by KENNETH SAUER

VOLUME 247. Neoglycoconjugates (Part B: Biomedical Applications)
Edited by Y. C. LEE AND REIKO T. LEE

VOLUME 248. Proteolytic Enzymes: Aspartic and Metallo Peptidases
Edited by ALAN J. BARRETT

VOLUME 249. Enzyme Kinetics and Mechanism (Part D: Developments in Enzyme Dynamics)
Edited by DANIEL L. PURICH

VOLUME 250. Lipid Modifications of Proteins
Edited by PATRICK J. CASEY AND JANICE E. BUSS

VOLUME 251. Biothiols (Part A: Monothiols and Dithiols, Protein Thiols, and Thiyl Radicals)
Edited by LESTER PACKER

VOLUME 252. Biothiols (Part B: Glutathione and Thioredoxin; Thiols in Signal Transduction and Gene Regulation)
Edited by LESTER PACKER

VOLUME 253. Adhesion of Microbial Pathogens
Edited by RON J. DOYLE AND ITZHAK OFEK

VOLUME 254. Oncogene Techniques
Edited by PETER K. VOGT AND INDER M. VERMA

VOLUME 255. Small GTPases and Their Regulators (Part A: Ras Family)
Edited by W. E. BALCH, CHANNING J. DER, AND ALAN HALL

VOLUME 256. Small GTPases and Their Regulators (Part B: Rho Family)
Edited by W. E. BALCH, CHANNING J. DER, AND ALAN HALL

VOLUME 257. Small GTPases and Their Regulators (Part C: Proteins Involved in Transport)
Edited by W. E. BALCH, CHANNING J. DER, AND ALAN HALL

VOLUME 258. Redox-Active Amino Acids in Biology
Edited by JUDITH P. KLINMAN

VOLUME 259. Energetics of Biological Macromolecules
Edited by MICHAEL L. JOHNSON AND GARY K. ACKERS

VOLUME 260. Mitochondrial Biogenesis and Genetics (Part A)
Edited by GIUSEPPE M. ATTARDI AND ANNE CHOMYN

VOLUME 261. Nuclear Magnetic Resonance and Nucleic Acids
Edited by THOMAS L. JAMES

VOLUME 262. DNA Replication
Edited by JUDITH L. CAMPBELL

VOLUME 263. Plasma Lipoproteins (Part C: Quantitation)
Edited by WILLIAM A. BRADLEY, SANDRA H. GIANTURCO, AND JERE P. SEGREST

VOLUME 264. Mitochondrial Biogenesis and Genetics (Part B)
Edited by GIUSEPPE M. ATTARDI AND ANNE CHOMYN

VOLUME 265. Cumulative Subject Index Volumes 228, 230–262

VOLUME 266. Computer Methods for Macromolecular Sequence Analysis
Edited by RUSSELL F. DOOLITTLE

VOLUME 267. Combinatorial Chemistry
Edited by JOHN N. ABELSON

VOLUME 268. Nitric Oxide (Part A: Sources and Detection of NO; NO Synthase)
Edited by LESTER PACKER

VOLUME 269. Nitric Oxide (Part B: Physiological and Pathological Processes)
Edited by LESTER PACKER

VOLUME 270. High Resolution Separation and Analysis of Biological Macromolecules (Part A: Fundamentals)
Edited by BARRY L. KARGER AND WILLIAM S. HANCOCK

VOLUME 271. High Resolution Separation and Analysis of Biological Macromolecules (Part B: Applications)
Edited by BARRY L. KARGER AND WILLIAM S. HANCOCK

VOLUME 272. Cytochrome P450 (Part B)
Edited by ERIC F. JOHNSON AND MICHAEL R. WATERMAN

VOLUME 273. RNA Polymerase and Associated Factors (Part A)
Edited by SANKAR ADHYA

VOLUME 274. RNA Polymerase and Associated Factors (Part B)
Edited by SANKAR ADHYA

VOLUME 275. Viral Polymerases and Related Proteins
Edited by LAWRENCE C. KUO, DAVID B. OLSEN, AND STEVEN S. CARROLL

VOLUME 276. Macromolecular Crystallography (Part A)
Edited by CHARLES W. CARTER, JR., AND ROBERT M. SWEET

VOLUME 277. Macromolecular Crystallography (Part B)
Edited by CHARLES W. CARTER, JR., AND ROBERT M. SWEET

VOLUME 278. Fluorescence Spectroscopy
Edited by LUDWIG BRAND AND MICHAEL L. JOHNSON

VOLUME 279. Vitamins and Coenzymes (Part I)
Edited by DONALD B. MCCORMICK, JOHN W. SUTTIE, AND CONRAD WAGNER

VOLUME 280. Vitamins and Coenzymes (Part J)
Edited by DONALD B. MCCORMICK, JOHN W. SUTTIE, AND CONRAD WAGNER

VOLUME 281. Vitamins and Coenzymes (Part K)
Edited by DONALD B. MCCORMICK, JOHN W. SUTTIE, AND CONRAD WAGNER

VOLUME 282. Vitamins and Coenzymes (Part L)
Edited by DONALD B. MCCORMICK, JOHN W. SUTTIE, AND CONRAD WAGNER

VOLUME 283. Cell Cycle Control
Edited by WILLIAM G. DUNPHY

VOLUME 284. Lipases (Part A: Biotechnology)
Edited by BYRON RUBIN AND EDWARD A. DENNIS

VOLUME 285. Cumulative Subject Index Volumes 263, 264, 266–284, 286–289

VOLUME 286. Lipases (Part B: Enzyme Characterization and Utilization)
Edited by BYRON RUBIN AND EDWARD A. DENNIS

VOLUME 287. Chemokines
Edited by RICHARD HORUK

VOLUME 288. Chemokine Receptors
Edited by RICHARD HORUK

VOLUME 289. Solid Phase Peptide Synthesis
Edited by GREGG B. FIELDS

VOLUME 290. Molecular Chaperones
Edited by GEORGE H. LORIMER AND THOMAS BALDWIN

VOLUME 291. Caged Compounds
Edited by GERARD MARRIOTT

VOLUME 292. ABC Transporters: Biochemical, Cellular, and Molecular Aspects
Edited by SURESH V. AMBUDKAR AND MICHAEL M. GOTTESMAN

VOLUME 293. Ion Channels (Part B)
Edited by P. MICHAEL CONN

VOLUME 294. Ion Channels (Part C)
Edited by P. MICHAEL CONN

VOLUME 295. Energetics of Biological Macromolecules (Part B)
Edited by GARY K. ACKERS AND MICHAEL L. JOHNSON

VOLUME 296. Neurotransmitter Transporters
Edited by SUSAN G. AMARA

VOLUME 297. Photosynthesis: Molecular Biology of Energy Capture
Edited by LEE MCINTOSH

VOLUME 298. Molecular Motors and the Cytoskeleton (Part B)
Edited by RICHARD B. VALLEE

VOLUME 299. Oxidants and Antioxidants (Part A)
Edited by LESTER PACKER

VOLUME 300. Oxidants and Antioxidants (Part B)
Edited by LESTER PACKER

VOLUME 301. Nitric Oxide: Biological and Antioxidant Activities (Part C)
Edited by LESTER PACKER

VOLUME 302. Green Fluorescent Protein
Edited by P. MICHAEL CONN

VOLUME 303. cDNA Preparation and Display
Edited by SHERMAN M. WEISSMAN

VOLUME 304. Chromatin
Edited by PAUL M. WASSARMAN AND ALAN P. WOLFFE

VOLUME 305. Bioluminescence and Chemiluminescence (Part C)
Edited by THOMAS O. BALDWIN AND MIRIAM M. ZIEGLER

VOLUME 306. Expression of Recombinant Genes in Eukaryotic Systems
Edited by JOSEPH C. GLORIOSO AND MARTIN C. SCHMIDT

VOLUME 307. Confocal Microscopy
Edited by P. MICHAEL CONN

VOLUME 308. Enzyme Kinetics and Mechanism (Part E: Energetics of Enzyme Catalysis)
Edited by DANIEL L. PURICH AND VERN L. SCHRAMM

VOLUME 309. Amyloid, Prions, and Other Protein Aggregates
Edited by RONALD WETZEL

VOLUME 310. Biofilms
Edited by RON J. DOYLE

VOLUME 311. Sphingolipid Metabolism and Cell Signaling (Part A)
Edited by ALFRED H. MERRILL, JR., AND YUSUF A. HANNUN

VOLUME 312. Sphingolipid Metabolism and Cell Signaling (Part B)
Edited by ALFRED H. MERRILL, JR., AND YUSUF A. HANNUN

VOLUME 313. Antisense Technology
(Part A: General Methods, Methods of Delivery, and RNA Studies)
Edited by M. IAN PHILLIPS

VOLUME 314. Antisense Technology (Part B: Applications)
Edited by M. IAN PHILLIPS

VOLUME 315. Vertebrate Phototransduction and the Visual Cycle (Part A)
Edited by KRZYSZTOF PALCZEWSKI

VOLUME 316. Vertebrate Phototransduction and the Visual Cycle (Part B)
Edited by KRZYSZTOF PALCZEWSKI

VOLUME 317. RNA–Ligand Interactions (Part A: Structural Biology Methods)
Edited by DANIEL W. CELANDER AND JOHN N. ABELSON

VOLUME 318. RNA–Ligand Interactions (Part B: Molecular Biology Methods)
Edited by DANIEL W. CELANDER AND JOHN N. ABELSON

VOLUME 319. Singlet Oxygen, UV-A, and Ozone
Edited by LESTER PACKER AND HELMUT SIES

VOLUME 320. Cumulative Subject Index Volumes 290–319

VOLUME 321. Numerical Computer Methods (Part C)
Edited by MICHAEL L. JOHNSON AND LUDWIG BRAND

VOLUME 322. Apoptosis
Edited by JOHN C. REED

VOLUME 323. Energetics of Biological Macromolecules (Part C)
Edited by MICHAEL L. JOHNSON AND GARY K. ACKERS

VOLUME 324. Branched-Chain Amino Acids (Part B)
Edited by ROBERT A. HARRIS AND JOHN R. SOKATCH

VOLUME 325. Regulators and Effectors of Small GTPases
(Part D: Rho Family)
Edited by W. E. BALCH, CHANNING J. DER, AND ALAN HALL

VOLUME 326. Applications of Chimeric Genes and Hybrid Proteins
(Part A: Gene Expression and Protein Purification)
Edited by JEREMY THORNER, SCOTT D. EMR, AND JOHN N. ABELSON

VOLUME 327. Applications of Chimeric Genes and Hybrid Proteins
(Part B: Cell Biology and Physiology)
Edited by JEREMY THORNER, SCOTT D. EMR, AND JOHN N. ABELSON

VOLUME 328. Applications of Chimeric Genes and Hybrid Proteins (Part C: Protein–Protein Interactions and Genomics)
Edited by JEREMY THORNER, SCOTT D. EMR, AND JOHN N. ABELSON

VOLUME 329. Regulators and Effectors of Small GTPases (Part E: GTPases Involved in Vesicular Traffic)
Edited by W. E. BALCH, CHANNING J. DER, AND ALAN HALL

VOLUME 330. Hyperthermophilic Enzymes (Part A)
Edited by MICHAEL W. W. ADAMS AND ROBERT M. KELLY

VOLUME 331. Hyperthermophilic Enzymes (Part B)
Edited by MICHAEL W. W. ADAMS AND ROBERT M. KELLY

VOLUME 332. Regulators and Effectors of Small GTPases (Part F: Ras Family I)
Edited by W. E. BALCH, CHANNING J. DER, AND ALAN HALL

VOLUME 333. Regulators and Effectors of Small GTPases (Part G: Ras Family II)
Edited by W. E. BALCH, CHANNING J. DER, AND ALAN HALL

VOLUME 334. Hyperthermophilic Enzymes (Part C)
Edited by MICHAEL W. W. ADAMS AND ROBERT M. KELLY

VOLUME 335. Flavonoids and Other Polyphenols
Edited by LESTER PACKER

VOLUME 336. Microbial Growth in Biofilms (Part A: Developmental and Molecular Biological Aspects)
Edited by RON J. DOYLE

VOLUME 337. Microbial Growth in Biofilms (Part B: Special Environments and Physicochemical Aspects)
Edited by RON J. DOYLE

VOLUME 338. Nuclear Magnetic Resonance of Biological Macromolecules (Part A)
Edited by THOMAS L. JAMES, VOLKER DÖTSCH, AND ULI SCHMITZ

VOLUME 339. Nuclear Magnetic Resonance of Biological Macromolecules (Part B)
Edited by THOMAS L. JAMES, VOLKER DÖTSCH, AND ULI SCHMITZ

VOLUME 340. Drug–Nucleic Acid Interactions
Edited by JONATHAN B. CHAIRES AND MICHAEL J. WARING

VOLUME 341. Ribonucleases (Part A)
Edited by ALLEN W. NICHOLSON

VOLUME 342. Ribonucleases (Part B)
Edited by ALLEN W. NICHOLSON

VOLUME 343. G Protein Pathways (Part A: Receptors)
Edited by RAVI IYENGAR AND JOHN D. HILDEBRANDT

VOLUME 344. G Protein Pathways (Part B: G Proteins and Their Regulators)
Edited by RAVI IYENGAR AND JOHN D. HILDEBRANDT

VOLUME 345. G Protein Pathways (Part C: Effector Mechanisms)
Edited by RAVI IYENGAR AND JOHN D. HILDEBRANDT

VOLUME 346. Gene Therapy Methods
Edited by M. IAN PHILLIPS

VOLUME 347. Protein Sensors and Reactive Oxygen Species (Part A: Selenoproteins and Thioredoxin)
Edited by HELMUT SIES AND LESTER PACKER

VOLUME 348. Protein Sensors and Reactive Oxygen Species (Part B: Thiol Enzymes and Proteins)
Edited by HELMUT SIES AND LESTER PACKER

VOLUME 349. Superoxide Dismutase
Edited by LESTER PACKER

VOLUME 350. Guide to Yeast Genetics and Molecular and Cell Biology (Part B)
Edited by CHRISTINE GUTHRIE AND GERALD R. FINK

VOLUME 351. Guide to Yeast Genetics and Molecular and Cell Biology (Part C)
Edited by CHRISTINE GUTHRIE AND GERALD R. FINK

VOLUME 352. Redox Cell Biology and Genetics (Part A)
Edited by CHANDAN K. SEN AND LESTER PACKER

VOLUME 353. Redox Cell Biology and Genetics (Part B)
Edited by CHANDAN K. SEN AND LESTER PACKER

VOLUME 354. Enzyme Kinetics and Mechanisms (Part F: Detection and Characterization of Enzyme Reaction Intermediates)
Edited by DANIEL L. PURICH

VOLUME 355. Cumulative Subject Index Volumes 321–354

VOLUME 356. Laser Capture Microscopy and Microdissection
Edited by P. MICHAEL CONN

VOLUME 357. Cytochrome P450, Part C
Edited by ERIC F. JOHNSON AND MICHAEL R. WATERMAN

VOLUME 358. Bacterial Pathogenesis (Part C: Identification, Regulation, and Function of Virulence Factors)
Edited by VIRGINIA L. CLARK AND PATRIK M. BAVOIL

VOLUME 359. Nitric Oxide (Part D)
Edited by ENRIQUE CADENAS AND LESTER PACKER

VOLUME 360. Biophotonics (Part A)
Edited by GERARD MARRIOTT AND IAN PARKER

VOLUME 361. Biophotonics (Part B)
Edited by GERARD MARRIOTT AND IAN PARKER

VOLUME 362. Recognition of Carbohydrates in Biological Systems (Part A)
Edited by YUAN C. LEE AND REIKO T. LEE

VOLUME 363. Recognition of Carbohydrates in Biological Systems (Part B)
Edited by YUAN C. LEE AND REIKO T. LEE

VOLUME 364. Nuclear Receptors
Edited by DAVID W. RUSSELL AND DAVID J. MANGELSDORF

VOLUME 365. Differentiation of Embryonic Stem Cells
Edited by PAUL M. WASSAUMAN AND GORDON M. KELLER

VOLUME 366. Protein Phosphatases
Edited by SUSANNE KLUMPP AND JOSEF KRIEGLSTEIN

VOLUME 367. Liposomes (Part A)
Edited by NEJAT DÜZGÜNEŞ

VOLUME 368. Macromolecular Crystallography (Part C)
Edited by CHARLES W. CARTER, JR., AND ROBERT M. SWEET

VOLUME 369. Combinational Chemistry (Part B)
Edited by GUILLERMO A. MORALES AND BARRY A. BUNIN

VOLUME 370. RNA Polymerases and Associated Factors (Part C)
Edited by SANKAR L. ADHYA AND SUSAN GARGES

VOLUME 371. RNA Polymerases and Associated Factors (Part D)
Edited by SANKAR L. ADHYA AND SUSAN GARGES

VOLUME 372. Liposomes (Part B)
Edited by NEJAT DÜZGÜNEŞ

VOLUME 373. Liposomes (Part C)
Edited by NEJAT DÜZGÜNEŞ

VOLUME 374. Macromolecular Crystallography (Part D)
Edited by CHARLES W. CARTER, JR., AND ROBERT W. SWEET

VOLUME 375. Chromatin and Chromatin Remodeling Enzymes (Part A)
Edited by C. DAVID ALLIS AND CARL WU

VOLUME 376. Chromatin and Chromatin Remodeling Enzymes (Part B)
Edited by C. DAVID ALLIS AND CARL WU

VOLUME 377. Chromatin and Chromatin Remodeling Enzymes (Part C)
Edited by C. DAVID ALLIS AND CARL WU

VOLUME 378. Quinones and Quinone Enzymes (Part A)
Edited by HELMUT SIES AND LESTER PACKER

VOLUME 379. Energetics of Biological Macromolecules (Part D)
Edited by JO M. HOLT, MICHAEL L. JOHNSON, AND GARY K. ACKERS

VOLUME 380. Energetics of Biological Macromolecules (Part E)
Edited by JO M. HOLT, MICHAEL L. JOHNSON, AND GARY K. ACKERS

VOLUME 381. Oxygen Sensing
Edited by CHANDAN K. SEN AND GREGG L. SEMENZA

VOLUME 382. Quinones and Quinone Enzymes (Part B)
Edited by HELMUT SIES AND LESTER PACKER

VOLUME 383. Numerical Computer Methods (Part D)
Edited by LUDWIG BRAND AND MICHAEL L. JOHNSON

VOLUME 384. Numerical Computer Methods (Part E)
Edited by LUDWIG BRAND AND MICHAEL L. JOHNSON

VOLUME 385. Imaging in Biological Research (Part A)
Edited by P. MICHAEL CONN

VOLUME 386. Imaging in Biological Research (Part B)
Edited by P. MICHAEL CONN

VOLUME 387. Liposomes (Part D)
Edited by NEJAT DÜZGÜNEŞ

VOLUME 388. Protein Engineering
Edited by DAN E. ROBERTSON AND JOSEPH P. NOEL

VOLUME 389. Regulators of G-Protein Signaling (Part A)
Edited by DAVID P. SIDEROVSKI

VOLUME 390. Regulators of G-Protein Signaling (Part B)
Edited by DAVID P. SIDEROVSKI

VOLUME 391. Liposomes (Part E)
Edited by NEJAT DÜZGÜNEŞ

VOLUME 392. RNA Interference
Edited by ENGELKE ROSSI

VOLUME 393. Circadian Rhythms
Edited by MICHAEL W. YOUNG

VOLUME 394. Nuclear Magnetic Resonance of Biological Macromolecules (Part C)
Edited by THOMAS L. JAMES

VOLUME 395. Producing the Biochemical Data (Part B)
Edited by ELIZABETH A. ZIMMER AND ERIC H. ROALSON

VOLUME 396. Nitric Oxide (Part E)
Edited by LESTER PACKER AND ENRIQUE CADENAS

VOLUME 397. Environmental Microbiology
Edited by JARED R. LEADBETTER

VOLUME 398. Ubiquitin and Protein Degradation (Part A)
Edited by RAYMOND J. DESHAIES

VOLUME 399. Ubiquitin and Protein Degradation (Part B)
Edited by RAYMOND J. DESHAIES

VOLUME 400. Phase II Conjugation Enzymes and Transport Systems
Edited by HELMUT SIES AND LESTER PACKER

VOLUME 401. Glutathione Transferases and Gamma Glutamyl Transpeptidases
Edited by HELMUT SIES AND LESTER PACKER

VOLUME 402. Biological Mass Spectrometry
Edited by A. L. BURLINGAME

VOLUME 403. GTPases Regulating Membrane Targeting and Fusion
Edited by WILLIAM E. BALCH, CHANNING J. DER, AND ALAN HALL

VOLUME 404. GTPases Regulating Membrane Dynamics
Edited by WILLIAM E. BALCH, CHANNING J. DER, AND ALAN HALL

VOLUME 405. Mass Spectrometry: Modified Proteins and Glycoconjugates
Edited by A. L. BURLINGAME

VOLUME 406. Regulators and Effectors of Small GTPases: Rho Family
Edited by WILLIAM E. BALCH, CHANNING J. DER, AND ALAN HALL

VOLUME 407. Regulators and Effectors of Small GTPases: Ras Family
Edited by WILLIAM E. BALCH, CHANNING J. DER, AND ALAN HALL

VOLUME 408. DNA Repair (Part A)
Edited by JUDITH L. CAMPBELL AND PAUL MODRICH

VOLUME 409. DNA Repair (Part B)
Edited by JUDITH L. CAMPBELL AND PAUL MODRICH

VOLUME 410. DNA Microarrays (Part A: Array Platforms and Web-Bench Protocols)
Edited by ALAN KIMMEL AND BRIAN OLIVER

VOLUME 411. DNA Microarrays (Part B: Databases and Statistics)
Edited by ALAN KIMMEL AND BRIAN OLIVER

VOLUME 412. Amyloid, Prions, and Other Protein Aggregates (Part B)
Edited by INDU KHETERPAL AND RONALD WETZEL

VOLUME 413. Amyloid, Prions, and Other Protein Aggregates (Part C)
Edited by INDU KHETERPAL AND RONALD WETZEL

VOLUME 414. Measuring Biological Responses with Automated Microscopy
Edited by JAMES INGLESE

VOLUME 415. Glycobiology
Edited by MINORU FUKUDA

VOLUME 416. Glycomics
Edited by MINORU FUKUDA

VOLUME 417. Functional Glycomics
Edited by MINORU FUKUDA

VOLUME 418. Embryonic Stem Cells
Edited by IRINA KLIMANSKAYA AND ROBERT LANZA

VOLUME 419. Adult Stem Cells
Edited by IRINA KLIMANSKAYA AND ROBERT LANZA

VOLUME 420. Stem Cell Tools and Other Experimental Protocols
Edited by IRINA KLIMANSKAYA AND ROBERT LANZA

VOLUME 421. Advanced Bacterial Genetics: Use of Transposons and Phage for Genomic Engineering
Edited by KELLY T. HUGHES

VOLUME 422. Two-Component Signaling Systems, Part A
Edited by MELVIN I. SIMON, BRIAN R. CRANE, AND ALEXANDRINE CRANE

VOLUME 423. Two-Component Signaling Systems, Part B
Edited by MELVIN I. SIMON, BRIAN R. CRANE, AND ALEXANDRINE CRANE

VOLUME 424. RNA Editing
Edited by JONATHA M. GOTT

VOLUME 425. RNA Modification
Edited by JONATHA M. GOTT

VOLUME 426. Integrins
Edited by DAVID CHERESH

VOLUME 427. MicroRNA Methods
Edited by JOHN J. ROSSI

VOLUME 428. Osmosensing and Osmosignaling
Edited by HELMUT SIES AND DIETER HAUSSINGER

VOLUME 429. Translation Initiation: Extract Systems and Molecular Genetics
Edited by JON LORSCH

VOLUME 430. Translation Initiation: Reconstituted Systems and Biophysical Methods
Edited by JON LORSCH

VOLUME 431. Translation Initiation: Cell Biology, High-Throughput and Chemical-Based Approaches
Edited by JON LORSCH

VOLUME 432. Lipidomics and Bioactive Lipids: Mass-Spectrometry–Based Lipid Analysis
Edited by H. ALEX BROWN

VOLUME 433. Lipidomics and Bioactive Lipids: Specialized Analytical Methods and Lipids in Disease
Edited by H. ALEX BROWN

VOLUME 434. Lipidomics and Bioactive Lipids: Lipids and Cell Signaling
Edited by H. ALEX BROWN

VOLUME 435. Oxygen Biology and Hypoxia
Edited by HELMUT SIES AND BERNHARD BRÜNE

VOLUME 436. Globins and Other Nitric Oxide-Reactive Protiens (Part A)
Edited by ROBERT K. POOLE

VOLUME 437. Globins and Other Nitric Oxide-Reactive Protiens (Part B)
Edited by ROBERT K. POOLE

VOLUME 438. Small GTPases in Disease (Part A)
Edited by WILLIAM E. BALCH, CHANNING J. DER, AND ALAN HALL

VOLUME 439. Small GTPases in Disease (Part B)
Edited by WILLIAM E. BALCH, CHANNING J. DER, AND ALAN HALL

VOLUME 440. Nitric Oxide, Part F Oxidative and Nitrosative Stress in Redox Regulation of Cell Signaling
Edited by ENRIQUE CADENAS AND LESTER PACKER

VOLUME 441. Nitric Oxide, Part G Oxidative and Nitrosative Stress in Redox Regulation of Cell Signaling
Edited by ENRIQUE CADENAS AND LESTER PACKER

VOLUME 442. Programmed Cell Death, General Principles for Studying Cell Death (Part A)
Edited by ROYA KHOSRAVI-FAR, ZAHRA ZAKERI, RICHARD A. LOCKSHIN, AND MAURO PIACENTINI

VOLUME 443. Angiogenesis: *In Vitro* Systems
Edited by DAVID A. CHERESH

VOLUME 444. Angiogenesis: *In Vivo* Systems (Part A)
Edited by DAVID A. CHERESH

VOLUME 445. Angiogenesis: *In Vivo* Systems (Part B)
Edited by DAVID A. CHERESH

VOLUME 446. Programmed Cell Death, The Biology and Therapeutic Implications of Cell Death (Part B)
Edited by ROYA KHOSRAVI-FAR, ZAHRA ZAKERI, RICHARD A. LOCKSHIN, AND MAURO PIACENTINI

VOLUME 447. RNA Turnover in Bacteria, Archaea and Organelles
Edited by LYNNE E. MAQUAT AND CECILIA M. ARRAIANO

VOLUME 448. RNA Turnover in Eukaryotes: Nucleases, Pathways and Analysis of mRNA Decay
Edited by LYNNE E. MAQUAT AND MEGERDITCH KILEDJIAN

VOLUME 449. RNA Turnover in Eukaryotes: Analysis of Specialized and Quality Control RNA Decay Pathways
Edited by LYNNE E. MAQUAT AND MEGERDITCH KILEDJIAN

VOLUME 450. Fluorescence Spectroscopy
Edited by LUDWIG BRAND AND MICHAEL L. JOHNSON

VOLUME 451. Autophagy: Lower Eukaryotes and Non-Mammalian Systems (Part A)
Edited by DANIEL J. KLIONSKY

VOLUME 452. Autophagy in Mammalian Systems (Part B)
Edited by DANIEL J. KLIONSKY

VOLUME 453. Autophagy in Disease and Clinical Applications (Part C)
Edited by DANIEL J. KLIONSKY

VOLUME 454. Computer Methods (Part A)
Edited by MICHAEL L. JOHNSON AND LUDWIG BRAND

VOLUME 455. Biothermodynamics (Part A)
Edited by MICHAEL L. JOHNSON, JO M. HOLT, AND GARY K. ACKERS (RETIRED)

VOLUME 456. Mitochondrial Function, Part A: Mitochondrial Electron Transport Complexes and Reactive Oxygen Species
Edited by WILLIAM S. ALLISON AND IMMO E. SCHEFFLER

VOLUME 457. Mitochondrial Function, Part B: Mitochondrial Protein Kinases, Protein Phosphatases and Mitochondrial Diseases
Edited by WILLIAM S. ALLISON AND ANNE N. MURPHY

VOLUME 458. Complex Enzymes in Microbial Natural Product Biosynthesis, Part A: Overview Articles and Peptides
Edited by DAVID A. HOPWOOD

VOLUME 459. Complex Enzymes in Microbial Natural Product Biosynthesis, Part B: Polyketides, Aminocoumarins and Carbohydrates
Edited by DAVID A. HOPWOOD

VOLUME 460. Chemokines, Part A
Edited by TRACY M. HANDEL AND DAMON J. HAMEL

VOLUME 461. Chemokines, Part B
Edited by TRACY M. HANDEL AND DAMON J. HAMEL

VOLUME 462. Non-Natural Amino Acids
Edited by TOM W. MUIR AND JOHN N. ABELSON

VOLUME 463. Guide to Protein Purification, 2nd Edition
Edited by RICHARD R. BURGESS AND MURRAY P. DEUTSCHER

VOLUME 464. Liposomes, Part F
Edited by NEJAT DÜZGÜNEŞ

VOLUME 465. Liposomes, Part G
Edited by NEJAT DÜZGÜNEŞ

VOLUME 466. Biothermodynamics, Part B
Edited by MICHAEL L. JOHNSON, GARY K. ACKERS, AND JO M. HOLT

VOLUME 467. Computer Methods Part B
Edited by MICHAEL L. JOHNSON AND LUDWIG BRAND

VOLUME 468. Biophysical, Chemical, and Functional Probes of RNA Structure, Interactions and Folding: Part A
Edited by DANIEL HERSCHLAG

VOLUME 469. Biophysical, Chemical, and Functional Probes of RNA Structure, Interactions and Folding: Part B
Edited by DANIEL HERSCHLAG

VOLUME 470. Guide to Yeast Genetics: Functional Genomics, Proteomics, and Other Systems Analysis, 2nd Edition
Edited by GERALD FINK, JONATHAN WEISSMAN, AND CHRISTINE GUTHRIE

VOLUME 471. Two-Component Signaling Systems, Part C
Edited by MELVIN I. SIMON, BRIAN R. CRANE, AND ALEXANDRINE CRANE

VOLUME 472. Single Molecule Tools, Part A: Fluorescence Based Approaches
Edited by NILS G. WALTER

VOLUME 473. Thiol Redox Transitions in Cell Signaling, Part A Chemistry and Biochemistry of Low Molecular Weight and Protein Thiols
Edited by ENRIQUE CADENAS AND LESTER PACKER

VOLUME 474. Thiol Redox Transitions in Cell Signaling, Part B Cellular Localization and Signaling
Edited by ENRIQUE CADENAS AND LESTER PACKER

VOLUME 475. Single Molecule Tools, Part B: Super-Resolution, Particle Tracking, Multiparameter, and Force Based Methods
Edited by NILS G. WALTER

VOLUME 476. Guide to Techniques in Mouse Development, Part A Mice, Embryos, and Cells, 2nd Edition
Edited by PAUL M. WASSARMAN AND PHILIPPE M. SORIANO

VOLUME 477. Guide to Techniques in Mouse Development, Part B Mouse Molecular Genetics, 2nd Edition
Edited by PAUL M. WASSARMAN AND PHILIPPE M. SORIANO

VOLUME 478. Glycomics
Edited by MINORU FUKUDA

VOLUME 479. Functional Glycomics
Edited by MINORU FUKUDA

VOLUME 480. Glycobiology
Edited by MINORU FUKUDA

VOLUME 481. Cryo-EM, Part A: Sample Preparation and Data Collection
Edited by GRANT J. JENSEN

VOLUME 482. Cryo-EM, Part B: 3-D Reconstruction
Edited by GRANT J. JENSEN

VOLUME 483. Cryo-EM, Part C: Analyses, Interpretation, and Case Studies
Edited by GRANT J. JENSEN

VOLUME 484. Constitutive Activity in Receptors and Other Proteins, Part A
Edited by P. MICHAEL CONN

VOLUME 485. Constitutive Activity in Receptors and Other Proteins, Part B
Edited by P. MICHAEL CONN

VOLUME 486. Research on Nitrification and Related Processes, Part A
Edited by MARTIN G. KLOTZ

VOLUME 487. Computer Methods, Part C
Edited by MICHAEL L. JOHNSON AND LUDWIG BRAND

VOLUME 488. Biothermodynamics, Part C
Edited by MICHAEL L. JOHNSON, JO M. HOLT, AND GARY K. ACKERS

VOLUME 489. The Unfolded Protein Response and Cellular Stress, Part A
Edited by P. MICHAEL CONN

VOLUME 490. The Unfolded Protein Response and Cellular Stress, Part B
Edited by P. MICHAEL CONN

VOLUME 491. The Unfolded Protein Response and Cellular Stress, Part C
Edited by P. MICHAEL CONN

VOLUME 492. Biothermodynamics, Part D
Edited by MICHAEL L. JOHNSON, JO M. HOLT, AND GARY K. ACKERS

VOLUME 493. Fragment-Based Drug Design
Tools, Practical Approaches, and Examples
Edited by LAWRENCE C. KUO

VOLUME 494. Methods in Methane Metabolism, Part A
Methanogenesis
Edited by AMY C. ROSENZWEIG AND STEPHEN W. RAGSDALE

VOLUME 495. Methods in Methane Metabolism, Part B
Methanotrophy
Edited by AMY C. ROSENZWEIG AND STEPHEN W. RAGSDALE

VOLUME 496. Research on Nitrification and Related Processes, Part B
Edited by MARTIN G. KLOTZ AND LISA Y. STEIN

VOLUME 497. Synthetic Biology, Part A
Methods for Part/Device Characterization and Chassis Engineering
Edited by CHRISTOPHER VOIGT

VOLUME 498. Synthetic Biology, Part B
Computer Aided Design and DNA Assembly
Edited by CHRISTOPHER VOIGT

VOLUME 499. Biology of Serpins
Edited by JAMES C. WHISSTOCK AND PHILLIP I. BIRD

VOLUME 500. Methods in Systems Biology
Edited by DANIEL JAMESON, MALKHEY VERMA, AND HANS V. WESTERHOFF

VOLUME 501. Serpin Structure and Evolution
Edited by JAMES C. WHISSTOCK AND PHILLIP I. BIRD

VOLUME 502. Protein Engineering for Therapeutics, Part A
Edited by K. DANE WITTRUP AND GREGORY L. VERDINE

VOLUME 503. Protein Engineering for Therapeutics, Part B
Edited by K. DANE WITTRUP AND GREGORY L. VERDINE

VOLUME 504. Imaging and Spectroscopic Analysis of Living Cells
Optical and Spectroscopic Techniques
Edited by P. MICHAEL CONN

SECTION ONE

TECHNIQUES

CHAPTER ONE

Laser-Induced Radiation Microbeam Technology and Simultaneous Real-Time Fluorescence Imaging in Live Cells

Stanley W. Botchway,* Pamela Reynolds,[†] Anthony W. Parker,* and Peter O'Neill[†]

Contents

1. Introduction	4
1.1. Principles of multiphoton absorption	6
1.2. Pulsed laser sources for multiphoton techniques	7
1.3. High-resolution cellular DNA damage upon NIR laser microbeam induction	8
2. Experimental Procedure	10
2.1. Construction of the NIR laser microbeam: The light source	11
2.2. The multiphoton laser microbeam scanning system	13
2.3. Simultaneous confocal imaging of microbeam-induced DNA damage	14
2.4. Cell culture and NIR multiphoton irradiation	15
2.5. DNA DSBs determination: comet assay–single-cell gel electrophoresis in NIR laser microbeam studies	20
2.6. Observation of DNA damage–repair dynamics following laser microbeam induction	22
2.7. Spectral detection of multiphoton microbeam-induced excited state chemical species	23
3. Conclusion/Forward Look	25
Acknowledgments	26
References	26

* Research Complex at Harwell, Central Laser Facility, STFC, Rutherford Appleton Laboratory, Harwell-Oxford, Didcot, Oxford, Oxfordshire, United Kingdom
[†] MRC/CRUK Gray Institute for Radiation Oncology and Biology, University of Oxford, Oxford, Oxfordshire, United Kingdom

Methods in Enzymology, Volume 504
ISSN 0076-6879, DOI: 10.1016/B978-0-12-391857-4.00001-X

© 2012 Elsevier Inc.
All rights reserved.

Abstract

The use of nano- and microbeam techniques to induce and identify subcellular localized energy deposition within a region of a living cell provides a means to investigate the effects of low radiation doses. Particularly within the nucleus where the propagation and processing of deoxyribonucleic acid (DNA) damage (and repair) in both targeted and nontargeted cells, the latter being able to study cell–cell (bystander) effects. We have pioneered a near infrared (NIR) femtosecond laser microbeam to mimic ionizing radiation through multiphoton absorption within a 3D femtoliter volume of a highly focused Gaussian laser beam. The novel optical microbeam mimics both complex ionizing and UV-radiation-type cell damage including double strand breaks (DSBs). Using the microbeam technology, we have been able to investigate the formation of DNA DSB and subsequent recruitment of repair proteins to the submicrometer size site of damage introduced in viable cells. The use of a phosphorylated H2AX (γ-H2AX a marker for DSBs, visualized by immunofluorescent staining) and real-time imaging of fluorescently labeling proteins, the dynamics of recruitment of repair proteins in viable mammalian cells can be observed. Here we show the recruitment of ATM, p53 binding protein 1 (53BP1), and RAD51, an integral protein of the homologous recombination process in the DNA repair pathway and Ku-80-GFP involved in the nonhomologous end joining (NHEJ) pathway as exemplar repair process to show differences in the repair kinetics of DNA DSBs. The laser NIR multiphoton microbeam technology shows persistent DSBs at later times post laser irradiation which are indicative of DSBs arising at replication presumably from UV photoproducts or clustered damage containing single strand breaks (SSBs) that are also observed. Effects of the cell cycle may also be investigated in real time. Postirradiation and fixed cells studies show that in G1 cells a fraction of multiphoton laser-induced DSBs is persistent for >6 h in addition to those induced at replication demonstrating the broad range of timescales taken to repair DNA damage.

1. INTRODUCTION

The mammalian deoxyribonucleic acid (DNA) is continuously under attack from both intracellular (such as from metabolic processes) and extracellular (such as from ionizing radiation) processes (Hall, 1994). The majority of the damage from metabolic processes that occurs with high frequency are mainly single strand breaks (SSBs) and base lesions and are repaired with high fidelity. However, DNA damage caused by ionizing radiation, including ultraviolet, X-rays, gamma rays, and energetic particles, may lead to various forms of base damage, SSB, clusters of lesions including double strand breaks (DSBs), base–base dimerization, and protein–DNA cross-links (Botchway et al., 1997; Jenner et al., 1998; Melvin et al., 1996; Ward, 1991). Such damage to the genomic DNA of cells can lead to severe errors in transcription and replication, if not repaired correctly, and spawn mutations, genomic instability, and even cell death (Kadhim *et al.*, 1992; Simpson and Savage, 1996). Cell viability studies including clonogenic assay under both aerobic and anaerobic conditions for sparsely ionizing radiation have shown that the survival fraction of cells as a quotient of total irradiated cells (from the plating efficiency of a particular cell line) shows a biphasic continuously curving response that may be expressed by Eq. (1.1):

$$S = e^{-(\alpha D + \beta D^2)} \qquad (1.1)$$

where S, surviving fraction at a specific dose D; α and β are constants.

The origin of this effect is still unknown, although in the case for high LET irradiation, the survival curve is a monoexponential on dose (Cox and Masson, 1979; deLara *et al.*, 1995). To study the mechanistic details at the molecular level requires a technique that allows targeting of subcellular components. Collectively such subcellular probes that use ionizing radiation to induce cellular damage are described as radiation microbeams (Daudin *et al.*, 2006; Folkard *et al.*, 1997; Pallon and Malmqvist, 1994; Prise *et al.*, 1994; Sheng *et al.*, 2009). These provide a means to, for example, deliver alpha particles through a single cell nucleus or placing a mask over the cell culture prior to ultrasoft X-ray low LET irradiation to induce localized damage (Nelms *et al.*, 1998; Fig. 1.1). However, these techniques irradiate several cellular components over a relatively large area or, in the case of the mask, perturb the sample and/or restrict visible observation of sample because of necessary safety measures for handling the hazardous ionizing radiation. Thus, any observed endpoint may be limited by a lag-time equivalent to the time taken to remove the sample from the radiation source and transported to the imaging device, usually a fluorescence microscope.

We have developed a near infrared (NIR) laser microbeam to overcome some of these limitations associated with ionizing radiation microbeam sources (NB: current developments of heavy ion beams and electrons produced in short pulses by accelerators will also be able to induce cellular DNA damage with simultaneous images). Although the pulsed NIR laser

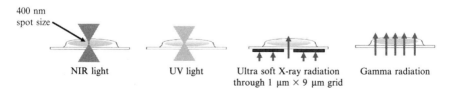

Figure 1.1 Schematic diagram showing the radiation path of different sources of radiation through a mammalian cell. (For color version of this figure, the reader is referred to the Web version of this chapter.)

microbeam is a complementary setup, it has several advantages for simultaneous cellular irradiation and imaging:

- Excitation with red light that is not directly absorbed by cellular materials
- Reduced cellular toxicity in biological studies
- Reduced photobleaching
- Deeper penetration of NIR light source into biological specimen
- Femtoliter volume excitation
- Single laser system can induce damage and image repair, without sample relocation
- Real-time observation of repair protein induction
- Flexible multimodal platform for cellular DNA damage and repair studies
- The ability to deliver UV-equivalent photon energies directly beneath UV-absorbing materials and molecules
- Ability to perform time-resolved studies due to short-pulsed light source

1.1. Principles of multiphoton absorption

The multiphoton absorption process was first described by Maria Goepert-Mayer in 1931 during her PhD thesis (Eq. (1.2); Goeppert-Mayer, 1931).

$$N_a = \frac{P_o^2 \delta}{\tau_p f_p^2} \left(\frac{A^2}{2hc\lambda}\right)^2 \tag{1.2}$$

N_a is number of photons absorbed by molecule, P_o is average power, out of optics δ is absorption cross section, τ_p is pulse duration in seconds, f_p is laser pulse repetition rate, h is planks constant, c is speed of light, λ is wavelength of light used, and A is numerical aperture of the lens (objective). The equation describes the square dependence for a two-photon absorption process.

The nonlinear quantum event of multiphoton absorption and excitation allows molecules such as DNA to be photoexcited/ionized by lower energy photons (longer wavelength) below the normal quantized energy levels (Fig. 1.2). The process can be envisaged as the electromagnetic field has sufficient intensity to perturb the electrons through a number of "virtual" states with residence time of about 10^{-15} s (femtoseconds) to reach an upper electronic state or drive the electrons a sufficient distance from their retaining atomic/molecular orbital that ionization occurs creating a solvated electron and a radical cation. The n-photon absorption depends on the nth power of the required intensity of light. The time average laser power (milliwatts) and subsequent peak power of the laser pulse is chosen so that sufficient photon intensity is found only within the central portion of the temporal profile of the Gaussian-shaped ultrafast laser pulse. The peak power is the key factor here, where power = energy (W)/time (s). Thus although the energy in a single laser pulse is very small, tens of nanojoules

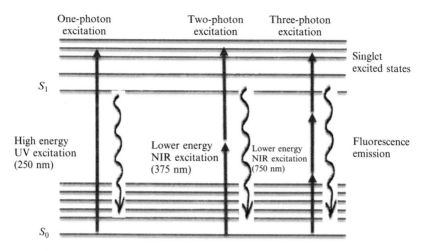

Figure 1.2 One-, two-, and three-photon Jablonski energy diagram.

(100×10^{-9} J) the ultrashort femtosecond (200×10^{-15} s) the peak power is megawatts (MW). This leads to the photon interaction occurring only in the most intense region of the laser spot, the center focused region, and this provides uniquely three-dimensional (3D) localization of the process beyond the diffraction limit (Denk et al., 1990; Meldrum et al., 2003), typically 300 nm spot. Exploitation of this enables high-resolution femtoliter 3D microbeam to induce DNA damage within individual mammalian cells such as depicted in Fig. 1.2. Although it is theoretically possible that the extremely high peak intensities required to drive the multiphoton process could lead to localized heating of the solvent, that is, water surrounding the cellular DNA, this heating process has been estimated to be negligible (no more than a rise of 10^{-5} K; Langford et al., 2001) since the rate of cooling and energy dissipation from 10 s of MHz of laser pulses is extremely fast compared with that of microjoule low repetition rate or continuous wave (CW) lasers. The multiphoton process is characterized and confirmed by a curvilinear plot of effect versus laser average power in the absence of any saturation process. A straight line graph is generated by creating a log–log plot, the slope of which gives the nth order of the observed effect.

1.2. Pulsed laser sources for multiphoton techniques

In the past few years, pulsed laser-based techniques developed in fundamental physical sciences, such as harmonic generation and multiphoton excitation, have found applications in many avenues of biophysical research. Photons are now routinely applied as noninvasive tools for imaging, cell and tissue surgery, manipulation of single molecules, cells, and tissues, with unprecedented accuracy (resolution) and dexterity (penetration depth) and energy (wavelength) across a wide range of timescales (seconds to femtoseconds; Grigaravicius et al., 2009; Hopt and Neher, 2001; Kawata et al., 2001; Konig, 2000; Konig et al., 2005; Tirlapur and Konig, 2002). Further, the available technology makes them budget able within grants and importantly reliable enough to be used routinely. Earlier research involving multiphoton processes used either CW lasers with very high average powers (100 mW or more) or nanosecond (10^{-9} s) lasers with medium to high pulse energy (micro–millijoules) to generate the required high peak intensities (Liu et al., 1995). However, both CW and nanosecond lasers have excessively high average powers and energies that cannot be tolerated by biological samples.

In practice, the high peak intensities required for the multiphoton processes (with low average power) are achieved by the use of high repetition rate ultrafast solid state lasers such as the self mode-locking Ti (titanium):sapphire laser (pump by a CW green laser) operating at 50–100 MHz with pulses as short as 10–200 fs (Konig, 2000). The addition of an OPO allows further tunability below 690 nm or above 1060 nm of the Ti:sapphire laser while maintaining the high repetition rate and ultrashort pulse nature of the laser. The OPO laser operates by a nonlinear optical process by generating photon pairs described as a signal (photons of interest) and an idler which can be made to lase. Tuning of the OPO laser depends on changing the Ti:sapphire laser wavelength as well as the temperature and angle of the nonlinear material, usually a beta barium borate crystal. The laser light is focused using a high numerical aperture (0.5–1.49) microscope objective. The sample is placed on the microscope stage (Fig. 1.3) and is irradiated with an average laser power of 1–30 mW (23–700 GJ cm^{-2}). For efficient multiphoton processes, the laser pulse width, pulse shape in time, and the collimation of the laser beam at the back aperture of the microscope objective lens need to be critically controlled using chirping methods to avoid optics-induced spreading of the broad range of wavelengths needed to generate femtosecond pulses. Note, a typical at 800 nm 100 fs laser pulse covers ca. 7 nm and these wavelengths have different transmission times through the optics (lenses), and hence this so-called *group velocity dispersion* induces a broadening of the required ultrashort pulse widths (Davis et al., 2002; Kong et al., 2009).

1.3. High-resolution cellular DNA damage upon NIR laser microbeam induction

A plethora of DNA damage can be induced by femtosecond high repetition lasers (Harper et al., 2008; Kong et al., 2009; König et al., 2001; Meldrum et al., 2003), and these lasers can also be applied to yield much needed information on the mechanism of cellular repair (Houtsmuller et al., 1999; Mari et al., 2006; Uematsu et al., 2007). The nature of DNA damage

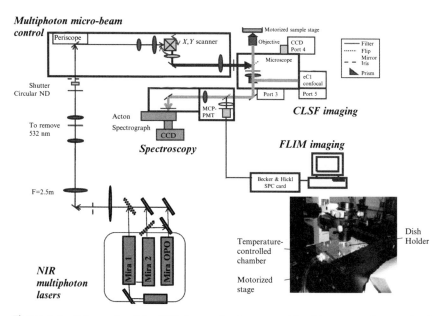

Figure 1.3 Schematic of the NIR laser microbeam combined laser scanning confocal and spectral detection setup. CLSFM, confocal laser scanning fluorescence microscopy; CW, continuous wave 1-P excitation lasers; DM, dichroic mirror; FLIM, advanced microscopy fluorescence lifetime image; KI, Kohler illumination for white light image; M, motorized computer-controlled stage; S, spectrograph for spectroscopy. (For color version of this figure, the reader is referred to the Web version of this chapter.)

generated by the NIR laser microbeam will depend on a number of conditions; the laser wavelength, material present, and molecular intermediates generated (singlet, triplet electronically excited states, ionization products). Such complexity means it is unclear which processes lead to the formation of DNA SSBs and DSBs although the majority of base modifications are likely to be via excited state and ionization products (Melvin et al., 1996; Tirlapur et al., 2001). As well as direct processes, indirect processes may also be possible as some reports have indicated the involvement of reactive oxygen species such as peroxides, hydroxyl radicals, and singlet oxygen (Tirlapur et al., 2001). For multiphoton processes to occur, a molecule must interact with a sufficient number of photons to generate the excited state (most likely a vibrationally excited electronic excited state; see Section 1.2). As the wavelength increases the photon's energy decreases (Table 1.1), for example, the energy of a single photon from a 750 nm source is a third of the energy of a photon from a 250 nm source. If three photons are, therefore, absorbed by a molecule in a stepwise fashion from a 750 nm source (NIR), it is equivalent in energy to that molecule absorbing a single photon from a 250 nm source (UV; Table 1.1).

Table 1.1 Multiphoton irradiation wavelength and photon equivalents (modified from Davis et al., 2002)

Energy/photon (eV)	One photon λ/n equivalent	Two photon λ/n equivalent	Three photon λ/n equivalent nm
1.73	715	359	238
1.70	730	360	248
1.65	750	375	250
1.59	780	390	260
1.55	800	400	267
1.38	900	450	300

It has been shown that the very high peak photon powers generated (10–12 W cm^{-2}) is capable of generating plasma, the forth state of matter which describes multielectron ionization and energetic electrons (Turcu and Dance, 1998). Such an environment would efficiently cause DNA damage. It has been shown previously that CW UV light source together with loading bromodeoxyuridine (BrdU) and the DNA minor grove interchelator dye Hoechst leads to efficient DNA DSB induction (König et al. 2001, Limoli and Ward, 1993). However, the yield of DNA DSBs is not enhanced following NIR irradiation in the presence of Hoechst and BrdU (Harper et al., 2008).

2. Experimental Procedure

Our NIR laser microbeam setup is an adapted commercial fluorescence laser scanning confocal microscope (Nikon TE2000-U and Nikon Ti-U-eC1), which has custom-built multiphoton laser scanning system (Botchway et al., 2008; Davis et al., 2002). The configuration allows us to operate both confocal and multiphoton techniques simultaneously so that cellular DNA can be damaged and *in situ* confocal microscopy imaging can be carried out simultaneously. Such a setup has several advantages over operating the multiphoton technique through the same commercial scanning system. For example, the ability to independently control the position of either the imaging laser or the DNA damage NIR laser is lost in a single scan-head system. However, constructing the laser microbeam around a commercial microscope allows us to utilize fully other options available on the microscope for conventional imaging studies such as bright-field Kohler illumination, epifluorescence, differential interference contrast, phase contrast, and confocal that is needed to allow selection of cellular components prior to irradiation. The several imaging ports on the standard research microscope (up to eight) also provide the means to add several imaging and detector devices for time-resolved measurements (single-photon counting), super-resolution techniques, and spectroscopy (see Fig. 1.3). Although commercialization of multiphoton microscopy is becoming widespread (at a significant cost), the majority in use are custom built and this still offer many key advantages as described here. Most setups have only one scan head for controlling the multiphoton laser and confocal laser and do not offer the ability to perform both nonlinear and linear microscopy easily and independently at the same time.

2.1. Construction of the NIR laser microbeam: The light source

The availability of Ti:sapphire lasers operating over the NIR region with ultrashort (<300 fs) and at ~ 100 MHz have paved the way for efficiently driving multiphoton processes. Tuning the laser to a desired wavelength produces a nonmonochromatic light. The ultrashort pulse nature of the Ti: sapphire laser has a corresponding broad range of wavelength of about 5 nm to 100s of nm. So that the exact spectral width depends on the precise shape of the pulse and is related to the temporal profile as Eq. (1.3)

$$\Delta \lambda \sim \frac{\lambda_o^2}{c\tau_p} \qquad (1.3)$$

where λ_o is the center wavelength, c is speed of light, and τ_p is the pulse length.

It is possible that the spectral width of a multiphoton laser pulse can extend beyond that of the excitation spectrum of the molecule of interest, for example, that of a 10 fs pulse width laser system. In our system, we have used a commercial Kerr-lens self mode-locking Ti:Sapphire laser (Mira900F) from coherent laser, with a reported pulse width of 180 fs directly out of the laser output. (It is worth noting that this laser system provides a valuable option to operate in either a pulsed or CW mode and is tunable from 690 to 1060 nm.) The addition of an OPO in our setup allows further tunability from 540–650 nm to 1080–1300 nm). Due to the broad spectral nature of the ultrashort pulse lasers, significant pulse broadening occurs when propagating through glass optical elements such as lenses, dichroics, filters, and prisms. This is particularly worse through a high numerical aperture microscope objective that contains several lens elements. We and others have determined this broadening to be as much as two to four times for a 180 fs pulse width. The problem is worse for the very short laser pulses <30 fs. Simply, as the speed and frequency of light varies through different media compared to air, this leads to a change in the group velocity, leading to a dispersion of the pulse (GVD). Since the multiphoton process is directly related to the pulse width, it is, therefore, necessary to compensate for this broadening by means of chirping. This can be efficiently achieved by the use of a prism pair or a matched grating such as the femtocontrol (from APE, Germany) as used in our setup. This allows delivering the original pulse length after passing through the optical train for laser excitation. For optimization of the setup and any quantification process, it is necessary to determine the pulse width at the microscope stage. This is achieved using a commercially available autocorrelator, which briefly works by splitting the beam into two, one going onto a delay line and then colliding these beams again while scanning the delay line. The sum frequency signal generated by the beams, measured using a standard two-photon detector, correlates with the movement of the delay line and hence the temporal width of the laser beams. The use of nonlinear crystal or crystalline material form (such as dried saturated solutions of urea or potassium dihydrogen phosphate, KDP) to detect the second harmonic generation (SHG) prior to detection provides an efficient method to optimize the multiphoton process and optimization procedure of the setup. Simply, the SHG is sensitive to fluctuations in the laser power, center wavelength, spectral width as well as pulse width. Since one of the main drivers of the multiphoton process is the laser power (peak intensity), this can be achieved by high average power or reduced pulse width. Fine control of the laser beam characteristics results in optimized sample irradiation and reducing unwanted secondary effects, such as photobleaching and interference from unwanted autofluorescence. The advantage of using laser pulse width of ~ 180 fs is that the spectral width is sufficiently low to avoid a drop in the spatial resolution produced by the microscope objective (comprising several optical elements)—a key requirement for the NIR laser microbeam technique. The average laser power at the sample needs to be monitored using a standard thermocouple laser power meter (Thorlabs or Molectron 500D) or a simple calibrated photodiode and maintained throughout the laser irradiation. The spectral characteristics may be monitored using a standard USB spectrometer (Ocean Optics).

The laser output of the Mira 900F laser system has a fairly large divergence of ~ 1.7 mrad (divergence over 1 m distance) from an original spot size of 0.7 mm. There is, therefore, a need to recollimate the beam before expansion into the microscope for the microbeam DNA damage. Without this step, the high numerical aperture of the microscope objective will not be fully utilized, and DNA damage is likely to be inefficient and/or spread over a large region. The collimation is achieved either by a Keplerian telescope (a pair of lenses, Fig. 1.4) or a single long-focal length lens placed one focal length from the focus of the laser, usually from inside of the laser. This method has a number of advantages as it reduces the number of optical elements that would contribute to the pulse group velocity dispersion and chromatic aberration. However, the position of this lens is critical to allow

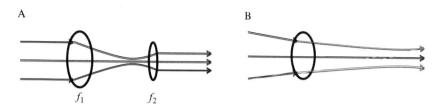

Figure 1.4 Kerperian telescope: (A) The lens separation needs to be the sum of the two focal lengths (f_1 and f_2) and positioned anywhere along the beam path. (B) A long-focal length lens acts as a pseudopair of a telescope.

the right amount of convergence or divergence at the backplane of the microscope objective. This is particularly important since the majority of lenses are designed for visible wavelengths (400–600 nm). For a custom-built multiphoton microbeam to work perfectly having a focus in the same plane as the confocal scanning system or Kohler illumination bright field, the right beam characteristics at the objective is vital. The magnification or demagnification can be obtained by selecting a pair of lenses whose focal length ratio equals the required outcome.

2.2. The multiphoton laser microbeam scanning system

The principles and operation of this part of the system is similar to that of the commercial confocal scanning system. The custom system consists of a pair of galvanometers (galvos, VM500, GSI Limonics, now Cambridge Technology) incorporating aluminum mirrors. The NIR laser is attenuated using a variable motorized, computer-controlled neutral density filter with the laser beam being expanded to 6 mm prior to entering the galvanometers (Fig. 1.5). The deflection and subsequent movement of the laser beam by the galvanometer is computer controlled through a sc2000 pc card. A second set of telescopes further expand the laser beam to 11 mm to allow for marginally overfilling of the back aperture (BA) of the microscope objective, usually ×60 water, numerical aperture 1.2, BA 10 mm. These same lenses act to image the galvos onto the BA of the objective to further expand the beam to the desired 11–12 mm and to act as beam steering optics so that the galvo movement is not translated off the back of the objective. During initial alignment of the system, care must be taken to avoid either "pin-cushion" or "barrel" distortions. This is kept in check by ensuring the laser beam travels only through the central axes of the optics (Fig. 1.5). Generally the NIR laser microbeam is not sensitive to polarization. Although the Ti:sapphire output is linearly polarized, there is a small degree of depolarization at the microscope objective. A quarter waveplate may be used together with polarizing optics to clearly define the polarization

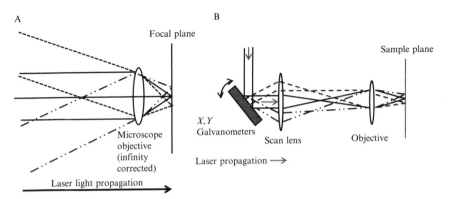

Figure 1.5 Schematics of the basic optics control and arrangements of the optical path and conjugate aperture plane to allow stationary laser beam scanning at the sample (A) at the infinity corrected microscope objective and (B) with scan lens and tube lens.

characteristics if required. The position of the polarizer optic is not too critical. In our setup, a dichroic mirror (660IK Comar or FF665-Di01, Semrock) is used to reflect the NIR laser beam into the objective. It is important to note that in this configuration, the multiphoton microbeam for cellular irradiation is identical to that used for confocal microscopy with the exception of a lack of a pinhole. A major advantage of the Nikon Ti and TE2000 series inverted microscopes is the large parfocus distance (60 mm) making it possible to utilize this space for other optical elements without compromising the beam path and characteristic of the objective. A set of dichroics were therefore stacked vertically to allow top-collected fluorescence emission, second harmonic generated photons onto a fast microchannel plate (MCP-PMT) for excited state lifetime measurements and the bottom dichroic for reflecting the laser light into the microscope objective. This arrangement allows weak fluorescence emission photons from the sample to be detected with high efficiency without first passing through any optics or descanned except being focused onto the detector (Fig. 1.3).

2.3. Simultaneous confocal imaging of microbeam-induced DNA damage

The multiphoton laser scanning microbeam was constructed around the back port of the inverted Nikon TE200 and Ti-U microscopes. This allowed attachment of a Nikon eC1 and eC1–Si at the side port (number 5). Thus, both confocal and multiphoton beams could arrive at the sample at the same time and can be controlled independently. The Nikon eC1 confocal is equipped with a supercontinuum laser as the excitation source for multicolor imaging while the eC1-Si has a three-laser attachment

(Diode laser 405 nm, Becker and Hickl, 488 nm, and 543 nm both from Melles Griot) as well as four detection capabilities, including transmission light. All lasers may be used simultaneously. The eC1 also offers the ability to perform spectral detection. However, a better, more sensitive spectral analysis is performed by utilizing a separate port on the inverted microscope and detector system. Our setup uses auxiliary port 3, fitted with a UV prism and the output directed to an Action spectrograph 275 and an Andor CCD IDus 420-B camera, providing Raman (not discussed here) as well as fluorescence spectroscopy. Emission from the one-photon confocal is collected by the objective, descanned through a selectable pinhole focused onto a light-guide fiber with detection using onboard photomultiplier tubes (PMTs). The microscopes were fitted with a computer-controlled motorized stages (IM 120, Marhauzer) and housed in an environmental-controlled casing (Solnent) controlling the temperature and gas flow for live cell studies.

2.4. Cell culture and NIR multiphoton irradiation

In this section, we discuss a standard method for cell culture, NIR laser irradiation to induce DNA DSB, real-time visualizing of repair protein as well as using antibodies that have been raised against an antigen of protein of interest followed by fluorescently tagging these antibodies with a fluorophore.

2.4.1. Required materials and device

A gyratory rocker (purchased from VWR) can be used to ensure an even distribution of reagents across the cells.

Additional materials

- Cells of interest: may be purchased from American Type Culture Collection (ATCC) or European Collection of Cell Cultures (ECACC), a Health Protection Cell Culture Collection.
- Growth medium: The growth medium to be used is specific to the cell type and can be found on the ATCC or ECCAC data sheet supplied with the cell line. Alternatively, the data sheet is also available on the ATCC or ECCAC website (http://www.lgcstandards-atcc.org; http://www.hpacultures.org.uk). Growth medium requires the addition of supplements 10% fetal calf serum and 2 mM L-glutamine. Antibiotics can also be added to the growth medium such as 100 units/ml penicillin and 100 µg ml^{-1} streptomycin.
- Primary antibodies raised against the antigen/protein of interest: Primary antibodies can be purchased from a number of companies including Abcam and Insight Biotechnology. Antibodies can vary between monoclonal and polyclonal. There are a number of advantages and

disadvantages of each; monoclonal antibodies recognize a specific target epitope reducing background staining. These antibodies may have reduced reactivity while polyclonal antibodies react with multiple epitopes allowing amplification of the signal although can cross react to give high background staining.
- Secondary antibodies: Secondary antibodies can be purchased from Invitrogen (Alexa Fluor secondary antibodies). Secondary antibodies must react with the species in which the primary antibody was raised in and must also be specific to the primary antibody immunoglobulin, that is, IgG.

Other reagents

- Phosphate buffer saline (PBS), pH 7.0 (1×)
- Triton X-100 (1% in PBS)
- Bovine serum albumin (BSA) 1% in PBS
- Fish skin gelatine (FSG) 1% in PBS
- Paraformaldehyde, 3% in PBS
- Vectashield® antifading medium

Disposables

- Pipettes (p1000, p200, p20, p10)
- Pipette tips (1 ml, 200 µl, 20 µl, 10 µl)
- 1.5 ml (Eppendorff) tubes
- 30 mm glass bottom dishes (glass bottom thickness number 1)

2.4.2. Plating adherent cells for NIR laser microbeam irradiation

V79-4 and CHO cells are plated at 1.5×10^5 cells/dish in 30 mm diameter glass walled, number 1 glass coverslip bottom dishes containing 3 ml of medium at 37 °C to obtain 70% confluency prior to irradiation. For other cell types, the plating density will need to be predetermined for ∼70% confluency on the day of irradiation. Incubate cells for 24 h prior to irradiation at 37 °C and 5% CO_2 humidified air to form a monolayer of adherent cells.

2.4.3. Laser microbeam irradiation

Plate cells as described in Section 2.4.2 on number 1 glass coverslip bottom dishes and incubate with 10 µg ml^{-1} Hoechst 33258 dye for 10 min prior to irradiation at 37 °C for real-time studies. Place a red bandpass filter (RG610) on top of the culture dish to prevent the Hoechst 33258 dye from absorbing UV light from ambient light.

For real-time analysis, maintain cells at 37 °C throughout the irradiation using a temperature control chamber such as the Warner TCM-1 and controller CL-100. Irradiate cells at a laser wavelength of 730 nm and a nominal power of 10 mW measured through the ×40 air, numerical aperture 0.95, or ×60 water immersion objective, numerical aperture 1.2 microscope objective. A custom-made computer software program written in LabVIEW^{TM} (or the proprietary software, Wincommander) is used in our laboratory to set experimental parameters for automated movement of the microscope stage (Märzhäuser Wetzlar GmbH & Co) in the x- and y-planes in colocalization studies. The parameters were set for irradiation of mammalian cells in a raster scanning pattern with the area set for an optimal step size of 12 μm to allow one irradiation per nucleus (Fig. 1.6). This will need to be optimized for other cell types as it is desirable to irradiate each cell nucleus only once so that saturation of the repair pathways does not occur. These parameters are best for V79-4 and CHO cells. Each movement of the stage can be controlled individually using the Wincommander software allowing for more precise irradiation of a small number of cells and faster image collection. Time zero is recorded immediately following irradiation of the cells (< 10 s), and images can be collected at predetermined times following irradiation using confocal microscopy (EC1, Nikon) equipped with an argon ion laser at 488 nm and HeNe (helium/neon) laser at 543 nm.

During colocalization studies, the cells should be cooled to 10 °C with 10 μg ml^{-1} Hoechst dye for 10 min prior to irradiation to slow cellular processes and hence DNA damage repair during laser microbeam irradiation. Irradiations are carried out at a laser wavelength of 730 nm, 25 mW nominal power, measured through the ×40 air, numerical aperture 0.95, microscope objective, and over a raster scanning area of 0.5 × 0.25 cm so that >100 cells are irradiated and imaged for statistical analysis (see above). Cells are irradiated in culture medium and maintained at 10 °C throughout the irradiation (approximately 12 min) using the temperature control chamber. Following irradiation of the cells in colocalization studies, the culture medium should be replaced with medium warmed to 37 °C. The

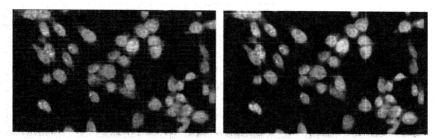

Figure 1.6 Raster scanning optimization figure: Left panel shows a cell population covered by a 5 μm raster grid, whereas right panel shows a cell population covered by a 25 μm raster grid. (For color version of this figure, the reader is referred to the Web version of this chapter.)

irradiated cells are incubated at 37 °C with 5% CO_2 in humidified air for the predetermined repair times. The cells are then fixed prior to immunofluorescent staining.

2.4.4. Visualization of proteins in real time

A significant advantage in using the NIR multiphoton microbeam technology to induce DNA damage is its ability to observe the cell nucleus in real time even before the damage occurs. While the use of other forms of ionizing radiation requires the samples to be first irradiated on ice or at 4 °C (to limit processing of the damage by the cell repair machinery), removed to the imaging system (the microscope stage), the sample warmed up to 37 °C before the imaging can begin. In our experience, this process (using ultrasoft X-rays, 1.5 keV at a dose of 10 Gy) results in a minimum initial time loss of 5 min. Any fast repair processes are therefore lost by the ionizing radiation techniques (particularly in X-ray and gamma ray irradiation experiments).

In the real-time NIR laser microbeam irradiation, cells containing a fluorescently tagged proteins (e.g., Ku80-EGFP and DNA-PKcs-YFP, key proteins involved in nonhomologous end joining (NHEJ) repair of DNA damage) are laser microbeam irradiated in the presence of Hoechst 33258 dye at 37 °C and simultaneously imaged using a Nikon Eclipse C1 confocal microscope coupled to a Nikon TE2000U microscope. An argon ion (488 nm) laser excites the EGFP-tagged and YFP-tagged DNA repair proteins in the fluorescently tagged cells. The pinhole and gain of the confocal microscope are adjusted to obtain optimal fluorescence without saturation of the pixels. The images can be collected using three Kalman filtered scans. Kalman filtered averages a number of scans thus increasing the signal-to-noise ratio, resulting in a less grainy image. Images should be collected at the predetermined times following initial radiation with a minimum of 10 cells visualized for analysis per experiment (Fig. 1.7). Since both the NIR light and visible confocal laser light are parfocal at the same focus region (~ 300 nm), the confocal imaging can begin prior to the NIR irradiation. Images obtained prior to the NIR irradiation is considered as time zero. The induction of GFP (green fluorescence protein)-tagged Ku70/80 involved in NHEJ was observed < 15 s post DNA damage induction as lines of fluorescence.

2.4.5. Cell preparation for immunofluorescent staining

One day prior to immunofluorescent staining for the protein of interest (e.g., RAD51 or γ-H2AX), mammalian cells are cultured in 30 mm glass bottom dishes in 3 ml fresh growth medium. The cells should reach $\sim 70\%$ confluency on the day of immunofluorescent staining.

On the day of immunofluorescent staining, prepare 1% Triton X-100 solution in PBS and blocking solution containing 1% FSG and 1% BSA in PBS.

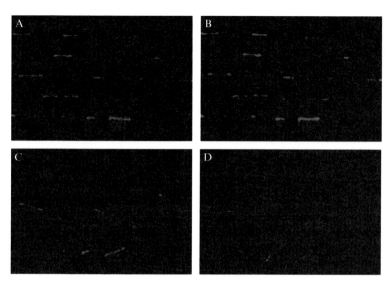

Figure 1.7 Real-time visualization of Ku80-EGFP following NIR multiphoton laser microbeam irradiation of exponentially growing Ku80-EGFP-tagged XR1V79B cells. (A) 50 s, (B) 3 min, (C) 30 min, and (D) 60 min postionizing radiation. (For color version of this figure, the reader is referred to the Web version of this chapter.)

2.4.5.1. Immunofluorescent staining Following incubation at 37 °C for the required repair times postirradiation with NIR laser microbeam, fix the cells in 1 ml of 3% paraformaldehyde in PBS at 4 °C for a minimum of 10 min. Aspirate the 3% paraformaldehyde into a waste bottle for hazardous waste disposal. The following steps should be carried out by placing the cells on a gyratory rocker. Wash the cells three times with 1 ml PBS for 5 min at room temperature. Permeabilize the cells with 500 µl of 1% Triton X-100 in PBS for a maximum of 10 min. Wash the cells three times with 1 ml PBS for 5 min at room temperature and block for 1 h at room temperature in 500 µl of 1% BSA and 1% FSG in PBS. Label the cellular proteins by adding 200 µl of the primary antibody, in 1% BSA and 1% FSG, for 1 h at room temperature or 4 °C overnight. Antibody concentrations should be optimized for each cell type (in CHO and V79-4 cells γ-H2AX at 1:200 dilution and RAD51 at 1:50 dilution). Wash the cells three times with 1 ml PBS for 5 min at room temperature followed by the addition of 200 µl of secondary antibody (diluted 1:1000) for 1 h at room temperature. Wash the cells three times with 1 ml PBS for 5 min at room temperature. Mount the cells in antifading medium by adding one drop of Vectashield® ± DAPI to each sample. Immediately cover the cells with a 25 mm diameter number 1 glass coverslip. Visualize the fluorescently tagged cellular proteins using confocal microscopy.

2.5. DNA DSBs determination: comet assay–single-cell gel electrophoresis in NIR laser microbeam studies

As yet, only a few studies have demonstrated DNA DSBs induced by a NIR multiphoton microbeam (Collins, 2004; Olive, 2009; Piperakis et al., 1999). This is primarily due to the difficulty in dealing with a relatively low number of cells irradiated with the multiphoton laser microbeam and the associated low copy of strand breaks. Analysis of a sample culture dish containing majority of unirradiated cell would lead to false results.

To our knowledge, the comet assay appears to be the main method to directly demonstrate and quantify strand breakage following NIR laser irradiation. Although an indirect method for strand breaks may be inferred from immunohistochemical staining of γH2AX as an indicator of DSB (see below). In the immunohistochemical staining method, exponentially growing mammalian cells in culture (Mat-tek 30 mm culture dish with #1.5 glass coverslip or specially constructed glass ring with #1.5 coverslip glued to the bottom) may be used for the cell culture. This allows for an even illumination field of view. Prior to irradiation, 10 $\mu g\ ml^{-1}$ Hoechst may be added to the cell culture and allowed to load into the DNA for 10 min for greater yield of DNA damage. The culture dishes are placed in a custom aluminum block jacket with circulating water. The temperature can be controlled between 2 and 50 °C. Low temperature studies (2–10 °C) allow cell studies where DNA repair functions are significantly minimized. By maintaining the temperature-controlled jacket or the microscope incubator at 37 °C and flushed with 5% CO_2 in air, cells may be kept at optimum growth condition for long periods of time. Comparing the relative quantities of DNA damage, strand breakage etc., induced by ionizing radiation with those produced by NIR microbeam irradiation (at defined laser power levels) is of great interest to develop our understanding of the damage thresholds that activate repair signaling pathways. Generally studies for DNA DSB analysis that used comet assay analysis were performed at 10 °C. Here the NIR laser beam is kept stationary (at an average laser power of 4, 10, or 25 mW) and the microscope stage was raster scanned over a 5 × 5 mm area. Raster scan steps of 12 μm were selected following optimization with different step sizes ranging from 5 to 25 μm. This optimization step ensures that the majority of cells (Chinese hamster V79-4 cells) within the scan area were exposed to the laser beam. However, a small fraction of cells (14%) is likely to receive two laser scans as a consequence of the cells having a distribution of cell areas (Fig. 1.6). A small percentage of cells (4%) may not show damage tracks mainly as a consequence of the variation in cell heights with respect to the position of the laser focal point within the cell. For the present studies, a raster scan area of 2.5 × 2.5 mm was used and the scan time was approximately 12 min. Further experimental control is also available using a custom-written software (LabVIEW^{TM}), which allows the identification and logging of cell positions during the laser irradiation with 100 nm accuracy for revisiting later. Following irradiation at 10 °C, the scanned area is washed in 1× PBS, carefully trypsinized and transferred to lysis buffer prior to cell lysis (see below for details).

2.5.1. Transfer of irradiated cells to microscope slide for comet assay

Part frosted microscope slides were coated with a layer of 1% low-melting point agarose (LMPA) to aid adhesiveness. Once the slides were dried, the agarose sandwich was added to precooled slides and 170 μl of 2% LMPA was placed onto the slide and immediately covered with a coverslip. The agarose is allowed to set for about 10 min on the coverslip then removed to leave the agarose layer on the slide. The V79-4 cells suspended in FCS were mixed 1:1 with 2% LMPA and a second layer of 170 μl containing the cells was placed on the slide and a second coverslip added. This coverslip was removed and a final layer of 200 μl 1% LMPA was added. Two slides were prepared per sample. The slides were incubated in ice-cold lysis buffer (2.5 M NaCl, 100 mM EDTA (pH 8.0), 1% Sarkosyl, 10 mM Tris–HCl (pH 8.0), 10% DMSO, and 10% Triton X-100) for 1 h and rinsed in ice-cold PBS for 15 min. A horizontal electrophoresis tank was placed on ice and filled with ice-cold alkali buffer (0.3 M NaOH, 1 mM EDTA, pH 13.0) and the slides were placed in the alkali buffer for 30 min prior to electrophoresis to allow the DNA to unwind. The slides were electrophoresed at a constant voltage of 22 V (500 mA) for 30 min at $\sim$4 °C then rinsed for 10 min in ice-cold neutralizing buffer (0.5 M Tris–HCl, pH 7.5) followed by 15 min in ice-cold PBS. The slides were stored in a moist environment overnight and rehydrated the following day in 500 ml ice-cold PBS for 10 min prior to staining.

2.5.2. Image capture and analysis of comets by confocal microscopy

The DNA was stained for analysis by flooding each rehydrated slide with 1 ml of 10 μg ml^{-1} of ethidium bromide for 2 min. Excess ethidium bromide was removed and a coverslip (#1) placed on top of the gel. Each slide was examined for DNA damage using the one-photon confocal microscope described above. Images were collected at ×20 magnification with a numerical aperture of 0.4 and an excitation wavelength of 543 or 568 nm (Fig. 1.8). A minimum of 25 comets per slide were captured to allow qualitative analysis of 50 comets per sample as quantitative analysis is not appropriate with nonhomogenous distributed damage (Davis et al., 2002).

Recent work using the NIR microbeam setup to activate damage in the absence or presence of Hoechst dye, 25 mW (peak intensity 580 GWcm^{-2})

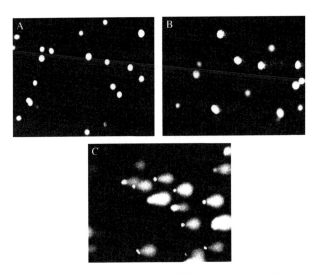

Figure 1.8 Induction of DNA damage upon NIR microbeam multiphoton excitation. Observation using the comet assay in (A) V79-4 control cells, (B) cells exposed to 750 nm multiphoton irradiation 25 mW at the sample at 10 °C, or (C) cells exposed to 10 μM H$_2$O$_2$.

of average laser power has been determined to be equivalent to 2 Gy dose of low LET radiation (70 DSBs along DNA damage track, i.e., ∼3 DSBs per mW; Botchway et al., 1997; Harper et al., 2008; Reynolds, 2009).

2.6. Observation of DNA damage–repair dynamics following laser microbeam induction

To understand the processes involved in the repair of DNA strand breaks in live cells, it is necessary to follow the dynamics of recruitment of the repair proteins to the sites of damage and their subsequent loss. This can be achieved by using GFP technology to tag key proteins of interest or by halting the repair process at specific time points, fixing the cells for immunohistochemical antibody staining. In the real-time studies, we use GFP-tagged cells stably expressing either Rad-51 or Ku-80 cells (gift from Dik van Gent).

2.6.1. Real-time studies

A significant advantage in using the NIR multiphoton microbeam technology to induce DNA damage is the ability to observe the cell nucleus in real time even before the DNA damage occurs. While the use of other forms of ionizing radiation requires the samples to be first irradiated on ice or at 4 °C to limit processing of damage during the irradiation period and uptake by diffusion, removed to the imaging system (the microscope stage, which has sensitive and radiation-damageable detectors) and the sample warmed up to 37 °C before imaging may begin. In our experience, this process (using ultrasoft X-rays, 1.5 keV at a dose of 5 Gy) results in a minimum initial time loss of 5 min. This means any immediate and fast repair processes are lost from observation in ionizing radiation (particularly in X-ray and gamma ray experiment) techniques.

In the NIR laser microbeam technique, cells expressing GFP-labeled repair proteins Ku70/80 or Rad51 are seeded at a density of 2×10^5 on either Mat-Tek petri dish with number 1.5 bottom cover glass or custom glass-bottom cell culture dish may be prepared by gluing (with araldite mixture) glass coverslips to glass rings with 30 mm (outer diameter with 27 mm internal diameter). The exponential growing cells are irradiated as described for the comet assay above. Here the cells may also be irradiated by raster scanning the NIR laser light. Since both the NIR light and visible confocal laser light are parfocal (same focus region ~ 300 nm), the confocal imaging can begin immediately after the NIR irradiation. Images obtained prior to the NIR irradiation was considered as the zero time point. The speediness of the technique is fully demonstrated by observing the induction of GFP-tagged Ku70/80 involved in NHEJ within 15 s of the DNA damage induction as lines of fluorescence. Analysis of a time course of fluorescent images with time gives a half-life for the Ku70/80 induction as 2.0 ± 0.25 min. The induction of GFP-tagged Rad51 involved in HR was observed within 5 min of the DNA damage. The half-life of this protein was determined to be ~ 60 min for the initial phase of the repair process using the florescence intensity change over time (Fig. 1.9).

2.7. Spectral detection of multiphoton microbeam-induced excited state chemical species

Generally, the left side port of the Nikon TE200-U is fitted with a 20/80 prism and labeled as number 2. The same side port labeled as number 3 (auxiliary) also contains an empty holder for a user-inserted optics. We have fitted this port with a 100% UV-reflecting prism. This allows directing fluorescence or harmonic generated photons from the sample and focusing ($f = 80$ mm) into a spectrograph (275 Action). An Andor iDUS CCD is fitted to the spectrograph for detection. Figure 1.10 shows cultured cells (Section 2.4) expressing GFP-tagged repair protein as well as labeled with Hoechst dye. The excitation wavelength at more than 900 nm, 180 fs, 75 MHz (multiphoton processes) shows the fluorescence spectra from GFP while excitation at the same pixel position with 730 nm (multiphoton) shows a complex spectra of autofluorescence, Hoechst, and GFP emission. More importantly, in this configuration, we are able to record fluorescence spectra from any pixel in the image.

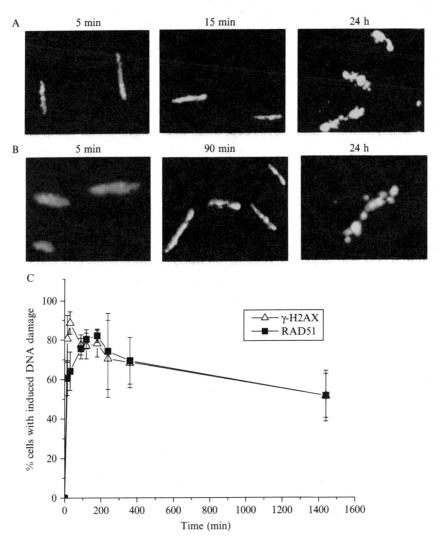

Figure 1.9 From Harper et al., 2008: (A) colocalization (yellow) of γ-H2AX (green) and 53BP1 (red), (B) colocalization (yellow) of γ-H2AX (green) and RAD51 (red) in a fixed exponential population of V79-4 cells following NIR laser microbeam irradiation. (C) Time course of the recruitment of γ-H2AX and RAD51 to sites of NIR laser microbeam-induced damage in exponentially growing V79-4 cells. The graph represents the time dependence of the percentage of cells (from the mean of at least three independent experiments ± SEM) that show damage induced by multiphoton laser excitation and visualized as lines. (For interpretation of the references to color in this figure legend, the reader is referred to the Web version of this article.)

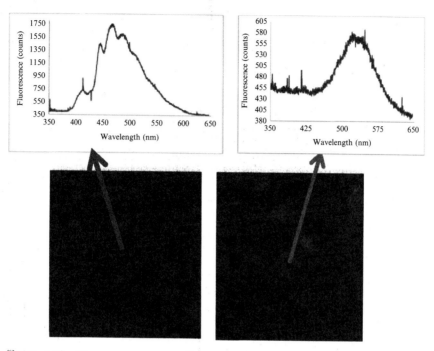

Figure 1.10 Fluorescence spectra from a single pixel of the image. Multiphoton excitation at (A) 730 nm or (B) 910 nm to reveal autofluorescence/Hoechst dye and GFP-tagged repair protein, respectively. (For color version of this figure, the reader is referred to the Web version of this chapter.)

3. CONCLUSION/FORWARD LOOK

This report has concentrated on primarily the development of realtime induction of DNA strand break in mammalian cells upon high repetition rate femtosecond NIR laser excitation. The technique is being driven and supported by a rapid growth in instrument availability, reliability all at reduced price of the ultrafast femtosecond laser sources suitable for biological science departments. Although the current literature shows a plethora of DNA lesions induced by the laser microbeam, the mechanisms leading to the observed biological end point remain to be better understood. However, the future role of ultrafast NIR lasers will provide exciting new ways to move this field forward with it offering offer excellent subcellular (submicron) localization of the DNA damage in four dimensions (3D and temporal) which has been difficult to obtain using more conventional high LET radiation.

While the very high peak intensities (GWcm^{-1}) generated within the femtoliter volume of the laser focus and subsequently, directly within the

DNA in a cell, is capable of generating either X-rays or fast energetic electrons for radiobiological experiments such as those previously reported (Turcu and Dance, 1998) or from UV (ns) pulses, these have not been discussed here since these are not laser microbeam, but nonetheless, offer new insights into the early (10–15 s) physical effects of radiation on DNA on the pico- and femtosecond timescales in living cells.

The NIR laser microbeam potentially offers the ability to investigate radiobiological studies such as genomic instability and bystander effects following subcellular targeted damage without the surrounding medium or neighboring effects. While the damage size produced followed NIR multiphoton microbeam so far reported is 300–1000 nm, it is possible to generate cellular DNA damage on the order of 100 nm (corresponding to $\sim$65 kb) as demonstrated for multiphoton polymerization processes (Kawata et al., 2001). Finally, the combination of the NIR laser microbeam together with simultaneous confocal and other imaging techniques offer the ability to investigate DNA damage and repair dynamics in real time; a significant advantage over other microbeam techniques.

ACKNOWLEDGMENTS

The work reported here was supported by the Science and Technology Facility council BioMed Network, access to the Central Laser Facility, Lasers for Science Facility imaging setup, and Oxford University, Gray Institute, MRC. Figures 1.1–1.3 have been modified and reprinted with permission from Rights link, Elsevier (2657001169267). Thanks to Dr Rahul Yadav for help with the manuscript.

REFERENCES

Botchway, S. W., Stevens, D. L., Hill, M. A., Jenner, T. J., and O'Neill, P. (1997). Induction and rejoining of DNA double-strand breaks in Chinese hamster V79-4 cells irradiated with characteristic Aluminum K and Copper L ultrasoft X rays. *Radiat. Res.* **148**(4), 317–324.

Botchway, S. W., Parker, A. W., Bisby, R. H., and Crisostomo, A. G. (2008). Real-time cellular uptake of serotonin using fluorescence lifetime imaging with two-photon excitation. *Microsc. Res. Tech.* **71**(4), 267–7344.

Collins, A. R. (2004). The comet assay for DNA damage and repair: Principles, applications, and limitations. *Mol. Biotechnol.* **26**(3), 249–262.

Cox, R., and Masson, W. K. (1979). Mutation and inactivation of cultured mammalian cells exposed to beams of accelerated heavy ions III. Human diploid fibroblasts. *Int. J. Radiat. Biol.* **36**(2), 149–1609.

Daudin, L., Carrière, M., Gouget, B., Hoarau, J., and Khodja, H. (2006). Development of a single ion hit facility at the Pierre Sue Laboratory: A collimated microbeam to study radiological effects on targeted living cells. *Radiat. Prot. Dosimetry* **122**, 310–312.

Davis, E. L., Jenner, J. T., O'Neill, P., Botchway, S. W., Conein, E., and Parker, A. W. (2002). Characterisation of DNA damage induced by near infrared multiphoton absorption. RAL-TR-(2003) **018**, 145–147.

deLara, C. M., Jenner, T. J., Townsend, K. M. S., Marsden, S. J., and O'Neill, P. (1995). The effect of dimethyl sulfoxide on the induction of DNA double-strand breaks in V79-4 mammalian cells by alpha particles. *Radiat. Res.* **144**(1), 43–49.

Denk, W., Strickler, J., and Webb, W. (1990). Two-photon laser scanning fluorescence microscopy. *Science* **248**, 73–76.

Folkard, M., Vojnovic, B., Prise, K. M., Bowey, A. G., Locke, R. J., Schettino, G., and Michael, B. D. (1997). A charged-particle microbeam. I. Development of an experimental system for targeting cells individually with counted particles. *Int. J. Radiat. Biol.* **72**, 375–385.

Goeppert-Mayer, M. (1931). Elementary process with two quantum jumps. *Ann. Phys.* **9**, 273–294.

Grigaravicius, P., Greulich, K. O., and Monajembashi, S. (2009). Laser microbeams and optical tweezers in ageing research. *Chem. Phys. Chem.* **10**, 79–85.

Hall, E. J. (1994). *Radiolobiology For The Radiologist*. 4th edn. J. B. Lippincott Co., Philadelphia.

Harper, J. V., Reynolds, P., Leatherbarrow, E. L., Botchway, S. W., Parker, A. W., and O'Neill, P. (2008). Induction of persistent double strand breaks following multiphoton irradiation of cycling and G1-arrested mammalian cells-replication-induced double strand breaks. *Photochem. Photobiol.* **84**, 506–1514.

Hopt, A., and Neher, E. (2001). Highly nonlinear photodamage in two-photon fluorescence microscopy. *Biophys. J.* **80**, 2029–2036.

Houtsmuller, A. B., Rademakers, S., Nigg, A. L., Hoogstraten, D., Hoeijmakers, J. H. J., and Vermeulen, W. (1999). Action of DNA repair endonuclease ERCC1/XPF in living cells. *Science* **284**, 958–961.

Jenner, T. J., Cunniffe, S. M. T., Stevens, D. L., and O'Neill, P. (1998). Induction of DNA-protein crosslinks in Chinese hamster V79-4 cells exposed to high- and low-linear energy transfer radiation. *Radiat. Res.* **150**(5), 593–599.

Kadhim, M. A., Macdonald, D. A., Goodhead, D. T., Lorimore, S. A., Marsden, S. J., and Wright, E. G. (1992). Transmission of chromosomal instability after plutonium alpha-particle irradiation. *Nature* **355**(6362), 738–740.

Kawata, S., Sun, H. B., Tanaka, T., and Takada, K. (2001). Finer features for functional microdevices. *Nature* **412**, 697–698.

Kong, X., Mohanty, S. K., Stephens, J., Heale, J. T., Gomez-Godinez, V., Shi, L. Z., Kim, J., Yokomori, K., and Berns, M. W. (2009). Comparative analysis of different laser systems to study cellular responses to DNA damage in mammalian cells. *Nucleic Acids Res.* **37**, 68.

Konig, K. (2000). Multiphoton microscopy in life sciences. *J. Microsc.* **200**, 83–104.

Konig, K., Riemanna, I., Strackea, F., and Le Harzica, R. (2005). Nanoprocessing with nanojoule near-infrared femtosecond laser pulses[[n]]Nanobearbeitung mit Nanojoule Nahinfrarot Femtosekunden Laserpulsen. *Med. Laser Appl.* **20**, 169–184.

König, K., Riemann, I., and Fritzsche, W. (2001). Nanodissection of human chromosomes with near-infrared femtosecond laser pulses. *Opt. Lett.* **26**(11), 819–821.

Langford, V. S., McKinley, A. J., and Quickenden, T. I. (2001). Temperature dependence of the visible-near-infrared absorption spectrum of liquid water. *J. Phys. Chem. A* **105**, 8916–8921.

Limoli, C. L., and Ward, J. F. (1993). A new method for introducing double-strand breaks into cellular DNA. *Radiat. Res.* **134**, 160–169.

Liu, G. J., Sonek, M., Berns, W., Koenig, K., and Tromberg, B. J. (1995). Two-photon fluorescence excitation in continuous-wave infrared optical tweezers. *Opt. Lett.* **20**, 2246–2248.

Mari, P. O., Florea, B. I., Persengiev, S. P., Verkaik, N. S., Bruggenwirth, H. T., Modesti, M., Giglia-Mari, G., Bezstarosti, K., Demmers, J. A., Luider, T. M., Houtsmuller, A. B., and van Gent, D. C. (2006). Dynamic assembly of end-joining complexes requires interaction between Ku70/80 and XRCC4. *Proc. Natl. Acad. Sci. USA* **103,** 18597–18602.

Meldrum, R. A., Botchway, S. W., Wharton, C. W., and Hirst, G. J. (2003). Nanoscale spatial induction of ultraviolet photoproducts in cellular DNA by three-photon near-infrared absorption. *EMBO Rep.* **4,** 1144–1149.

Melvin, T., Botchway, S. W., Parker, A. W., and O'Neill, P. (1996). Induction of strand breaks in single-stranded polyribonucleotides and DNA by photoionization: One electron oxidized nucleobase radicals as precursors. *J. Am. Chem. Soc.* **118,** 10031–10036.

Nelms, B. E., Maser, R. S., MacKay, J. F., Lagally, M. G., and Petrini, J. H. J. (1998). In situ visualization of DNA double-strand break repair inhuman fibroblasts. *Science* **280,** 590–592.

Olive, P. L. (2009). Impact of the comet assay in radiobiology. *Mutat. Res.* **681,** 13–23.

Pallon, J., and Malmqvist, K. (1994). New applications of the nuclear microprobe for biological samples. *Scanning Microsc. Suppl.* **8,** 317–324.

Piperakis, S. M., Visvardis, E.-E., and Tassiou, A. M. (1999). Comet assay for nuclear DNA damage. *Methods Enzymol.* **300,** 184–194.

Prise, K. M., Belyakov, O. V., Folkard, M., Ozols, A., Schettino, G., Vojnovic, B., and Michael, B. D. (1994). Investigating the cellular effects of isolated radiation tracks using microbeam techniques: Meeting Report: Microbeam Probes of Cellular Radiation Response, 4th L.H. Gray Workshop, 8–10 July 1993, Int. J. Radiat. Biol. **65,** pp. 503–508.

Reynolds, P. (2009). Ph.D. Thesis, Reading University, UK.

Sheng, L., Song, M., Zhang, X., Yang, X., Gao, D., He, Y., Zhang, B., Liu, J., Sun, Y., Dang, B., Li, W., Su, H., et al. (2009). Design of the IMP microbeam irradiation system for 100 MeV/u heavy ions. *Chinese Phys. C* **33,** 315–320.

Simpson, P. J., and Savage, J. R. (1996). Dose–response curves for simple and complex chromosome aberrations induced by X-rays and detected using fluorescence in situ hybridization. *Int. J. Radiat. Biol.* **69,** 429–436.

Tirlapur, U. K., and Konig, K. (2002). Targeted transfection by femtosecond laser. *Nature* **418,** 290–291.

Tirlapur, U. K., Konig, K., Peuckert, C., Krieg, R., and Halbhuber, K. J. (2001). Femtosecond near infrared laser pulses elicit generation of reactive oxygen species in mammalian cells leading to apoptosis-like death. *Exp. Cell Res.* **263,** 88–97.

Turcu, E., and Dance, J. (1998). *X-rays from Laser Plasmas: Generation and Applications.* Wiley, New York.

Uematsu, N., Weterings, E., Yano, K., Morotomi-Yano, K., Jakob, B., Taucher-Scholz, G., Mari, P. O., van Gent, D. C., Chen, B. P. C., and Chen, D. J. (2007). Autophosphorylation of DNA-PKcs regulates its dynamics at DNA double-strand breaks. *J. Cell Biol.* **177,** 219–229.

Ward, J. F. (1991). The early effects of radiation on DNA. In "NATO ASI, Series H," (E. M. Fielden and P. O'Neill, eds.), Vol. 54, pp. 1–16. Springer-Verlag, Berlin.

CHAPTER TWO

A Cell Biologist's Guide to High Resolution Imaging

Graeme Ball, Richard M. Parton, Russell S. Hamilton, *and* Ilan Davis

Contents

1. Introduction	30
2. Physical Limitations on the Resolution of Conventional Microscopy	31
2.1. The Abbe diffraction limit and optical resolution	31
2.2. Photon detection and signal-to-noise	32
2.3. Fluorophore properties	34
2.4. Temporal resolution limits	34
3. Preparations for High Resolution Fluorescence Imaging	35
3.1. Microscope setup	35
3.2. Sample preparation	37
4. High Resolution Image Data Acquisition and Processing	39
4.1. High resolution techniques	39
4.2. Improving the axial resolution	40
4.3. RESOLFT microscopy	40
4.4. Structured illumination microscopy (SIM)	41
4.5. Localization microscopy techniques	42
4.6. Three-dimensional super-resolution microscopy techniques	42
5. Processing	43
5.1. Denoising methods	44
5.2. Deblurring and deconvolution	46
6. Analysis of High Resolution Image Data	47
6.1. Intensity and molecular quantification	47
6.2. Accurate localization of fluorophores	47
6.3. Segmentation and particle tracking	47
6.4. Analysis of distribution and colocalization	48
6.5. Motility statistics and motion models	48
7. Conclusions	51
Acknowledgments	51
References	51

Department of Biochemistry, The University of Oxford, Oxford, United Kingdom

Methods in Enzymology, Volume 504
ISSN 0076-6879, DOI: 10.1016/B978-0-12-391857-4.00002-1

© 2012 Elsevier Inc.
All rights reserved.

Abstract

Fluorescence microscopy is particularly well suited to the study of cell biology, due to its noninvasive nature, high sensitivity detection of specific molecules, and high spatial and temporal resolution. In recent years, there has been an important transition from imaging the static distributions of molecules as a snapshot in time in fixed material to live-cell imaging of the dynamics of molecules in cells: in essence visualizing biochemical processes in living cells. Furthermore, in the last 5 years, there have been important advances in so-called "super-resolution" imaging methods that have overcome the resolution limits imposed by the diffraction of light in optical systems. Live-cell imaging is now beginning to deliver in unprecedented detail, bridging the resolution gap between electron microscopy and light microscopy. We discuss the various factors that limit the spatial and temporal resolution of microscopy and how to overcome them, how to best prepare specimens for high resolution imaging, and the choice of fluorochromes. We also summarize the pros and cons of the different super-resolution techniques and introduce some of the key data analysis tasks that a cell biologist employing high resolution microscopy is typically interested in.

1. INTRODUCTION

Light microscopy in all its varied and sophisticated forms, has become one of the most important and ubiquitous techniques in cell biology. Fluorescence microscopy, in particular, has become of central importance for making quantitative measurements of molecular behavior and for testing mechanistic hypotheses. The discovery of green fluorescent protein (GFP; Chalfie *et al.*, 1994) has been well recognized as a key milestone in the development of microscopy methods, resulting in the award of a Nobel prize (Nobelprize.org, 2008). The introduction of GFP and a plethora of related proteins and techniques have certainly revolutionized the field of cell biology, enabling the tagging of specific protein or mRNA components in living cells. The past 20 years have seen a stream of equally exciting developments. Perhaps, the simplest way to classify these milestones in fluorescence microscopy is into four areas: first, specimen preparation, which is highly dependent on the model system; second, the development of fluorochromes and methods of introducing them into cells; and third, the ever diversifying range of microscopes ranging from commercial to bespoke systems built for specific kinds of experiments. Finally, and equally important, is the post-acquisition analysis of the images, with the aim of extracting quantitative information about the dynamics and relationships of molecules inside cells. This chapter is structured around these distinct four parts of imaging technology and its major goal is to provide a practical guide to pushing the resolution limits of fluorescence microscopy and the benefits this affords.

Fluorescence microscopy suffers from major limitations to both spatial and temporal resolution. Temporal resolution is limited by the speed and sensitivity of the instrument, but perhaps most importantly by the photon budget of the specimen, that is, the total number of photons that can be obtained before photobleaching. All impose limits on the questions a biologist is able to answer using fluorescence microscopy; the major issues being that poor spatial resolution prevents accurate identification and colocalization of features in a cell, and poor temporal resolution prevents accurate characterization of dynamic cellular processes. We describe current approaches used to address these problems: starting from sample preparation, through instrumentation to the reconstruction and processing methods used to achieve maximum resolution.

2. Physical Limitations on the Resolution of Conventional Microscopy

2.1. The Abbe diffraction limit and optical resolution

As discovered by Ernest Abbe in 1873, light from a point source with a wavelength λ produces an observable image spot feature with a resolved lateral (X, Y) radius given by Eq. (2.1) below:

$$\text{Resolution}_{x,y} = \lambda/2\text{NA}, \qquad (2.1)$$

where NA is the numerical aperture of the lens. Abbe obtained a corresponding expression for axial (Z) resolution given by Eq. (2.2) below:

$$\text{Resolution}_z = 2\lambda/\text{NA}^2. \qquad (2.2)$$

There are several different resolution criteria including the Rayleigh criterion, the Sparrow resolution limit, and the full width at half maximum (FWHM) criterion; which give similar resolution limits. Since the NA of a lens can reach ~ 1.4 with current designs, this introduces a diffraction limit for the ability to resolve two point sources of light that is around half of the wavelength of the light. At the blue end of the visible spectrum, this implies a spatial resolution limit of ~ 200 nm, increasing to ~ 250 nm for green light, and ~ 300 nm at the red end of the spectrum. The shapes of objects at, or below, the resolution limit are indistinguishable, resulting in an identical system-dependent point spread function (PSF) of the light. In the case of confocal laser scanning microscopy (CLSM), the resolution is theoretically improved by a factor of $\sqrt{2}$ as an infinitely small pinhole is approached and increasing amounts of out-of-focus light are excluded, but in practice this maximum gain in resolution is never achieved. As we describe below, there are several practical ways to evade this limit known as super-resolution techniques, but they all come at a cost in terms of speed and/or photodamage.

The optical resolution is a fundamental limit of the optical components of an imaging system. However, the actual resolution achieved in the captured image data is also dependent upon how this data is sampled by the detector—so-called sampling theory. The criterion that must be considered in determining an appropriate level of sampling is the Nyquist frequency. This is defined as slightly more than twice the highest frequency that must be sampled. In the context of microscope images, this refers to the spatial frequency of the minimum optically resolvable object given by Eq. (2.3).

$$\text{Resolution}_{x,y} = 0.61\lambda/\text{NA}. \qquad (2.3)$$

In the case of a charge-coupled device (CCD) camera, this determines the magnification at which optimum resolution is achieved for a given NA and pixel size on the detector.

2.2. Photon detection and signal-to-noise

An image is made up of detected photons. Ultimately, the resolution that can be usefully obtained from any imaging system is limited by the number of photons actually detected. This is commonly described as the "signal-to-noise" limit of imaging, or the signal in relation to the uncertainty or variation in that signal. The accuracy with which the number of photons (brightness) can be determined for a signal is limited by the "shot noise" or statistical variation in that signal defined according to the Poisson statistics of photon counting ($\sqrt{n}$). Together with the optical limits, the signal-to-noise level imposes a practical restriction on achievable resolution. The more photons that can be detected, the closer the result is to the true optical limit. Recently, advances have been made in processing techniques which attempt to "recover" data limited by the number of photons detected, as discussed in Section 6.

Modern imaging systems rely on a variety of photon detectors to collect the image data. The two most important families of detectors reflect the two different approaches to capturing microscope image data. CCD detectors and complementary metal-oxide semiconductor (CMOS) sensors consist of an array of photodetectors in an integrated circuit for simultaneous detection of photons from the whole image in a wide-field or multifocal scanning (e.g., spinning disc confocal) microscopy system. Photomultiplier tube (PMT) and avalanche photodiode (APD) detectors, on the other hand, are the devices most commonly used to detect the single stream of photons from a CLSM. The sensitivity of all types of photon detectors depends on two factors: the detection limit, and the quantum efficiency (electron output/photon input).

The detection limit of a photodetector corresponds to the lowest photon signal that produces an electronic response distinguishable from system noise: this limit is termed the noise equivalent power (NEP), and the smaller it is, the better the detector. The most important sources of noise vary depending on the type of detector, but one that is unavoidable is the stochastic nature of photon arrivals, which results in a Poisson distribution of detected photons and therefore a Poisson or Shot noise equal to the square root of the number of detected photons.

In the case of CCD cameras, readout noise is the second major contribution and increases in proportion to the square root of the readout speed (Goldman et al., 2009). Thermal noise makes a second, less important contribution, and can be reduced by cooling. Electron-multiplying CCD (EM-CCD) cameras use multiple multiplication steps prior to readout, making them better suited to low light levels, and reduces the importance of readout noise. However, EM-CCD devices are affected more by thermal, charge, and excess noise as a consequence of the extended multiplication process (Robbins and Hadwen, 2003). PMT devices have excellent noise characteristics, but suffer from poor quantum efficiency at higher wavelengths (Pawley, 1995). APD detectors, on the other hand, introduce a greater amount of noise during the signal multiplication process, but have generally superior quantum efficiency compared to PMTs.

Finally, quantum efficiency is simply a measure of light collection efficiency and depends on the physical setup and properties of the detector. On this measure, CCD and CMOS detectors tend to give good results: ~ 45–50% maximum efficiency (wavelength dependent) for front-illuminated detectors, and as much as $\sim 95\%$ efficiency, or more typically $\sim 85\%$, for back-thinned detectors. PMT detectors typically have a quantum efficiency $< 25\%$, and APD detectors $\sim 70\%$ (Pawley, 1995).

The dynamic range of the detector is also extremely important and the linearity of its response to photons. CCD and CMOS detectors amplify the photon signal in a relatively linear fashion and are not particularly prone to saturation where increased signal does not register, with the exception of EM-CCD devices; whereas PMTs and APDs have a much less linear response and saturation must be carefully avoided by adjusting the gain that adjusts the degree to which the electrical output signal is amplified by the detector relative to the photon input (Pawley, 1995). However, some PMTs and APDs can be operated in a photon-counting (Geiger) mode which negates the nonlinearity of the amplification response provided that the photon flux is low enough that readout can be achieved between photon arrival events.

2.3. Fluorophore properties

Photobleaching and photodamage are the ultimate roadblocks for most strategies to overcome the diffraction and sensitivity limits that reduce the resolution of fluorescence microscopy. Photobleaching and photodamage are why sensitivity limitations cannot be simply bypassed by increasing excitation intensity, and why applying techniques such as Three-dimensional (3D)-SIM and STED to live-cell imaging remains a challenge.

The fluorescence phenomenon involves excitation of the fluorophore by a photon at the excitation wavelength, which produces an excited electronic state in the molecule, followed by spontaneous relaxation back to the ground state with the emission of another photon at the emission wavelength. Fluorescence lifetimes for typical fluorophores are typically <20 ns, so this is not currently a limitation. However, there are various competing pathways for relaxation of the excited state, most notably conversion to a long-lived triplet state that is not fluorescent. These excited states are also chemically reactive, explaining the photobleaching and much of the photodamage caused by fluorescence imaging. A typical fluorophore will undergo thousands of excitation/emission cycles before bleaching occurs, however. In addition to the excitation and emission spectra and photostability, two of the most important properties of a fluorophore are the quantum yield which, in this context, is a measure of the number of photons emitted for the number absorbed; and the extinction coefficient, which is a measure of how readily the fluorophore absorbs excitation photons.

There is a choice to be made between the various small organic molecule fluorophores such as the cyanine dyes, Alexa dyes, and Atto dyes, versus the recombinant fluorescent proteins derived from GFP and other natural fluorophores (Shaner et al., 2005). Small molecule dyes have two major drawbacks: the difficulty of labeling for live-cell imaging, and generally increased phototoxicity. However, some of the Alexa fluor and Atto dyes in particular have excellent brightness and photostability characteristics, and are the main fluorophores that have been successfully used in photon-intensive super-resolution techniques like STED (see Section 4.4).

Finally, an important class of fluorophores for super-resolution imaging are the photoactivatable, photoconvertible, and photoswitchable molecules that are particularly applicable to the localization microscopy techniques described in Section 4.5 (Endesfelder et al., 2011).

2.4. Temporal resolution limits

As mentioned above, the fluorescence lifetime is the ultimate limiting factor on temporal resolution, but modern fluorescence microscopy has a long way to go before reaching this limit. Both wide-field and laser scanning methods are currently limited to temporal resolutions of milliseconds (Pawley, 1995). In the case of laser scanning methods, PMT and APD detectors have very rapid response times of a few nanoseconds, but the necessity to raster scan the sample area results in much slower (milliseconds to seconds) temporal resolution. In the case of CCD and the faster CMOS cameras, readout times are of the order of milliseconds and the entire image field is imaged at once, which makes them the detectors of choice for fast 3D imaging. Other hardware limitations include shutter speeds and/or laser switching times.

3. Preparations for High Resolution Fluorescence Imaging

In any form of imaging, but especially for high resolution microscopy, the importance of correctly setting up and aligning the imaging system and of optimizing sample preparation for imaging cannot be overemphasized. As the spatial resolution is increased, so do the pitfalls and potential artifacts that can arise from optical aberrations and the inhomogeneity of the refractive index (RI) of live specimens.

3.1. Microscope setup

A badly serviced and aligned microscope cannot achieve optimal performance. The best defense against this is regular calibration using appropriate test samples, usually fluorescent bead slides, which are easily prepared. We find three types of bead slide test samples particularly useful: very sparsely distributed 100 nm beads, either multicolor or single color; moderately distributed 200 nm beads, ideally with a wide excitation and emission range; very crowded 100 or 200 nm beads as a "bead lawn." Section 3.1.1 describes the preparation of these different test slides.

3.1.1. Protocol 1: preparation of test slides

1. It is important to select good quality, clean slides, and coverslips to prepare your sample slides. Generally we use No. 1.5 coverslips with a nominal thickness of 170 μm (0.16–0.19 mm).
2. Beads should be mounted directly onto the coverslip, not the glass slide. Arrange your coverslips on sheets of filter paper to make it easier to work with them and use coarse forceps to aid handling.
3. It is helpful to poly-lysine treat your coverslip prior to mounting beads to localize and adhere them. Take neat poly-D-lysine solution and apply 100 μl to the center of a 22 × 22 mm coverslip. Leave for about 5 min and then remove with a pipette and wash off the residue with distilled water from a wash bottle. Tilt the coverslip and blot dry at the edge only with filter paper. Allow to dry fully.

4. Make up your bead suspension. Often it helps to briefly sonicate the beads to help disperse clumps, you may also want to very briefly centrifuge the bead sample and take off the supernatant. You will need to dilute appropriately depending on the density of your stock solution, this is best determined by a dilution series. Beads can be diluted into water or ethanol, although the latter should be avoided for storage of beads. Depending upon the density of beads required, dilutions of between 1:1 and 1:10,000 may be used.
5. There are different ways to achieve a useful coating of beads:
   a. For very sparse dispersions we recommend using a poly-lysine-coated coverslip and applying about 20 µl of bead suspension. This should be allowed to settle for between 5 and 30 min. Ideally there should be some slight drying of the edge of the drop of beads. This should achieve a dense ring of beads where the suspension has been dried down, and a sparsely coated inner region. The dense ring facilitates location of beads when imaging. After settling, the excess can be very gently and briefly washed off with distilled water from a wash bottle and allowed to dry.
   b. For denser applications of beads or a "bead lawn," bead suspensions in ethanol can be dried down onto the coverslip. As all the beads are concentrated on the glass surface, greater initial dilutions of the stock are usually required. To assist drying, the coverslip may be placed briefly on a 50 °C heat block. When beads are dried down from ethanol, it is usually not necessary to assist adherence with poly-lysine.
6. The prepared bead-coated coverslips may then be mounted on glass slides with a drop of mountant. About 10–15 µl is usually appropriate for a 22 × 22 mm coverslip. To avoid air bubbles, place the drop of mountant on the center of the glass slide and angle the coverslip down onto it using a needle to gently lower one side until it is resting flat. Allow the mountant time to spread and cover the whole area of the coverslip before sealing.
7. When using mountants such as Prolong Gold, which require a curing period to harden before sealing, it is best to leave overnight on a level surface in the dark. Note that the time of curing of these reagents will affect the eventual RI of the mountant.
8. Always remember to label slides with the characteristics of the beads, including the density and the type of mountant used.
9. Slides survive longest when they are double varnished and stored level in appropriate slide boxes at 4 °C. Slides stored on their sides tend to leak the mountant, especially if excess mountant was used in the preparation.

Sparsely prepared bead slides of 100 nm beads should be used to determine the PSF of your imaging system. The PSF is an image set of a single subresolution fluorescent bead. To obtain a PSF, a solitary bead should be selected and imaged through Z from about 5 μm above the focal plane to 5 μm below the focal plane. The resultant image set represents the optical characteristics of the imaging system for the specific lens used. The PSF provides information about the alignment of the system and aberrations present. The more symmetrical the PSF, the better the system is set up and the better the imaging will be.

A moderate field of beads is useful for assessing spherical aberration (SA). In this case, the preparation may be modified such that beads are mounted directly on the glass slide as well as on the coverslip. By using different volumes of mountant applied to a 22 × 22 mm coverslip, it is possible to determine the effects of imaging at different defined depths into a specimen (calculated from the known volume and area of the coverslip). Mountants of different RI, for example, different dilutions of glycerol may be useful to match the conditions in live tissue.

When imaging multiple channels of data, it is essential to be able to assess the correct registration of those channels. This is especially important when systems with multiple-independent detectors are used. Tetra spec or broad spectrum beads are imaged in multiple channels and the relative position of the beads are used to correct for XYZ shifts, rotation, and magnification. It is important to have a moderately even distribution of beads across the whole field of view. With modern algorithms, it is possible to automatically register channels to subpixel accuracy. This becomes even more important where registration of super-resolution images is being considered. More advanced alignment protocols can perform nonlinear transformations across the field of view.

A dense bead lawn is very useful for assessing flatness of the field of view to determine if the excitation illumination is properly aligned. It can also be useful in assessing the success of super-resolution techniques. It should be possible to resolve the packing of subresolution 20–100 nm beads. For structured illumination techniques, a bead lawn is useful for assessing and optimizing the illumination pattern which is applied to the specimen, the quality of which determines the ability to increase resolution.

When assessing the performance of an imaging system, it is important to bear in mind that individual objective lenses, even of the same type, can vary considerably in how well they perform. This will be manifested in the PSF. In purchasing objectives it is worth comparing several.

3.2. Sample preparation

Sample preparation has always been an important aspect of microscopy from the inception of the method, but as the instruments used have increased in complexity and sophistication, so has the importance and diversity of the methods used to prepare samples for fixed and live imaging. This is of particular importance when considering super-resolution imaging.

For example, additional resolution is only useful on fixed specimens if the fixation has preserved structural integrity to the appropriate degree. Further, as the spatial resolution is increased, so do the pitfalls and potential artifacts that can arise from optical aberrations and inhomogeneities in the RI of live specimens. The following paragraph discusses some of the key considerations for specimen preparation of both fixed and live tissue, when using super-resolution imaging modalities.

SA poses a significant problem when imaging biological specimens. It is usually caused by mismatches in RI along the optical axis from the lens into the specimen. The consequence of this is reduced resolution in XY and particularly in Z, and decreased signal strength at the point of focus. The effects of SA are most easily seen as a distortion in the PSF when viewing subresolution beads manifesting as an elongation or "smearing" of the signal along the optical axis and an uneven distribution of signal above and below the plane of focus. The best way to avoid this problem is to match as far as possible the RI of the mounting medium and immersion medium for the lens with the RI of the glass coverslip. With fixed material, it is helpful to deposit beads (see Section 3.1.1) at the edge of the coverslip before mounting the specimen. By using either the "coverslip thickness correction" adjustment of the objective (where one exists) or by using immersion oils of systematically varying RI it is possible to correct for SA. In this way, the beads at the corner of the coverslip are imaged to monitor the extent of SA, and the conditions can be modified until it is fully corrected. SA increases with the depth of mismatched media through which the features of interest are imaged. The best imaging is, therefore, achieved within < 15 µm of the coverslip inner surface.

When imaging deep into thick specimens cannot be avoided, SA can still be assessed and ameliorated by immersion media of appropriate RI, using thinner coverslips or, particularly with live imaging, using objectives designed to work with immersion media which more closely match that of the cell or tissues being imaged. Water, glycerol, and the new silicon immersion objectives (from Olympus) are all designed to reduce the affects of SA. The latter we have found particularly effective for high resolution imaging as it is available as a 1.3-NA $\times 60$ lens. A significant consideration when using immersion media and mounting media that differ significantly from that of the glass coverslip, is that if the latter is at an incline, this will introduce aberrations apparent as a distortion to the PSF. More recently, it has become possible to correct for imaging depth-dependent SA in a much more dynamic way. Adaptive optics using lens arrangements or deformable mirrors (Kner *et al.*, 2010), which correct the phase of the light, can be calibrated to allow SA correction specific to the depth of imaging. So far commercial systems are limited such as the mSAC produced by Intelligent Imaging Innovations.

Another significant limitation to achieving optimal resolution is that of limited signal strength (Section 2). Sample preparation can significantly help in this regard. In addition to SA correction, discussed in the preceding section, using the best labels and mounts which include antifade reagents can significantly improve the signal-to-noise and hence to achievable resolution. With fixed material, it is possible to introduce antibodies (Ab) coupled to stable high quantum efficiency dyes (Section 2). Even when GFP-tagged proteins are to be imaged, it is possible to boost this signal by using dye-conjugated Ab (Guizetti *et al.*, 2011). With live material, it is often possible to enhance photostability by introducing antifades based upon vitamin E derivatives such as Trolox (Roche).

4. High Resolution Image Data Acquisition and Processing

The first step in acquiring useful high resolution image data is to carefully define the problem and then decide which of the various methods is compatible with the sample and will deliver the necessary resolution without compromising the biology. This means answering questions such as: What *XY* resolution is required? What *Z* resolution is required? What time resolution is required? What contrast-to-noise ratio (CNR) is required? How sensitive is the sample to photobleaching and photodamage? It is also essential to ensure that all necessary instrument calibrations have been properly carried out (e.g., checking registration of channels).

4.1. High resolution techniques

When the guidelines above are followed, the maximum resolution of widefield and confocal microscopy is given in Section 2.1, and temporal resolution has been discussed in Section 2.4. Some other important high resolution techniques that have not been mentioned until this section are light sheet fluorescence microscopy (LSFM), reviewed by Reynaud *et al.* (2008); two-photon microscopy (Denk *et al.*, 1990); and total internal reflection microscopy (TIRFM; Axelrod, 1989). Light sheet microscopy (Reynaud *et al.*, 2008; Santi, 2011; Verveer *et al.*, 2007) uses a light sheet perpendicular to the objective for excitation illumination, affording good sectioning capability and reduced photobleaching and photodamage. The multiple views acquired to build up a 3D LSFM image can also be used in multiview deconvolution (Planchon *et al.*, 2011; Verveer *et al.*, 2007), resulting in improved, isotropic resolution. Two-photon microscopy involves high photon fluxes of low wavelength light which produce some two-photon absorption events that can excite a transition equal to the sum of the two photons' energy, that is, at a shorter wavelength. It also offers improved contrast when working with deep samples. TIRFM on the other hand is limited to imaging samples within around 100 nm of the coverslip, but this is also the advantage of the technique since only a thin slice is observed, and the Z position is known to a high level of precision. The 100-nm limitation stems from the rapid decay of the evanescent field with distance from the coverslip.

There already exist a number of excellent reviews of the super-resolution techniques (Davis, 2009; Galbraith and Galbraith, 2011; Gustafsson, 1999; Huang et al., 2009; Ji et al., 2008; Schermelleh et al., 2010; Toomre and Bewersdorf, 2010; Toprak et al., 2010). Far-field super-resolution techniques fall into two broad categories: those that employ spatial patterning of the excitation illumination; and those that detect single molecules to enable high resolution spatial localization. The super-resolution microscopy, or nanoscopy field is in its infancy though (Evanko, 2009; Hell, 2009; Lippincott-Schwartz and Manley, 2009) and it is still important to corroborate findings using other techniques such as conventional fluorescence microscopy and electron microscopy in a correlative light-EM scheme for example.

4.2. Improving the axial resolution

As mentioned previously, the axial resolution of a typical microscopy is normally significantly poorer than the lateral resolution (Eqs. (2.1) and (2.2) above). The first successful far-field attempt to overcome this problem was the 4Pi microscope (Hell and Stelzer, 1992; Hell and Wichmann, 1994), which is a CLSM that uses a pair of objectives on either side of the sample to allow addition of the wavefronts observed from opposite sides to capture a greater proportion of the full spherical wavefront. The increase in axial resolution depends on the exact setup used (Hell and Wichmann, 1994), but is around fivefold, meaning that an axial resolution of about 100 nm is achievable. The I5M microscope (Gustafsson et al., 1999) is a wide-field setup that similarly uses two objectives, and can also achieve an axial resolution of around 100 nm. A comparison of 4Pi and I5M microscopes (Bewersdorf et al., 2006) shows that I5M is more sensitive, but suffers from more problematic artifacts due to pronounced and complicated side-lobe structures, which must be removed by processing. Both methods obviously suffer from the requirement to sandwich the sample between two objectives, meaning that they are tricky to set up and incompatible with many biological specimens, particularly those that are thick.

4.3. RESOLFT microscopy

One way to overcome the diffraction limit is to use a pair of excitation lasers that interact with the fluorophore photophysics in a nonlinear fashion, effectively engineering a smaller PSF. The generic name for this approach is REversible Saturable or switchable OpticaL Fluorescence Transitions (RESOLFT; Hell, 2003). The first among this family of techniques to be reported was STtimulated Emission Depletion (STED; Hell and Wichmann, 1994; Klar et al., 2000, 2001). STED uses an excitation laser beam combined with a second "de-excitation" laser beam at a second wavelength with a ring-like de-excitation profile. Although the de-excitation beam is also diffraction limited, the dependence of the de-excitation process (stimulated emission) upon de-excitation intensity is nonlinear. This means that a small spot at the center of focus remains excited and fluorescence from this spot can be detected after the de-excitation beam. There are two major disadvantages of this approach: firstly, the high photon flux required results in rapid photobleaching and photodamage. Secondly, although the lateral (XY) resolution can be improved to 20 nm or less using STED, the axial resolution remains unaltered in a basic STED microscope, producing a needle-like PSF which is very elongated in the Z-axis. In spite of these caveats though, a commercial STED system is sold by Leica and live-cell STED imaging has been reported a number of times (Hein et al., 2008; Nägerl et al., 2008).

Ground state depletion (GSD) microscopy (Bretschneider et al., 2007) also uses a nonlinear response to excitation photons. In this case, the same laser may be used for both excitation, and by a second excitation event, for GSD to a long-lived dark state. GSD can achieve a similar level of resolution to STED (Rittweger et al., 2009). The disadvantage of GSD vis-a-vis STED is that fluorophores in the triplet state are less photostable, and photobleaching is a major challenge.

4.4. Structured illumination microscopy (SIM)

SIM is a wide-field method that makes use of illumination stripes of varying phase and angle to create low resolution Moiré patterns that contain higher resolution information (Gustafsson, 2000), producing a potential doubling of the resolution to around 100 nm. The high resolution images must be computationally reconstructed from the raw data, unlike the RESOLFT methods described in the previous section, which is a disadvantage in that artifacts may be introduced and the reconstruction may fail if the signal-to-noise ratio is not adequate. However, SIM requires less excitation illumination in total than RESOLFT methods. SIM has also been successfully combined with TIRFM for live super-resolution imaging (Kner et al., 2009).

Further improvement in the resolution of wide-field structured illumination methods requires using the same sort of nonlinear photophysical responses as employed by the RESOLFT methods. saturated structured illumination microscopy (SSIM; Gustafsson, 2005) and saturated patterned excitation microscopy (SPEM; Heintzmann et al., 2002) have been reported, but these techniques require very high photon fluxes.

4.5. Localization microscopy techniques

The localization microscopy techniques employ a completely different principle to the patterned excitation methods discussed above. Localization microscopy techniques rely on detection of photons from a single fluorophore in isolation, and fitting of the PSF pattern in order to localize the fluorophore to a high level of precision (Moerner, 2006; Toprak et al., 2010). The differences between the various methods are in the way in which the isolation of active fluorophores is achieved. The localization precision of the methods used to determine fluorophore positions is discussed in Section 6.2.

Prior to reports of actual localization microscopy, fluorescence imaging to 1 nm accuracy (FIONA) was reported (Yildiz and Selvin, 2005). Actual super-resolution imaging was then reported in 2006 by a number of groups using different activation/switching mechanisms. Photoactivation light microscopy (PALM; Betzig et al., 2006) and fluorescence PALM (FPALM; Hess et al., 2006) use a brief pulse of activation wavelength light in the presence of a photoactivatable fluorophore to activate a small population that is subsequently imaged until bleached. Stochastic optical reconstruction microscopy (STORM; Rust et al., 2006) and direct STORM (dSTORM; Endesfelder et al., 2010) on the other hand, use photoswitchable fluorophores to achieve a limited population for sampling, and have also been shown to achieve an accuracy of 20 nm. Until recently, the localization microscopy techniques suffered from the disadvantage that acquisition was necessarily long in order to separate fluorophores temporally, but rapid photoswitching (Endesfelder et al., 2010), and improved algorithms for reconstruction of images at higher fluorophore densities such as DAOSTORM (Holden et al., 2011) mean that live-cell imaging is becoming increasingly feasible using this super-resolution technique.

4.6. Three-dimensional super-resolution microscopy techniques

Each of the classes of super-resolution microscopy introduced above has been successfully extended to a 3D version. isoSTED (Schmidt et al., 2008, 2009) uses two opposing lenses to achieve improved axial resolution. SIM has similarly been implemented in a system using two opposing lenses to improve the axial resolution, called I5S (Shao et al., 2008). A more practical 3D-SIM method using only one lens has also been created (Gustafsson et al., 2008; Schermelleh et al., 2008; Fig. 2.1). In the case of the localization microscopy techniques, 3D STORM has been reported (Huang et al., 2008) as well as two-photon PALM (York et al., 2011). Finally, a promising double-helical PSF (DH-PSF) technique (Pavani et al., 2009), to improve axial resolution, was recently reported in localization microscopy experiments studying nucleoid-associated proteins in bacteria (Lee et al., 2011).

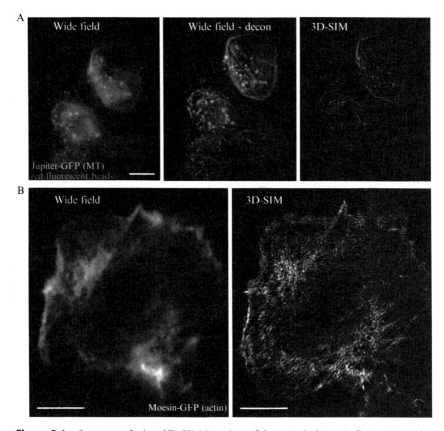

Figure 2.1 Super-resolution 3D-SIM imaging of the cytoskeleton in living *Drosophila* macrophages captured with an OMX-Blaze microscope (manufactured and distributed by Applied Precision, a GE Healthcare Company). (A) Microtubules (Jupiter GFP) and endocytosed red fluorescent beads. A single time point is shown from a time lapse movie of 11 frames: Conventional wide-field, deconvolved, and SI-reconstructed images derived from the same data set are compared. Each 3D-SIM image requires 240 images/μm of Z depth (15 images/slice, 8 slices/μm) captured at a rate of about 1 s/μm. (B) Actin cytoskeleton in a well-spread living macrophage (Moesin GFP). Wide-field and SI-reconstructed images derived from the same data set are compared as a projected Z series of 0.5 μm from a single time point. Resolution in 3D-SIM was about 120 nm, wide-field 275 nm, and deconvolved wide-field 250 nm. (See Color Insert.)

5. PROCESSING

Image processing has two basic aims: firstly, to remove any artifacts present in the images, and secondly, to present the data in a way that enables the human eye to make sense of the information therein.

5.1. Denoising methods

Denoising or noise filtering aims to remove the effects of noise contamination to recover a better estimate for the true features an image represents, that is, a classical inverse problem. A thorough treatment of image denoising can be found elsewhere (Buades et al., 2005). Instead, our aim here is to make a few important points and draw the reader's attention to some of the most useful and versatile tools for dealing with the noise filtering problem.

It is very important to remember that noise contamination is essentially irreversible from the point of view of the information content of the data. Sometimes there may be particular pieces of prior knowledge that can be used to filter out noise, such as knowledge about the imaging system or sample; but in the general case, the best one can hope for is to make the underlying information content of the data more apparent to the human eye. This is a reasonable goal, since human vision has a number of limitations including sensitivity, acuity, and the spatial and temporal intervals over which intensity information is integrated. Where true quantitative measurements based on the image data are the goal, however; it is generally better to fit the raw data taking account of noise contamination.

The simplest denoising algorithms are spatial smoothing filters such as the mean filter, median filter, and Gaussian blur (convolution with a Gaussian kernel). These rely on the fact that in an oversampled microscope image, rapid intensity changes from one pixel to the next can only have their origin in noise or other artifacts. All result in some amount of smoothing and loss of resolution however. 3D and 4D versions of these filters for data sets with multiple Z planes and time points can actually be surprisingly effective considering their simplicity. An equivalent approach is to filter out the highest frequency information in Fourier space, and the Wiener filter (Gonzalez and Woods, 2002) is a moderately sophisticated version of this basic idea employing a statistical approach to identify the frequency spectrum of the noise.

There exist a number of more intelligent spatial filters for denoising, which weigh the averaging process according to the estimated likelihood that a given pair of pixels are equivalent, reducing the degree of oversmoothing. The equivalence of pixels is usually determined on the basis of the similarity of patches of pixels surrounding the two pixels under consideration (Buades et al., 2005; Kervrann and Boulanger, 2008). Figure 2.2 shows the results of denoising using the patch-based algorithm of Kervrann and Boulanger (2008), which varies the neighborhood window adaptively. This algorithm has been used to good effect in 4D live-cell imaging sequences at extremely low levels of illumination (Carlton et al., 2010). An important point to note is that this class of denoising approach relies on redundancy in the image or image sequence. In the complex microscope images typically recorded by a cell biologist, truly repeating patterns within

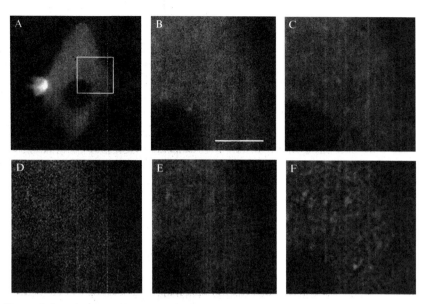

Figure 2.2 The effects of denoising and deblurring. Data are EB1:GFP in a stage 8 *Drosophila* oocyte, marking microtubule plus-ends; acquired using a Deltavision Core wide-field microscope (Applied Precision, a GE Healthcare Company) with EM-CCD camera. (A) Raw data at high excitation illumination intensity: Z-plane 2/3, time point 30/121. (B) A blowup of the region highlighted in (A); scale-bar is 5 μm. (C) The result of iterative constrained deconvolution for the image plane shown in (B) using soft-WoRx 4.5, "trailed" for 5 time points. (D) Raw data at low excitation illumination intensity for the same oocyte: Z-plane 2/3, time point 30/121, sequence taken immediately prior to the high excitation data shown in (A–C). (E) Result of denoising using the nD-SAFIR algorithm with a neighborhood size of $5 \times 5 \times 5 \times 3$ ($XYZT$). (F) Result of deconvolution following denoising in (E), "trailed" for 5 time points.

a single image are unlikely, and it is instead the immediately adjacent pixels and neighboring planes in Z and time that are likely to provide multiple samples of equivalent pixels. The Cramer-Rao lower bound (Kay, 1993) for a normally distributed variable states that N observations of a variable of unknown mean lead to a reduction of the error in the estimate σ by factor of $\sqrt{N}$; and while we do not suggest that the noise in a typical microscope image is normally distributed, in our experience this gives a reasonable idea of the improvement that can be expected from effective denoising. For this reason, we find that denoising is a technique that lends itself to well-sampled 4D image sequences where the signal-to-noise ratio is often modest as a consequence of the need to keep total illumination to a minimum to reduce photobleaching. Where it is possible to make additional assumptions about the nature of the features in an image, signal-to-noise in feature-enhanced images can be further increased (Yang *et al.*, 2010).

There are many similarities between the denoising problem and the image compression problem: both seek to uncover redundancy in an image, the former to identify and remove noise, the latter to produce a more compact representation of the information. Some of the most effective noise filters have much in common with image compression techniques, including the bilateral filter (Tomasi and Manduchi, 1998), K-SVD (Elad and Aharon, 2006), and BM3D (Dabov et al., 2007). Other popular denoising tools such as nonlinear anisotropic diffusion (Perona and Malik, 1990) and total variation minimization (Rudin et al., 1992) make use of the gradients in an image to preserve feature boundaries while smoothing the noise.

5.2. Deblurring and deconvolution

Blurring in raw microscope image data is a consequence of the PSF of the imaging system as well as inevitable aberrations due to imperfections of the optics. The most important and problematic aberrations are: SA, chromatic aberration, and astigmatism. A detailed discussion of the origins of the various types of aberration and their removal is beyond the scope of this chapter, but suffice it to say that one should check for their presence using a diagnostic standard sample such as that described in Section 3.1.1.

The blurring that results from the PSF of the imaging system can be corrected by measuring this PSF using a standard sample and then performing deconvolution to recover an image in which the light has been reassigned to the true source. This is of much greater importance for wide-field microscope images than for confocal images in which a large part of the out-of-focus light has already been excluded by the pinhole, and spinning disk confocal images lie between the two extremes.

One of the most popular, successful deconvolution algorithms is constrained iterative deconvolution (Agard et al., 1989), the results of which are shown in Fig. 2.2. If the PSF of the system is carefully measured, and there are no other aberrations or noise, then deconvolution problem is an easy one. In this case, an inverse filter can be used to arrive at a solution for the original object image. Unfortunately, however, microscope images are invariably contaminated with unpredictable noise. This renders inverse filtering ineffective. Instead, constrained iterative deconvolution utilizes a noise filter such as the Wiener filter to suppress noise, and makes successive estimates of the original image before reblurring these using knowledge of the PSF, and then subtracting these out-of-focus blur contributions to arrive at a better estimate for the original image. Note that only true 3D deconvolution preserves and indeed improves the distribution of intensity information in a microscope image, whereas 2D deblurring operations such as the unsharp mask and nearest neighbor deconvolution merely improve the appearance, and actually worsen the fidelity. For a more detailed practical guide to the different deconvolution methods, see Wallace et al. (2001).

6. Analysis of High Resolution Image Data

6.1. Intensity and molecular quantification

Rigorous interpretation of intensity information is not limited to high resolution imaging *per se*, but fluorescence intensity is of much greater utility when it can be assigned with confidence to a discrete structure in the cell. Hence, the better the spatial and temporal resolution one can achieve, the more useful it becomes to relate intensity to the number of fluorescently labeled molecules in a structure.

A protocol for the quantification of fluorescently labeled molecules can be found in a book chapter written by members of the Singer lab (Goldman *et al.*, 2009), who have pioneered many of the techniques to study mRNA molecules *in vivo*. In this protocol, they recommend wide-field fluorescence microscopy combined with a CCD camera due to the higher sensitivity compared to confocal methods. For a series of experiments, they point out that it is essential to use the same lens and camera settings, and that constrained iterative deconvolution should be used to reassign out-of-focus light to the correct position without introducing intensity artifacts. The basic procedure is to create a series of standard slides with known fluorophore concentration, and image these in order to build up a calibration curve of fluorescence intensity versus concentration that can be used to interpret the real data. It is also possible by careful calibration to determine the actual number of photon counts (Chen *et al.*, 2003; Thompson *et al.*, 2010).

6.2. Accurate localization of fluorophores

Accurate localization of individual fluorophores is key to the localization microscopy techniques introduced in Section 4.5. This involves fitting the observed distribution of detected photons to determine a more accurate centroid position, producing a localization error that decreases with the square root of the number of measured photons (Thomann *et al.*, 2002; Thompson *et al.*, 2002). A resolution of 1 nm or better, as in the FIONA technique (Churchman *et al.*, 2005; Yildiz and Selvin, 2005) was shown to be possible using this method, and more recently, subnanometer single molecule localization has been achieved (Pertsinidis *et al.*, 2010). The limits of the resolution of localization microscopy and the fitting techniques used can be found in a recent series of chapters introduced here (Larson, 2010).

6.3. Segmentation and particle tracking

Tracking moving objects in live cells is often a difficult undertaking, but the results can be extremely rewarding in terms of the information they provide about the underlying cellular processes. A sufficiently high temporal resolution is crucially important for successful tracking, and this must be kept in mind when considering the trade-off with signal-to-noise and spatial resolution. In order to ensure that the tracking problem is tractable, one must achieve a temporal resolution such that the majority of particle displacements from one frame to the next are less than the average interparticle distance. According to the Rose criterion (Rose, 1948), a CNR of 5 or more should be the aim to ensure reliable detection. High spatial resolution on the other hand, reduces the number of particle crossing events during which the individual particles cannot be resolved.

There have been a number of reviews and comparisons of tracking algorithms (Cheezum et al., 2001; Meijering et al., 2006), and several powerful algorithms are freely available (Jaqaman et al., 2008; Sbalzarini and Koumoutsakos, 2005). The segmentation and tracking problem is covered in detail elsewhere in this issue, and we refer the reader there.

6.4. Analysis of distribution and colocalization

Fluorescence microscopy is the most important tool for the analysis of the distribution of macromolecules in live cells, and the higher the resolution achieved, the greater the utility and reliability of the technique. Two common questions posed are: Does the detected distribution of molecules indicate order of some kind, or is it entirely random? Where two different kinds of molecules are detected, is there any correlation between their distributions?

Both problems belong to the domain of spatial statistics, particularly the treatment of spatial point patterns (Cressie, 1993), the analysis of which is most easily accomplished using the R statistical computing language (Bivand et al., 2008). In the case of the first problem, deciding whether a spatial distribution is random or not, researchers typically calculate statistics such as the nearest neighbor distance function G (Andrey et al., 2010) and Ripley's K function (Lee et al., 2011).

Colocalization is a concept that has received much interest from cell biologists from the inception of modern fluorescence microscopy, and important developments were the adoption of Pearson's correlation coefficient (Manders et al., 1992) and the creation of Manders' overlap coefficients M1 and M2. These overlap coefficients quantify the dependence of each channel on the other, since any dependence may not be mutual. For a review of the topic, see Bolte et al. (Bolte and Cordelières, 2006).

6.5. Motility statistics and motion models

Motility statistics are an excellent way to interrogate dynamic cellular processes in order to understand the underlying mechanisms. Two of the most important applications include the study of molecular motor-driven transport processes and diffusion processes in membranes. Although modeling underlying processes with reference to the data is a rewarding approach for understanding a biological system, a much simpler method is to compare a parameter from two populations for which a different phenotype is suspected, and ask the question: Do the two populations show a statistically significant difference? Table 2.1 provides a summary of common statistical tests and the situations in which they are appropriate.

The most important motility statistics are speed, directionality, position, and for saltatory movements, run length and duration, as well as the duration of the pauses. Average values of these motility statistics can be valuable, but histograms of their distributions are even more informative, and are able to reveal whether there are multiple populations within the sample. Another useful graphical representation of motion that can aid analysis is a kymograph (meaning "wave drawing"), which displays the evolution of the one-dimensional position of a particle over time. In the case of directionality information about motility on the other hand, the most useful representation is the display of an overall "flow" or "wind map."

Table 2.1 Choosing an appropriate statistical test

Statistical distributions	Description/uses
Normal distribution	(Gaussian distribution) a distribution of random variables whose means cluster around a single value.
Extreme value distribution	The generalized extreme value distribution is a family of distributions describing normal distributions weighted in one tail. This occurs when selecting a particular feature from a population. For example, the heights of the tallest person in each city.
Binomial distribution	A discrete probability distribution of the successes on a series of events, for example, tossing a coin.
Poisson distribution	A distribution of discrete random variables from events in time, distance, area, or volume.
t-Test	Comparing distributions of speeds and run lengths.
Regression analysis	Used to fit a line to a data set. The R statistic then gives an indication of how well the line describes the data.
Chi-squared test	To test whether enumerated data are from a particular category.
Rayleigh test	Test for the significance of the mean direction in a circular distribution.
von Mises distribution	A circular distribution, like a circular version of the normal distribution.
Watson's test	Tests goodness of fit to the von Mises circular distribution.

We previously described an open source software package called ParticleStats that can produce these plots, as well as performing some of the relevant statistical tests. This is illustrated in Fig. 2.3 below. ParticleStats is described in Hamilton *et al.* (2010) and Oliveira *et al.* (2010), and can be accessed from www.particlestats.com.

A common goal when tracking particle motility is to fit the observed movements to a specific motion model. What are the known motion models, and how does one go about deciding whether the data is fit?

The process of diffusion normally results in a mean-squared displacement that increases linearly with time (Saxton, 1994). However, this simple case is unlikely in a living cell, where up to ∼40% of the total volume is taken up by various biomacromolecules. A plot of mean-squared displacement versus time will reveal the anomalous subdiffusion that results from

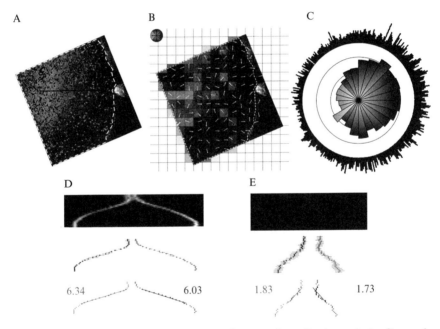

Figure 2.3 ParticleStats: An open source software package for the analysis of intracellular particle motility. Panels A–C summarize EB1:GFP tracks that mark growing microtubule plus-end directionality in a stage 8 *Drosophila* oocyte. (A) ParticleStats: Directionality track plot, where colors represent directionality as for the wind map. (B) "Wind map" representing sum of track directions in each square, where colors correspond to the four sectors shown top left. (C) "Rose diagram" summarizing overall directionality, where each outer dot represents a track. (D) ParticleStats:Kymograph used to create a kymograph summarizing the movement of eight aligned kinetochore pairs during meiosis in a *Drosophila* embryo. (E) Kymograph for a mutant with poor kinetochore synchrony. (See Color Insert.)

the crowded intracellular environment, as well as the existence of corralled diffusion if this is the case. If, on the other hand, an active transport process is taking place, then super-diffusion will result in a mean-squared displacement that increases more rapidly than a linear rate with time.

7. Conclusions

This is an exciting time to be involved in high resolution imaging in cell biology. Each new improvement in the resolution of microscopy has delivered a wealth of scientific discoveries in diverse fields of research, leading to the ability to bypass old limits. New developments in super-resolution imaging, combined with advances in detector sensitivity and speed, efficient methods for image restoration, and new improved fluorophores mean that fluorescence microscopy is on the cusp of resolving intracellular interactions live, in real time, at the molecular level of detail. This promises to deliver profound benefits to research in cell biology, greatly expediting the dissection of complicated pathway interactions by visualizing them live in action.

ACKNOWLEDGMENTS

The authors would like to thank R. A. Oliveria for permission to include the *Drosophila* embryo kinetochore data shown in Fig. 2.3. We would also like to acknowledge Charles Kervrann and INRIA Rennes for providing the nD-SAFIR denoising algorithm, and John Sedat's lab for the implementation in the Prism image processing and analysis suite. This work was supported by a Wellcome Trust Senior Fellowship to I.D. (Grant number 081858). G. B. is employed by the Micron Oxford Advanced Imaging Facility, which is supported by a Wellcome Trust Strategic Award.

REFERENCES

Agard, D. A., Hiraoka, Y., et al. (1989). Fluorescence microscopy in three dimensions. *Methods Cell Biol.* **30,** 353–377.

Andrey, P., Kiêu, K. K., et al. (2010). Statistical analysis of 3D images detects regular spatial distributions of centromeres and chromocenters in animal and plant nuclei. *PLoS Comput. Biol.* **6**(7), (e1000853).

Axelrod, D. (1989). Total internal reflection fluorescence microscopy. *Methods Cell Biol.* **30,** 245–270.

Betzig, E., Patterson, G. H., et al. (2006). Imaging intracellular fluorescent proteins at nanometer resolution. *Science* **313**(5793), 1642–1645.

Bewersdorf, J., Schmidt, R., et al. (2006). Comparison of I5M and 4Pi-microscopy. *J. Microsc.* **222**(2), 105–117.

Bivand, R. S., Pebesma, E. J., et al. (2008). Applied Spatial Data Analysis with R. Springer, New York.

Bolte, S., and Cordelières, F. P. (2006). A guided tour into subcellular colocalization analysis in light microscopy. *J. Microsc.* **224**(Pt. 3), 213–232.
Bretschneider, S., Eggeling, C., et al. (2007). Breaking the diffraction barrier in fluorescence microscopy by optical shelving. *Phys. Rev. Lett.* **98**(21), 218103.
Buades, A., Coll, B., et al. (2005). A review of image denoising algorithms, with a new one. *Multiscale Model. Simul.* **4**(2), 490–530.
Carlton, P. M., Boulanger, J., et al. (2010). Fast live simultaneous multiwavelength four-dimensional optical microscopy. *Proc. Natl. Acad. Sci. USA* **107**(37), 16016–16022.
Chalfie, M., Tu, Y., et al. (1994). Green fluorescent protein as a marker for gene expression. *Science* **263**(5148), 802–805.
Cheezum, M. K., Walker, W. F., et al. (2001). Quantitative comparison of algorithms for tracking single fluorescent particles. *Biophys. J.* **81**(4), 2378–2388.
Chen, Y., Wei, L.-N., et al. (2003). Probing protein oligomerization in living cells with fluorescence fluctuation spectroscopy. *Proc. Natl. Acad. Sci. USA* **100**(26), 15492–15497.
Churchman, L. S., Okten, Z., et al. (2005). Single molecule high-resolution colocalization of Cy3 and Cy5 attached to macromolecules measures intramolecular distances through time. *Proc. Natl. Acad. Sci. USA* **102**(5), 1419–1423.
Cressie, N. (1993). Statistics for Spatial Data. Wiley-Blackwell, Hoboken, NJ.
Dabov, K., Foi, A., et al. (2007). Image denoising by sparse 3-D transform-domain collaborative filtering. *IEEE Trans. Image Process.* **16**(8), 2080–2095.
Davis, I. (2009). The 'super-resolution' revolution. *Biochem. Soc. Trans.* **37**(Pt. 5), 1042–1044.
Denk, W., Strickler, J. H., et al. (1990). Two-photon laser scanning fluorescence microscopy. *Science* **248**(4951), 73–76.
Elad, M., and Aharon, M. (2006). Image denoising via sparse and redundant representations over learned dictionaries. *IEEE Trans. Image Process.* **15**(12), 3736–3745.
Endesfelder, U., van de Linde, S., et al. (2010). Subdiffraction-resolution fluorescence microscopy of myosin–actin motility. *Chemphyschem* **11**(4), 836–840.
Endesfelder, U., Malkusch, S., et al. (2011). Chemically induced photoswitching of fluorescent probes—A general concept for super-resolution microscopy. *Molecules (Basel, Switzerland)* **16**(4), 3106–3118.
Evanko, D. (2009). Primer: Fluorescence imaging under the diffraction limit. *Nat. Methods* **6**(1), 19–20.
Galbraith, C. G., and Galbraith, J. A. (2011). Super-resolution microscopy at a glance. *J. Cell Sci.* **124**(10), 1607–1611.
Goldman, R. D., Swedlow, J. R., et al. (2009). Live Cell Imaging: A Laboratory Manual. Cold Spring Harbor Laboratory Press, Woodbury, NY.
Gonzalez, R. C., and Woods, R. E. (2002). Digital Image Processing. Prentice Hall, New Jersey.
Guizetti, J., Schermelleh, L., et al. (2011). Cortical constriction during abscission involves helices of ESCRT-III-dependent filaments. *Science* **331**(6024), 1616–1620.
Gustafsson, M. G. L. (1999). Extended resolution fluorescence microscopy. *Curr. Opin. Struct. Biol.* **9**(5), 627–628.
Gustafsson, M. G. (2000). Surpassing the lateral resolution limit by a factor of two using structured illumination microscopy. *J. Microsc.* **198**(Pt. 2), 82–87.
Gustafsson, M. G. L. (2005). Nonlinear structured-illumination microscopy: Wide-field fluorescence imaging with theoretically unlimited resolution. *Proc. Natl. Acad. Sci. USA* **102**(37), 13081–13086.
Gustafsson, M. G., Agard, D. A., et al. (1999). I5M: 3D widefield light microscopy with better than 100 nm axial resolution. *J. Microsc.* **195**(Pt. 1), 10–16.
Gustafsson, M. G. L., Shao, L., et al. (2008). Three-dimensional resolution doubling in wide-field fluorescence microscopy by structured illumination. *Biophys. J.* **94**(12), 4957–4970.

Hamilton, R. S., Parton, R. M., et al. (2010). ParticleStats: Open source software for the analysis of particle motility and cytoskeletal polarity. *Nucleic Acids Res.* **38**(Web Server), W641–W646.

Hein, B., Willig, K. I., et al. (2008). Stimulated emission depletion (STED) nanoscopy of a fluorescent protein-labeled organelle inside a living cell. *Proc. Natl. Acad. Sci. USA* **105**(38), 14271–14276.

Heintzmann, R., Jovin, T. M., et al. (2002). Saturated patterned excitation microscopy: A concept for optical resolution improvement. *J. Opt. Soc. Am. A* **19**(8), 1599–1609.

Hell, S. W. (2003). Toward fluorescence nanoscopy. *Nat. Biotechnol.* **21**(11), 1347–1355.

Hell, S. W. (2009). Microscopy and its focal switch. *Nat. Methods* **6**(1), 24–32.

Hell, S., and Stelzer, E. H. K. (1992). Properties of a 4Pi confocal fluorescence microscope. *J. Opt. Soc. Am. A* **9**(12), 2159–2166.

Hell, S. W., and Wichmann, J. (1994). Breaking the diffraction resolution limit by stimulated emission: Stimulated-emission-depletion fluorescence microscopy. *Opt. Lett.* **19**(11), 780–782.

Hess, S. T., Girirajan, T. P. K., et al. (2006). Ultra-high resolution imaging by fluorescence photoactivation localization microscopy. *Biophys. J.* **91**(11), 4258–4272.

Holden, S. J., Uphoff, S., et al. (2011). DAOSTORM: An algorithm for high-density super-resolution microscopy. *Nat. Methods* **8**(4), 279–280.

Huang, B., Jones, S. A., et al. (2008). Whole-cell 3D STORM reveals interactions between cellular structures with nanometer-scale resolution. *Nat. Methods* **5**(12), 1047–1052.

Huang, B., Bates, M., et al. (2009). Super-resolution fluorescence microscopy. *Annu. Rev. Biochem.* **78**, 993–1016.

Jaqaman, K., Loerke, D., et al. (2008). Robust single-particle tracking in live-cell time-lapse sequences. *Nat. Methods* **5**(8), 695–702.

Ji, N., Shroff, H., et al. (2008). Advances in the speed and resolution of light microscopy. *Curr. Opin. Neurobiol.* **18**(6), 605–616.

Kay, S. M. (1993). *Fundamentals of Statistical Signal Processing: Estimation Theory, V. 1*. Prentice Hall, New Jersey.

Kervrann, C., and Boulanger, J. (2008). Local adaptivity to variable smoothness for exemplar-based image regularization and representation. *Int. J. Comput. Vis.* **79**(1), 45–69.

Klar, T. A., Jakobs, S., et al. (2000). Fluorescence microscopy with diffraction resolution barrier broken by stimulated emission. *Proc. Natl. Acad. Sci. USA* **97**(15), 8206–8210.

Klar, T. A., Engel, E., et al. (2001). Breaking Abbe's diffraction resolution limit in fluorescence microscopy with stimulated emission depletion beams of various shapes. *Phys. Rev. E Stat. Nonlin. Soft Matter Phys.* **64**(6 Pt. 2), 066613.

Kner, P., Chhun, B. B., et al. (2009). Super-resolution video microscopy of live cells by structured illumination. *Nat. Methods* **6**(5), 339–342.

Kner, P., Sedat, J. W., et al. (2010). High-resolution wide-field microscopy with adaptive optics for spherical aberration correction and motionless focusing. *J. Microsc.* **237**(2), 136–147.

Larson, D. R. (2010). The economy of photons. *Nat. Methods* **7**(5), 357–359.

Lee, S. F., Thompson, M. A., et al. (2011). Super-resolution imaging of the nucleoid-associated protein HU in Caulobacter crescentus. *Biophys. J.* **100**(7), L31–L33.

Lippincott-Schwartz, J., and Manley, S. (2009). Putting super-resolution fluorescence microscopy to work. *Nat. Methods* **6**(1), 21–23.

Manders, E. M., Stap, J., et al. (1992). Dynamics of three-dimensional replication patterns during the S-phase, analysed by double labelling of DNA and confocal microscopy. *J. Cell Sci.* **103**(Pt. 3), 857–862.

Meijering, E., Smal, I., et al. (2006). Tracking in molecular bioimaging. *IEEE Signal Process. Mag.* **23**(3), 46–53.
Moerner, W. E. (2006). Single-molecule mountains yield nanoscale cell images. *Nat. Methods* **3**(10), 781–782.
Nägerl, U. V., Willig, K. I., et al. (2008). Live-cell imaging of dendritic spines by STED microscopy. *Proc. Natl. Acad. Sci. USA* **105**(48), 18982–18987.
Nobelprize.org (2008). *Nobel Prize in Chemistry 2008.* http://nobelprize.org/nobel_prizes/chemistry/laureates/2008/.
Oliveira, R. A., Hamilton, R. S., et al. (2010). Cohesin cleavage and Cdk inhibition trigger formation of daughter nuclei. *Nat. Cell Biol.* **12**(2), 185–192.
Pavani, S. R. P., Thompson, M. A., et al. (2009). Three-dimensional, single-molecule fluorescence imaging beyond the diffraction limit by using a double-helix point spread function. *Proc. Natl. Acad. Sci. USA* **106**(9), 2995–2999.
Pawley, J. (1995). *Handbook of Biological Confocal Microscopy.* Springer, Springer-Verlag London Ltd.
Perona, P., and Malik, J. (1990). Scale-space and edge detection using anisotropic diffusion. *IEEE Trans. Pattern Anal. Mach. Intell.* **12**(7), 629–639.
Pertsinidis, A., Zhang, Y., et al. (2010). Subnanometre single-molecule localization, registration and distance measurements. *Nature* **466**(7306), 647–651.
Planchon, T. A., Gao, L., et al. (2011). Rapid three-dimensional isotropic imaging of living cells using Bessel beam plane illumination. *Nat. Methods* **8,** 417–423.
Reynaud, E. G., Krzic, U., et al. (2008). Light sheet-based fluorescence microscopy: More dimensions, more photons, and less photodamage. *HFSP J.* **2**(5), 266–275.
Rittweger, E., Han, K. Y., et al. (2009). STED microscopy reveals crystal colour centres with nanometric resolution. *Nat. Photon.* **3**(3), 144–147.
Robbins, M. S., and Hadwen, B. J. (2003). The noise performance of electron multiplying charge-coupled devices. *IEEE Trans. Electron Devices* **50**(5), 1227–1232.
Rose, A. (1948). The sensitivity performance of the human eye on an absolute scale. *J. Opt. Soc. Am.* **38**(2), 196–208.
Rudin, L. I., Osher, S., et al. (1992). Nonlinear total variation based noise removal algorithms. *Phys. D* **60**(1–4), 259–268.
Rust, M. J., Bates, M., et al. (2006). Sub-diffraction-limit imaging by stochastic optical reconstruction microscopy (STORM). *Nat. Methods* **3**(10), 793–796.
Santi, P. A. (2011). Light sheet fluorescence microscopy. *J. Histochem. Cytochem.* **59**(2), 129–138.
Saxton, M. J. (1994). Single-particle tracking: Models of directed transport. *Biophys. J.* **67**(5), 2110–2119.
Sbalzarini, I. F., and Koumoutsakos, P. (2005). Feature point tracking and trajectory analysis for video imaging in cell biology. *J. Struct. Biol.* **151**(2), 182–195.
Schermelleh, L., Carlton, P. M., et al. (2008). Subdiffraction multicolor imaging of the nuclear periphery with 3D structured illumination microscopy. *Science (New York, N.Y.)* **320**(5881), 1332–1336.
Schermelleh, L., Heintzmann, R., et al. (2010). A guide to super-resolution fluorescence microscopy. *J. Cell Biol.* **190**(2), 165–175.
Schmidt, R., Wurm, C. A., et al. (2008). Spherical nanosized focal spot unravels the interior of cells. *Nat. Methods* **5**(6), 539–544.
Schmidt, R., Wurm, C. A., et al. (2009). Mitochondrial cristae revealed with focused light. *Nano Lett.* **9**(6), 2508–2510.
Shaner, N. C., Steinbach, P. A., et al. (2005). A guide to choosing fluorescent proteins. *Nat. Methods* **2**(12), 905–909.
Shao, L., Isaac, B., et al. (2008). I5S: Wide-field light microscopy with 100-nm-scale resolution in three dimensions. *Biophys. J.* **94**(12), 4971–4983.

Thomann, D., Rines, D. R., et al. (2002). Automatic fluorescent tag detection in 3D with super-resolution: Application to the analysis of chromosome movement. *J. Microsc.* **208**(Pt. 1), 49–64.

Thompson, R. E., Larson, D. R., et al. (2002). Precise nanometer localization analysis for individual fluorescent probes. *Biophys. J.* **82**(5), 2775–2783.

Thompson, M. A., Lew, M. D., et al. (2010). Localizing and tracking single nanoscale emitters in three dimensions with high spatiotemporal resolution using a double-helix point spread function. *Nano Lett.* **10**(1), 211–218.

Tomasi, C., and Manduchi, R. (1998). Bilateral filtering for gray and color images. Proceedings of the Sixth International Conference on Computer Vision, *IEEE Comput. Soc.* 839–846.

Toomre, D., and Bewersdorf, J. (2010). A new wave of cellular imaging. *Annu. Rev. Cell Dev. Biol.* **26,** 285–314.

Toprak, E., Kural, C., et al. (2010). Super-accuracy and super-resolution getting around the diffraction limit. *Methods Enzymol.* **475,** 1–26.

Verveer, P. J., Swoger, J., et al. (2007). High-resolution three-dimensional imaging of large specimens with light sheet-based microscopy. *Nat. Methods* **4**(4), 311–313.

Wallace, W., Schaefer, L. H., et al. (2001). A workingperson's guide to deconvolution in light microscopy. *Biotechniques* **31**(5), 1076–1078. (1080, 1082 passim-1076–1078, 1080, 1082 passim).

Yang, L., Parton, R., et al. (2010). An adaptive non-local means filter for denoising live-cell images and improving particle detection. *J. Struct. Biol.* **172**(3), 233–243.

Yildiz, A., and Selvin, P. R. (2005). Fluorescence imaging with one nanometer accuracy: Application to molecular motors. *Acc. Chem. Res.* **38**(7), 574–582.

York, A. G., Ghitani, A., et al. (2011). Confined activation and subdiffractive localization enables whole-cell PALM with genetically expressed probes. *Nat. Methods* **8**(4), 327–333.

CHAPTER THREE

Applications of Fluorescence Lifetime Spectroscopy and Imaging to Lipid Domains *In Vivo*

André E. P. Bastos,* Silvia Scolari,* Martin Stöckl,[‡] *and*
Rodrigo F. M. de Almeida*,[†]

Contents

1. The Challenge of Studying Lipid Domains *In Vivo* Defied by Fluorescence Lifetimes — 58
2. Importance of Studies *In Vitro* to Design and Rationalize Studies *In Vivo* — 61
   2.1. Fundamental issues: Study lipid domains away from the complexity and dynamics of a living cell — 61
   2.2. Practical issues: Selecting a probe for *in vivo* studies — 62
3. Labeling Cell Membranes *In Vivo*: Lipophilic Probes Versus GFP-Tagged Membrane Proteins — 67
   3.1. Fluorescent lipid analogues — 67
   3.2. Chemical dyes sensitive to lipid composition — 68
   3.3. Fluorescent protein engineering — 69
4. Cuvette Lifetimes and FLIM: *In Vivo* Applications — 70
   4.1. Cuvette advantages and applications — 70
   4.2. FLIM advantages and applications — 73
Acknowledgments — 78
References — 79

Abstract

Lipid domains are part of the current description of cell membranes and their involvement in many fundamental cellular processes is currently acknowledged. However, their study in living cells is still a challenge. Fluorescence lifetimes have and will continue to play an important role in unraveling the properties and function of lipid domains, and their use *in vivo* is expected to increase in the near future, since their extreme sensitivity to the physical

* Centro de Química e Bioquímica, Faculdade de Ciências da Universidade de Lisboa, Lisboa, Portugal
[‡] Nanobiophysics, MESA+ Institute for Nanotechnology, University of Twente, Enschede, The Netherlands
[†] Corresponding Author

properties of the membrane and the possibility of optical imaging are particularly suited to deal with the hurdles that are met by researchers. In this review, a practical guide on the use of fluorescence lifetimes for the study of this subject is given. A section is devoted to studies *in vitro*, particularly membrane model systems, and how they are used to better design and correctly interpret results obtained in living cells. Criteria are presented for selecting suitable probes to solve each problem, drawing attention to factors sometimes overlooked and which may affect the fluorescence lifetime such as subcellular distribution and concentration of the probe. The principal groups of lifetime probes for lipid domains: (i) fluorescent lipid analogues, (ii) other lipophilic probes, and (iii) fluorescent proteins, and respective applications are briefly described and lab tips about the labeling of living cells are provided. The advantages and complementarities of spectroscopy (cuvette) work and fluorescence lifetime imaging microscopy are presented and illustrated with three selected case studies: (i) the finding of a new type of lipid rafts in yeast cells; (ii) the detection of liquid ordered type heterogeneity in animal cells below optical resolution; and (iii) establish a role for the transmembrane domain of influenza virus hemagglutinin with cholesterol-enriched domains in mammalian cells.

1. The Challenge of Studying Lipid Domains *In Vivo* Defied by Fluorescence Lifetimes

Lipid domains are lateral heterogeneities in cellular membranes that are formed on the basis of differential lipid–lipid interactions. They were first described several decades ago, but only recently a detailed comprehension about the mechanisms of formation and functions of those domains has started to emerge (Elson *et al.*, 2010). This has been in great part due to the intensive research on lipid rafts, which are a special type of lipid domains, formed through preferential interactions between sphingolipids and certain sterols involved and which are involved in signal transduction and sorting of proteins and lipids (Pike, 2006). Several have been the challenges on the study of this subject. One of them is that proteins which are also major components of biomembranes can strongly modulate membrane lateral organization, and therefore, it is not easy to discriminate whether differential lipid–lipid interactions are the major driving force for the formation of a certain type of heterogeneity or not. Therefore, the more general term membrane microdomain is often used, referring to areas of the membrane with distinct lipid and protein composition (Malinsky *et al.*, 2010; Pedroso *et al.*, 2009). Another difficulty has been the nature of lipid rafts in mammalian cells, since they are small and transient domains (Pike, 2006) which usually only become more stable and can be observed by fluorescence microscopy under certain stimuli, such as T-cell activation (Gaus *et al.*, 2005) or ceramide generation under stress stimuli (Grassme *et al.*, 2003).

However, in yeast, spatiotemporally stable membrane domains can be observed, namely, membrane compartment occupied by Can1p (MCC) domains which are ergosterol enriched, and membrane compartment occupied by Pma1p (MCP) domains, among others (Malinsky et al., 2010). This is a major difference in membrane organization when compared to mammalian cells. In the latter, there are special cases such as the apical and basolateral domains of polarized epithelial cells which have different lipid and protein compositions and are stable over time, but in this case, physical barriers, namely, tight junctions play an important role in the maintenance of lipid and protein segregation (Schuck and Simons, 2004). Another challenge stems from the fact that membranes are highly and rapidly adaptable entities, for example, as described for *Saccharomyces cerevisiae* cellular adaptation to a steady-state dose of hydrogen peroxide (Folmer et al., 2008; Pedroso et al., 2009).

Therefore, noninvasive, highly sensitive techniques with minimal perturbation have to be used in the study of lipid domains *in vivo*. Due to their complexity, the combination of imaging (which allows real-time visualization of the cell membrane) with spectroscopic information has been a recent trend in the field. Fluorescence offers ideal approaches where such combination can be achieved, since living cell imaging is now carried out chiefly by fluorescence microscopy, and at the same time, the fluorescence properties of fluorescent markers can be determined, namely, spectral shifts, fluorescence lifetimes, and fluorescence anisotropy, retrieving information about the physical properties of the membranes, such as order, charge distribution, polarity, and hydration (examples are given along the text). Fluorescence presents the advantages of being intrinsically a very sensitive technique, requiring very low probe concentrations.

Time-resolved fluorescence gives information on the kinetics of fluorescence emission and excited-state lifetimes, that is, the average time the fluorophore spends in the excited state attained after light absorption and prior to fluorescence emission or deactivation by other processes. Since the emission probability decreases with time after excitation, in time-resolved fluorescence, a fluorescence intensity decay is obtained (see, e.g., Fig. 3.1A). The simplest kinetics of fluorescence emission is a first order law (exponential decay) and the fluorescence (or excited-state) lifetime is the reverse of the decay rate constant. However, in most cases, the fluorescence decay is more complex and is usually described by a sum of exponentials,

$$I(t) = \sum_{i=1}^{n} \alpha_i \exp(-t/\tau_i) \tag{3.1}$$

or by a lifetime distribution. In Eq. (3.1), α_i and τ_i are the normalized amplitude and lifetime of component i, respectively. Multicomponent decays or width of lifetime distribution are indications of heterogeneity

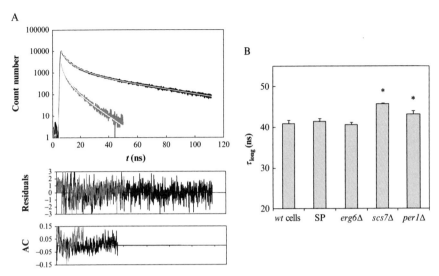

Figure 3.1 (A) Global analysis of fluorescence intensity decay of a suspension of LC4 cells (0.6 absorbance at 600 nm) in mid-exponential phase at 24 °C labeled with 2 μM t-PnA (black) or unlabeled (gray). Top panel: experimental decays and best fitting function with a sum of exponentials (white lines). Middle panel: random distribution of weighted residuals of the fitting. Bottom panel: autocorrelation of the residuals (AC). Fluorescence measurements were carried out by the single photon counting technique on a Horiba Jobin Yvon FL-1057 Tau 3 spectrofluorometer. A nanoLED N-320 (Horiba Jobin Yvon) was used for the excitation of t-PnA (nominal wavelength 315 nm), and emission wavelength was 404 nm. Colloidal silica in water was used as the scatterer to obtain the instrument response function. Data analysis was performed with TRFA Data Processor v.1.4 software (Minsk, Belarus). The global analysis method is described in the text. Plasmid Ylp211 (a kind gift of W. Tanner, University of Regensburg, Germany) was transformed into BY4741 (wt) strain to obtain strain LC4 expressing CAN1-GFP. (B) Long component lifetime of t-PnA obtained from the fluorescence intensity decay of the probe at 24 °C as described in (A). The analysis was performed for wt (BY4147) intact cells and spheroplasts (SP) with cell-wall digested by zymolyase, and $erg6\Delta$, $scs7\Delta$, and $per1\Delta$ in mid-exponential phase. The values are the mean ± standard deviation of at least four independent experiments. $\star P < 0.001$ versus wt cells; the *S. cerevisiae* strains were obtained from EUROSCARF. While $erg6\Delta$ cells have a very different sterol profile, but similar sphingolipid composition when compared to wt cells, and τ_{long} remains unchanged, in $scs7\Delta$ the opposite change in lipids occurs, and τ_{long} is significantly different from the wt. In $per1\Delta$, the remodeling of GPI-anchors is impaired, and τ_{long} is also significantly different from that of the wt. This shows that sphingolipids and GPI-anchored proteins, but not sterols, are important components of the lipid domains that give raise to the long lifetime component of t-PnA. See Aresta-Branco et al. (2011) and text for further details.

and, at the same time, give information on the biophysical properties and relative abundance of the lipid domains that are responsible for the appearance of each lifetime. Multiexponentiality in membrane probes and the cautions that are necessary in data analysis and interpretation were discussed elsewhere (de Almeida et al., 2009).

Fluorescence lifetimes, usually in the order of nanoseconds, are sensitive to biomolecule dynamics, solvent relaxation, and presence of other molecules (concentration and diffusion coefficient) that decrease the excited-state lifetime (Berezin and Achilefu, 2010; Stöckl and Herrmann, 2010). To this, we can add Förster resonance energy transfer (FRET), another process that reduces the fluorescence lifetime and is dependent of the distance between the donor (the fluorophore) and an acceptor, with a maximum sensitivity on the length scale of nanometer. Therefore, fluorescence lifetime spectroscopy and imaging are among the most promising techniques to unravel the elusive nature of lipid domains *in vivo*.

2. Importance of Studies *In Vitro* to Design and Rationalize Studies *In Vivo*

2.1. Fundamental issues: Study lipid domains away from the complexity and dynamics of a living cell

Some of the difficulties and limitations of studying lipid domains *in vivo* can be circumvented by using membrane model systems, which are artificial structures composed by well-defined and controlled number and proportion of lipids (and proteins). The study of lipid domains in membrane model systems can be made with essentially the same techniques as for *in vivo* studies (de Almeida *et al.*, 2009; Stöckl and Herrmann, 2010). Therefore, not only these studies have been crucial to establish the state of the art in lipid domains, but they can be also extremely helpful to design and interpret data that is obtained *in vivo* (next sections). In general, cuvette (spectroscopic) experiments (Section 4.1) are performed with suspensions of multilamellar vesicles (MLVs) or large unilamellar vesicles (LUVs), whereas fluorescence lifetime imaging spectroscopy (FLIM) experiments (Section 4.2) are carried out with giant unilamellar vesicles (GUVs) or planar bilayers. This is because only the latter two have sizes that reach tens of micrometers, and thus can be adequately imaged by optical microscopic techniques. In all these systems, the lipids are organized in closed-shell bilayers (one in the case of LUV and GUV, or several in the case of MLV) that separate inner and outer aqueous media such as in biological membranes.

Studies in model systems have revealed that a small number of components is sufficient to give rise to a variety of different types of lateral heterogeneities, with sizes ranging from only a small number of molecules up to micrometer-sized domains, and that they can be dynamic composition/critical fluctuations or may correspond to stable-phase separation situations (Elson *et al.*, 2010). Therefore, the different phases formed by lipids are considered to be good models for the lipid domains and other lateral heterogeneities that may be found in biological membranes *in vivo*. Three types of lipid phases have been considered the most relevant in this context:

the gel or solid ordered (s_o) phase where the lipids have highly ordered acyl chains and slow lateral diffusion, fluid or liquid disordered (l_d) phase where there is a high degree of disorder in the acyl chains and a fast lateral diffusion, and the liquid ordered (l_o) phase that only forms in the presence of certain sterols, which have been described as raft-forming sterols such as cholesterol and ergosterol. This phase is characterized by an acyl chain order similar to the gel phase, but lateral diffusion on the same order of magnitude as the l_d phase (though slightly slower) (Korlach et al., 1999), and thus also considered fluid. All these phases have the organization of the lipids as a fully hydrated bilayer as a common feature.

There were several fundamental studies using fluorescence lifetimes in ternary lipid systems that mimic lipid rafts *in vivo*. Such models contain one lipid with a low gel/fluid transition (melting) temperature (T_m), typically, an unsaturated phospholipid, one high T_m lipid (either a saturated phospholipid or a sphingolipid), and cholesterol. It was possible to determine the range of compositions where coexistence of l_d (enriched in the low T_m lipid) and l_o phase (enriched in the high T_m lipid and cholesterol) occurs, which would correspond in cells to the raft and nonraft fractions of the membrane (de Almeida et al., 2003). A FLIM study provided the first direct evidence for the presence of gel, l_o and l_d domains in a single GUV (de Almeida et al., 2007). Factors governing lipid raft size were pointed out by a FRET study using fluorescence lifetimes (de Almeida et al., 2005).

2.2. Practical issues: Selecting a probe for *in vivo* studies

As described above, lipid phases and lateral heterogeneities that occur in membrane model systems are representative of lipid organization in cells. Therefore, it is expected that the fluorescence properties of a probe *in vivo* will resemble those in model systems that mimic the lipid domains preferentially stained by the probe in living cells. However, there are, in general, no probes that label exclusively a certain type of domain. Even when using techniques that combine imaging with fluorescence lifetime (Section 4.2), it is common that the domains are smaller than the spatial resolution of the technique. Therefore, the fluorescence signal is most of the time a composition of the fluorescence from the probe incorporated in different types of domains. The contribution of each type of domain to the total signal depends on three factors in model systems and four in the case of living cells:

(i) The relative abundance of the lipid domain;
(ii) The preference of the probe for each type of domain (that can be expressed as one or several partition coefficients between two lipid phases);
(iii) The fluorescence quantum yield of the probe in each type of domain;

(iv) In case of cells, the intracellular distribution of the probe has to be considered; this will be discussed and exemplified along the text.

Consequently, a systematic study of the probe's fluorescence behavior in model systems over a wide range of compositions that allow establishing the relevant parameters for factors (ii) and (iii) are often advisable (examples can be found in de Almeida et al., 2009).

Studies in model systems may therefore provide important criteria for the selection of fluorescent probes for the study of lipid domains *in vivo* using fluorescence lifetimes. In addition to general criteria for the choice of a fluorescent probe, such as excitation and emission wavelengths, photostability, and brightness, the following points should be considered:

(a) The typical lifetime range in each lipid phase (e.g., if the lifetime in different phases in model systems is similar, the probe will not be useful for studies *in vivo*);
(b) The preference of the probe for each lipid phase;
(c) Other factors that may affect strongly the fluorescence lifetime (e.g., probe concentration) and probe aggregation that often leads to alterations of fluorescence lifetime or even membrane properties, including the appearance of artifactual domains.

For probes that do not partition preferentially into a certain type of lipid domain (in fact, very few probes prefer the more ordered and compact phases), it is still possible to obtain information on those domains from fluorescence lifetimes, if the average lifetime of the probe is significantly affected by the small fraction of probe that is emitting from those domains. When resolution of multiple exponentials is possible, the lifetime components together with their amplitudes (Eq. (3.1)) contain information on the type and relative abundance of domains. Examples are given in following sections and Tables 3.1 and 3.2.

Probes located in more compact phases have expectedly longer fluorescence lifetimes. In those phases, processes that compete with fluorescence emission, such as collisions, torsions, or isomerizations are, in principle, less effective. An example is the long lifetime component (τ_{long}) of *trans*-parinaric (*t*-PnA) acid in sphingolipid-enriched domains of *S. cerevisiae* (Fig. 3.1B). In wild-type (wt) cells, the very long chain fatty acids (VLFA) of sphingolipids are mostly 2-hydroxylated, and therefore the acyl chains are not as tightly packed as in *scs7Δ* cells (Aresta-Branco et al., 2011), which lack the sphingolipid 2-hydroxylase. However, this is not always the case. The fluorescence lifetime of BODIPY-PC (2-(4,4-difluoro-5,7-dimethyl-4-bora-3a,4a-diaza-*s*-indacene-3-pentanoyl)-1-hexadecanoyl-*sn*-glycero-3-phosphocholine) is shorter in the gel phase (Ariola et al., 2006) and Rhod-DOPE (*N*-(lyssamine rhodamine B sulfonyl)-dioleoylphosphatidylethanolamine) in the l_o phase (de Almeida et al., 2007)

Table 3.1 Representative probes and respective excitation/emission wavelengths used for the study of lipid domains through FLIM and cuvette fluorescence lifetimes *in vivo* and/or *in vitro* (model systems and isolated/reconstituted cell membranes)

Probe	Exc/em (nm)	Problem/system addressed		Lifetime components or average lifetime (ns)			Technique	References
DPH	325/430	Isolated synaptic plasma membrane (PM) from mice		10.7 ± 0.1	3.8 ± 1.5		Cuvette	Colles et al. (1995), de Almeida et al. (2003), Ravichandra and Joshi (1999)
		Model system	POPC/PSM/chol. (l_d)	7.4				
			POPC/PSM/chol. (l_o)	9.7				
		PM of PC—12 living cells	Control	8.83 ± 0.22				
			+ Bovine brain gangliosides (BBG)	≈8.83 ± 0.22				
TMA-DPH	337/415	PM of PC—12 living cells	Control	6.22 ± 0.29			Cuvette	Ravichandra and Joshi (1999), Sinha et al. (2003)
			200 μg/ml BBG	4.71 ± 0.11				
		Model system	Gel DPPC (25 °C)	1.99 ± 0.25	6.66 ± 0.09			
			l_d DPPC (50 °C)	3.51 ± 0.23	–			
			l_o DPPC + 30 mol% chol. (50 °C)	–	6.84 ± 0.41			
		PM isolated from 87 MG cells	Detergent-resistant fraction	–	6.54 ± 0.32			
			PM	1.56 ± 0.08	5.66 ± 0.11			
			PM chol. depleted	1.66 ± 0.24	4.07 ± 0.03			
Oregon Green (OG)	470/>500	Isolated synaptic vesicles from mouse brains	Control	4.01 ± 0.02			Cuvette	Zeigler et al. (2011)
			Glutamate-loaded	3.84 ± 0.05				
			OG in water	3.85 ± 0.05				
CrFP	860/480	Living plant protoplasts	CDC48A-CrFP	2.67 ± 0.05			FRET-FLIM	Aker et al. (2006)
			CDC48A-CrFP/SERK1-YFP	2.23 ± 0.10				
Nile-Red	575/620	Bovine hippocampal membranes	Native	5.63	3.65	0.85	Cuvette	Mukherjee et al. (2007)
			Lipid extract	5.5	2.57	0.5		

Table 3.1 (Continued)

Probe	Exc/em (nm)	Problem/system addressed		Lifetime components or average lifetime (ns)		Technique	References
C6-NBD-PC/PS	468/540	Model system	POPC/PSM/chol. (25 °C)	7/2/1 (l_d)	8.4 ± 0.1 —	FLIM	Stöckl et al. (2008)
		PM of living cells	HepG2 cells at 25 °C	2/2/6 (l_o)	7.0 ± 0.4 11.4 ± 0.1		
Rhod-DOPE	860/620	Model system	GUVs	DOPC (fluid) DPPC (gel) DPPC/chol. 3:1 (l_o)	2.64 2.60 2.46	FLIM	de Almeida et al. (2007)
	575/620		MLVs, LUVs	DOPC (fluid) DPPC (gel) DPPC/chol. 3:1 (l_o)	2.66 2.64 2.11	Cuvette	
Laurdan	358/434	Native bovine hippocampal membranes		20 °C 50 °C	5.69 2.89	Cuvette	Mukherjee and Chattopadhyay (2005)
Bodipy-PC	775/525	GUVs		DOPC (l_d) DPPC (gel)	2.16 ± 0.07 9.4 ± 0.02 1.9 ± 0.07 4.1 ± 0.02	FLIM	Ariola et al. (2006)

Abbreviations: Bodipy-PC: 2-(4,4-difluoro-5,7-dimethyl-4-bora-3a,4a-diaza-s-indacene-3-pentanoyl)-1-hexadecanoyl-sn-glycero-3-phosphocholine; chol.: cholesterol; C6-NBD-PC/PS: 1-palmitoyl-2-[6-[(7-nitro-2-1,3-benzoxadiazol-4-yl)amino]-hexanoyl]-sn-glycero-3-phosphatidylcholine/serine; CrFP: cerulean fluorescent protein; DOPC: dioleoylphosphatidylcholine; DPPC: dipalmitoylphosphatidylcholine; GUV: giant unilamellar vesicle; LUV: large lamellar vesicle; MLV: multilamellar vesicle; Nile-Red: 9-diethylamino-5H-benzo[α]phenoxazine-5-one; POPC: 1-palmitoyl-2-oleoyl-sn-glycero-3-phosphocholine; PSM: N-palmitoyl-D-sphingomyelin; Rhod-DOPE: N-(lyssamine rhodamine B sulfonyl)-dioleoyl-phosphatidylethanolamine. See the other abbreviations in the text.

Table 3.2 *trans*-Parinaric acid fluorescence lifetimes labeling the membranes of living cells and isolated/reconstituted cell membranes and conclusions about membrane biophysical properties/lipid domains retrieved thereof

Problem/system addressed		Lifetime components or average lifetime (ns)		Main conclusion	References
		Average lifetime	Long component τ_4		
PM of *Sacharomyces cerevisiae*	Living cells (wt)	25.7	40.9	PM of *S. cerevisiae* contains highly ordered lipid domains and the presence of proteins limits the rigidity of the domains	Aresta-Branco et al. (2011)
	Isolated PM	40.8	46.6		
	Liposomes from isolated PM lipids	40.6	49.7		
Isolated rat-liver PMs	0 mM Ca^{2+}	3.1 ± 0.6	12.6 ± 1.7	Coexistence of fluid and solid domains that may be regulated by Ca^{2+}	Schroeder and Soler-Argilaga (1983), Schroeder (1983)
	2.4 mM Ca^{2+}	2.8 ± 0.4	19.9 ± 0.7		
Isolated PMs from cultured human fibroblasts	Male	4.8 ± 0.1	18.5 ± 0.5	Sexual-related differences in PM properties	Schroeder, F. (1984)
	Female	5.6 ± 0.6	20.7 ± 1.9		
Fragments of the PM of human platelets	27	15.5 ± 0.3	6.0 ± 0.3 1.5 ± 0.3	Confirm the existence of the thermotropic lipid phase separation at temperatures of 37 °C and lower	Mateo et al. (1991)
	35	12.5 ± 0.3	5.4 ± 0.3 1.3 ± 0.3		
	43	9.0 ± 0.3	4.1 ± 0.3 0.9 ± 0.3		
Mouse brain synaptic PMs	Dolichol −	18.3 ± 0.4	2.0 ± 0.2	Dolichol can dramatically alter the structure and dynamics of lipid motion	Schroeder et al. (1987)
	+	14.7 ± 0.7	1.4 ± 0.7		

than in the l_d phase (Table 3.1). This is due to a self-quenching process more effective in the condensed phases. If so, it is expected that reducing total probe concentration will increase the lifetime of the probe in those phases, which may reduce the lifetime contrast with l_d domains.

3. LABELING CELL MEMBRANES *IN VIVO*: LIPOPHILIC PROBES VERSUS GFP-TAGGED MEMBRANE PROTEINS

Microscopy and fluorescence represent nowadays almost a binomial condition necessary to study living systems. There exist several ways to perform sample labeling, some more straightforward that rely on the addition of the fluorescent agent externally (Sections 3.1 and 3.2) and some require genetic manipulation of the matter of study (e.g., GFP and derivatives fusion proteins—Section 3.3). Finally, some systems combine both the external addition of the fluorescent agent together with the manipulation of the subject (e.g., SNAP tag). With this technique developed by New England Biolabs, a short DNA sequence specifically reactive to benzylguanine (BG) and benzylchloropyrimidine (CP) derivatives is fused to the protein of interest. Labeling occurs upon covalent attachment of the tag to a synthetic fluorescent probe (Klein *et al.*, 2011).

In Table 3.1, a series of probes that have been used to study lipid domains both *in vivo* and *in vitro*, in cuvette and FLIM experiments, are presented, together with some important characteristics of those probes. It should be noted that this is not an exhaustive list of all studies available in the literature, but rather a collection of illustrative examples, showing, for example, excitation/emission wavelengths for which there are known fluorophores.

3.1. Fluorescent lipid analogues

In this first part of the section, the use of lipid-based probes for the labeling of cellular membranes is discussed. In such molecules, the chromophore can constitute the head of the structure, such as in Rhod-DOPE; can be attached to the end of one of the acyl chains of the lipid moiety (most NBD (Nitro-2-1,3-BenzoxaDiazol-4-yl) analogues); or could be opposite to the hydrophilic group as in the case of BODIPY-cholesterol. Examples are given in Table 3.1.

Lipid bilayers are the common structural basis of all membranes in a cell, but they distribute differently not only within compartments but also within layers (Pomorski *et al.*, 2001). Lipids carrying a chromophore represent the most "natural" method for labeling cells since they cause only a marginal disturbance of the membrane dynamics. Further, if the properties of the lipid consent its transport through the membrane by means of lipid transporters, such analogues may be suitable as well for a precise visualization of intracellular compartments.

In general, chromophores like NBD change their photophysical properties such as fluorescence polarization or lifetime, because they are sensitive to the environment in which they are embedded (Mukherjee et al., 2004; Stöckl et al., 2008). Further, these lipid-like fluorophores can preferentially segregate into areas of the membrane which differ in their lipid composition (Baumgart et al., 2007), although in living animal cells, a net separation of membrane compartments is not visible. However, depending on the length of their acyl chains, these analogues can be more rapidly or slowly incorporated into the membranes. For example, the use of N-Rho-DOPE for labeling intact biological membranes is difficult due to its long acyl chains (18:1).

Labeling of biological samples with such dyes is not trivial. As above mentioned, being lipids, these probes tend to be rapidly incorporated into the membranes and even translocated into inner compartments. Therefore, if the goal of the study is the plasma membrane, or even the external leaflet of the bilayer, cells are often labeled at 4 °C (on ice) and incubation time should be adjusted according to the efficiency of labeling. However, temperature variations may strongly affect lipid domain organization both *in vivo* (Magee et al., 2005) and *in vitro* (de Almeida et al., 2003). If measurements are carried out at room temperature, they should be performed in the shortest time possible in order to avoid plasma membrane recycling and for some probes flip-flop. It is recommended washing after labeling to remove the excess of dye from the cell surrounding that would create high background disturbance, always bearing in mind that each manipulation may affect the lipid domains (e.g., if the experiment is to be performed at 30 °C, the buffers should be warmed to 30 °C before washing the cells, to avoid temperature variations). Other lipid-based fluorophores such as BODIPY-cholesterol need to be incubated even overnight or for a more rapid absorption should be first complexed with Methyl-β-Cyclodextrin (MBCD) (Holtta-Vuori et al., 2008).

In any case, in the range of concentration usually employed (on the order of micromolar or less), no toxicity effects have been observed with such dyes.

3.2. Chemical dyes sensitive to lipid composition

In addition to lipid fluorescent analogues, nonlipid fluorescent probes are sensitive to the environment, thus changing their properties (lifetime, emission wavelength, anisotropy). As for labeling with fluorescent lipid analogues, washing after the incubation with such molecules is suggested to avoid high background noise. In some exceptional cases, such as that of *t*-PnA and 1,6-diphenyl-1,3,5-hexatriene (DPH; Section 3.1), the partition into the membranes is so strong and fast and the quantum yield in water is so low that washing can be avoided.

Many are the molecules able to effectively label cell membranes in a very fast and economic way. Within these probes, di-4-ANEPPDHQ (aminonaphthylethenylpyridinium derivative), born as a voltage sensitive probe, is now used also to sense the lipid organization of membranes since its fluorescence lifetime changes with the cholesterol content of the bilayer (Owen et al., 2006).

Laurdan is a very interesting dye since it changes its physical properties according to the hydration of the membrane, that is, it is affected by the solvent relaxation effect undergoing a typical shift of its emission spectrum and change in anisotropy (Harris et al., 2002). One application of Laurdan fluorescence lifetimes is given in Table 3.1.

Depending on the structure of the molecule, the translocation to inner compartments might be slower or faster and therefore they exhibit fluorescence parameters that report on membrane domains of different cellular compartments. Labeling of Chinese hamster ovary (CHO)-K1 cells with di-4-ANEPPS required incubation of cells for 30 min at room temperature or eventually at 37 °C and more than 60 min for labeling of inner compartments such as ER (endoplasmic reticulum) and Golgi apparatus, whereas in S. cerevisiae cells, inner membranes were labeled within a few minutes (S. Scolari, unpublished observations).

DPH, one of the most popular fluorescent probes to study membrane order and fluidity in model systems, is rapidly internalized when used to label cell membranes. Nevertheless its cationic derivative, trimethylammonium (TMA)-DPH, remains at the cell membrane surface sufficient time, which allows to obtain a fluorescence decay before internalization (Abe and Hiraki, 2009). However, virtually all probes become internalized after some time due to endocytic processes.

3.3. Fluorescent protein engineering

Cell labeling can also occur indirectly through the fusion of fluorescent proteins (FPs) to the protein of interest. In this way, one can follow the destiny of the protein under study, that is, intracellular trafficking, subcellular localization, colocalization with other proteins, and many other processes. When indeed the matter of study is not the lipid bilayer but the specific function exerted by a protein, it is advantageous to directly label the subject and thus directly follow its destiny. However, for the production of FP-tagged constructs cloning and plasmid design is required and biochemical/structural analysis is necessary to prove that no alterations in the functionality/structure have occurred. Constructs need then to be transfected into the cells for expression of the protein. Transfection can be carried out by many different methods, including electroporation and DNA complex formation with a lipid carrier (e.g., Lipofectamin^{TM}) or with polymers (e.g., polyethylenimine (PEI)). The efficiency of transfection depends strongly on the cell type, size of the plasmid, and cell state (e.g., cell confluency). Indeed, some cell lines do not tolerate very well the introduction of exogenous DNA, therefore transfection causes high mortality that can be due either to the overexpression of the FP or to the treatment with the transfection reagent. It has to be pointed out that, in general, transfection of tumor cell lines is more successful than transfection on primary cells. Although the use of FP tags does not require the addition of organic solvents or solubilizing agents in which the chemical dyes need to be resuspended and which may alter membrane properties, in any case also transfection agents may be cause of high toxicity as above discussed.

4. Cuvette Lifetimes and FLIM: *In Vivo* Applications

4.1. Cuvette advantages and applications

The cuvette experiment *in vivo* is performed with cells in suspension and is quite similar to the acquisition of a fluorescence intensity decay for *in vitro* systems, such as a suspension of liposomes labeled with a certain membrane probe. Extra care has to be taken due to problems that may arise *in vivo* related to the instability of the cells and their sensitivity to radiation that is used to excite the fluorophore.

An important factor to take into account is also that the culture medium is in general much more complex than the solvent used for *in vitro* studies and some of its components can have strong absorption and/or emission at the wavelengths of interest (an illustrious example is phenol red), and in this case, it is necessary to change the medium prior to fluorescence measurements. Because cuvette experiments are performed in cellular suspensions, usually they are not carried out with highly adhesive cells. Gentle stirring is also recommended, as cells may tend to deposit on the bottom of the cuvette. However, since fluorescence lifetimes are independent of total fluorescence intensity, this is not a major problem with time-resolved fluorescence, provided that a pronounced decrease in fluorescence intensity is not registered during the acquisition. Such fluorescence intensity fluctuations should be avoided mainly because for *in vivo* studies the acquisition time should be minimized and kept as much as possible constant from sample to sample and experiment to experiment within the same study. This will ensure that no variability is being introduced in cell behavior. Photobleaching is also a problem that affects much less time-resolved fluorescence data in comparison to experiments that rely on total fluorescence intensity.

Fluorescent probes with virtually any excitation wavelength can be used. This is a very important feature of fluorescence lifetime spectroscopy in the study of lipid domains because polyene probes such as DPH and t-PnA can be used, since UV excitation and emission is possible and photobleaching is a minor problem. In addition, probes absorbing/emitting in the visible can also be utilized (see Section 2). Polyene probes are a very important class of membrane probes because they align parallel to the acyl chains of phospholipids/sphingolipids in lipid bilayers and therefore their fluorescence properties reflect very faithfully structural and dynamic properties of the membranes. In addition, they are cylindrically shaped and introduce minimal perturbation in the system. Different polyene probes have preference for different types of lipid phases or regions of the bilayer, and more than one probe can be used to obtain an unbiased or quantitative description of the lipid domains behavior (Aresta-Branco et al., 2011; Castro et al., 2009).

In cuvette experiments, it is not possible to study a single cell. The typical number of cells used is on the order of 10^6–10^7ml^{-1} (Aresta-Branco et al., 2011; Ravichandra and Joshi, 1999). Therefore, the fluorescence decay obtained is an average of a large number of cells. Whereas this may be disadvantageous in some cases (see Section 4.2), it is a way of obtaining in one measurement an average over the variability that a biological sample always contains. In addition, it is not the same cell that is being continuously irradiated, which could limit severely the acquisition time due to the effects that exposure to radiation may itself have. Thus, decays with a large number of counts can be obtained and therefore resolution of a high number of components is possible. The benefits from such feature will become clear in the "case study" below.

In sum, cuvette lifetime measurements are highly suitable to study lipid domain *in vivo* of non adherent cells when spatial resolution or single cell manipulation are not mandatory issues.

An important limitation in cell suspension studies is that a significant fraction of the probes may be incorporated into intracellular membranes; the signal obtained is a composite signal from all the membranes that are labeled, as discussed in Section 3.2. Isolating the several subcellular membrane fractions and labeling them with the fluorescent probe of interest is a strategy to overcome this limitation. Although such studies are not *in vivo*, they have proven to be quite useful:

(i) They revealed important differences in the biophysical properties of different subcellular fractions.
(ii) It is sometimes necessary to obtain lipid extracts from those fractions and reconstitute them into liposomes in order to understand which may be the influence of proteins in the behavior of the membranes, and finally if the domains detected are due to lipid–lipid interactions ("true" lipid domains) or if they are highly influenced by the presence of proteins. It was found, using fluorescence lifetimes, that the highly ordered sphingolipid-enriched domains in the plasma membrane of *S*.

cerevisiae become even more compact in the absence of proteins (Aresta-Branco *et al.*, 2011; Table 3.2), whereas proteins have very little influence on the heterogeneity of bovine hippocampal membranes (Mukherjee *et al.*, 2007).

(iii) *In vitro* studies, with isolated cell membrane fractions and lipid extracts reconstituted into liposomes, or with model systems, allow certain types of studies not possible *in vivo*. One such case is to perform large temperature variations to determine initial and final melting temperatures of ordered domains. Also the addition/extraction of specific membrane components and the evaluation of the effects produced on lipid domain organization and properties (Marquês *et al.*, 2011) are carried out in a more controlled manner. Examples of these studies are presented in Tables 3.1 and 3.2.

4.1.1. Case study 1: Using *trans*-parinaric acid to detect gel domains in the plasma membrane of *S. cerevisiae* living cells

S. cerevisiae cells in mid-exponential phase under physiological conditions were labeled with *t*-PnA, and the fluorescence intensity decay obtained and analyzed. A surprising result was obtained, this being the presence of a very long lifetime of ca. 41 ns (Aresta-Branco *et al.*, 2011; Fig. 3.1B). This lifetime is known from a series of studies in model systems to correspond to gel or s_o domains (reviewed in de Almeida *et al.* (2009); see also Castro *et al.* (2009)) which were up till now considered absent in growing cells in the absence of stress or pathological situations. In fact, the maximum lifetime component of *t*-PnA detected in mammalian cells or membranes was clearly below the 30 ns threshold of the gel phase (see Table 3.2). A detailed study combining *in vivo* experiments using spheroplasts and intact cells of different mutant strains (Fig. 3.1B) with *in vitro* experiments using isolated plasma membrane, lipid extracts reconstituted into liposomes (Table 3.2), and model systems was conducted. Such study allowed to conclude that the domains responsible for the 41 ns fluorescence lifetime are indeed gel, presenting a gel/fluid transition at a temperature much higher than the physiological one; they contain glycosyl-phosphadityl-inositol (GPI)-anchored proteins and are mostly in the plasma membrane where they present a significant abundance; they are largely sterol-free, contradicting the current paradigm of lipid rafts (Aresta-Branco *et al.*, 2011). This study can be used to illustrate several practical aspects concerning the study of lipid domains *in vivo* by time-resolved fluorescence spectroscopy.

One of the practical problems was cell light absorption/autofluorescence, since in the case of yeast cells, in addition to proteins, and other components such as Nicotinamide adenine dinucleotide (NADH), in its reduced form and Flavin adenine dinucleotide (FAD), also sterols have significant absorption/fluorescence (Woods, 1971). Therefore, a global analysis method was used, which consists in acquiring a fluorescence intensity decay for the labeled and the unlabeled sample in exactly the same conditions, and analyzed them as described next (Fig. 3.1A), instead of using the more common approach of directly subtracting the decay of the unlabeled to the labeled sample (blank subtraction). The following steps were undertaken:

1. Analyze separately the fluorescence decay of the labeled and unlabeled samples.
2. Establish the minimum number of lifetime components to obtain a good description of such decays.
3. Realize that there are lifetime components that are the same (or very similar) in both decays.
4. Establish a global analysis procedure: the labeled sample will have a number of components usually larger than the unlabeled sample; some of those components will be common to the unlabeled sample; in the global analysis, the two decays are analyzed simultaneously and the common components will be linked, that is, they will be forced to be the same in both decays.
5. To obtain the real decay of the probe in the labeled samples, subtract the preexponential (nonnormalized) of the unlabeled sample corresponding to a certain linked component to the preexponential (nonnormalized) of the same component in the labeled sample. When the resulting corrected preexponential is not significantly different from zero, that component should be discarded, because it is attributed completely to cell autofluorescence.

In the case of *t*-PnA labeling *S. cerevisiae* membranes, the unlinked component was a very long lifetime component assigned to the presence of a gel phase. One advantage of global analysis is that the number of components statistically meaningful is generally larger than in single-decay analysis.

4.2. FLIM advantages and applications

Many processes in cellular membranes do not affect the composition or properties of the membrane as a total, but take place in a localized fashion. A pivotal role is hereby played by the microdomains formed by lipids with different properties, in which a local enrichment of target molecules, for example, cellular receptors is found. Although overall changes in the lipid environment can be also detected by cuvette experiments, the observed changes cannot be observed *in situ*. For example, identifying protein–protein interactions between two labeled protein species due to a drop in donor fluorescence lifetime will only give information that this process is occurring, but will not yield the membrane sites at which the interaction occurs. Therefore, FLIM techniques have been developed, which not only measure fluorescence lifetimes, but also allow visualizing their spatial distribution (Fig. 3.2). In order to record the fluorescence image, various fluorescence microcopy techniques, including wide-field illumination, confocal laser scanning, or total internal reflection can be used. Thus, the achievable spatial resolution is solely determined by the used technique, and normally in the range of a several hundred nanometers in the x–y plane (for a review on FLIM techniques, see Periasamy and Clegg (2009), Stöckl and Herrmann (2010)).

The main advantage of spatial resolved images recorded by FLIM is that in contrast to the cuvette experiments, where always whole cells are measured, spots of interest can be analyzed separately. On the one hand, this can be exploited to specifically study processes occurring at localized spots at the membrane, while unaffected parts of the membrane may serve as a direct internal control region. On the other hand, a specific labeling of the membrane compartment is often not possible. As different cellular membranes have a different composition, in experiments where environment sensitive probes are used, these may yield various lifetimes according to their localization. Even FPs may show a different lifetime behavior, according to their subcellular localization (Konig et al., 2008; Fig. 3.2A). While this has to be accounted for in cuvette experiments (Section 4.1), using FLIM it is possible to include only the membranes of interest in the analysis, which strongly facilitates it (Fig. 3.2B and C).

When studying process in living cells, the fact that single cells can be distinguished in the FLIM images is a big advantage over cuvette experiments. As it is possible to measure individual cells, the degree of variation between the fluorescence lifetimes from cell to cell can be directly measured. Also fluorescent entities, like dying or dead cells, cellular debris, or other artifacts, which would normally contribute to the fluorescence signal can be easily excluded from the measurements. While most of these objects can be easily identified by a visual inspection of the images, the state of health of the cells can be assessed using assays based on fluorescent markers.

To determine the cell-to-cell variability is of particular interest if the impact of artificially expressed proteins is to be measured in cells, as the expression level may vary strongly from cell to cell. Especially in experiments measuring interactions between partner molecules using FRET, the donor–acceptor ratio may influence the measured donor lifetimes, if an ensemble of free and bound donor molecules is measured. Also in experiments where the protein of interest has an impact on the organization of the lipid membranes, the knowledge of the amount of protein present at the membrane is desirable in order to relate the measured effect to the protein dose. Also a differential enrichment of the target protein at the membrane is accessible. Protein expression and thus cell labeling is not possible to be externally controlled since the insertion of a specific promoter would be

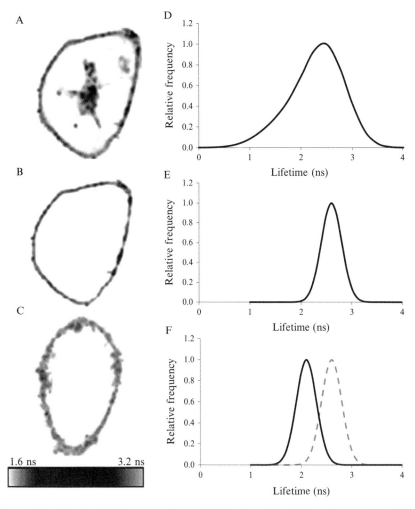

Figure 3.2 (A–C) FLIM images taken at 25 °C of one CHO-K1 cell-expressing GPI-CFP (A, B) and one cell coexpressing GPI-CFP and TMD-HA-YFP (C). In (B) and (C), only the region of interest (ROI) of the cells is shown (plasma membrane). The pseudo-color scale at the bottom refers to the average lifetime. It is clear that in the presence of FRET acceptor (YFP), the CFP (donor) lifetime is clearly reduced in most of the plasma membrane. The instrumental setup consisted of an inverted FluoView 1000 microscope (Olympus, Tokyo, Japan) equipped with a time-resolved LSM Upgrade kit (PicoQuant, Berlin, Germany) and a ×60 (1.35 numerical aperture) oil immersion objective. The donor was excited at 440 nm using a pulsed laser diode and fluorescence emission detected by a single photon avalanche photodiode at 470 ± 15 nm. Analysis of the 512 × 512 pixels FLIM images was performed using SymPhoTime software (PicoQuant), taking into account the instrument response function. See text and Scolari et al. (2009) for further details. (D–E) Schematic representation of the histograms of average lifetime distribution in the images, corresponding to the situation in (A), that is, cell expressing only donor, but no ROI selected; (B) the same cell, but after ROI selection, the distribution becomes narrower; and (C) presence of donor and acceptor and ROI selection, the lifetime distribution is shifted to shorter values due to the FRET process. (For color version of this figure, the reader is referred to the Web version of this chapter.)

required in order to downregulate GFP-tagged proteins production. In any case, this kind of regulation does not allow choosing a defined amount of protein. In general, plasmids in which an FP and derivative sequences are cloned contain a CMV (Cytomegalovirus) promoter that leads to the overexpression of the fluorescent molecule. In order to gain results that are the most reproducible, it is important to always keep microscopy parameters such as gain, high voltage, and pinhole constant. Once the experiment setup has been done, similar cells should be selected, that is, cells clearly alive and showing a reasonable fluorescence level that would not comport a change in the standard setting. If for any reasons (low expression, few transfected cells) settings have to be changed, it is important to correct the retrieved data for the standard parameters so that experiments can be compared (Engel *et al.*, 2010).

However, recording spatially resolved images in which single cells and different cellular compartments can be discerned comes with the cost of greatly reduced statistical significance. While a cuvette sample usually contains about $\sim 10^6$ cells, in FLIM experiments normally only tens to hundreds of cells are averaged. This is owed to the effort which is necessary for the selection of regions of interest and individual data analysis.

In order to analyze FLIM data, after selection of a region of interest, normally in a first step the fluorescence signal is summed up and the underlying rate constants and the corresponding amplitudes are determined by fitting of the exponential decays (Eq. 3.1). If short lifetimes or relative amplitudes between different lifetime components have to be determined accurately, during fitting, a reconvolution with the measured instrument response function is required. The calculated fluorescence lifetimes yield the information about the ensemble of fluorescent molecules present in the region of interest. However, often the lateral heterogeneity of fluorescence lifetimes in cellular membranes in correlation with secondary markers (morphological features, enrichment of a probe) yields information about the organization of the membrane. In the case of a recorded multiexponential fluorescence decay, it is often possible to associate lifetime components with certain properties of the lipid membrane (l_d vs. l_o, high vs. low cholesterol), as the behavior of the probe can be studied in an *in vitro* model system in which, for example, the lipid composition can be deliberately chosen. In order to visualize and quantify the distribution of these lifetime components within the sample, the lifetime histograms can be calculated, by applying the fitting procedure on single pixels. However, this requires the intensity of single pixels to be high enough to yield statistically sound results, otherwise a binning of pixels has to be used (reducing lateral resolution). This especially can be a problem if photolabile probes are used, as due to photobleaching, only a finite amount of photons can be recorded per cell.

4.2.1. Case study 2: Using NBD-labeled lipids to compare lipid domains of artificial and natural membranes

The organization of the lipids in the plasma membrane of cells is still not known in detail. The necessary experiments are not straightforward as the entities in question are on the one hand short lived and on the other hand have a size (10–200 nm) which is below the resolution limit of optical microscopes (Pike, 2006). However, FLIM experiments should be able to detect even such short-lived lipid domains as the fluorescence lifetime of incorporated fluorescent probes sensitive to the lipid environment still should report their presence.

A thorough characterization of the behavior of NBD-labeled lipids in GUV showing phase separated domains in the micromolar range showed that the fluorescence of C6-NBD-PC decays follows a biexponential decay law. The cardinal lifetime component is thereby environment sensitive and changes from about 7 ns in an l_d environment compared to 12 ns in a l_o domain (Stöckl et al., 2008). Similar values could be retrieved using lipid compositions which yield GUVs showing l_d–l_o coexistence (Table 3.1), albeit with domain sizes below the resolution limit. Although no different lipid domains could be observed, the domain-specific lifetimes could be recovered from the decay curves.

These model systems consist only of two to three defined lipid species, therefore as a more physiologically relevant model system, giant plasma membrane vesicles can be used which contain the components of the plasma membrane of the cells they are derived from. Also in these vesicles, a phase separation into l_d- and l_o-domains occurs. However, the difference between the fluorescence lifetimes is much smaller in this case (l_d: 6.7 ns; l_o: 9.6 ns; Stöckl et al., 2008). This indicates that the domains are much more similar in their properties from a physicochemical view.

The labeling of HepG2 cells with C6-NBD-PC showed two different lifetime distributions which are typical for intracellular membrane compartments and the plasma membrane. In this view, the intracellular lifetimes are distinctively shorter than the lifetimes measured for the plasma membrane. This can be explained by the high fraction of cholesterol present in the plasma membrane leading to a more ordered lipid environment. For the plasma membrane, no distinct lifetime typical for l_d domains could be observed (Table 3.1). This would suggest that the plasma membrane, at least mostly, in its properties resembles a l_o phase. The measured lifetimes in the range of 12 ns are also much longer than the lifetime measured for giant plasma membrane vesicles, a strong hint that the interactions between the proteins embedded in the plasma membrane and the cellular scaffolds have a strong ordering effect on the lipids. This strong connection is also reflected in the absence of a large-scale phase separation of lipid domains in living cells (Stöckl et al., 2008).

4.2.2. Case study 3: Hemagglutinin transmembrane domain of influenza virus and cholesterol-enriched domains using FPs

As well as with chemical dyes or lipid fluorescent analogues, also with FP-tagged molecules it is possible to use two or more different colors at the same time, so that more information can be retrieved at a time. Using cyan fluorescent protein (CFP) and yellow fluorescent protein (YFP) tagged proteins, it was demonstrated through FLIM-FRET that viral spike proteins and raft markers cluster into cholesterol-enriched domains (Scolari et al., 2009). In order to show that hemagglutinin (HA) transmembrane domain has a preferential affinity for lipid microdomains, as a "raft reporter," a GPI-anchored CFP molecule was produced (Fig. 3.2A), since it is well established that GPI-anchored proteins are in general associated to lipid rafts (Schuck and Simons, 2004). Choosing the so-called raft-marker was important to then design the architecture of the protein under study, that was made substituting the HA ectodomain with a YFP protein. CFP and YFP fluorophores were selected since their emission and excitation spectra, respectively, overlap, and thus they represent suitable proteins for using FRET. It is important to point out that the phenomenon occurs only if the participating donor and acceptor molecules are in very close proximity, in the order of 10 nm. However, detecting energy transfer is not so trivial, since small acceptor emissions indeed due to energy transfer are not easily distinguishable from the normal spillover of the donor fluorescence into the acceptor channel. Thus, it was chosen to detect energy transfer by FLIM (Fig. 3.2C). In this way, even small events of energy transfer reflected in changes into donor fluorescence lifetime can be detected and analyzed. FRET with an efficiency in the order of 10–12% was observed that could be translated into a distance within proteins in the order of 60 Å (Scolari et al., 2009), which is in fact a quite short distance considering that the β-barrel of the GFP has a diameter of ~3 nm (Yang et al., 1996). Further, they could demonstrate that disruption of lipid microdomains by means of cholesterol extraction from the plasma membrane of cells expressing both GPI-CFP and HA-YFP lowered/abolished energy transfer. It was therefore concluded that the transmembrane domain of HA preferentially partitions into those regions of the plasma membrane enriched in cholesterol and in general equated with lipid rafts.

ACKNOWLEDGMENTS

Funding from F.C.T., Portugal, through grant no. PTDC/QUI-BIQ/104311/2008, including a research fellowship to A. E. P. B., is acknowledged. L. Cyrne is gratefully acknowledged for the kind gift of LC4 cells. M. S. is sponsored by a German Academic Exchange Service (DAAD) postdoctoral fellowship (D/09/50722).

REFERENCES

Abe, F., and Hiraki, T. (2009). Mechanistic role of ergosterol in membrane rigidity and cycloheximide resistance in Saccharomyces cerevisiae. *Biochim. Biophys. Acta* **1788,** 743–752.
Aker, J., Borst, J. W., Karlova, R., and de Vries, S. (2006). The *Arabidopsis thaliana* AAA protein CDC48A interacts *in vivo* with the somatic embryogenesis receptor-like kinase 1 receptor at the plasma membrane. *J. Struct. Biol.* **156,** 62–71.
Aresta-Branco, F., Cordeiro, A. M., Marinho, H. S., Cyrne, L., Antunes, F., and de Almeida, R. F. (2011). Gel domains in the plasma membrane of Saccharomyces cerevisiae: Highly ordered, ergosterol-free, and sphingolipid-enriched lipid rafts. *J. Biol. Chem.* **286,** 5043–5054.
Ariola, F. S., Mudaliar, D. J., Walvick, R. P., and Heikal, A. A. (2006). Dynamics imaging of lipid phases and lipid-marker interactions in model biomembranes. *Phys. Chem. Chem. Phys.* **8,** 4517–4529.
Baumgart, T., Hunt, G., Farkas, E. R., Webb, W. W., and Feigenson, G. W. (2007). Fluorescence probe partitioning between Lo/Ld phases in lipid membranes. *Biochim. Biophys. Acta* **1768,** 2182–2194.
Berezin, M. Y., and Achilefu, S. (2010). Fluorescence lifetime measurements and biological imaging. *Chem. Rev.* **110,** 2641–2684.
Castro, B. M., Silva, L. C., Fedorov, A., de Almeida, R. F. M., and Prieto, M. (2009). Cholesterol-rich fluid membranes solubilize ceramide domains: Implications for the structure and dynamics of mammalian intracellular and plasma membrane. *J. Biol. Chem.* **284,** 22978–22987.
Colles, S., Wood, W. G., Myers-Payne, S. C., Igbavboa, U., Avdulov, N. A., Joseph, J., and Schroeder, F. (1995). Structure and polarity of mouse brain synaptic plasma membrane: Effects of ethanol *in vitro* and *in vivo*. *Biochemistry* **34,** 5945–5959.
de Almeida, R. F. M., Fedorov, A., and Prieto, M. (2003). Sphingomyelin/phosphatidylcholine/cholesterol phase diagram: Boundaries and composition of lipid rafts. *Biophys. J.* **85,** 2406–2416.
de Almeida, R. F., Loura, L. M., Fedorov, A., and Prieto, M. (2005). Lipid rafts have different sizes depending on membrane composition: A time-resolved fluorescence resonance energy transfer study. *J. Mol. Biol.* **346,** 1109–1120.
de Almeida, R. F., Borst, J., Fedorov, A., Prieto, M., and Visser, A. J. (2007). Complexity of lipid domains and rafts in giant unilamellar vesicles revealed by combining imaging and microscopic and macroscopic time-resolved fluorescence. *Biophys. J.* **93,** 539–553.
de Almeida, R. F. M., Loura, L. M. S., and Prieto, M. (2009). Membrane lipid domains and rafts: Current applications of fluorescence lifetime spectroscopy and imaging. *Chem. Phys. Lipids* **157,** 61–77.
Elson, E. L., Fried, E., Dolbow, J. E., and Genin, G. M. (2010). Phase separation in biological membranes: Integration of theory and experiment. *Annu. Rev. Biophys.* **39,** 207–226.
Engel, S., Scolari, S., Thaa, B., Krebs, N., Korte, T., Herrmann, A., and Veit, M. (2010). FLIM-FRET reveal association of influenza virus haemagglutinin with membrane rafts. *Biochem. J.* **425,** 567–573.
Folmer, V., Pedroso, N., Matias, A. C., Lopes, S. C. D. N., Antunes, F., Cyrne, L., and Marinho, H. S. (2008). H_2O_2 induces rapid biophysical and permeability changes in the plasma membrane of Saccharomyces cerevisiae. *Biochim. Biophys. Acta* **1778,** 1141–1147.
Gaus, K., Chklovskaia, E., Fazekas de St, G. B., Jessup, W., and Harder, T. (2005). Condensation of the plasma membrane at the site of T lymphocyte activation. *J. Cell Biol.* **171,** 121–131.
Grassme, H., Jendrossek, V., Riehle, A., Von, K. G., Berger, J., Schwarz, H., Weller, M., Kolesnick, R., and Gulbins, E. (2003). Host defense against Pseudomonas aeruginosa requires ceramide-rich membrane rafts. *Nat. Med.* **9,** 322–330.

Harris, F. M., Best, K. B., and Bell, J. D. (2002). Use of laurdan fluorescence intensity and polarization to distinguish between changes in membrane fluidity and phospholipid order. *Biochim. Biophys. Acta* **1565**, 123–128.
Hölttä-Vuori, M., Uronen, R. L., Repakova, J., Salonen, E., Vattulainen, I., Panula, P., Li, Z., Bittman, R., and Ikonen, E. (2000). BODIPY-cholesterol: A new tool to visualize sterol trafficking in living cells and organisms. *Traffic* **9**, 1839–1849.
Klein, T., Loschberger, A., Proppert, S., Wolter, S., van de Linde, S., and Sauer, M. (2011). Live-cell dSTORM with SNAP-tag fusion proteins. *Nat. Methods* **8**, 7–9.
Konig, I., Schwarz, J. P., and Anderson, K. I. (2008). Fluorescence lifetime imaging: Association of cortical actin with a PIP3-rich membrane compartment. *Eur. J. Cell Biol.* **87**, 735–741.
Korlach, J., Schwille, P., Webb, W. W., and Feigenson, G. W. (1999). Characterization of lipid bilayer phases by confocal microscopy and fluorescence correlation spectroscopy. *Proc. Natl. Acad. Sci. USA* **96**, 8461–8466.
Magee, A. I., Adler, J., and Parmryd, I. (2005). Cold-induced coalescence of T-cell plasma membrane microdomains activates signalling pathways. *J. Cell Sci.* **118**, 3141–3151.
Malinsky, J., Opekarova, M., and Tanner, W. (2010). The lateral compartmentation of the yeast plasma membrane. *Yeast* **27**, 473–478.
Marquês, J. T., Viana, A. S., and de Almeida, R. F. (2011). Ethanol effects on binary and ternary supported lipid bilayers with gel/fluid domains and lipid rafts. *Biochim. Biophys. Acta* **1808**, 405–414.
Mateo, C. R., Lillo, M. P., Gonzhlez-Rodriguez, J., and Acufia, A. U. (1991). Lateral heterogeneity in human platelet plasma membrane and lipids from the time-resolved fluorescence of *trans-parinaric* acid. *Eur. Biophys. J.* **20**, 53–59.
Mukherjee, S., and Chattopadhyay, A. (2005). Monitoring the organization and dynamics of bovine hippocampal membranes utilizing Laurdan generalized polarization. *Biochim. Biophys. Acta* **1714**, 43–55.
Mukherjee, S., Raghuraman, H., Dasgupta, S., and Chattopadhyay, A. (2004). Organization and dynamics of N-(7-nitrobenz-2-oxa-1,3-diazol-4-yl)-labeled lipids: A fluorescence approach. *Chem. Phys. Lipids* **127**, 91–101.
Mukherjee, S., Kombrabail, M., Krishnamoorthy, G., and Chattopadhyay, A. (2007). Dynamics and heterogeneity of bovine hippocampal membranes: Role of cholesterol and proteins. *Biochim. Biophys. Acta* **1768**, 2130–2144.
Owen, D. M., Lanigan, P. M., Dunsby, C., Munro, I., Grant, D., Neil, M. A., French, P. M., and Magee, A. I. (2006). Fluorescence lifetime imaging provides enhanced contrast when imaging the phase-sensitive dye di-4-ANEPPDHQ in model membranes and live cells. *Biophys. J.* **90**, L80–L82.
Pedroso, N., Matias, A. C., Cyrne, L., Antunes, F., Borges, C., Malhó, R., de Almeida, R. F., Herrero, E., and Marinho, H. S. (2009). Modulation of plasma membrane lipid profile and microdomains by H2O2 in Saccharomyces cerevisiae. *Free Radic. Biol. Med.* **46**, 289–298.
Periasamy, A., and Clegg, R. M. (eds.), (2009). FLIM Microscopy in Biology and Medicine, Chapman & Hall/CRC Press, Boca Raton, FL.
Pike, L. J. (2006). Rafts defined: A report on the Keystone Symposium on lipid rafts and cell function. *J. Lipid Res.* **47**, 1597–1598.
Pomorski, T., Hrafnsdottir, S., Devaux, P. F., and van Meer, G. (2001). Lipid distribution and transport across cellular membranes. *Semin. Cell Dev. Biol.* **12**, 139–148.
Ravichandra, B., and Joshi, P. G. (1999). Gangliosides asymmetrically alter the membrane order in cultured PC-12 cells. *Biophys. Chem.* **76**, 117–132.
Schroeder, F. (1983). Domains in plasma membranes from rat liver. *Eur. J. Biochem.* **132**, 509–516.

Schroeder, F. (1984). Sex and age alter plasma membranes of cultured fibroblast. *Eur. J. Biochem.* **142,** 183–191.

Schroeder, F., and Soler-Argilaga, C. (1983). Calcium modulates fatty acid dynamics in rat liver plasma membranes. *Eur. J. Biochem.* **132,** 517–524.

Schroeder, F., Gorka, C., Wdliamson, L. S., and Wood, W. G. (1987). The influence of dolichols on fluidity of mouse synaptic plasma membranes. *Biochim. Biophys. Acta* **902,** 385–393.

Schuck, S., and Simons, K. (2004). Polarized sorting in epithelial cells: Raft clustering and the biogenesis of the apical membrane. *J. Cell Sci.* **117,** 5955–5964.

Scolari, S., Engel, S., Krebs, N., Plazzo, A. P., de Almeida, R. F., Prieto, M., Veit, M., and Herrmann, A. (2009). Lateral distribution of the transmembrane domain of influenza virus hemagglutinin revealed by time-resolved fluorescence imaging. *J. Biol. Chem.* **284,** 15708–15716.

Sinha, M., Mishra, S., and Joshi, P. G. (2003). Liquid-ordered microdomains in lipid rafts and plasma membrane of U-87 MG cells: A time-resolved fluorescence study. *Eur. Biophys. J.* **32,** 381–391.

Stöckl, M. T., and Herrmann, A. (2010). Detection of lipid domains in model and cell membranes by fluorescence lifetime imaging microscopy. *Biochim. Biophys. Acta* **1798,** 1444–1456.

Stöckl, M., Plazzo, A. P., Korte, T., and Herrmann, A. (2008). Detection of lipid domains in model and cell membranes by fluorescence lifetime imaging microscopy of fluorescent lipid analogues. *J. Biol. Chem.* **283,** 30828–30837.

Woods, R. A. (1971). Nystatin-resistant mutants of yeast: Alterations in sterol content. *J. Bacteriol.* **108,** 69–73.

Yang, F., Moss, L. G., and Phillips, G. N., Jr. (1996). The molecular structure of green fluorescent protein. *Nat. Biotechnol.* **14,** 1246–1251.

Zeigler, M. B., Allen, P. B., and Chiu, D. T. (2011). Probing rotational viscosity in synaptic vesicles. *Biophys. J.* **100,** 2846–2851.

CHAPTER FOUR

Detecting and Tracking Nonfluorescent Nanoparticle Probes in Live Cells

Gufeng Wang[*] and Ning Fang[†,‡]

Contents

1. Introduction 84
2. Techniques and Tools 84
   2.1. Scattering-based microscopy 85
   2.2. Absorption-based microscopy 87
   2.3. Interferometric detection techniques 88
   2.4. Differential interference contrast microscopy 88
   2.5. Nonlinear optical microscopy 89
3. Biological Applications 90
   3.1. Single particle tracking and single particle orientation and rotational tracking 90
   3.2. Biosensors for disease diagnosis and therapy 99
4. Cytotoxicity of Nanoparticle Probes 100
5. Conclusions and Future Perspective 101
Acknowledgments 102
References 102

Abstract

Precisely imaging and tracking dynamic biological processes in live cells are crucial for both fundamental research in life sciences and biomedical applications. Nonfluorescent nanoparticles are emerging as important optical probes in live-cell imaging because of their excellent photostability, large optical cross sections, and low cytotoxicity. Here, we provide a review of recent development in optical imaging of nonfluorescent nanoparticle probes and their applications in dynamic tracking and biosensing in live cells. A brief discussion on cytotoxicity of nanoparticle probes is also provided.

[*] Department of Chemistry, North Carolina State University, Raleigh, North Carolina, USA
[†] Ames Laboratory, U.S. Department of Energy, Ames, Iowa, USA
[‡] Department of Chemistry, Iowa State University, Ames, Iowa, USA

1. INTRODUCTION

Studying dynamic biological processes at the single molecule level inside live cells is a challenging yet rewarding task in life science. Recent advances in single molecule and single particle techniques gave great opportunities to examine time-dependent biological processes of target molecules, one at a time, in unprecedented detail (Joo et al., 2008; Saxton and Jacobson, 1997). In particular, the transient intermediates and individual time-dependent pathways that are difficult, if not impossible, to measure in ensemble experiments can be resolved.

To date, the majority of single molecule fluorescence experiments are limited to *in vitro* studies due to high fluorescence background and high bleaching propensity of fluorophores in the cellular environment. However, the fast development of nanoparticle probes greatly advanced the study of dynamic biological processes (Levi and Gratton, 2007; Miura, 2005; Saxton and Jacobson, 1997). Of particular interest are nonfluorescent nanoparticles. They have large cross sections so that they can be localized with high spatial precision and temporal resolution. Compared to single fluorophores or fluorophore-doped nanoparticles, nonfluorescent nanoparticle probes have excellent photostability, allowing them to be dynamically tracked for arbitrarily long time without blinking or bleaching. As no fluorophore is required to be present in the particle, these nonfluorescent nanoparticles are usually nontoxic. Combined with far-field optical imaging techniques, which require no contact thus minimum interference with the biological processes, nonfluorescent nanoparticles are suitable for visualizing dynamic processes in live cells. In addition, nanoparticles can serve as carriers for functional molecules. They can be manufactured as multifunctional centers. For these reasons, nanoparticles have found widespread applications in live cells including *in vivo* biosensing, targeted drug/gene delivery, biomedical diagnostics and therapies, etc. (Love et al., 2008; Murphy et al., 2008; Sperling et al., 2008; Wax and Sokolov, 2009).

In this chapter, we provide a review on recent technological development in detecting and tracking nonfluorescent nanoparticle probes in live cells, and the achievements in biological studies. Nanoparticles doped with fluorescent dyes, quantum dots, and particles giving surface-enhanced Raman scattering are not covered in this chapter. Interested readers are referred to other chapters in this volume or recent reviews (Levi and Gratton, 2007; Michalet et al., 2005; Miura, 2005).

2. TECHNIQUES AND TOOLS

The detection and tracking of individual nanoparticles with a size of $1 \sim 100$ nm require high-sensitivity imaging techniques. The detection schemes reported in recent literature include both existing techniques that are tailored and new techniques to fulfill this special requirement.

2.1. Scattering-based microscopy

Detecting Rayleigh scattering is the most used approach in detecting nonfluorescent particles. In live-cell imaging, detecting nanoparticles is not trivial because a cell itself is a highly scattering medium due to the presence of many particular and tubular structures. The key for sensitive detection of nanoparticle probes in the cellular environment is to increase the signal yield from nanoparticles. Noble metal nanoparticles are excellent probes in this sense because of their large scattering cross sections at or near their plasmon resonance frequencies. By adjusting the particle size and shape, the plasmon resonance wavelength can be tuned in the visible to near-infrared (NIR) region. This feature is especially attractive because biological tissues are relatively transparent in the NIR region.

The optical absorption and scattering cross sections of noble metal nanospheres were first estimated by Gustav Mie in 1908 (Mie, 1908). In the limit where the particle size is much smaller than the wavelength of the illuminating light, the cross sections σ's are given by (Born and Wolf, 1999):

$$\sigma_{\text{absorption}} = \frac{8\pi^2}{\lambda} R^3 \text{Im}\left\{\frac{m^2 - 1}{m^2 + 2}\right\}, \qquad (4.1)$$

and

$$\sigma_{\text{scattering}} = \frac{128\pi^5}{3\lambda^4} R^6 \left|\frac{m^2 - 1}{m^2 + 2}\right|^2, \qquad (4.2)$$

where λ is the wavelength; R is the radius of the particle; m is the ratio of refractive indices of the particle and the medium, which is a complex number; Im denotes the imaginary part of the complex number. Rayleigh scattering decreases with the sixth power of the particle radius. It is possible to detect particles as small as a few nanometers under ideal conditions, but practically only particles with a diameter larger than 30 nm are used in complex biological environments (Wayne, 2008).

Microscopic scattering images can be obtained in the wide field with a conventional dark-field microscope, in which oblique illumination is applied through a high numerical aperture (N.A.) condenser with the center portion blocked by a disc stop, leaving only the outer ring of light for illumination (Fig. 4.1A). The Rayleigh scattering of light, which is spatially separated from the zeroth order (direct) illuminating light, is collected with an objective with a smaller N.A. The acquired sample image only shows the

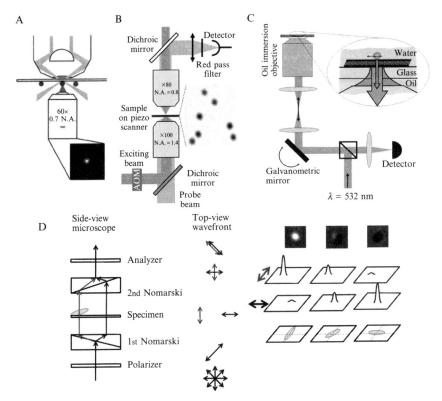

Figure 4.1 Schematics of different types of microscopy. (A) Scattering-based dark-field microscopy. Figure reprinted from Hu *et al.* (2008). Reproduced by permission of The Royal Society of Chemistry. (B) Absorption-based Laser-Induced Scattering around a NanoAbsorber (LISNA) technique. Figure reprinted from Lasne *et al.* (2006). Reproduced by permission of Elsevier. (C) Interference-based optical microscopy. Reprinted with permission from Ewers *et al.* (2007). Copyright 2011 American Chemical Society. (D) Differential interference contrast microscopy. (For color version of this figure, the reader is referred to the Web version of this chapter.)

scattering centers as bright spots on a dark background, similar to a fluorescence microscopic image.

In recent years, other types of microscopy were developed to collect images based on scattering signals. Louit *et al.* (2009) developed a confocal Rayleigh scattering spectroscopic and imaging system using a white-light continuum femtosecond laser as illumination source. The advantage of this system is that a focused laser beam is used so that there is enough scattering signal to monitor the plasmon resonance band shift of gold nanoparticle probes, allowing dynamic tracking of the changes on or near the nanoparticle surfaces thus the dynamic biological events occurring around the gold nanoparticle probes.

The scattering signal can also be generated with evanescent wave illumination using a modified objective-type total internal reflection microscope (He and Ren, 2008), in which a perforated mirror replaces the dichroic mirror to spatially separate the scattering signal from the reflected laser beam background. Using evanescent wave illumination, only a thin layer (~200 nm) of the sample is illuminated, giving higher signal-to-noise ratios of the scattering image. Also, in this evanescent wave illumination setup, all optics are on the same side of the sample, leaving free space above the specimen for *in situ* manipulation of live-cell samples, such as incubation, microinjection, patch clamping, etc. Ueno *et al.* used a similar system to track the translational displacement of a gold nanoparticle attached to the center unit of a rotary motor F_1-ATPase. Nanometer spatial precision and microsecond temporal resolution were realized, allowing the visualization of the stepping behavior of the rotary motor (Ueno *et al.*, 2010).

2.2. Absorption-based microscopy

2.2.1. Bright-field microscopy

Bright-field microscopy is one of the simplest optical microscopy. In bright-field microscopy, illumination light is transmitted through the sample and the contrast is generated by the absorption of light in dense areas of the specimen. The limitations of bright-field microscopy include low contrast for weakly absorbing samples and low resolution due to the blurry appearance of out-of-focus material.

Colloidal gold nanoparticles can serve as labels in bright-field microscopy due to their large absorption and scattering cross sections. When applied in live cells, they usually appear as dark spots on a blurred, bright image of the host cell. In the 1980s, with the development of video-enhancement techniques, small gold nanoparticles (20–40 nm) became visible and can be imaged at video rates (~30 Hz) through digital enhancement of the image contrast (Saxton and Jacobson, 1997). This advancement resulted in a wave of live-cell single particle tracking (SPT) studies of cellular activities including receptor-mediated endocytosis, motion of cell surface molecules, intracellular transport, etc. Bright-field microscopy is still being used frequently today, despite its high background level (Kural *et al.*, 2007).

2.2.2. Photothermal effect-based detection

The photothermal effect-based detection takes advantage of photoinduced change in the refractive index of the environment surrounding the probe (van Dijk *et al.*, 2006). In general, nanoparticles that strongly absorb light, for example, noble metal nanoparticles, carbon nanotubes, semiconductor nanocrystals, etc., are likely candidates for detection with photothermal effect-based methods. Importantly, this technique is insensitive to scattering background, making it ideal for live-cell studies. Cognet *et al.* developed a technique named Laser-Induced Scattering around a NanoAbsorber (LISNA) based on photothermal effect that is able to detect 5-nm gold nanoparticles in live cells (Cognet *et al.*, 2008; Lasne *et al.*, 2006) (Fig. 4.1B). Two laser beams, one for pumping and the other for probing, need to be focused onto the nanoparticle probe. Thus, this technique requires raster scanning of the samples to collect an image and the temporal resolution is sacrificed.

2.3. Interferometric detection techniques

Since the scattering of a nanoparticle probe decays as the 6th order of the particle dimension (Eq. (4.2)), the scattering signal vanishes quickly as the particle size decreases. One way to amplify the signal is to let the scattering signal interfere with the illumination background:

$$I = |E_i + E_s|^2 = |E_i|^2 \{1 + s^2 - 2s\sin\phi\}, \qquad (4.3)$$

where I is the interference signal, E_i and E_s are the amplitudes of the electronic vectors of the incidence and scattering light; s and φ are the modulus and phase of the complex scattering electronic vector, where s is a function of R^3 (Wang *et al.*, 2010a). The three terms at the right-hand side of Eq. (4.3) represent the background intensity, the scattering signal that is a function of R^6, and the cross term that scales to R^3. As the particle size decreases, the scattering term becomes negligible and the cross term (scales to R^3) dominates the interference signal. Thus, it can be deemed that the scattering is magnified by superimposing with the incidence beam and the detectability is greatly enhanced for small nanoparticles.

Based on this mechanism, Sandoghdar and coworkers developed an interferometric optical detection system that can detect 5-nm individual gold nanoparticles at the water–glass interface with an integration time of 2 ms (Ewers *et al.*, 2007; Jacobsen *et al.*, 2006) (Fig. 4.1C). Focused laser beam was used to generate large scattering signal. This method, in combination with fluorescence microscopy, was later demonstrated to track both position and orientation of quantum dot-labeled, virus-like nanoparticles on artificial membranes (Kukura *et al.*, 2009).

2.4. Differential interference contrast microscopy

Differential interference contrast (DIC) microscopy in principle is also an interferometric detection technique (Sun *et al.*, 2009). Due to its unique design, conventional incandescent lamps that are incoherent in nature can be used for illumination. Because of this and its usefulness in biology studies, DIC microscopy is discussed in this separate section.

DIC microscopy was first developed in the 1950s and has several different versions (Mehta and Sheppard, 2008). Most currently available DIC microscopes belong to the Nomarski type and adopt a two-prism configuration (Fig. 4.1D). The first Nomarski prism shears the illumination beam into two orthogonally polarized beams that are separated by a sub-wavelength distance—the shear distance. Two identical, intermediate bright-field images are generated behind the microscope objective and then laterally shifted back by the second Nomarski prism to overlay and generate the interference DIC image. The first Nomarski prism guarantees that the two intermediate images at the overlapped region are coherent, allowing incoherent light sources, for example, incandescent lamps, to be used for illumination.

One major advantage of DIC microscopy is that it uses the full objective and condenser apertures, which provides the user with the highest lateral and axial resolutions. Contrast of a DIC image is generated primarily within the thin focal plane with minimal influence from objects outside of the focal plane, thereby supplying DIC with higher contrast and better z-resolution than what are achievable with bright- or dark-field microscopy. A comparison of live-cell images was performed by Tsunoda *et al.* using DIC, bright-, and dark-field microscopies (Tsunoda *et al.*, 2008). The bright-field images of the cell are the worst, blurred by the cell features from outside of the focal plane. In the dark-field images, the edges of the cell produced intense scattering that strongly interferes with the observation of organelles inside the cell. In addition, the cell features appear to be larger due to the use of a small N.A. objective. DIC microscopy gives the best image quality in terms of contrast, lateral, and axial resolutions.

Another major advantage of DIC microscopy is that it can image nanoprobes and live cells simultaneously for an extended period of time without the need of staining. For example, Sun *et al.* (2008) used a DIC microscope to monitor the whole endocytosis process of mesoporous silica nanoparticles by live A549 human lung cancer cells. Notably, the cell morphology changes were recorded near the entry spot on cell membrane without staining the sample. Such morphology information of the cell during a dynamic process is not readily available with other modes of microscopy, such as dark-field or fluorescence-based microscopy.

2.5. Nonlinear optical microscopy

Some nanoparticles display unique nonlinear optical (NLO) properties. For example, noble metal nanoparticles can enhance the electric field of a light wave under plasmon resonance conditions, making them well suited for NLO imaging. Several research groups are engaged in developing new detection techniques based on NLO properties of nanoparticle probes. Noncentrosymmetric $BaTiO_3$ nanocrystals have been demonstrated as second harmonic generation probes (Hsieh *et al.*, 2009). Third harmonic signals from individual gold nanospheres (Lippitz *et al.*, 2005) and gold nanorods (Schwartz and Oron, 2009) show potential as single biomolecule labels for sensitive, background-free imaging of biological samples. Masia *et al.* (2009) demonstrated high-contrast, photostable imaging of Golgi structures labeled with 5–10 nm gold nanoparticles in HepG2 cells by detecting plasmon resonance-enhanced four-wave mixing.

Weak photoluminescence of noble metal nanoparticles can also be enhanced by many orders of magnitude under plasmon resonance conditions (Mohamed *et al.*, 2000; Sonnichsen *et al.*, 2002). Plasmon resonance two-photon luminescence of gold nanorods was recently used to serve as contrast agent in cancer cell studies (Durr *et al.*, 2007) and to monitor cellular uptake, transportation, excretion, and degradation of surface-modified gold nanorods (Huff *et al.*, 2007; Wang *et al.*, 2005).

Coherent anti-Stokes Raman scattering (CARS) is a third-order NLO process. Improvements in the instrumentation during the past few years have prepared CARS imaging for biological and biomedical applications. For example, 200-nm polystyrene particles were used as CARS probes in the studies of endocytosis and intracellular trafficking of folate-targeted liposomes in KB cells (Cheng and Xie, 2004; Tong *et al.*, 2007). CARS can also be used as a label-free technique to noninvasively probe the spatial and temporal distribution of biomolecules, for example, protein, lipids, and nucleic acids in live cells (Pliss *et al.*, 2010; Rago *et al.*, 2011).

Due to the NLO nature of excitation, these methods require the use of pulsed and tightly focused laser(s). The images are acquired by raster scanning of the sample or the laser beam.

3. BIOLOGICAL APPLICATIONS

3.1. Single particle tracking and single particle orientation and rotational tracking

Diffusion of membrane proteins and lipids is essential to many membrane processes including signal transduction, domain formation, protein sorting, protein–lipid complex formation, etc. (Dix and Verkman, 2008; Fan *et al.*, 2010; Kahya and Schwille, 2006; Niemela *et al.*, 2010). Tracking individual lipid or protein molecule movement on cell membranes either with fluorescent molecular probes (Douglass and Vale, 2005; Groc *et al.*, 2004; Kusumi *et al.*, 2005; Murase *et al.*, 2004; Ohsugi *et al.*, 2006; Reister and Seifert, 2005; Ritchie *et al.*, 2005; Srivastava and Petersen, 1998; Vrljic *et al.*, 2002) or with nonfluorescent particle probes (Bates *et al.*, 2006; Dietrich *et al.*, 2002; Fujiwara *et al.*, 2002; Sako *et al.*, 1998; Schuler *et al.*, 1999; Suzuki *et al.*, 2005) will help us understand fundamental membrane processes. In addition, numerous biological nanomachines perform various functions that keep a creature to live and prosper. Tracking their motions helps us understand their dynamics and working mechanisms for the human health and bioengineering purposes. SPT techniques provide us with an opportunity to investigate these cellular processes with high temporal and spatial resolutions.

3.1.1. Membrane diffusion

Various techniques have been used, including bright-field, dark-field, DIC, photothermal- (Leduc *et al.*, 2011), and interferometry-based techniques to study phospholipids, protein, and exogenous particles (Ewers *et al.*, 2007) diffusing on synthetic and live-cell membranes.

It should be noted that in these studies, localizing the target molecules/particles with nanometer precision is important. For dark- and bright-field microscopies that have a symmetric point spread function, the localization precision can be as high as ~ 1 nm by nonlinear least squares fitting of the particle image profile with an approximated Gaussian function. In DIC microscopy, the localization procedure is more complicated. Due to DIC's reliance on the principles of interference, nanoparticles show a unique type of polarized images: the half-bright and half-dark regions superimposed on the nonsignal gray background. This image pattern is instrument specific and does not represent any simple functions, excluding the possibility of localizing the particle through simple fitting. Sheetz *et al.* (1989) developed a method that utilized correlation coefficient mapping to establish the coordinates of a moving particle. They subsequently applied this method to study the motion of motor molecules and membrane proteins. The lateral localization precision obtained with this method was a few nanometers, comparable to those obtained from fluorescence and dark-field techniques.

Another important parameter in membrane diffusion studies is high temporal resolution. Diffusion of biomolecules is extremely fast even in viscous cell membranes, where conventional microscopic imaging at video rate is not sufficient to resolve the fast dynamics. Kusumi *et al.* (2005) developed assays with high temporal resolution to study the trajectories of unsaturated phospholipid L-α-dioleoylphosphatidylethanolamine (DOPE) tagged with 40-nm colloidal gold particles. A temporal resolution of 25 μs was achieved. At this enhanced time resolution, the gold nanoparticle-tagged DOPE molecules appeared to undergo short-term confined diffusion within individual compartments and long-term hopping movement between the compartments. This study depicts the plasma membrane as a compartmentalized fluid, in which compartmentalization is caused by the fence effects of the membrane skeleton (actin filaments) as well as the hydrodynamic slowing effects of transmembrane proteins anchored on the membrane skeleton fence (Fujiwara *et al.*, 2002).

3.1.2. Motor proteins

The complete understanding of the working mechanisms of motor proteins is imperative in order to realize the ultimate goal of the applications of intracellular transport for human health and medicines purposes. Advances in single particle tracking techniques and mechanical manipulation revealed the stepping mechanism of linear motors working *in vitro*: dynein and kinesin taking predominant 8-nm steps on microtubules and myosin taking 37-nm steps on actin filaments (Block, 2007; Coy et al., 1999; Gennerich and Vale, 2009; Hancock and Howard, 1998; Khalil et al., 2008; Reck-Peterson et al., 2006; Ross et al., 2008; Schnitzer and Block, 1997; Sindelar and Downing, 2010; Svoboda et al., 1993). However, how these motor proteins work *in vivo* (e.g., do they work in coordination or fight a tug of war?) is still a mystery (Kural et al., 2005; Miller and Heidemann, 2008; Ross et al., 2008). This information is crucial to the understanding of the motor–regulator or motor–motor interactions, which is currently under intense investigation (Miller and Heidemann, 2008; Ross et al., 2008). The most compelling evidence would be from resolving individual steps of cargoes carried by molecular motors in live cells. Recently, high-temporal-resolution and high-spatial-precision imaging were attempted *in vivo* using both fluorescence (Kural et al., 2005; Nan et al., 2005; Watanabe and Higuchi, 2007) and nonfluorescence techniques (Kural et al., 2007; Nan et al., 2008; Sims and Xie, 2009).

Selvin and coworkers used bright-field microscopy to study the intracellular transport of melanosomes, membrane organelles filled with a dark, nonfluorescent pigment melanin in cells known as Xenopus melanophores, with a spatial precision of 2 nm and a temporal resolution of 1 ms (Kural et al., 2007). They observed 8-nm step size of kinesin-2 and 35-nm steps of myosin V, and found that motors of actin filaments and microtubules work on the same cargo nearly simultaneously, indicating that a diffusive step is not needed between the two systems of transport. Xie and coworkers developed a method to track organelle movement within live cells that is based on dark-field imaging of targeted gold nanoparticles (150 nm in diameter) with a quadrant diode detector. The achieved temporal resolution was 25 µs and the spatial localization precision was 1.5 nm (Nan et al., 2008). They discovered that cargo-carrying kinesin moves only in 8-nm steps, whereas cargo-carrying dynein can move in 8-, 12-, 16-, 20-, and 24-nm steps. Xie and coworkers further demonstrated simultaneous optical manipulation and tracking of endogenous lipid droplets as actively transported cargoes in a living mammalian cells with submillisecond time resolution (Sims and Xie, 2009). They were able to detect steps of dynein- and kinesin-driven cargoes under known force loads thus to distinguish single and multiple motor-driven cargoes.

3.1.3. Rotation and orientation tracking

Resolving rotational motion is as important as resolving translational motion in understanding dynamics and working mechanism of protein macromolecule nanomachines in live cells. However, tracking orientation of nano-objects and their rotational motion generated by a cell organelle in live cells was not as successful as tracking translational movements. The difficulty is that the geometric shapes and orientation of objects with a dimension below the diffraction limit of light (~200 nm) cannot be resolved in far-field optical microscopy. Using special techniques or smart experiment designs, the orientation and rotational motion of a nano-object can be partially resolved. For example, in *in vitro* experiments, using single-molecule fluorescence polarization (Forkey *et al.*, 2003; Ha *et al.*, 1999; Sase *et al.*, 1997; Toprak *et al.*, 2006; Xiao *et al.*, 2010) or tracking translational movement of a large object (Adachi *et al.*, 2007; Itoh *et al.*, 2004; Kukura *et al.*, 2009; Nitzsche *et al.*, 2008; Yajima *et al.*, 2008; Yasuda *et al.*, 1998) can resolve the rotational motion of a nano-object. However, neither method can be readily applied to live-cell studies. The single molecule fluorescence polarization-based methods suffer from high autofluorescence background and high bleaching propensity of single fluorophores in cellular environment. The translation movement-based methods require specific geometries, for example, fixed rotation center or axis, and unhindered space in the pathway of the rotating object, and thus are also not suitable for imaging inside live cells.

Several research groups pioneered in exploring tracking the rotational motion of nonfluorescent nano-objects (Chang *et al.*, 2010; Sonnichsen and Alivisatos, 2005; Spetzler *et al.*, 2006; Wang *et al.*, 2010b; Xiao *et al.*, 2010). All of these groups used plasmonic gold nanorods (Jana *et al.*, 2001) as orientation probes. This is because gold nanorods show anisotropic absorption and scattering of polarized light due to their geometry-confined plasmon resonance modes (Gans, 1915), allowing their orientation in the 3D space to be determined. In addition, gold nanorods are small, having a size comparable to protein assemblies, and can be delivered into cells with minimum disruption of cell activity. What's more, gold nanorods are nontoxic, have large optical cross sections and excellent photostability, and can be synthesized reproducibly.

Sonnichsen *et al.* tracked the 2D orientation of individual gold nanorods confined at the interface of glass and water at video rate by monitoring polarized light scattering using a modified dark-field microscope (Sonnichsen and Alivisatos, 2005). Frasch *et al.* also used dark-field microscopy but with a point detector to achieve much higher temporal resolution (up to 2.5 µs) (Spetzler *et al.*, 2006). They were able to resolve the 120° rotation steps of the membrane-bound molecular motors the F_0F_1-ATP synthase and the dwell time between rotation steps (Ishmukhametov *et al.*, 2010). Xiao *et al.* employed a defocused approach, through deciphering the field distribution pattern in the slightly defocused images, to resolve the 3D orientation of single gold nanorods on a standard optical dark-field microscope (Xiao *et al.*, 2010). The beauty of this method is that the orientation of individual nanorods can be resolved in the complete 3D space without degeneracy. Link *et al.* demonstrated that the orientation of single gold nanorods can be determined from not only the longitudinal but also the transverse surface plasmon resonance mode by using polarization-sensitive photothermal imaging (Chang et al., 2010).

The Fang group for the first time successfully applied the gold nanorod rotational probes into live-cell imaging (Wang et al., 2010b). A platform—single particle orientation and rotational tracking (SPORT)—was developed with the combined use of Nomarski-type DIC microscopy and anisotropic gold nanorods. As discussed earlier, a Nomarski-type DIC microscope can be viewed as a two-beam interferometer and two orthogonally polarized beams are employed to illuminate the sample. Two *independent* intermediate images are generated, laterally shifted by ~100 nm, and then overlapped to generate the final interference image. In this way, *isotropic* samples, for example, polymeric nanospheres, will have two *identical* intermediate images, leading to evenly bright and dark DIC images. However, optically *anisotropic* samples such as gold nanorods will have two *different* intermediate images that are phase delayed to different extent depending on the relative orientation of the nanorod to the polarization direction of the illumination beams. The result is that the DIC images of gold nanorods appear as diffraction-limited spots with disproportionate bright and dark parts. Figure 4.2 shows the DIC images of two 25 nm × 73 nm gold nanorods placed at different orientations and illuminated at the transverse (540 nm) and longitudinal (720 nm) surface plasmon resonance wavelengths, respectively. It was demonstrated that the 2D and 3D orientation of individual gold nanorods can be determined through the bright and dark intensity pattern of their DIC images. When a nanorod is confined in the 2D space, the relative brightness $\Delta I/\Delta I_{max}$ and darkness $\Delta I'/\Delta I'_{max}$ adopt the $\sin^4$ and $\cos^4$ relationship, respectively, with their orientation angle ϕ (defined in Fig. 4.2B):

$$\phi = \operatorname{asin}(\Delta I/\Delta I_{max})^{1/4}, \quad (4.4)$$

$$\phi = \operatorname{acos}(\Delta I'/\Delta I'_{max})^{1/4}. \quad (4.5)$$

Ha et al. (2011) defined a parameter, namely, DIC polarization anisotropy, to calculate the orientation angle ϕ which is more robust to intensity fluctuations and give more reliable angle measurements in dynamic studies.

By realizing that the bright and dark intensities are the projections of the dipole of the nanorod onto the two polarization directions, Fang and coworkers demonstrated that the 3D orientation can be determined in each single DIC image. In reality, the errors of the 3D orientation angles are directly related to the signal-to-noise ratio of a DIC image. For live-cell imaging at video rates (32 frames/s) using commercially available instruments, such as a Nikon Eclipse 80i upright microscope with an Andor EMCCD

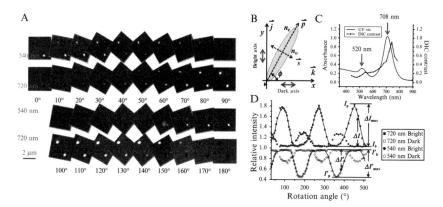

Figure 4.2 Gold nanorod orientations in 2D space. (A) DIC images of two 25 nm × 73 nm gold nanorods at different orientations in 2D space. The same nanorods were illuminated at their transverse (540 nm) or longitudinal plasmonic resonance wavelengths (720 nm). The gold nanorods were positively charged and physically adsorbed on a negatively charged glass slide and submerged in deionized water. The glass slide was fixed on a rotating stage that allows 360° rotation. (B) Definition of the 2D orientation (*azimuth angle* ϕ of a nanorod with respect to the polarization directions of the two illumination beams in a DIC microscope. (C) UV–vis spectrum of the gold nanorod suspension in deionized water (red) and DIC spectrum of an immobilized, randomly orientated gold nanorod (blue). The DIC contrast is defined as the difference between the maximum and the minimum intensities divided by the average local background intensity. The two DIC peaks are red-shifted compared to their plasmon resonance wavelengths. (D) Periodic changes of the bright/dark intensities of a gold nanorod when rotated under a DIC microscope and illuminated at the two DIC peak wavelengths. All intensities are relative to the background level. The periodic patterns at these two illumination wavelengths are shifted by ∼90°, consistent with the relative orientation between the transverse and the longitudinal plasmonic resonance modes. Reprinted with permission from Wang *et al.* (2010b). Copyright 2011 American Chemical Society. (See Color Insert.)

camera, the estimated standard deviation of the angle determination can be less than 3° for a nanorod lying flat in the focal plane.

In a later work, the characteristic bright and dark pattern of DIC images are correlated to the absolute orientation of gold nanorods acquired using transmission electron microscopy (Stender *et al.*, 2010), proving the mathematic model used in estimating the nanorod orientation. In addition, the optical responses of several common yet important gold nanorod configurations: isolated nanorods, uncoupled nanorods with an interparticle distance below the diffraction limit, and coupled nanorod dimers are recorded and analyzed. The results show that a dimer can exhibit a multipole plasmon at wavelengths that are close to the dipole plasmon of single nanorods in the sample. These configurations may show similar images under conventional optical microscopes. However, using DIC microscopy, one can distinguish these absolute geometries by simply rotating the sample under the microscope. As noble metal nanoparticles are employed to increasingly sophisticated environments like live cells, these findings establish an important step toward being able to characterize nanorods in dynamic environments without the use of electron microscopy.

With these understandings, the SPORT technique was employed to report rotational motions of nano-objects in both engineered environments and live cells. In one example, Wang et al. tracked the rotational motion of nanocargos when they were transported on the gliding microtubules on the microfabricated substrates. The kinesin motor proteins were coated on the substrate surface and served as the driving force to propel the reconstituted microtubules gliding on the substrate surface upon ATP hydrolyzation. For microtubules that are composed of non-13 protofilaments, they will self-rotate along their long axis during the translational movement (Ray et al., 1993). Apparently, such self-rotation is not readily identifiable using conventional microscopy. Using gold nanorods as probes, this rotation can be easily identified. This example demonstrates that the SPORT imaging technique can effectively disclose the rotational motion of nano-objects in an engineered environment.

Wang et al. further showed that SPORT technique can be used to track dynamic orientation information in live cells. In this application, the surface-modified gold nanorods (25 nm × 73 nm) were passively delivered into A549 human lung cancer cells and their subsequent transport on cytoskeleton tracks was tracked. The SPORT technique shows that when the nanorod-containing vesicles are being transported linearly on a microtubule track over a long distance (several microns), they tend to keep their orientation throughout, giving nearly constant bright/dark DIC intensities during the time course. More interestingly, it was observed that when the nanorod's moving direction changed $\sim 90°$, the orientation of the gold nanorod also changed $\sim 90°$. These observations showed for the first time the dynamic orientation tracking of nano-objects in live-cell transport processes and suggested that the relative orientations of the cargos to the underneath microtubule tracks are tightly controlled. The result is a conveyor-like movement of cargo vesicles on the intracellular microtubule network. The new approach may shed new light on many live-cell transport mysteries, such as the question whether molecular motors with different directionalities work in coordination or fight a "tug of war" to reverse the transport direction (Gennerich and Vale, 2009; Gross et al., 2002; Kural et al., 2005; Muller et al., 2008).

More recently, the rotational dynamics of drug and gene delivery vectors on live-cell membranes were studied by the SPORT technique (Gu et al., 2011). Using surface-modified gold nanorods as models, Gu et al. tracked the rotational patterns of functional nanorods continuously at 200 frames/s. The rotational dynamics, for example, fast or slow rotation (Fig. 4.3), of the

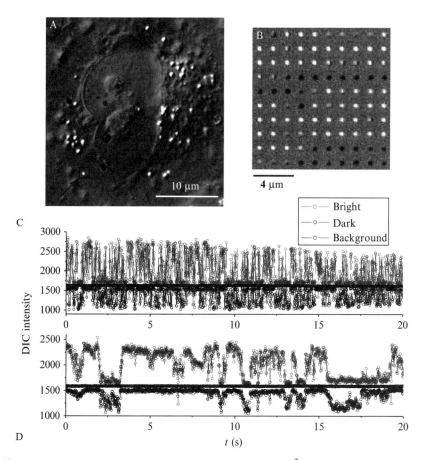

Figure 4.3 (A) DIC image of an A549 cell with a PEG–CO_2^{2-}-modified gold nanorod highlighted in the red square. (B) Composite of 100 consecutive images of the gold nanorod. (C, D) Typical DIC intensity traces as functions of time for (C) fast and (D) slow rotations. The rectangle in (C) distinguishes the intensities of the 100 DIC images shown in (B). The fast-rotation trace was recorded right after the nanorod landed on the cell membrane, and the slow-rotation trace was recorded 7 min later for the same nanorod. Reprinted with permission from Gu et al. (2011). Copyright 2011 American Chemical Society. (For interpretation of the references to color in this figure legend, the reader is referred to the Web version of this chapter.)

gold nanorod probes can be semiquantified using the autocorrelation function of the stochastic DIC intensity fluctuations. It was observed that the characteristic rotational behavior of gold nanorod vectors is strongly related to their surface charges. For example, the strongly positively charged polyethyleneimine-coated gold nanorods lost their rotation freedom almost immediately after appearing on the cell membrane. The gold nanorods modified with cell penetration peptides from HIV-1 Tat protein have a weaker positive charge (zeta potential $\zeta = +22.3$ mV) and their rotational Brownian motion slowed down gradually. As a contrast, the gold nanorods modified with neutral polyethylene glycol (PEG) or slightly negatively charged PEG–CO_3^{2-} showed much longer durations of active rotation. Their rotation rate fluctuated, showing the struggle between the interactions with the cell membrane and the thermal activities of the nanorods and their surrounding environments. Specific surface functional groups also contribute to the rotational dynamics significantly. The transferrin-coated gold nanorods have a negative charge ($\zeta = -11.2$ mV) similar to that of PEG–CO_3^{2-}-modified gold nanorods. But their active rotation lasted much shorter before they were endocytosed by the cell possibly because the membrane receptor binding triggers the initiation of endocytosis. This study will lead to a better understanding of drug delivery process and provide guidance in designing surface modification strategies for efficient drug delivery under different circumstances.

In summary, the SPORT technique allows the real-time rotational dynamics of single nano-objects in cellular environments to be studied for the first time. Compared to single molecule fluorescence polarization imaging, the SPORT technique is more robust in terms of photostability and resistance to background noise and photobleaching. The study of rotational dynamics of gold nanoparticles on live cells will have a general impact on the understanding of cellular processes, including but not limited to endocytosis, exocytosis, intracellular transport, and cell–cell communication, and thus bring us further on the exploration into the cell kingdom.

3.1.4. Plasmonic ruler for measuring minute displacement

Due to diffraction of light, optical microscopy in the visible range usually cannot resolve two particles fall in a distance of ~200 nm. However, plasmonic nanoparticles show a special attribute of coupling: the interparticle-distance-dependent red-shift of plasmon resonance frequency. Thus, noble metal particle pairs become attractive as a distance probe, namely, plasmonic ruler. Recently, there has been an increasing interest in applying these plasmonic rulers to understanding transient cellular processes.

Sonnichsen et al. (2005) introduced the concept of plasmonic ruler by showing that the plasmon resonance frequency shift of closely spaced gold nanoparticle dimers linked via streptavidin-biotin binding is strongly correlated with their distance. The distance dependence of plasmon coupling has been systematically calibrated, and the quantitative relationship was derived by El-Sayed and coworkers (Jain and El-Sayed, 2007). The universal scaling model, correlating the plasmon resonance frequency to the interparticle distance in terms of the particle size, is potentially useful to track nanometer distance or displacement in biological systems. Rong et al. applied the plasmon ruler concept to probe individual encounters of fibronectin–integrin complexes labeled with gold nanoparticles through a two-wavelength ratiometric detection approach in live HeLa cells. They were able to resolve the interparticle distance down to ~15 nm (Rong et al., 2008). The same group also used polymer-tethered silver nanoparticle pairs to track the distance and orientation of the nanoruler simultaneously using polarized illumination. This study leads to the understanding of the structural details of individual membrane compartments and the lateral heterogeneity of cell membranes on nanometer length scales (Rong et al., 2010).

3.2. Biosensors for disease diagnosis and therapy

3.2.1. Cancer diagnosis and therapy

An important application of functionalized nanoparticles is their potential in cancer diagnosis and photothermal therapies (Sokolov et al., 2003). Plasmonic nanoparticles are especially of interest because their plasmon resonance band can be tuned to the IR region by simply adjusting their material, shape, and aspect ratio. It has been demonstrated that nanospheres and nanorods conjugated with anti-epidermal growth factor receptor antibodies bind specifically to the surface of malignant cancer cells, allowing the differentiation of malignant and normal cells (Aaron et al., 2007; Huang et al., 2006). Functionalized gold nanoparticles were successfully applied to label tumors at the tissue level (Huang et al., 2010b) and in live animals (Kopwitthaya et al., 2010). More recent studies focused on the efficiency and cell death pathways upon photothermal therapy (Huang et al., 2010a; Kang et al., 2010). These studies establish critical steps toward optical diagnosis and subsequent surgery or photothermal therapy of cancers.

3.2.2. Multiplexed detection

For many diseases, there is not a single biomarker but rather a disease is associated with the abnormal levels of a group of biomarkers. It is important to establish and measure the disease-specific "marker signatures" for the purpose of diagnosis and treatment. The most cost-effective way is to detect the group of biomolecules in a single assay—namely, multiplexed detection. Sun et al. (2009) developed a spectrally resolved DIC microscopy technique that is able to differentiate gold and silver nanoparticles. In a later study, Luo et al. (2010) exploited the advantage of this technique and systematically investigated the suitability of 19 kinds of nanoparticles of different materials and/or sizes. A unique DIC contrast spectrum was identified for each kind of nanoparticle. It was demonstrated that four different nanoparticles can be differentiated on live-cell membranes in a single assay while providing high-contrast images of both the nanoprobes and the cell features.

The multiplexed detection of noble metal nanoparticles can also be realized in the dark-field detection schemes. For example, Yu et al. (2007) demonstrated that up to three kinds of gold nanorod markers can be identified on live human breast epithelial cells. Similarly, Hu et al. (2009)

relied on silver nanospheres and gold nanorods to simultaneously target different markers inside pancreatic cancer cells.

3.2.3. Single particle sensors in live cells

Many analytes of biological importance emerging in highly dynamic processes need to be detected *in situ*. For example, the intra- and intercellular signaling molecules only appear transiently at specific locations inside a living organism. As an alternative to genetically encoded fluorescent proteins, plasmonic nanoparticle probes can be delivered to the regions of interest and serve as *in situ* sensors. The conditions for generating a surface plasmon mode are highly dependent on the chemical composition, size, geometry, and local environment of the plasmonic structure. In particular, changes in local environment, such as the refractive index change caused by binding of biomolecules, can significantly affect the plasmon resonance properties including the resonance wavelength, magnitude, spectral width, etc.

4. Cytotoxicity of Nanoparticle Probes

With the increasingly fast development of nanotechnology, human and environmental exposure to engineered nanomaterials becomes inevitable. However, the consequences of the application of engineered nanomaterials are not well understood, and the public are becoming increasingly concerned about the potential toxicity on human health and environmental sustainability. It is important that the nanotoxicology research uncovers how physiochemical properties of nanomaterials influence their toxicity so that their undesirable properties can be avoided. It has become necessary to integrate the optical imaging tools with the use of nanoparticle probes to reveal the underlying mechanisms of nanoparticle uptake, transportation, targeting, cell organ transformation under the influence of nanoparticles, etc.

A huge amount of literature on nanotoxicity is available today. This section is only intended to bring readers' attention to this important issue by summarizing a few selected reports on the cytotoxicity of nanoparticles that are useful for optical imaging.

Xu and coworkers successfully applied *in vivo* tracking of single silver and gold nanoparticles while monitoring cell development under the influence of nanoparticles (Lee *et al.*, 2007). Their results show that gold nanoparticles are much more biocompatible with (or less toxic to) the embryos than the silver nanoparticles (Browning *et al.*, 2009). These results provide powerful evidences that the biocompatibility of nanoparticles is highly dependent on their chemical compositions. They also found that gold nanoparticles can be located in both normally developed and deformed zebrafish, suggesting that cytotoxicity might be a stochastic event during embryonic development (Browning *et al.*, 2009). Murphy and coworkers studied the origin of cytotoxicity of the growth-directing surfactant, cetyltrimethylammonium bromide (CTAB)-coated gold nanorods using human colon carcinoma cells (HT-29) (Alkilany *et al.*, 2009). They found that the apparent cytotoxicity is caused by free CTAB in solution and overcoating the nanorods with polymers substantially reduces cytotoxicity. They also showed that the surface properties of nanoparticles change significantly after they are introduced to biological media, such as cell culture media containing bovine serum albumin, due to surface adsorption. Such changes need to be considered when examining the biological properties or environmental impacts of nanoparticles.

5. Conclusions and Future Perspective

The research at the interface of live-cell imaging and the optical detection of nonfluorescent nanoparticles has been particularly active in recent years. The number of fundamental biological studies using nonfluorescent techniques should continue to increase during the next few years. Due to the high cellular autofluorescence background, single-molecule experiments in live cells remain challenging. The large optical cross sections and nontoxic nature of gold nanoparticles make them ideal probes for studies inside live cells at the molecular level. We anticipate increased collaboration among the fields of biology, medicine, and materials science for clinical applications of these nonfluorescent nanoparticles.

Despite the rapid development in this field of study, there are still several issues associated with the application of nanoparticles.

(1) The need for smaller nanoparticle probes. In order to be optically imaged, most of current frequently used nanoparticles are synthesized to be large (20–100 nm) as compared to their targets, for example, protein molecules with a diameter of ~ 10 nm. Apparently, there are adverse effects for these oversized labels to probe much smaller targets. Particles with a dimension one order of magnitude smaller (< 10 nm) are highly desired. In a rare, successful attempt of using single nanoparticle probes of several nanometers in diameter for live-cell sensing, Xu and coworkers synthesized stable silver nanoparticles with an average diameter of ~ 2.6 nm, which exhibited higher dependence on optical properties (a 29-nm red-shift of the plasmon resonance wavelength) on surface functional groups, for tracking the binding reaction quantitatively in real time at the single molecule level on the single particle sensor surface (Huang *et al.*, 2008).

(2) The need for fast 3D tracking techniques. A cell can be thought of as a complicated 3D matrix. While precise 2D SPT has become a mature technique, the localization precision in the axial direction is still more than an order of magnitude worse than the lateral precision, and dynamic axial tracking remains challenging. As such, there currently exists a demand for a new method to localize nanoparticles in the z-direction within a single image.

(3) The need for multimodality imaging. Cellular processes usually involve dynamic assembly and disassembly of multiple biomolecules. Despite the efforts of combinatorial microscopy techniques (Axelrod and Omann, 2006) and multiplexed detection schemes, there still exists a need for combining nonluminescent nanoparticle imaging with single-molecule fluorescence microscopy in order to provide complimentary information. It also needs to include other types of imaging schemes, for example, photoacoustic imaging, magnetic resonance imaging, etc., to maximize the yield of information obtained from single label (Liu et al., 2011).

ACKNOWLEDGMENTS

This work was supported by start-up funds from Iowa State University and North Carolina State University and a grant from U.S. Department of Energy, Office of Basic Energy Sciences, Division of Chemical Sciences, Geosciences, and Biosciences through the Ames Laboratory. The Ames Laboratory is operated for the U.S. Department of Energy by Iowa State University under contract no. DE-AC02-07CH11358.

REFERENCES

Aaron, J., Nitin, N., Travis, K., Kumar, S., Collier, T., Park, S. Y., Jose-Yacaman, M., Coghlan, L., Follen, M., Richards-Kortum, R., and Sokolov, K. (2007). Plasmon resonance coupling of metal nanoparticles for molecular imaging of carcinogenesis in vivo. *J. Biomed. Opt.* **12,** 034007.

Adachi, K., Oiwa, K., Nishizaka, T., Furuike, S., Noji, H., Itoh, H., Yoshida, M., and Kinosita, K. (2007). Coupling of rotation and catalysis in F-1-ATPase revealed by single-molecule imaging and manipulation. *Cell* **130,** 309–321.

Alkilany, A. M., Nagaria, P. K., Hexel, C. R., Shaw, T. J., Murphy, C. J., and Wyatt, M. D. (2009). Cellular uptake and cytotoxicity of gold nanorods: Molecular origin of cytotoxicity and surface effects. *Small* **5,** 701–708.

Axelrod, D., and Omann, G. M. (2006). Combinatorial microscopy. *Nat. Rev. Mol. Cell Biol.* **7,** 944–952.

Bates, I. R., Hebert, B., Luo, Y. S., Liao, J., Bachir, A. I., Kolin, D. L., Wiseman, P. W., and Hanrahan, J. W. (2006). Membrane lateral diffusion and capture of CFTR within transient confinement zones. *Biophys. J.* **91,** 1046–1058.

Block, S. M. (2007). Kinesin motor mechanics: Binding, stepping, tracking, gating, and limping. *Biophys. J.* **92,** 2986–2995.

Born, M., and Wolf, E. (1999). Principles of Optics: Electromagnetic Theory of Propagation, Interference and Diffraction of Light. 7th edn. Cambridge University Press, New York.

Browning, L. M., Lee, K. J., Huang, T., Nallathamby, P. D., Lowman, J. E., and Xu, X. H. N. (2009). Random walk of single gold nanoparticles in zebrafish embryos leading to stochastic toxic effects on embryonic developments. *Nanoscale* **1**, 138–152.

Chang, W. S., Ha, J. W., Slaughter, L. S., and Link, S. (2010). Plasmonic nanorod absorbers as orientation sensors. *Proc. Natl. Acad. Sci. USA* **107**, 2781–2786.

Cheng, J. X., and Xie, X. S. (2004). Coherent anti-Stokes Raman scattering microscopy: Instrumentation, theory, and applications. *J. Phys. Chem. B* **108**, 827–840.

Cognet, L., Octeau, V., Lasne, D., Berciaud, S., and Lounis, B.Ieee (2008). Photothermal detection and tracking of individual non-fluorescent nano-objects in live cells, 2008 Digest of the Leos Summer Topical Meetings. 2008 Digest of the Leos Summer Topical Meetings. Ieee, New York, pp. 65–66.

Coy, D. L., Wagenbach, M., and Howard, J. (1999). Kinesin takes one 8-nm step for each ATP that it hydrolyzes. *J. Biol. Chem.* **274**, 3667–3671.

Dietrich, C., Yang, B., Fujiwara, T., Kusumi, A., and Jacobson, K. (2002). Relationship of lipid rafts to transient confinement zones detected by single particle tracking. *Biophys. J.* **82**, 274–284.

Dix, J. A., and Verkman, A. S. (2008). Crowding effects on diffusion in solutions and cells. *Annu. Rev. Biophys.* **37**, 247–263.

Douglass, A. D., and Vale, R. D. (2005). Single-molecule microscopy reveals plasma membrane microdomains created by protein-protein networks that exclude or trap signaling molecules in T cells. *Cell* **121**, 937–950.

Durr, N. J., Larson, T., Smith, D. K., Korgel, B. A., Sokolov, K., and Ben-Yakar, A. (2007). Two-photon luminescence imaging of cancer cells using molecularly targeted gold nanorods. *Nano Lett.* **7**, 941–945.

Ewers, H., Jacobsen, V., Klotzsch, E., Smith, A. E., Helenius, A., and Sandoghdar, V. (2007). Label-free optical detection and tracking of single virions bound to their receptors in supported membrane bilayers. *Nano Lett.* **7**, 2263–2266.

Fan, J., Sammalkorpi, M., and Haataja, M. (2010). Formation and regulation of lipid microdomains in cell membranes: Theory, modeling, and speculation. *FEBS Lett.* **584**, 1678–1684.

Forkey, J. N., Quinlan, M. E., Shaw, M. A., Corrie, J. E. T., and Goldman, Y. E. (2003). Three-dimensional structural dynamics of myosin V by single-molecule fluorescence polarization. *Nature* **422**, 399–404.

Fujiwara, T., Ritchie, K., Murakoshi, H., Jacobson, K., and Kusumi, A. (2002). Phospholipids undergo hop diffusion in compartmentalized cell membrane. *J. Cell Biol.* **157**, 1071–1081.

Gans, R. (1915). The state of ultramicroscopic silver particles. *Ann. Phys.* **47**, 270–284.

Gennerich, A., and Vale, R. D. (2009). Walking the walk: How kinesin and dynein coordinate their steps. *Curr. Opin. Cell Biol.* **21**, 59–67.

Groc, L., Heine, M., Cognet, L., Brickley, K., Stephenson, F. A., Lounis, B., and Choquet, D. (2004). Differential activity-dependent regulation of the lateral mobilities of AMPA and NMDA receptors. *Nat. Neurosci.* **7**, 695–696.

Gross, S. P., Welte, M. A., Block, S. M., and Wieschaus, E. F. (2002). Coordination of opposite-polarity microtubule motors. *J. Cell Biol.* **156**, 715–724.

Gu, Y., Sun, W., Wang, G. F., and Fang, N. (2011). Single particle orientation and rotation tracking discloses distinctive rotational dynamics of drug delivery vectors on live cell membranes. *J. Am. Chem. Soc.* **133**, 5720–5723.

Ha, T., Laurence, T. A., Chemla, D. S., and Weiss, S. (1999). Polarization spectroscopy of single fluorescent molecules. *J. Phys. Chem. B* **103**, 6839–6850.

Ha, J. W., Sun, W., Wang, G. F., and Fang, N. (2011). Differential interference contrast polarization anisotropy for tracking rotational dynamics of gold nanorods. *Chem. Commun.* **47,** 7743–7745.

Hancock, W. O., and Howard, J. (1998). Processivity of the motor protein kinesin requires two heads. *J. Cell Biol.* **140,** 1395–1405.

He, H., and Ren, J. C. (2008). A novel evanescent wave scattering imaging method for single gold particle tracking in solution and on cell membrane. *Talanta* **77,** 166–171.

Hsieh, C. L., Grange, R., Pu, Y., and Psaltis, D. (2009). Three-dimensional harmonic holographic microcopy using nanoparticles as probes for cell imaging. *Opt. Express* **17,** 2880–2891.

Hu, M., Novo, C., Funston, A., Wang, H. N., Staleva, H., Zou, S. L., Mulvaney, P., Xia, Y. N., and Hartland, G. V. (2008). Dark-field microscopy studies of single metal nanoparticles: Understanding the factors that influence the linewidth of the localized surface plasmon resonance. *J. Mater. Chem.* **18,** 1949–1960.

Hu, R., Yong, K. T., Roy, I., Ding, H., He, S., and Prasad, P. N. (2009). Metallic nanostructures as localized plasmon resonance enhanced scattering probes for multiplex dark-field targeted imaging of cancer cells. *J. Phys. Chem. C* **113,** 2676–2684.

Huang, X. H., El-Sayed, I. H., Qian, W., and El-Sayed, M. A. (2006). Cancer cell imaging and photothermal therapy in the near-infrared region by using gold nanorods. *J. Am. Chem. Soc.* **128,** 2115–2120.

Huang, T., Nallathamby, P. D., and Xu, X. H. N. (2008). Photostable single-molecule nanoparticle optical biosensors for real-time sensing of single cytokine molecules and their binding reactions. *J. Am. Chem. Soc.* **130,** 17095–17105.

Huang, X. H., Kang, B., Qian, W., Mackey, M. A., Chen, P. C., Oyelere, A. K., El-Sayed, I. H., and El-Sayed, M. A. (2010a). Comparative study of photothermolysis of cancer cells with nuclear-targeted or cytoplasm-targeted gold nanospheres: Continuous wave or pulsed lasers. *J. Biomed. Opt.* **15,** 058002.

Huang, X. H., Peng, X. H., Wang, Y. Q., Wang, Y. X., Shin, D. M., El-Sayed, M. A., and Nie, S. M. (2010b). A reexamination of active and passive tumor targeting by using rod-shaped gold nanocrystals and covalently conjugated peptide ligands. *ACS Nano* **4,** 5887–5896.

Huff, T. B., Hansen, M. N., Zhao, Y., Cheng, J. X., and Wei, A. (2007). Controlling the cellular uptake of gold nanorods. *Langmuir* **23,** 1596–1599.

Ishmukhametov, R., Hornung, T., Spetzler, D., and Frasch, W. D. (2010). Direct observation of stepped proteolipid ring rotation in E. coli FoF1-ATP synthase. *Embo J.* **29,** 3911–3923.

Itoh, H., Takahashi, A., Adachi, K., Noji, H., Yasuda, R., Yoshida, M., and Kinosita, K. (2004). Mechanically driven ATP synthesis by F-1-ATPase. *Nature* **427,** 465–468.

Jacobsen, V., Stoller, P., Brunner, C., Vogel, V., and Sandoghdar, V. (2006). Interferometric optical detection and tracking of very small gold nanoparticles at a water-glass interface. *Opt. Express* **14,** 405–414.

Jain, P. K., and El-Sayed, M. A. (2007). Universal scaling of plasmon coupling in metal nanostructures: Extension from particle pairs to nanoshells. *Nano Lett.* **7,** 2854–2858.

Jana, N. R., Gearheart, L., and Murphy, C. J. (2001). Wet chemical synthesis of high aspect ratio cylindrical gold nanorods. *J. Phys. Chem. B* **105,** 4065–4067.

Joo, C., Balci, H., Ishitsuka, Y., Buranachai, C., and Ha, T. (2008). Advances in single-molecule fluorescence methods for molecular biology. *Annu. Rev. Biochem.* **77,** 51–76.

Kahya, N., and Schwille, P. (2006). Fluorescence correlation studies of lipid domains in model membranes (review). *Mol. Membr. Biol.* **23,** 29–39.

Kang, B., Mackey, M. A., and El-Sayed, M. A. (2010). Nuclear targeting of gold nanoparticles in cancer cells induces DNA damage, causing cytokinesis arrest and apoptosis. *J. Am. Chem. Soc.* **132**(5), 1517–1519.

Khalil, A. S., Appleyard, D. C., Labno, A. K., Georges, A., Karplus, M., Belcher, A. M., Hwang, W., and Lang, M. J. (2008). Kinesin's cover-neck bundle folds forward to generate force. *Proc. Natl. Acad. Sci. USA* **105,** 19247–19252.

Kopwitthaya, A., Yong, K. T., Hu, R., Roy, I., Ding, H., Vathy, L. A., Bergey, E. J., and Prasad, P. N. (2010). Biocompatible PEGylated gold nanorods as colored contrast agents for targeted in vivo cancer applications. *Nanotechnology* **21,** 315101.

Kukura, P., Ewers, H., Muller, C., Renn, A., Helenius, A., and Sandoghdar, V. (2009). High-speed nanoscopic tracking of the position and orientation of a single virus. *Nat. Methods* **6,** U923–U985.

Kural, C., Kim, H., Syed, S., Goshima, G., Gelfand, V. I., and Selvin, P. R. (2005). Kinesin and dynein move a peroxisome in vivo: A tug-of-war or coordinated movement? *Science* **308,** 1469–1472.

Kural, C., Serpinskaya, A. S., Chou, Y. H., Goldman, R. D., Gelfand, V. I., and Selvin, P. R. (2007). Tracking melanosomes inside a cell to study molecular motors and their interaction. *Proc. Natl. Acad. Sci. USA* **104,** 5378–5382.

Kusumi, A., Nakada, C., Ritchie, K., Murase, K., Suzuki, K., Murakoshi, H., Kasai, R. S., Kondo, J., and Fujiwara, T. (2005). Paradigm shift of the plasma membrane concept from the two-dimensional continuum fluid to the partitioned fluid: High-speed single-molecule tracking of membrane molecules. *Annu. Rev. Biophys. Biomol. Struct.* **34,** U351–U354.

Lasne, D., Blab, G. A., Berciaud, S., Heine, M., Groc, L., Choquet, D., Cognet, L., and Lounis, B. (2006). Single nanoparticle photothermal tracking (SNaPT) of 5-nm gold beads in live cells. *Biophys. J.* **91,** 4598–4604.

Leduc, C., Jung, J. M., Carney, R. R., Stellacci, F., and Lounis, B. (2011). Direct investigation of intracellular presence of gold nanoparticles via photothermal heterodyne imaging. *ACS Nano* **5,** 2587–2592.

Lee, K. J., Nallathamby, P. D., Browning, L. M., Osgood, C. J., and Xu, X. H. N. (2007). In vivo imaging of transport and biocompatibility of single silver nanoparticles in early development of zebrafish embryos. *ACS Nano* **1,** 133–143.

Levi, V., and Gratton, E. (2007). Exploring dynamics in living cells by tracking single particles. *Cell Biochem. Biophys.* **48,** 1–15.

Lippitz, M., van Dijk, M. A., and Orrit, M. (2005). Third-harmonic generation from single gold nanoparticles. *Nano Lett.* **5,** 799–802.

Liu, L. W., Ding, H., Yong, K. T., Roy, I., Law, W. C., Kopwitthaya, A., Kumar, R., Erogbogbo, F., Zhang, X. H., and Prasad, P. N. (2011). Application of gold nanorods for plasmonic and magnetic imaging of cancer cells. *Plasmonics* **6,** 105–112.

Louit, G., Asahi, T., Tanaka, G., Uwada, T., and Masuhara, H. (2009). Spectral and 3-dimensional tracking of single gold nanoparticles in living cells studied by Rayleigh light scattering microscopy. *J. Phys. Chem. C* **113,** 11766–11772.

Love, S. A., Marquis, B. J., and Haynes, C. L. (2008). Recent advances in nanomaterial plasmonics: Fundamental studies and applications. *Appl. Spectrosc.* **62,** 346a–362a.

Luo, Y., Sun, W., Gu, Y., Wang, G. F., and Fang, N. (2010). Wavelength-dependent differential interference contrast microscopy: Multiplexing detection using nonfluorescent nanoparticles. *Anal. Chem.* **82,** 6675–6679.

Masia, F., Langbein, W., Watson, P., and Borri, P. (2009). Resonant four-wave mixing of gold nanoparticles for three-dimensional cell microscopy. *Opt. Lett.* **34,** 1816–1818.

Mehta, S. B., and Sheppard, C. J. R. (2008). Partially coherent image formation in differential interference contrast (DIC) microscope. *Opt. Express* **16,** 19462–19479.

Michalet, X., Pinaud, F. F., Bentolila, L. A., Tsay, J. M., Doose, S., Li, J. J., Iyer, G., and Weiss, S. (2005). Peptide-coated semiconductor nanocrystals for biomedical applications. *Genet. Eng. Opt. Probes Biomed. Appl. III* **5704,** 57–68.

Mie, G. (1908). Contributions to the optics of turbid media, particularly of colloidal metal solutions. *Ann. Phys.* **25,** 377–445.

Miller, K. E., and Heidemann, S. R. (2008). What is slow axonal transport? *Exp. Cell Res.* **314,** 1981–1990.
Miura, K. (2005). Tracking Movement in Cell Biology, Microscopy Techniques. Springer-Verlag Berlin, Berlin(267–295).
Mohamed, M. B., Volkov, V., Link, S., and El-Sayed, M. A. (2000). The 'lightning' gold nanorods: Fluorescence enhancement of over a million compared to the gold metal. *Chem. Phys. Lett.* **317,** 517–523.
Muller, M. J. I., Klumpp, S., and Lipowsky, R. (2008). Tug-of-war as a cooperative mechanism for bidirectional cargo transport by molecular motors. *Proc. Natl. Acad. Sci. USA* **105,** 4609–4614.
Murase, K., Fujiwara, T., Umemura, Y., Suzuki, K., Iino, R., Yamashita, H., Saito, M., Murakoshi, H., Ritchie, K., and Kusumi, A. (2004). Ultrafine membrane compartments for molecular diffusion as revealed by single molecule techniques. *Biophys. J.* **86,** 4075–4093.
Murphy, C. J., Gole, A. M., Stone, J. W., Sisco, P. N., Alkilany, A. M., Goldsmith, E. C., and Baxter, S. C. (2008). Gold nanoparticles in biology: Beyond toxicity to cellular imaging. *Acc. Chem. Res.* **41,** 1721–1730.
Nan, X. L., Sims, P. A., Chen, P., and Xie, X. S. (2005). Observation of individual microtubule motor steps in living cells with endocytosed quantum dots. *J. Phys. Chem. B* **109,** 24220–24224.
Nan, X. L., Sims, P. A., and Xie, X. S. (2008). Organelle tracking in a living cell with microsecond time resolution and nanometer spatial precision. *Chemphyschem* **9,** 707–712.
Niemela, P. S., Miettinen, M. S., Monticelli, L., Hammaren, H., Bjelkmar, P., Murtola, T., Lindahl, E., and Vattulainen, I. (2010). Membrane proteins diffuse as dynamic complexes with lipids. *J. Am. Chem. Soc.* **132,** 7574–7575.
Nitzsche, B., Ruhnow, F., and Diez, S. (2008). Quantum-dot-assisted characterization of microtubule rotations during cargo transport. *Nat. Nanotechnol.* **3,** 552–556.
Ohsugi, Y., Saito, K., Tamura, M., and Kinjo, M. (2006). Lateral mobility of membrane-binding proteins in living cells measured by total internal reflection fluorescence correlation spectroscopy. *Biophys. J.* **91,** 3456–3464.
Pliss, A., Kuzmin, A. N., Kachynski, A. V., and Prasad, P. N. (2010). Biophotonic probing of macromolecular transformations during apoptosis. *Proc. Natl. Acad. Sci. USA* **107,** 12771–12776.
Rago, G., Bauer, B., Svedberg, F., Gunnarsson, L., Ericson, M. B., Bonn, M., and Enejder, A. (2011). Uptake of gold nanoparticles in healthy and tumor cells visualized by nonlinear optical microscopy. *J. Phys. Chem. B* **115,** 5008–5016.
Ray, S., Meyhofer, E., Milligan, R. A., and Howard, J. (1993). Kinesin follows the microtubules protofilament axis. *J. Cell Biol.* **121,** 1083–1093.
Reck-Peterson, S. L., Yildiz, A., Carter, A. P., Gennerich, A., Zhang, N., and Vale, R. D. (2006). Single-molecule analysis of dynein processivity and stepping behavior. *Cell* **126,** 335–348.
Reister, E., and Seifert, U. (2005). Lateral diffusion of a protein on a fluctuating membrane. *Europhys. Lett.* **71,** 859–865.
Ritchie, K., Shan, X. Y., Kondo, J., Iwasawa, K., Fujiwara, T., and Kusumi, A. (2005). Detection of non-Brownian diffusion in the cell membrane in single molecule tracking. *Biophys. J.* **88,** 2266–2277.
Rong, G. X., Wang, H. Y., Skewis, L. R., and Reinhard, B. M. (2008). Resolving sub-diffraction limit encounters in nanoparticle tracking using live cell plasmon coupling microscopy. *Nano Lett.* **8,** 3386–3393.
Rong, G. X., Wang, H. Y., and Reinhard, B. M. (2010). Insights from a nanoparticle minuet: Two-dimensional membrane profiling through silver plasmon ruler tracking. *Nano Lett.* **10,** 230–238.

Ross, J. L., Ali, M. Y., and Warshaw, D. M. (2008). Cargo transport: Molecular motors navigate a complex cytoskeleton. *Curr. Opin. Cell Biol.* **20,** 41–47.

Sako, Y., Nagafuchi, A., Tsukita, S., Takeichi, M., and Kusumi, A. (1998). Cytoplasmic regulation of the movement of E-cadherin on the free cell surface as studied by optical tweezers and single particle tracking: Corralling and tethering by the membrane skeleton. *J. Cell Biol.* **140,** 1227–1240.

Sase, I., Miyata, H., Ishiwata, S., and Kinosita, K. (1997). Axial rotation of sliding actin filaments revealed by single-fluorophore imaging. *Proc. Natl. Acad. Sci. USA* **94,** 5646–5650.

Saxton, M. J., and Jacobson, K. (1997). Single-particle tracking: Applications to membrane dynamics. *Annu. Rev. Biophys. Biomol. Struct.* **26,** 373–399.

Schnitzer, M. J., and Block, S. M. (1997). Kinesin hydrolyses one ATP per 8-nm step. *Nature* **388,** 386–390.

Schuler, J., Frank, J., Saenger, W., and Georgalis, Y. (1999). Thermally induced aggregation of human transferrin receptor studied by light-scattering techniques. *Biophys. J.* **77,** 1117–1125.

Schwartz, O., and Oron, D. (2009). Background-free third harmonic imaging of gold nanorods. *Nano Lett.* **9,** 4093–4097.

Sheetz, M. P., Turney, S., Qian, H., and Elson, E. L. (1989). Nanometer-level analysis demonstrates that lipid flow does not drive membrane glycoprotein movements. *Nature* **340,** 284–288.

Sims, P. A., and Xie, X. S. (2009). Probing dynein and kinesin stepping with mechanical manipulation in a living cell. *Chemphyschem* **10,** 1511–1516.

Sindelar, C. V., and Downing, K. H. (2010). An atomic-level mechanism for activation of the kinesin molecular motors. *Proc. Natl. Acad. Sci. USA* **107,** 4111–4116.

Sokolov, K., Follen, M., Aaron, J., Pavlova, I., Malpica, A., Lotan, R., and Richards-Kortum, R. (2003). Real-time vital optical imaging of precancer using anti-epidermal growth factor receptor antibodies conjugated to gold nanoparticles. *Cancer Res.* **63,** 1999–2004.

Sonnichsen, C., and Alivisatos, A. P. (2005). Gold nanorods as novel nonbleaching plasmon-based orientation sensors for polarized single-particle microscopy. *Nano Lett.* **5,** 301–304.

Sonnichsen, C., Franzl, T., Wilk, T., von Plessen, G., Feldmann, J., Wilson, O., and Mulvaney, P. (2002). Drastic reduction of plasmon damping in gold nanorods. *Phys. Rev. Lett.* **88,** 077402.

Sonnichsen, C., Reinhard, B. M., Liphardt, J., and Alivisatos, A. P. (2005). A molecular ruler based on plasmon coupling of single gold and silver nanoparticles. *Nat. Biotechnol.* **23,** 741–745.

Sperling, R. A., Rivera gil, P., Zhang, F., Zanella, M., and Parak, W. J. (2008). Biological applications of gold nanoparticles. *Chem. Soc. Rev.* **37,** 1896–1908.

Spetzler, D., York, J., Daniel, D., Fromme, R., Lowry, D., and Frasch, W. (2006). Microsecond time scale rotation measurements of single F-1-ATPase molecules. *Biochemistry* **45,** 3117–3124.

Srivastava, M., and Petersen, N. O. (1998). Diffusion of transferrin receptor clusters. *Biophys. Chem.* **75,** 201–211.

Stender, A. S., Wang, G. F., Sun, W., and Fang, N. (2010). Influence of gold nanorod geometry on optical response. *ACS Nano* **4,** 7667–7675.

Sun, W., Fang, N., Trewyn, B. G., Tsunoda, M., Slowing, I. I., Lin, V. S. Y., and Yeung, E. S. (2008). Endocytosis of a single mesoporous silica nanoparticle into a human lung cancer cell observed by differential interference contrast microscopy. *Anal. Bioanal. Chem.* **391,** 2119–2125.

Sun, W., Wang, G. F., Fang, N., and Yeung, E. S. (2009). Wavelength-dependent differential interference contrast microscopy: Selectively imaging nanoparticle probes in live cells. *Anal. Chem.* **81,** 9203–9208.

Suzuki, K., Ritchie, K., Kajikawa, E., Fujiwara, T., and Kusumi, A. (2005). Rapid hop diffusion of a G-protein-coupled receptor in the plasma membrane as revealed by single-molecule techniques. *Biophys. J.* **88**, 3659–3680.

Svoboda, K., Schmidt, C. F., Schnapp, B. J., and Block, S. M. (1993). Direct observation of kinesin stepping by optical trapping interferometry. *Nature* **365**, 721–727.

Tong, L., Lu, Y., Lee, R. J., and Cheng, J. X. (2007). Imaging receptor-mediated endocytosis with a polymeric nanoparticle-based coherent anti-stokes Raman scattering probe. *J. Phys. Chem. B* **111**, 9980–9985.

Toprak, E., Enderlein, J., Syed, S., McKinney, S. A., Petschek, R. G., Ha, T., Goldman, Y. E., and Selvin, P. R. (2006). Defocused orientation and position imaging (DOPI) of myosin V. *Proc. Natl. Acad. Sci. USA* **103**, 6495–6499.

Tsunoda, M., Isailovic, D., and Yeung, E. S. (2008). Real-time three-dimensional imaging of cell division by differential interference contrast microscopy. *J. Microsc.* **232**, 207–211.

Ueno, H., Nishikawa, S., Iino, R., Tabata, K. V., Sakakihara, S., Yanagida, T., and Noji, H. (2010). Simple dark-field microscopy with nanometer spatial precision and microsecond temporal resolution. *Biophys. J.* **98**, 2014–2023.

van Dijk, M. A., Tchebotareva, A. L., Orrit, M., Lippitz, M., Berciaud, S., Lasne, D., Cognet, L., and Lounis, B. (2006). Absorption and scattering microscopy of single metal nanoparticles. *Phys. Chem. Chem. Phys.* **8**, 3486–3495.

Vrljic, M., Nishimura, S. Y., Brasselet, S., Moerner, W. E., and McConnell, H. M. (2002). Translational diffusion of individual class II MHC membrane proteins in cells. *Biophys. J.* **83**, 2681–2692.

Wang, H. F., Huff, T. B., Zweifel, D. A., He, W., Low, P. S., Wei, A., and Cheng, J. X. (2005). In vitro and in vivo two-photon luminescence imaging of single gold nanorods. *Proc. Natl. Acad. Sci. USA* **102**, 15752–15756.

Wang, G. F., Stender, A. S., Sun, W., and Fang, N. (2010a). Optical imaging of non-fluorescent nanoparticle probes in live cells. *Analyst* **135**, 215–221.

Wang, G. F., Sun, W., Luo, Y., and Fang, N. (2010b). Resolving rotational motions of nano-objects in engineered environments and live cells with gold nanorods and differential interference contrast microscopy. *J. Am. Chem. Soc.* **132**, 16417–16422.

Watanabe, T. M., and Higuchi, H. (2007). Stepwise movements in vesicle transport of HER2 by motor proteins in living cells. *Biophys. J.* **92**, 4109–4120.

Wax, A., and Sokolov, K. (2009). Molecular imaging and darkfield microspectroscopy of live cells using gold plasmonic nanoparticles. *Laser Photonics Rev.* **3**, 146–158.

Wayne, R. O. (2008). Light and Video Microscopy. Academic Press, New York.

Xiao, L. H., Qiao, Y. X., He, Y., and Yeung, E. S. (2010). Three dimensional orientational imaging of nanoparticles with darkfield microscopy. *Anal. Chem.* **82**, 5268–5274.

Yajima, J., Mizutani, K., and Nishizaka, T. (2008). A torque component present in mitotic kinesin Eg5 revealed by three-dimensional tracking. *Nat. Struct. Mol. Biol.* **15**, 1119–1121.

Yasuda, R., Noji, H., Kinosita, K., and Yoshida, M. (1998). F-1-ATPase is a highly efficient molecular motor that rotates with discrete 120 degrees steps. *Cell* **93**, 1117–1124.

Yu, C. X., Nakshatri, H., and Irudayaraj, J. (2007). Identity profiling of cell surface markers by multiplex gold nanorod probes. *Nano Lett.* **7**, 2300–2306.

CHAPTER FIVE

Fluorescence Lifetime Microscopy of Tumor Cell Invasion, Drug Delivery, and Cytotoxicity

Gert-Jan Bakker,[*,1] Volker Andresen,[†,‡,1] Robert M. Hoffman,[§,¶] and Peter Friedl[*,†]

Contents

1. Introduction	110
2. Preparation of Mouse Mammary Tumor Cell Spheroids in a Collagen Matrix	112
2.1. Cell culture and reagents	112
2.2. Generation of multicellular spheroids	112
2.3. Incorporation of spheroids into 3D collagen matrix	113
3. Simultaneous Acquisition of Fluorescence Lifetime and Intensity to Monitor DOX Uptake	113
3.1. Microscope setup	114
3.2. FLIM detector	114
3.3. FLIM benchmarking	116
3.4. Testing the FLIM two-photon system in a biological setting	116
3.5. Time-lapse fluorescence lifetime and intensity detection	118
4. Monitoring Drug Uptake Kinetics by FLIM	119
4.1. Monitoring nuclear DOX accumulation by exponential-fit lifetime analysis	121
4.2. Phasor analysis for multiple fluorescent species	122
5. Conclusions and Outlook	123
Acknowledgment	123
References	124

[*] Microscopical Imaging of the Cell, Department of Cell Biology, Nijmegen Center for Molecular Life Science, Radboud University Nijmegen Medical Centre, Nijmegen, The Netherlands
[†] Rudolf-Virchow Center for Experimental Biomedicine and Department of Dermatology, Venerology and Allergology, University of Würzburg, Würzburg, Germany
[‡] LaVision BioTec GmbH, Bielefeld, Germany
[§] AntiCancer, Inc., San Diego, California, USA
[¶] Department of Surgery, University of California San Diego, San Diego, California, USA
[1] Equal author contribution.

Abstract

Fluorescence lifetime imaging microscopy (FLIM) enables detection of complex molecular assemblies within a single voxel for studies of cell function and communication with subcellular resolution in optically transparent tissue. We describe a fast FLIM technique consisting of a novel time-correlated single-photon counting (TCSPC) detector that features 80 MHz average count rate and the phasor analysis for efficient data acquisition and evaluation. This method in combination with multiphoton microscopy enables acquisition of a lifetime image every 1–2 s in 3D live organotypic tissue culture. 3D time-lapse fluorescence lifetime data were acquired over up to 20 h and analyzed by using exponential fitting and phasor analysis. By correlating specific areas in the phasor plot to the actual image, we obtained direct insight into cancer-cell invasion into a 3D collagen matrix, the differential uptake of doxorubicin by cells, and the consequences on cell invasion and apoptosis induction. Based on the fast acquisition and simplified image postprocessing and quantification, time-lapse 3D FLIM is a versatile approach for monitoring the 3D topography, kinetics, and biological output of structurally and spectrally complex cell and tissue models.

1. INTRODUCTION

Multiphoton laser-scanning microscopy (MPM; Denk et al., 1990) is an established approach for the visualization of morphology, dynamics, (mal)functioning, and underlying molecular pathways of cells and tissues during physiologic and disease processes in vitro and in vivo (Germain et al., 2008; Helmchen et al., 1999). Multiphoton excitation of a fluorophore relies on the quasi-simultaneous absorption of two or more photons by a molecule in a single quantitized event. Since the energy of a photon is inversely proportional to its wavelength, the two absorbed photons have a wavelength about twice than that required for a single-photon excitation process. Key advantages of MPM using pulsed near-infrared and infrared radiation are reduced photobleaching and phototoxicity at out-of-focus planes and reduced scattering and absorption of the laser light in tissue allowing larger imaging depths (Helmchen and Denk, 2005; Theer et al., 2003). Besides the excitation of fluorophores, nonlinear microscopy delivers additional contrast mechanisms including second and third harmonic generation (SHG, THG; Aptel et al., 2010; Friedl et al., 2007; Zipfel et al., 2003). Consequently, MPM has emerged as a key approach for detecting the structural and molecular basis of cell dynamics and communication in dense tissues in histocultures, intact organs, and in vivo.

To understand molecular processes in live cells, including protein interaction and posttranslational modification and function, the Förster Resonant Energy Transfer (FRET) is exploited. FRET is a mechanism describing the nonradiative energy transfer of an excited molecule to an acceptor fluorophore in its immediate proximity through dipole–dipole coupling (Förster, 1948; Fruhwirth et al., 2011). FRET measurements have provided direct topographic and molecular insight into cell functions, including membrane channel function for surface receptor signaling, pH and calcium regulation, as well as the delivery and function of drugs (Levitt et al., 2009; Timpson et al., 2011; Verveer et al., 2000). To study FRET in optically transparent, nonscattering environments in vitro, both intensity and ratiometric measurement techniques allow detection of intensity changes of the acceptor or donor fluorescence and estimate their proximity and interaction qualitatively (Elangovan et al., 2003).

For molecular imaging in living samples, complex fluorophore composition and overlapping emission spectra hamper the reliable separation of individual fluorophores. Because all fluorophores display distinct lifetimes due to their energetically unique composition, the inclusion of fluorescence lifetime imaging microscopy (FLIM) allows detection of two or multiple fluorophores in the same voxel and thus can reliably assess expression dynamics and compound distribution. Ratiometric techniques require the detection of both donor and acceptor fluorescence at different emission wavelengths. Both intensity and ratiometry FRET measurements have been successfully applied in intravital imaging, for example, for analyzing the performance of a live olfactory bulb (Ducros et al., 2009). Therefore, FLIM became the method of choice to quantify cellular function via FRET in live tissue and organs.

For live-cell and intravital studies in three-dimensional (3D) tissues, sufficiently fast and accurate fluorescence lifetime identification with cellular or subcellular resolution of FRET and FLIM measurements are hampered by several technical challenges. Widefield frequency-domain FLIM is fast enough but reliable only for monoexponential fluorescence decays, whereas multiexponential signals in complex tissues and multicolor models are poorly accommodated. Field detection time-domain approaches based on gated image intensifiers enable very fast FLIM in nonscattering in vitro settings (Soloviev et al., 2007; Straub and Hell, 1998), but the need for high fluorescence signals and topographically precise photon detection limits their usability for live microscopy in scattering 3D samples. Consequently, widefield FLIM techniques are less sensitive and have less spatial resolution for deep-tissue imaging as they lack the ability to suppress scattered photons. Likewise, fluorescence anisotropy measurements, based on usually slight signal modifications, provide calibration-free access to cell–cell interactions at the molecular level, but are limited by noise intrinsic to 3D tissue imaging. As alternative, single-point laser-scanning combined with time-correlated single-photon counting (TCSPC) allows detection of the lifetime of scattered photons while maintaining diffraction-limited spatial resolution. However, state-of-the-art TCSPC FLIM-systems feature an effective count rate of only 10 MHz which limits the analysis of fast processes and the efficient acquisition of 3D volumes in a time-resolved manner. Titanium:Sapphire lasers, typically used for MPM, feature a repetition rate of 80 MHz which is technically adequate for improved FLIM detection of fast processes in scattering samples, provided there are matching average TCSPC rates at the detector side.

We present here a FLIM technique that is fast enough to allow the acquisition of 3D lifetime stacks with subcellular spatial resolution to track changes due to interaction of motile cells. The method is based on single-point laser-scanning and TCSPC and allows the acquisition of a 512 × 512 pixels image in less than a second. In combination with the recently published phasor evaluation (Digman *et al.*, 2008), it allows dynamic deep-tissue imaging and online evaluation.

2. Preparation of Mouse Mammary Tumor Cell Spheroids in a Collagen Matrix

3D spheroids cultured in collagen lattices were used as a model for tumor outgrowth and matrix manipulation for live-cell imaging.

2.1. Cell culture and reagents

Mouse mammary tumor (MMT) cells, stably expressing different fluorescent proteins in the nucleus (Histone-2B-enhanced green-fluorescent protein (eGFP) and cytoplasm (DsRed2), were generated, as described (Yamamoto *et al.*, 2004). Cells were cultured in Roswell Park Memorial Institute 1640 medium (Gibco RPMI 1640 supplemented with L-glutamine; Invitrogen), supplemented with 10% FBS, 1% pyruvate, 1% glucose, and 1% penicillin/streptomycin. Cells were cultured up to 80% confluence and propagated three times a week after detachment from the culture dish with 0.075% trypsin/1 mM EDTA. Doxorubicin (DOX; Sigma Aldrich) was stored as stock solution (2 mg/ml in PBS at −20 °C) and, prior to the experiment, diluted with culture medium to a final concentration of 10 µg/ml.

2.2. Generation of multicellular spheroids

Multicellular spheroids with a controlled number of cells were made in hanging drops of methylcellulose, resulting in cell accumulation and aggregation at the bottom of the drop. The following steps were taken to make multicellular spheroids:

- A cell suspension with a final concentration of 40,000 cells/ml was prepared containing 4.5 ml medium + 1 ml low-viscosity methylcellulose (RPMI1640 supplemented with 10% methylcellulose).

- 25 μl drops (containing 1000 cells) were deposited on the hydrophobic lid of a 15-cm culture dish.
- The lid was carefully placed on the dish, and hanging drops containing cell suspension were cultured overnight in an incubator (37 °C and 5% CO_2, humidified atmosphere).
- Plates were carefully positioned upside down and spheroids were inspected by brightfield microscopy.
- To collect spheroids, 10-ml culture medium was added and spheroids were moved to the center of the dish by gently moving the culture dish in a circle.
- Spheroids were taken out of the dish with a 1-ml pipette. To avoid disintegration of spheroids, the top of the pipette tip was cut off to have a more gentle flow.
- Spheroids were transferred into an Eppendorf tube and rinsed with 1 ml culture medium.

2.3. Incorporation of spheroids into 3D collagen matrix

Design and preparation of the imaging chamber were as described (Friedl and Bröcker, 2004).

Preparation of rat tail collagen matrix (4 mg/ml):

- Collagen stock solution (10 mg/ml, BD 354249) and Minimal Essential Eagle's Medium (MEM) were maintained on ice.
- 30 μl MEM 10× (Sigma) and 30 μl sodium bicarbonate solution (7.5%, GIBCO) were mixed (4 °C).
- 242 μl collagen stock solution and 238 μl culture medium were added to the MEM and bicarbonate buffer (BIC), and the mixture was kept on ice.
- Spheroids in 60-μl medium were carefully aspirated and gently mixed with the collagen solution.
- The collagen/spheroid solution was pipetted into the glass chambers (200 μl gel per chamber) and allowed to polymerize in an incubator (5% CO_2, 37 °C, humidified atmosphere) in a vertical position.
- Polymerization of the gel is complete when homogeneous opacity is reached.
- As the last step, the chambers were filled with culture medium and sealed with paraffin.

3. SIMULTANEOUS ACQUISITION OF FLUORESCENCE LIFETIME AND INTENSITY TO MONITOR DOX UPTAKE

DOX is a chemotherapeutic clinically used agent which intercalates with nuclear DNA and causes DNA damage, secondary arrest of the cell cycle, and ultimately apoptosis-induced cell death (Thoronton, 2008).

DOX emits red fluorescence which is exploited for monitoring cellular uptake using fluorescence microscopy (Dai *et al.*, 2008; de Lange *et al.*, 1992; Hovorka *et al.*, 2010). FLIM and spectral unmixing have been successfully applied to monitor DOX accumulation during live-cell experiments (Hovorka *et al.*, 2010). Using DOX uptake in MMT cells expressing different fluorescent proteins in the nucleus (Histone-2B-eGFP) and cytoplasm (DsRed2) and a live-cell 3D tumor invasion model in a 3D collagen matrix, we demonstrate how to simultaneously acquire fluorescence lifetime and intensity data in order to coregister cell invasion, drug uptake, and apoptosis at subcellular and high temporal resolution.

3.1. Microscope setup

The setup is based on a commercially available multiphoton scanhead (TriM Scope II, LaVision BioTec), coupled to a tunable fs-pulsed Ti:Sa laser (680–1080 nm, 120 fs, 80 MHz, Ultra II, Coherent) and a fast TCSPC detector (FLIM-X_{16}, LaVision BioTec; Fig. 5.1A). In backward direction, the FLIM data were separated from the red-fluorescent intensity signal by a dichroic mirror (DM1, reflection 400–700 nm). The SHG and eGFP intensity signals were retrieved in the forward direction by collection of transmitted light with a 1.4NA oil immersion condenser (U-AAC, Olympus). Thereby, fluorescence and second-harmonic-generation intensity signals were acquired simultaneously with the TCSPC data.

3.2. FLIM detector

A novel FLIM detector was designed to collect photons from a single spot within the sample and spread them homogeneously across the surface of a multi-anode photo multiplier tube (PMT) detector by means of diffusion optics (Fig. 5.1A). The diffusion optics consists of an aspheric focusing lens followed by a hollow light tunnel for multiple reflection of the signals. The PMT contains 16 separate channels that are synchronously read out by analogue electronics. Two 8-channel time-to-digital converters (TDCs) measure the arrival times of all photons and a fast histogram module builds up the temporal distribution. The average count rate per channel is 5 MHz resulting in an average count rate of 80 MHz for the complete device. The overall transmission of the detection and beam homogenization optics for the visible spectra (400–660 nm) was determined to be 67%. The multialkali photocathodes of the PMT have a quantum efficiency of 17%.

The technical parameters of the detector are optimized for sensitive and fast readout. The electronic dead time of a single channel is only 5.5 ns, allowing a maximal burst count rate of 182 MHz per PMT channel. Given typical photon rates of 0.1–100 MHz used for biomedical applications, the parallelized detection efficiently prevents pile-up effects. Pile-up arises

Fluorescence Lifetime Microscopy

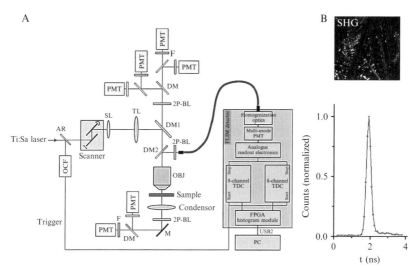

Figure 5.1 Design of microscope and TCSPC detector. (A) The beam of the Ti:Sa laser was passed through a shaping device (not shown) to control its diameter, collimation, pulse length, and power. Within the scanhead, a pair of galvanometric scanners was used to raster the beam across the sample. The objective lens (20× IR, NA 0.95, Olympus) was corrected for a wide wavelength range from 430 to 1450 nm and provided 2 mm working distance for deep-tissue penetration. To detect the FLIM signal, the emission was spectrally separated with a dichroic mirror directly after the objective (DM1), guided through a laser line blocking filter, and collimated into a liquid lightguide coupled to the TCSPC detector (*for details see text*). To detect the red fluorescence intensity signal transmitted by DM1, the beam was guided through a laser blocker filter (et700sp, Chroma) and a bandpass filter (ET620/60, Chroma) and focused onto a sensitive GaAsP PMT (H7422-40, Hamamatsu). Transmitted light was collected with a 1.4 NA oil immersion condenser (U-AAC, Olympus) and cleared from laser light by a blocking filter (et700sp, Chroma). EGFP and SHG signals were separated by a dichromatic mirror (t495lpxxr, Chroma), purified by appropriate bandpass filters (ET525/50 and ET420/50, respectively, Chroma), and guided onto PMTs (H6780-01 and H6780-20, respectively, Hamamatsu). (B) Intensity image (top) and lifetime (bottom) of second-harmonic signal elicited by urea crystals for estimating the instrument response function. (For color version of this figure, the reader is referred to the Web version of this chapter.)

when several photons are generated by a single laser pulse. After detection of the first photon, additional photons are lost due to the dead time of the detector. This is especially a problem when SHG/THG and fluorescence signals are emitted from the same spot, as the measured fluorescence lifetime will always be short and within the refractory phase of the PMT.

A fast photodiode (OCF-401, Becker & Hickl) is used to synchronize the counting electronics of the FLIM detector with the laser pulse train and derive electrical trigger pulses from optical pulses. Due to the constant-fraction trigger method, the delay is independent of the pulse amplitude.

This is of importance as the electrical trigger provided by Ti:Sa lasers shifts in time and amplitude when the wavelength is changed.

3.3. FLIM benchmarking

The performance of the FLIM detector was benchmarked, based on the instrument response function (IRF), accuracy, and acquisition speed. The maximal average count rate was determined by detecting bright fluorescence signals, yielding $\sim$400 counts at a dwell time of 5 μs from the brightest pixels, which is equivalent to an 80 MHz count rate.

The accuracy of a fluorescence lifetime value is given by the full-width-half-maximum (FWHM) of the lifetime distribution of a specific species. It was measured using Rh6G dissolved in water and resulted in a 164 ps distribution width and a lifetime value of 4233 ps. This fits the expected fluorescence lifetime of Rh6G in aqueous solution which is 4080 ps. The maximal accuracy given by the time jitter of the detector is 80 ps.

The IRF was measured using urea crystals that exhibit a strong SHG signal after irradiation with 800 nm pulsed light. SHG is a nonlinear light scattering effect based on the polarization of matter by electromagnetic radiation. Therefore, it arises instantaneously and $\sim$120 fs short light pulses reach the detector where they are detected as convoluted with the IRF. The FWHM of the resulting distribution of photon arrival times of SHG reflects the IRF which is $\sim$290 ps (Fig. 5.1B).

3.4. Testing the FLIM two-photon system in a biological setting

To test and optimize the performance of the FLIM two-photon system in a biological setting, TCSPC data of the emission spectrum ranging from 400 to 800 nm were acquired from MMT/dual-color cell spheroids in a 3D collagen lattice, before and after incubation with DOX (Fig. 5.2A). Since DOX, eGFP, DsRed2, and SHG have distinctive fluorescent lifetimes, this model served to estimate their relative contribution to each individual pixel. The sample was excited at 800 nm and the excitation power under the objective was 35 mW. For a pixel dwell time of 200 μs, a count rate of $\sim$5000 was achieved for the brightest areas of the image, and a single image of 250 × 250 pixels was acquired in <15 s. Regions of interest (ROIs; Fig. 5.2A, encircled regions) were selected within a central z-plane for estimating the baseline lifetimes of DsRed and eGFP, the lifetime of the nucleus after DOX treatment, and the lifetime distribution obtained from a DOX stock solution (Fig. 5.2B, upper panel). The lifetimes were 2.2 ± 0.2 ns (eGFP), 1.5 ± 0.1 ns (DsRed2), and 0.99 ± 0.03 ns (DOX), using a single-exponential model function to fit the decay curves of the lifetime distributions (Fig. 5.2B). The values of eGFP and DOX are

Fluorescence Lifetime Microscopy 117

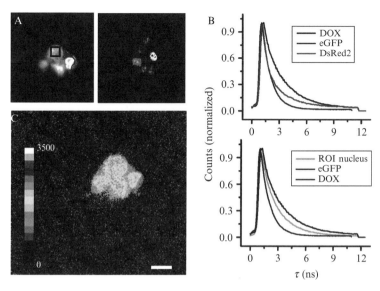

Figure 5.2 Fluorescence lifetime analysis of dual-color MMT (MMT-DC) breast cancer cells cultured in a 3D collagen matrix. (A) Summed intensity images of cells before (left) and after (right) 90 min incubation with DOX (5 μg/ml). (B, top panel) Photon-arrival-time histograms of eGFP-H2B in the nucleus and DsRed2 in the cytosol of MMT-DC cells using ROIs depicted in (A). (B, bottom panel) Lifetime histogram from the nucleus of MMT-DC cells after addition of DOX using the ROI in (A), right image. As a reference, a DOX stock solution in PBS (2 mg/ml) was used for the DOX lifetime (blue lines). (C) TCSPC lifetime image from (A) generated by fitting the decay curve at every pixel of the image using a single-exponential function. Lifetimes are presented using false-color coding (ns). (A, C) A single section from a 3D TCSPC datastack. Scale bar: 20 μm. (See Color Insert.)

consistent with published values (Hess *et al.*, 2003; Hovorka *et al.*, 2010, Jakobs *et al.*, 2000), whereas the lifetime for DsRed2 is ∼1 ns below previously found values (Hess *et al.*, 2003). It has been found that the fluorescent lifetime of DsRed is highly dependent on the protein maturation state (Jakobs *et al.*, 2000), rendering the lifetime of DsRed2 variable, depending on the cellular background and model system. With accumulation of DOX in the nucleus, a decrease of the lifetime inside the nucleus was obtained (down to 1.75 ± 0.06 ns), consistent with increasing proportions of DOX over eGFP (Fig. 5.2B, bottom). Assuming a mixture of eGFP and DOX emission, a biexponential model function was used, with fixed fluorescent lifetime parameters obtained for isolated components (Fig. 5.2B, top). Compared with a single exponential model function, a biexponential model function indeed resulted in a better fit (Chi squared = 0.04 and 0.012 respectively). According to the obtained fit parameters, the relative contribution of the DOX component to the lifetime in cell nuclei after 90 min of incubation was 55%. For display of differences in lifetimes per pixel, a false-color display represents each individual pixel after fitting the decay curve (Fig. 5.2C) and provides insight into subcellular lifetime distributions. The long lifetime of the eGFP in the nucleus was distinct from the shorter DsRed2 lifetime in the cytosol and the instantaneous lifetime of SHG (Fig. 5.2C). All fitting procedures have been performed with Inspector Pro (version 4.0.207), the acquisition and analysis software accompanied with the multiphoton microscope (LaVision BioTec).

3.5. Time-lapse fluorescence lifetime and intensity detection

To detect the time-course of DOX uptake by tumor cells and address whether invading and noninvading cells show similar uptake rates, large spheroids of MMT-DC cells were monitored for fluorescence intensity and lifetime by sequential imaging over up to 20 h. Spheroids containing ~1000 cells were embedded in a 3D collagen matrix (4 mg/ml) and monitored by time-lapse MPM starting at 24 h to allow for the formation of invasion zones (Fig. 5.3). For time-lapse microscopy, samples were placed on a temperature-controlled stage and kept at 37 °C (THO 60–16 Warm Stage and DC60 controller, Linkam; TC-HLS-1 flexible objective heater, Bioscience Tools). A spheroid at 600 μm distance from the collagen-medium

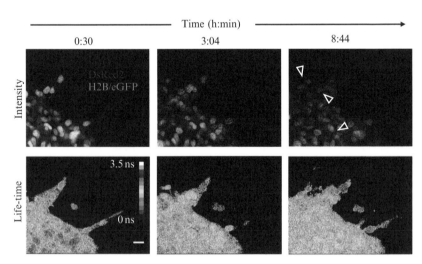

Figure 5.3 Simultaneous acquisition of fluorescence intensity (top) and TCSPC lifetime images based on single-exponential fit from a live-cell culture. Time-lapse microscopy of a spheroid of MMT-DC cells invading a 3D collagen matrix before (left panel) and after treatment (middle and right panels) with DOX (see Supplementary Movie S3, http://www.elsevierdirect.com/companions/9780123918574) followed by per-pixel analysis of the spectrally unseparated TCSPC stack. False-color range in bottom panel: 0 to 3.5 ns. Scale bar: 20 μm. (See Color Insert.)

interface was selected and lifetime stacks, as well as eGFP, SHG, and red fluorescence intensity images, were acquired simultaneously before, during and after DOX treatment. (see Supplementary Movie S1, http://www.elsevierdirect.com/companions/9780123918574). The image acquisition time was 3 s and the pixel or voxel ($0.8 \times 0.8 \times 5$ μm^3) dwell time was 34 μs. The excitation power in the focus of the objective lens was 30 mW at 800 nm. The counts per pixel were 150–2000 (background subtracted from the minimum and maximum intensity of the summed TCSPC image), which is sufficient for single-exponential fitting of the lifetime decay curve per pixel. First, a single 3D z-stack followed by single-section imaging over 66 time-points were acquired for an observation period of 124 min to determine the drug-free baseline (Supplementary Movies S1 and S4, http://www.elsevierdirect.com/companions/9780123918574). Then, the medium in the chamber was replaced by medium containing DOX (final concentration after equilibrium: 5 μg/ml) followed by simultaneous time-lapse recording of lifetime and intensity channels from a central plane with a 2-min interval (Fig. 5.3; Supplementary Movie S3, http://www.elsevierdirect.com/companions/9780123918574). With penetration of DOX into the spheroid, the intensity images show a shift toward enhanced red fluorescence in the nuclei and, accordingly, reduced mean lifetimes after single-exponential fitting (Fig. 5.3). In some cells, within few hours after DOX uptake, nuclear condensation and fragmentation were observed, indicating the onset of apoptosis (Fig. 5.3, arrowheads). In addition, the lifetime in the cytosol increased, consistent with the appearance of a recently characterized DOX degradation product which is retained in the cytoplasm (Hovorka et al., 2010). Finally, a 3D lifetime and intensity stack was measured after the time-lapse recording, at the same position as the 3D stack prior to DOX treatment (Supplementary Movie S2, http://www.elsevierdirect.com/companions/9780123918574). The changes in the average fluorescence life-time in the nuclei occurred to the same degree in regions that were not imaged repetitively; thus the observed changes cannot predominantly be due to the laser-induced bleaching of eGFP. Imaging a spheroid without addition of DOX, but under similar conditions, did not result in any observable changes in the fluorescence lifetime of the nucleus (Supplementary Movie S4, http://www.elsevierdirect.com/companions/9780123918574).

4. MONITORING DRUG UPTAKE KINETICS BY FLIM

We include here two numerical approaches to analyze time-resolved FLIM data acquired with the time-domain system, the lifetime based on single-exponential fitting and the phasor approach. The traditional method of fitting the decay curve at every pixel of the image (Fig. 5.2C), using single- or

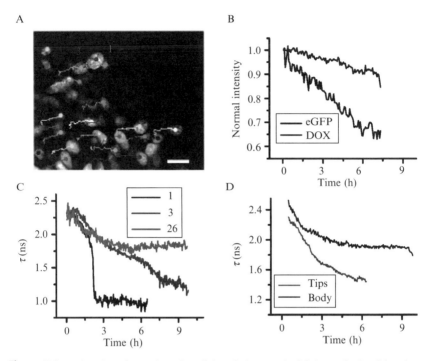

Figure 5.4 Migration dynamics of nuclei and changes in lifetime obtained by time-lapse imaging. (A) Trajectories of moving nuclei, obtained by manual tracking of the H2B-eGFP intensity from the nucleus (Fiji distribution of ImageJ software; http://fiji.sc/). Scale bar: 20 μm. (B) Bleaching of individual fluorophores, detected as mean green and red intensities of all nuclei as a function of time. Images were obtained after saturation with DOX (26 h after addition to the culture medium). (C) Fast, intermediate, and slow uptake of DOX by individual nuclei. Mean lifetimes from individual nuclei in the spheroid tip (cells 1 and 3 in (A)) and spheroid body (cell 26 in (A)). (D) The average lifetime of all nuclei present in collective invasion zones, compared with nuclei located in the main mass of the spheroid. Images were recorded and analyzed at 5-min intervals. (See Color Insert.)

multiexponential functions, allows identification of the relative occurrences of distinct fluorophores in the same voxel and changes thereof over time (Fig. 5.4). However, the long computation time required for multiexponential fitting at each pixel, the high photon counts required for robust multiexponential fitting, and the high level of expertise needed to identify the fluorophore species and their relative occurrence limit the utility of lifetime fitting. Phasor analysis allows identification of pixel-based lifetime characteristics without prior knowledge of the exact contribution of different components (Digman et al., 2008). Thereby, the phasor approach significantly simplifies the quantification of FLIM data and back-gating enables identification of lifetime subsets at subcellular resolution (Fig. 5.5).

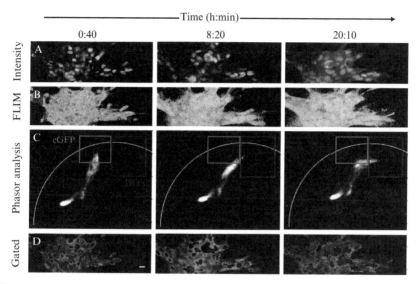

Figure 5.5 Matching intensity with single-exponential FLIM sequence with phasor analysis and reverse gating, in order to identify functional subregions. DOX uptake by an MMT-DC cell spheroid, embedded in a 3D collagen matrix, was monitored for 17 h (see Supplementary Movie S5, http://www.elsevierdirect.com/companions/9780123918574). (A) Intensity images, eGFP (green), DsRed2 and DOX (red), collagen (SHG; blue). (B) TCSPC image based on single-exponential fit from the spectrally unseparated TCSPC stack (false-color range 0 to 3.5 ns). (C) Phasor plot derived from the TCSPC stack (LaVision Phasor analysis software, V4.0.207). Green ROI includes longer lifetimes of eGFP and a DOX degradation product; red ROI includes shorter lifetimes from native DOX. (D) Gated image from the phasor population analysis. Remaining pixels represent the summed TCSPC (intensity) image. Scale bar: 20 μm. (See Color Insert.)

4.1. Monitoring nuclear DOX accumulation by exponential-fit lifetime analysis

The fluorescence lifetimes of the individual nuclei were traced over time along the nuclear trajectories (Fig. 5.4). To accommodate for the positional changes of invading cells over time, trajectories of moving nuclei were obtained by manual tracking of the eGFP signal (Fig. 5.4A). Then, the lifetime images of the time series obtained by single-exponential-fit lifetime analysis per pixel (Fig. 5.3; Supplementary Movie S3, http://www.elsevierdirect.com/companions/9780123918574) were spatially averaged using a mean filter with a radius of 5 pixels (Fiji distribution of ImageJ software; http://fiji.sc/). Finally, the time-dependent changes in lifetime of every individually tracked nucleus were obtained by taking the values from the averaged lifetime images at the coordinates of the center of each tracked nucleus along its trajectory. The last step was performed using a dedicated script programmed in Matlab (software release 7.01). To control for bleaching, which may affect retrieved lifetime values caused by multiple fluorophores, the eGFP and DOX intensities in individual nuclei were monitored during repetitive excitation, starting 26 h after addition of DOX to the culture. At this late time-point, DOX intercalation with DNA was in equilibrium, as detected by steady-state red-fluorescent intensity, and yielded twofold better photostability compared with eGFP, shown by consecutive scanning over 7 h (Fig. 5.4B). To understand the relative uptake in invading versus noninvading cells, the lifetimes were obtained from individual nuclei in invasion zones and the spheroid body (Fig. 5.4C) and averaged for both zones (Fig. 5.4D). With addition of DOX to the culture medium, a stepwise reduction of fluorescence lifetime was detected in all nuclei, consistent with intranuclear accumulation of DOX (Fig. 5.4C); however, each nucleus featured a distinct DOX uptake profile, detected from the uptake kinetics over time (Fig. 5.4C). Cells in the invasion zone showed more rapid uptake, compared with cells in the main body of the spheroid (Fig. 5.4D). Ultimately, with addition of DOX to the culture medium, a stepwise reduction of fluorescence lifetime was detected in all nuclei, consistent with intranuclear accumulation of DOX (Fig. 5.4C).

Besides changes in the relative amount of each fluorophore per ROI, interfering parameters need to be considered that may impact changes in lifetime from regions containing multiple fluorophores, including differential bleaching, quenching, and heterogeneous absorption and scattering of tissues at the cellular scale. Taking differential bleaching of DOX and eGFP during repeated laser-scanning into account (Fig. 5.4B), the noncorrected lifetime decay curves shown in Fig. 5.4C and D overestimate DOX accumulation by a factor of 2, whereby the uptake kinetics and internuclear variability at the end point are likely correctly represented. Further, DOX might quench eGFP fluorescence by intercalating with DNA directly adjacent to H2B-eGFP and thus limit its emission. However, the relative contribution of quenching cannot be retrieved from biological data samples but requires independent controls using isolated fluorophores with defined concentration in isolated nuclei. In conclusion, mono- or biexponential fitting permits addressing time-resolved changes of lifetimes in cell subregions to qualitatively estimate the uptake kinetics and its divergence between cells. However, normalization for photobleaching and quenching are required to obtain semiquantitative data sets on the frequency of each fluorophore.

4.2. Phasor analysis for multiple fluorescent species

To visualize all photon arrival times at a single pixel in a histogram, a transformation into a phasor is obtained using the following equations:

$$g_i(\omega) = \int_0^\infty I(t)\cos(\omega t)dt / \int_0^\infty I(t)dt$$
$$s_i(\omega) = \int_0^\infty I(t)\sin(\omega t)dt / \int_0^\infty I(t)dt$$

After sine–cosine transformations, the values are represented as a polar-plot two-dimensional histogram (Fig. 5.5A–C). The phasor space thus allows determining of clustering of pixels in specific regions of the phasor plot without the requirement for exponential lifetime fitting (Digman et al., 2008). Phasor transformation and back-gating require little computing time and therefore provide instantaneous discrimination of lifetime species characteristics during acquisition of images. By gating on circumscribed voxel "clouds," cell and tissue subregions with specific lifetime "fingerprints" are obtained, representing molecular information (Fig. 5.5C, D; rectangular gates and gates images; phasor module of the ImSpector Software V4.0). Thus, phasor analysis, followed by back-gating onto the ungated lifetime image (Fig. 5.5D grayscale), identifies the relative proportions of eGFP and DOX with subcellular resolution (Fig. 5.5D, color-coded regions). Gating subregions may further serve as ROIs for secondary exponential fitting and to confirm the numeric lifetime contribution of each fluorophore (not shown). Thus, the phasor approach is useful for the rapid and sensitive discrimination of lifetime families, in a time- and space-resolved manner.

5. Conclusions and Outlook

The combination of single-beam MPM, fluorescence lifetime detection with 80 MHz average count rate, and phasor evaluation allows rapid acquisition and quantitative processing of lifetime data in live cells and intravital imaging in 3D tissues, without *a priori* knowledge of the lifetime components. The approach is useful in determining compound uptake in tumor cells, the coregistration of cell invasion, proliferation, and apoptosis, together with 3D microenvironmental information. To reduce the impact of interfering parameters, including photobleaching, future FLIM approaches will benefit from including infrared excitation and the use of far-red probes, which greatly reduces photobleaching and enhances photon retrieval due to reduced light scattering in 3D tissues (Andresen et al., 2009). In conclusion, fast TCSPC detectors and phasor analysis will enhance experimental biology and biomedicine by delivering comprehensive and visual topographically controlled readout of molecular properties in complex samples.

ACKNOWLEDGMENT

This work was supported by the Dutch Cancer Foundation (KWF 2008-4031).

REFERENCES

Andresen, V., Alexander, S., Heupel, W. M., Hirschberg, M., Hoffman, R. M., and Friedl, P. (2009). Infrared multiphoton microscopy: Subcellular-resolved deep tissue imaging. *Curr. Opin. Biotechnol.* **20**, 54–62.
Aptel, F., Olivier, N., Deniset-Besseau, A., Legeais, J. M., Plamann, K., Schanne-Klein, M. C., and Beaurepaire, E. (2010). Multimodal nonlinear imaging of the human cornea. *Invest. Ophthalmol. Vis. Sci.* **51**, 2459–2465.
Dai, X., Yue, Z., Eccleston, M. E., Swartling, J., Slater, N. K., and Kaminski, C. F. (2008). Fluorescence intensity and lifetime imaging of free and micellar-encapsulated doxorubicin in living cells. *Nanomedicine* **4**, 49–56.
de Lange, J. H., Schipper, N. W., Schuurhuis, G. J., ten Kate, T. K., van Heijningen, T. H., Pinedo, H. M., Lankelma, J., and Baak, J. P. (1992). Quantification by laser scan microscopy of intracellular doxorubicin distribution. *Cytometry* **13**, 571–576.
Denk, W., Strickler, J. H., and Webb, W. W. (1990). Two-photon laser scanning fluorescence microscopy. *Science* **248**, 73–76.
Digman, M. A., Caiolfa, V. R., Zamai, M., and Gratton, E. (2008). The phasor approach to fluorescence lifetime imaging analysis. *Biophys. J.* **94**, L14–L16.
Ducros, M., Moreaux, L., Bradley, J., Tiret, P., Griesbeck, O., and Charpak, S. (2009). Spectral unmixing: Analysis of performance in the olfactory bulb in vivo. *PLoS One* **4**, e4418.
Elangovan, M., Wallrabe, H., Chen, Y., Day, R. N., Barroso, M., and Periasamy, A. (2003). Characterization of one- and two-photon excitation fluorescence resonance energy transfer microscopy. *Methods* **29**, 58–73.
Förster, T. (1948). Zwischenmolekulare Energiewanderung und Fluoreszenz. *Ann. Physik* **437**, 55.
Friedl, P., and Bröcker, E. B. (2004). Reconstructing leukocyte migration in 3D extracellular matrix by time-lapse videomicroscopy and computer-assisted tracking. *Methods Mol. Biol.* **239**, 77–90.
Friedl, P., Wolf, K., von Andrian, U. H., and Harms, G. (2007). Biological second and third harmonic generation microscopy. *Curr. Protoc. Cell Biol.* **4**, (Unit 4.15).
Fruhwirth, G. O., Fernandes, L. P., Weitsman, G., Patel, G., Kelleher, M., Lawler, K., Brock, A., Poland, S. P., Matthews, D. R., Keri, G., Barber, P. R., Vojnovic, B., et al. (2011). How Forster resonance energy transfer imaging improves the understanding of protein interaction networks in cancer biology. *Chemphyschem* **12**, 442–461.
Germain, R. N., Bajenoff, M., Castellino, F., Chieppa, M., Egen, J. G., Huang, A. Y., Ishii, M., Koo, L. Y., and Qi, H. (2008). Making friends in out-of-the-way places: How cells of the immune system get together and how they conduct their business as revealed by intravital imaging. *Immunol. Rev.* **221**, 163–181.
Helmchen, F., and Denk, W. (2005). Deep tissue two-photon microscopy. *Nat. Methods* **2**, 932–940.
Helmchen, F., Svoboda, K., Denk, W., and Tank, D. W. (1999). In vivo dendritic calcium dynamics in deep-layer cortical pyramidal neurons. *Nat. Neurosci.* **2**, 989–996.
Hess, S. T., Sheets, E. D., Wagenknecht-Wiesner, A., and Heikal, A. A. (2003). Quantitative analysis of the fluorescence properties of intrinsically fluorescent proteins in living cells. *Biophys. J.* **85**, 2566–2580.
Hovorka, O., Subr, V., Větvička, D., Kovář, L. L., Strohalm, J., Strohalm, M., Benda, A., Hof, M., Ulbrich, K., and Rihova, B. (2010). Spectral analysis of doxorubicin accumulation and the indirect quantification of its DNA intercalation. *Eur. J. Pharm. Biopharm.* **76**, 514–524.

Jakobs, S., Subramaniam, V., Schönle, A., Jovin, T. M., and Hell, S. W. (2000). EFGP and DsRed expressing cultures of Escherichia coli imaged by confocal, two-photon and fluorescence lifetime microscopy. *FEBS Lett.* **479,** 131–135.

Levitt, J. A., Matthews, D. R., Ameer-Beg, S. M., and Suhling, K. (2009). Fluorescence lifetime and polarization-resolved imaging in cell biology. *Curr. Opin. Biotechnol.* **20,** 28–36.

Soloviev, V. Y., Tahir, K. B., McGinty, J., Elson, D. S., Neil, M. A., French, P. M., and Arridge, S. R. (2007). Fluorescence lifetime imaging by using time-gated data acquisition. *Appl. Opt.* **46,** 7384–7391.

Straub, M., and Hell, S. W. (1998). Fluorescence lifetime 3D-microscopy with picosecond precision using a multifocal multiphoton microscope. *Appl. Phys. Lett.* **73,** 1769–1771.

Theer, P., Hasan, M. T., and Denk, W. (2003). Two-photon imaging to a depth of 1000 μm in living brains by use of a Ti:Al2o3 regenerative amplifier. *Opt. Lett.* **28,** 1022–1024.

Thoronton, K. (2008). Chemotherapeutic management of soft tissue sarcoma. *Surg. Clin. North Am.* **88**(viii), 647–660.

Timpson, P., McGhee, E. J., Morton, J. P., von Kriegsheim, A., Schwarz, J. P., Karim, S. A., Doyle, B., Quinn, J. A., Carragher, N. O., Edward, M., Olson, M. F., Frame, M. C., *et al.* (2011). Spatial regulation of RhoA activity during pancreatic cancer cell invasion driven by mutant p53. *Cancer Res.* **71,** 747–757.

Verveer, P. J., Wouters, F. S., Reynolds, A. R., and Bastiaens, P. I. (2000). Quantitative imaging of lateral ErbB1 receptor signal propagation in the plasma membrane. *Science* **290,** 1567–1570.

Yamamoto, N., Jiang, P., Yang, M., Xu, M., Yamauchi, K., Tsuchiya, H., Tomita, K., Wahl, G. M., Moossa, A. R., and Hoffman, R. M. (2004). Cellular dynamics visualized in live cells *in vitro* and *in vivo* by differential dual-color nuclear-cytoplasmic fluorescent-protein expression. *Cancer Res.* **64,** 4251–4256.

Zipfel, W. R., Williams, R. M., Christie, R., Nikitin, A. Y., Hyman, B. T., and Webb, W. W. (2003). Live tissue intrinsic emission microscopy using multiphoton-excited native fluorescence and second harmonic generation. *Proc. Natl. Acad. Sci. USA* **100,** 7075–7080.

CHAPTER SIX

Measuring Membrane Protein Dynamics in Neurons Using Fluorescence Recovery after Photobleach

Inmaculada M. González-González,* Frédéric Jaskolski,*,[1] Yves Goldberg,*,[2] Michael C. Ashby,*,† and Jeremy M. Henley*

Contents

1. Introduction	128
1.1. Fluorescent probes	128
1.2. Photobleaching and recovery	130
2. Experimental Parameters and Preparation	130
2.1. Buffer composition	130
2.2. Temperature	132
2.3. Osmolarity	132
2.4. Health and viability of cells	132
3. Equipment	133
4. Establishing FRAP Conditions	134
4.1. The bleach mode	135
4.2. The imaging mode	135
5. A Basic FRAP Protocol	136
6. Analysis of FRAP	137
6.1. Processing raw fluorescence data	137
6.2. Normalizing data	138
6.3. Calculating the recovery half-time and mobile fraction	139
6.4. FRAP in the soma	140
6.5. Cross-linking to investigate the role of lateral mobility	142
6.6. Combined FRAP and FLIP	142
6.7. Perspective	145
Acknowledgments	145
References	145

* MRC Centre for Synaptic Plasticity, School of Biochemistry, Medical Sciences Building, University of Bristol, Bristol, United Kingdom
† School of Physiology and Pharmacology, University of Bristol, Bristol, United Kingdom
[1] Permanent address: Grenoble - Institut des Neurosciences, Grenoble, France
[2] Present address: Ecole Privée Maso, Perpignan, France

Methods in Enzymology, Volume 504
ISSN 0076-6879, DOI: 10.1016/B978-0-12-391857-4.00006-9

© 2012 Elsevier Inc.
All rights reserved.

Abstract

The use of genetically encoded fluorescent tags such as green fluorescent protein (GFP) as reporters to monitor processes in living cells has transformed cell biology. One major application for these tools has been to analyze protein dynamics in neurons. In particular, fluorescence recovery after photobleach (FRAP) of surface expressed fluorophore-tagged proteins has been instrumental to addressing outstanding questions about how neurons orchestrate the synaptic delivery of proteins. Here, we provide an overview of the methodology, equipment, and analysis required to perform, analyze, and interpret these experiments.

1. Introduction

Cells are highly dynamic structures and their constituent protein components are in constant motion within and between cellular compartments, domains, and microdomains. Membrane spanning proteins such as receptors and ion channels oscillate between confined and free Brownian motion (Meier *et al.*, 2001; Serge *et al.*, 2002). These processes are important because they participate in the localization of proteins to the appropriate regions of the cell surface. This is especially critical in neurons, which are highly polarized and morphologically complex cells that transmit and process information via synaptic transmission. Each neuron can receive information via thousands of synapses and each Individual synapse requires the precise, activity-dependent, and coordinated delivery, retention, and removal of specialized sets of proteins.

1.1. Fluorescent probes

Green fluorescent protein (GFP) and subsequent derivatives and alternatives (Pakhomov and Martynov, 2008; Tsien, 1998) are bright, stable, nontoxic fluorophores that allow prolonged imaging of protein trafficking in live cells. Crucially, virtually any protein can be tagged with GFP resulting in a fusion that usually retains the same targeting and functional properties as the parent protein when expressed in cells. A potential limitation of most GFP-derived or -related fluorophores, however, is that they are fluorescent as soon as the protein is folded and remain fluorescent until degraded. Thus, fluorescent molecules can be found at all stages of the protein production pathway from early after synthesis until degradation. Therefore, an important experimental consideration is that the compartmental localization of the fluorescent signal corresponds to the main site of residence of the mature protein. To circumvent these complications, GFP family proteins have been engineered to generate a range of mutants with specifically altered properties such as shifts in emission wavelength or pH sensitivity (Table 6.1).

Among these, superecliptic pHluorin (SEP), a pH-sensitive derivative of eGFP, has been extensively used for live cell imaging of plasma membrane proteins (Ashby et al., 2004b; Sankaranarayanan et al., 2000). Protonation of SEP decreases photon absorption and therefore eliminates fluorescence emission at low pH. This allows the selective imaging of tagged proteins exposed to neutral pH environment. This is useful because most stages of the secretory pathway in neurons and other cells occur in acidic compartments, so a SEP-tagged membrane protein absorbs and emits fluorescence only when inserted in the plasma membrane (Ashby et al., 2004b).

Quantification of the pH-dependence of SEP fluorescence suggests that it is $21\times$ brighter at pH 7.4 than at 5.5 (Sankaranarayanan et al., 2000). Thus, if only $\sim 5\%$ of the SEP is surface expressed and exposed to pH 7.4 and 95% is intracellular and exposed to pH 5.5, there will be the same levels of fluorescence signal from the cell surface and inside the cell. This is an important issue to be aware of and control for with acid and ammonium chloride wash protocols (see Section 5) that specifically identify the contributions of surface expressed and intracellular proteins. Fortunately, our experience in neurons is that recombinantly expressed SEP-tagged surface membrane proteins produce bright signals with a very good signal to noise

Table 6.1 Properties of GFP derivatives

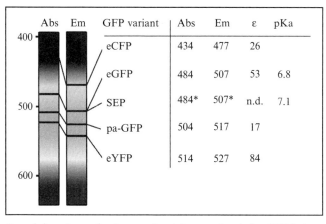

GFP variant	Abs	Em	ε	pKa
eCFP	434	477	26	
eGFP	484	507	53	6.8
SEP	484*	507*	n.d.	7.1
pa-GFP	504	517	17	
eYFP	514	527	84	

Compiled maximum absorbance and emission wavelength for eCFP (enhanced cyan fluorescent protein), eGFP (enhanced green fluorescent protein), SEP (superecliptic pHluorin), pa-GFP (photoactivable GFP), and eYFP (enhanced yellow fluorescent protein). On the left, values are pointed on spectral bands. Also given are the molar extinction coefficient and pKa.
Abs, maximum absorbance in nm; Em, maximum emission in nm; ε, molar extinction coefficient in $10^3\ M^{-1}\ cm^{-1}$.
* denotes when fluorescence not eclipsed by acidic pH.
Data taken from Ashby et al. (2004a); Miesenbock et al. (1998); Pakhomov and Martynov (2008); Sankaranarayanan et al. (2000).

ratio, meaning that surface expressed fluorophore can be readily distinguished (Ashby et al., 2004a, 2006;Bouschet et al., 2005; Jaskolski et al., 2009; Martin et al., 2008).

1.2. Photobleaching and recovery

Like other fluorescent dyes, when illuminated at high intensity, GFP derivatives can be irreversibly bleached without detectably damaging intracellular structures (Patterson et al., 1997; Swaminathan et al., 1997; Wiedenmann et al., 2009). As SEP only absorbs photons when exposed to neutral pH, it allows the selective bleaching of only SEP-tagged plasma membrane proteins (Ashby et al., 2004a, 2006; Bouschet et al., 2005; Jaskolski et al., 2009; Martin et al., 2008). Fluorescence recovery after photobleach (FRAP) relies on the high mobility of cellular proteins under physiological conditions within the plasma membrane; proteins undergo various types of motion from free diffusion to flow motion and/or anchoring (see Jaskolski and Henley, 2009). When fluorescently tagged proteins are bleached in a region of interest (ROI), the recovery in fluorescence occurs due to the movement of unbleached SEP-tagged protein from areas outside the ROI into the bleached region. At the same time, bleached SEP-tagged protein within the ROI moves out of the ROI (Fig. 6.1A). Thus, FRAP is a convenient and powerful technique to assess protein movement, and the fact that SEP is fluorescent predominantly when located only at the plasma membrane allows experiments to focus on the properties of surface expressed proteins.

2. Experimental Parameters and Preparation

2.1. Buffer composition

Bicarbonate-based solutions that are typically used for culturing cells are not usually used for short-term imaging experiments because of the practical difficulty of maintaining CO_2/O_2 balance. Further, it is also advisable to avoid the use of phenol red and serum since both are sources of fluorescence.

The buffer we use for cell culture is GIBCO™ Neurobasal™ Medium (1×) supplemented with 2% of B-27 and 0.5 mM glutamine or Glutamax. The composition of imaging buffer is NaCl 119 mM; Hepes 25 mM; Glucose 10 mM; NaHCO3 1–2 mM; KCl 2.5 mM; NaH2PO4 1 mM; CaCl2 1.8–2.5 mM; MgSO4 0.8–1.3 mM; adjust the pH to 7.4. For low pH solution, HEPES was replaced with equimolar MES and pH was adjusted to 6.0.

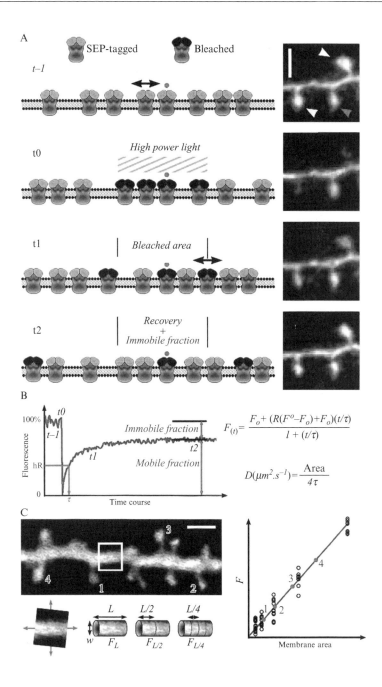

2.2. Temperature

Temperature affects the mobility of both soluble and membrane-associated molecules. Because viscosity is highly dependent on temperature, the effects on diffusion can be striking (Reits and Neefjes, 2001). Thus the temperature of the cells being imaged needs to be constant, for example, on a controlled temperature stage, ideally at a physiological temperature. It should also be noted that temperature fluctuations of the stage and objective can cause focus drift. To limit the impact of this variable, the imaging set up should be switched on and warmed up for 30 min prior to the experiment.

2.3. Osmolarity

Alterations in cell volume caused by hypo/hyperosmolarity can profoundly alter cell function (Sabirov and Okada, 2009). Changes in the osmolarity can occur during the experiment due to evaporation of the medium and/or differences in the composition of the culture media and the recording media. It is therefore necessary to determine the osmolarity of the culture media and adjust the osmolarity of the recording media accordingly and evaporation should be minimized by using a humidified environment.

2.4. Health and viability of cells

Our FRAP experiments are performed using dispersed cultures of hippocampal neurons (Ashby *et al.*, 2004a, 2006; Jaskolski *et al.*, 2009; Martin *et al.*, 2008). Cell morphology can be assessed using transmitted-light

Figure 6.1 Fluorescence recovery after photobleaching, theory, and analysis. (A) Schematic representing successive steps of a FRAP experiments using SEP-tagged proteins. A fraction of tagged proteins undergoes motion in the plasma membrane, while another fraction is clustered and thus immobilized (red dot). Left column, simulated FRAP images, t−1 is the prebleach step, t0 is the bleach step (see black bleached proteins), t1 is the initial recovery due to lateral diffusion (black double arrow), t2 corresponds to the steady state late recovery where only immobile proteins (red dot) remains within the bleached area. The right-hand panels show corresponding stages of FRAP in dendritic spines. In the top panel, the white arrows indicate the bleached spines and the red arrow a control spine. The scale bar is 1 μm. (B) A typical FRAP recording and formulas for curve fitting (according to Feder *et al.* 1996) and diffusion coefficient calculus (see analysis section). (C) Complex membrane area approximation using calibration curves. Dendrites with spines of membrane-anchored eGFP-expressing neuron. A nonspiny region of the shaft is cropped and aligned, and fluorescence is measured for various lengths (L). Using the width (w) of the shaft, the area of a corresponding cylinder is computed. Left, fluorescence is plotted versus calculated area to produce a calibration curve. Membrane area of labeled spines (1–4) can be read on the curve. (See Color Insert.)

microscopy techniques to identify cells that are stressed, dying, or dead before starting the experiment. The formation of irregular plasma membrane bulges, large vacuoles, and detachment from the tissue-culture plate are all indicators of cell stress. In addition, clustering of the fluorescent signal is a strong indication that cells are under stress (Samhan-Arias *et al.*, 2009) (Fig. 6.2). Another key indicator of stressed neurons is compromised membrane integrity, which is routinely assessed in the FRAP experiment using SEP by brief acid and ammonium chloride washes (see Section 5).

3. Equipment

FRAP experiments to monitor surface expressed fluorophore-tagged proteins in cultured neurons can be performed on most live imaging microscope setups as long as the appropriate excitation light source is available

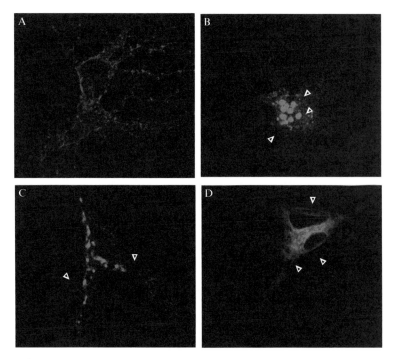

Figure 6.2 Examples of damaged neurons. These are 21 DIV hippocampal neurons expressing a SEP-tagged membrane protein. (A) An example of a neuron suitable for FRAP; (B–D) examples of stressed neurons that should not be imaged showing blebbing (B), intracellular aggregation of fluorescent proteins (C), and vacuoles (D). (For colour version of this figure, the reader is referred to the Web version of this chapter.)

(e.g., 488 nm for SEP) and the temperature can be controlled. Ideally, a perfusion system with at least three ports is required for pH calibration and additional ports are necessary for any pharmacological treatment.

For image capture in FRAP on relatively flat specimens, where depth is not much larger than the optical resolution, the more light collected from the area of interest, the better, so there is no need to restrict the depth of the focus. Thus, when imaging neuronal dendrites or dendritic spines, a confocal microscope is not required, although confocal is necessary when studying the somatic plasma membrane.

Most major microscope suppliers provide customized built-in software that often includes a FRAP module to control key parameters including excitation power, imaging rate, and designated ROI. Some also encompass analysis modules, although the freely available software, ImageJ, is widely used and entirely suitable (http://rsbweb.nih.gov/ij/ see below). The experimental guidelines below largely reflect our use of a Zeiss confocal microscope and the ImageJ software, but the steps can be generalized to different setups.

4. Establishing FRAP Conditions

Laborious, but essential, initialization experiments are required to define the parameters for successful FRAP. Each FRAP experiment consists of four different steps: prebleach, bleach, postbleach, and recovery (Fig. 6.1A and B). Rapid bleaching is used to decrease the initially bright prebleach fluorescence in the targeted region, ideally to values close to zero. The postbleach and recovery phases then proceed as a recovery of fluorescence in the ROI that is characteristic of the movement of unbleached molecules from outside the bleached area. Photobleaching experiments require a rapid switching between a low-intensity illumination mode during the pre/postbleaching and the recovery phases and a high-intensity mode during the bleach phase.

Here, we provide an overview protocol for FRAP of SEP-tagged plasma membrane-expressed proteins in dispersed hippocampal cultures. It is important to emphasize, however, that there is no universal protocol for FRAP experiments. The conditions for the imaging mode, the bleach mode, and the frequency of acquisition need to be established and optimized according to the specimen (e.g., cell monolayer, organotypic culture, etc.), the equipment (e.g., widefield, confocal, multiphoton photon microscope, etc.), and the kinetics of the molecule under investigation. For example, for proteins with rapid kinetics, the acquisition speed is a critical parameter. Each series of experiments needs to be optimized to bleach rapidly and effectively and to minimize diffusion during bleaching. For slower molecules, an important consideration is to minimize focus drift by, for example, the use of autofocus and/or microscopes with closed chambers.

4.1. The bleach mode

Photobleaching is photodestruction of the fluorophore. Every fluorescent molecule has characteristic maximum number of absorption/emission cycles; when this value is exceeded, the fluorophore is irrevocably bleached, that is, no longer fluorescent. An ideal bleaching should be instantaneous, but in practice, it should not exceed one-tenth of the half-time of the recovery. More specifically, typical bleaching conditions require a 100- to 1000-fold increase in laser illumination power (decrease in attenuation) for 1–5 bleach iterations ($\sim$0.01–0.5 s) for many ROIs. A higher number of iterations may affect the recovery (Weiss, 2004). Other parameters that influence bleaching are the zoom and the volume of the ROI. Often, when using a scanner, software-based zooming can cause an increase in the overlapping of illumination scans that tends to increase the speed of the bleaching. However, a problem arises when switching to the unzoomed mode as it can delay the acquisition sequence. This is especially undesirable when analyzing rapid kinetics. The degree of bleaching is often a critical issue in setting the acquisition conditions. Complete bleaching is nearly impossible to obtain, but a 70–80 % bleach of the signal is sufficient to explore the membrane dynamics in FRAP experiments.

4.2. The imaging mode

A critical issue stems from the fact that repeated illumination, even at low power, can cause unintended photobleaching resulting in a slow fade of fluorescence. This "acquisition" photobleaching needs to be minimized because it will influence measurement of recovery rates. In practical terms, pilot experiments are required to optimize imaging parameters to ensure there is minimal loss of fluorescence during acquisition since $\geq 10\%$ will have significant impact on the analysis. The dosage of excitation light determined by the light intensity and the exposure window must be optimized to minimize slow bleach by repeated illumination. Reducing either the excitation intensity or the exposure time not only decreases potential phototoxicity but also leads to decreased fluorescence and limits the signal to noise ratio causing a loss in spatial resolution. Thus, the imaging settings, like the bleaching parameters, require careful optimization and the following measures should be considered:

1. The detection signal to noise ratio should be optimized to ensure that as much light as possible is collected. The simultaneous excitation of multiple dyes requiring stringent separation of emitted wavelengths should be avoided whenever possible. When it is necessary to use multiple fluorophores, they should be imaged sequentially to avoid narrowing the collection wavelength window. However, this approach

does cause a time delay between images that will slow the imaging rate. A convenient way to enhance light collection is to open pinhole where resolution is not a major issue.
2. Ensure the most efficient light path of the microscope. Where possible, avoid using a Wollaston prism, beam splitter, narrow band dichroic filters, or anything else that can cause photon loss and decrease acquisition rate. If transmission light imaging or simultaneous dual probe imaging are required, the weight of such biases need to be considered during the analysis step.
3. Reduce the pixel dwell time by minimizing frame or line averaging and increase the scan speed (or reduce the exposure time in CCD and spinning disk systems). One convenient method to speed up the imaging sequence is to image a small format like 512 × 512. Further, in most cases, there is no need to perform z scanning because thin neuronal processes are smaller than the z resolution (i.e., the minimal focus depth is larger than thin dendrites).

During the recovery phase, the acquisition frequency should be adjusted to resolve the dynamic range of the recovery with good temporal resolution. High rate imaging is needed to follow the initial rapid recovery phase, but less frequent imaging is required for the late phase of recovery and to define the steady state. In practice, initial experiments should be conducted to define the time at which no noticeable further increase in fluorescence intensity is detected. Once this is established, the imaging sequence should be designed as a succession of modules with different imaging rates and thus illumination rates appropriate to the phase of the recovery.

5. A Basic FRAP Protocol

This protocol uses a Zeiss LSM or equivalent confocal microscope.

1. Prewarm the recording buffer and the imaging stage to 37 °C and warm up the microscope and laser(s) following manufacturer's instruction. Prepare the neurones in the imaging chamber with recording medium, wash the remaining culture medium twice to prevent contamination by residual light-sensitive material (phenol, serum, etc.).
2. Identify and focus on the cell of interest. Acquire an image of the whole cell at low excitation light intensity. Modify filters, pinhole, zoom, and detector gain for maximal fluorescence with minimal laser power.
3. Within the selected neuron, define a ROI for the photobleach and save the coordinates.
4. Input the photobleaching conditions (i.e., laser power, zoom, and the minimal number of laser iterations required for photobleaching) and

save the configuration as bleach mode. Empirically determine the photobleaching conditions before the experiment so that, after photobleaching, the fluorescent signal of the photobleached ROI decreases close to background levels.
5. Input the parameters for the imaging mode (i.e., laser power, zoom, scan speed, line/frame averaging, and format), and save the configuration as imaging mode. Empirically determine imaging conditions that do not significantly photobleach the cell outside of the bleach ROI during the experiment.
6. Input the frequency of acquisition during the different steps. As a starting reference, use millisecond intervals for the initial recovery step and second intervals for the prebleach and late recovery steps. Adjust the intervals once you approximate half-recovery values. Continue to image until the recovery process has reached a steady state, being careful to account for any anomalous diffusion that causes a very slow recovery to the steady state.
7. Define the experimental sequence by combining the frequency of acquisition (5) and imaging acquisition mode (4) during the pre/post bleach steps, the bleach mode (3) with no acquisition and the frequency of acquisition during the recovery step (5) with the imaging mode (4).
8. Collect at least 10–20 data sets for each fluorescently labeled protein and experimental protocol for statistical analysis. It maybe necessary to discard a fraction of data sets because of problems that potentially bias results (e.g., bleach was not complete, the focal plane shifted, or the phototoxicity damaged the cell affecting the recovery).
9. Distinguish the surface-expressed protein from intracellular fluorescence by briefly washing the cell with low pH (5.5) buffer to reversibly eclipse the fluorescence only from the surface protein.
10. Assess the maximal SEP fluorescence using ammonium chloride (NH_4Cl) wash to transiently increase intracellular pH and reveal total SEP signal. This peak value is used to determine the proportion of protein at the surface of the cell (Ashby *et al.*, 2004b).

6. ANALYSIS OF FRAP

6.1. Processing raw fluorescence data

ImageJ plug-ins are available to open image files generated with proprietary software from major microscope manufacturers. For example, the LOCI suite http://www.loci.wisc.edu/software/bio-formats can handle virtually all image formats used in the biosciences. Users of Zeiss microscopes will also find the LSM toolbox useful (http://www.image-archive.org/).

The metadata associated with the acquired images include measurements of real-time values, as recorded with each frame. The LSM Toolbox "Apply stamps" command can be used to report all the time values (t-stamps) from the Zeiss metadata into a self-generated text file, which is then copied into a spreadsheet and will provide the time coordinate of the experiment.

In addition, the exact contours of the bleached ROI can be extracted from the .lsm file. Using the LOCI importer, ROI coordinates that have saved within the Zeiss interface can be directly downloaded into the ImageJ ROI manager (in the Analyze > Tools menu). The time evolution of mean fluorescence per pixel within the ROI can then be obtained using the Image > Stacks > Plot z-axis profile sequence of commands. This generates a list of values that can be directly copied into the spreadsheet, next to the time coordinate.

One way to limit the noise contribution is to subtract the maximum noise level, as estimated from an empty region of the same plane, from mean fluorescence values of the ROI. An ImageJ selection tool can be used to delineate an empty area, preferably close to the ROI if background is not homogeneous or if the ROI fluorescence is weak. Use Image > Stacks > Plot z-axis profile as above to obtain the time series of mean fluorescence values and respective standard deviations in the empty area. After copying this list into the spreadsheet, calculate mean plus twice the standard deviation for each timepoint. This provides an estimate of the likely maximum noise at each timepoint, which is then subtracted from the mean fluorescence of the ROI at the same timepoint.

6.2. Normalizing data

To allow comparison between experiments, data must be normalized. After removing noise, if necessary, all fluorescence values in the time course are divided by the mean prebleach value. To average out random fluctuations, the prebleach value is the mean of the last 2–4 values prior to bleaching. As discussed, despite optimization, a slow fade of fluorescence can appear during the total time course of the experiment and these recordings should be excluded from the analysis. Sometimes, protein density or the pharmacological treatment can cause significant decrease in fluorescence intensity and, when combined with a repeated illumination for imaging, the slow fade is unavoidable. This problem can be overcome by correcting for nonspecific photobleaching. Delineate 3–4 control (unbleached) regions with shapes similar to that of the ROI, but as far as possible from the bleached region itself, and calculate the normalized fluorescence time course for these control regions. Fit the control time course with a monoexponential decay to estimate the rate of nonspecific fluorescence decay during the experiment. Use the inverse of this to correct the recordings, and apply the same correction to all compared recordings even if they do not display any apparent slow fluorescence fade. This correction should be made specifically for each individual experiment unless identical expression and imaging parameters are used across experiments, as the rate is likely to vary. The prebleach fluorescence level (F^0) is normalized to 1, the value immediately after bleach (F_0) typically lies between 0.2 and 0.3, and time-dependent recovery ($F(t)$) tends to a plateau value ≤ 1.

6.3. Calculating the recovery half-time and mobile fraction

The asymptotic nature of the plateau value and the dispersion of fluorescence values along the time course require the half-time and mobile fraction to be calculated by fitting the recovery time series to a theoretical curve, generated by an equation that describes the recovery process (Fig. 6.1B) (Weiss, 2004). Therefore, to analyze the recovery, the end of bleaching is designated as time zero, and the successive timepoints are adjusted accordingly. Curve fitting and parameter calculation can be performed directly within ImageJ (using the curve fitting tool in the "Analyze" menu). Alternatively other data analysis software (Igor pro, Qtiplot, Matlab, etc.) with accurate fitting menus can be used. The quality of fit is evaluated by the correlation coefficient between experimental and theoretical points.

Several equations have been used in the literature to fit the FRAP time course for SEP-tagged AMPA receptor subunits (SEP-GluR) in dendritic spines of cultured hippocampal neurons (Ashby *et al.*, 2006; Jaskolski *et al.*, 2009; Makino and Malinow, 2009). These reflect different approximations of the mixture of diffusion and binding that determine the kinetics of fluorescence recovery. For simplicity, here we present only an equation derived from a two-dimensional diffusion model equation (for an alternative dual exponential model, see Makino and Malinow, 2009).

Two-dimensional diffusion model (Axelrod *et al.*, 1976; Feder *et al.*, 1996):

$$F(t) = \frac{F_0 + (R(F^0 - F_0) + F_0)(t/\tau)}{1 + (t/\tau)},$$

F^0 is the fluorescence level at baseline, F_0 is the fluorescence level after the bleaching step, R is the mobile fraction, and τ is the half-time to fluorescence return. For proteins that undergo anomalous subdiffusion, the formula can be implemented with a time exponent (Feder *et al.*, 1996). Here the recovery kinetic is determined by the rate of diffusion of fluorescent protein into the bleached domain. The presence of immobilized or persistently bound protein is accounted for by the nondiffusing fraction $(1-R)$. This equation was derived (as a first-order approximation to the exact analytical solution) in the idealized case of free diffusion from all directions into a circular photobleached spot on a planar membrane

(Axelrod et al., 1976; Feder et al., 1996). As this clearly is not the geometry of a bleached dendritic spine, the calculated parameters do not have absolute value. However, the resulting theoretical curves provide remarkably close fits to both experimental FRAP time series and simulated receptor diffusion along spine head and neck (Ashby et al., 2006; Jaskolski et al., 2009). The curve parameters are therefore useful to compare between spines, provided the bleached regions have similar shape. In particular, R (the mobile fraction) and thus 1−R (the immobile fraction) are particularly informative parameters to compare experimental conditions. With respect to diffusion, τ should be computed with the bleach membrane area to obtain the diffusion coefficient:

$$D \text{ in } \mu m^2 s^{-1} = \frac{\text{Area}}{4\tau}.$$

This allows comparison of experimental conditions where the bleached areas may fluctuate. Measuring the plasma membrane area of complex structures like dendritic spines is not trivial, so values can be empirically derived from calibration curves using cell membrane segments with more regular shapes (Fig. 6.1C) (Jaskolski et al., 2009). Nonspiny straight dendrites can roughly be considered as cylinders. Slicing such a cylinder in subsections of known length and diameter (measured in the image) provides cylindrical sections from which fluorescence can be measured. The plot of fluorescence versus calculated membrane area provides a calibration curve with which fluorescence emitted from an arbitrarily shaped region can be used to estimate the corresponding membrane area (Fig. 6.1C).

6.4. FRAP in the soma

Targeting FRAP to the soma of neurons has been used to estimate diffusion rates of proteins at extrasynaptic locations, and indeed, the same experimental principles apply to the cell body of other cell types. The typically ovoid or spherical shape of the soma dictates that, by varying the focus, very different looking images can be obtained through optical sectioning (such as with confocal microscopy). This can impact the analysis of FRAP as differing bleaching profiles (i.e., the shape of the bleached area) can be obtained. If focused through the center of the cell body, the plasma membrane appears as an annulus around the edge of the cytoplasm whereas focusing at the top (or bottom) gives a circular image largely containing plasma membrane (Fig. 6.3). Bleaching profiles in these two planes therefore have different shapes. One consequence of this is that diffusing molecules underlying FRAP are likely to have different routes of entry into the FRAP region. When bleaching a section of the annular profile (Fig. 6.3, Optical section A), FRAP largely proceeds from the edges of the FRAP

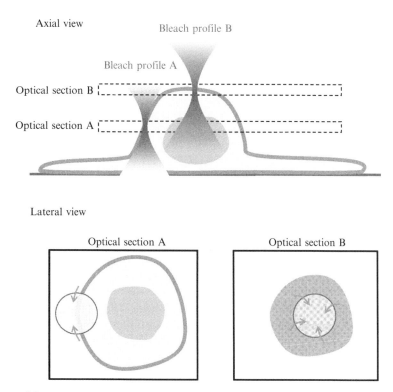

Figure 6.3 FRAP in the soma. The upper schematic shows the axial view of two bleach profiles when focused on the soma at different optical planes. When focused at optical section A, the image cuts through the vertically oriented somatic plasma membrane, which is imaged as an annulus (shown in the corresponding lateral view below). In this case, bleaching occurs at one section of the membrane and FRAP can be assumed to largely occur laterally from within the optical plane (one-dimensional). When focused on the top of the soma (optical section B), the image is circular (see lateral view below) and FRAP can occur laterally from all around the bleach ROI (two-dimensional). (For colour version of this figure, the reader is referred to the Web version of this chapter.)

region as molecules move in from the sides. This is because the optical section is usually smaller in depth than the bleaching profile, meaning that there is not much fluorescence recovered directly from above or below the plane of focus. Under these conditions, it is sensible to restrict analysis to a subregion of the FRAP area that contains the plasma membrane and to analyze FRAP curves based on diffusion along a one-dimensional line.

When bleaching the top of the soma, unbleached molecules can move into the FRAP region from any lateral direction. This is similar to early FRAP experiments on planar fluorescent layers that assumed diffusion occurring on a two-dimensional plane (Axelrod et al., 1976). Varying rates of FRAP can arise from differences in optical properties of the experiment and topological features of the structure studied. Thus, calculated diffusion coefficients from membranes with different shapes should be compared with caution. Ideally, detailed and accurate modeling of membrane topology and diffusion could provide solutions for comparative calculations; indeed, many variations have been produced (Klonis et al., 2002). However, it is likely that accurate measurement and modeling of membranes may be impractical, so a pragmatic approach is that only structures of relatively similar shape are compared directly (Reits and Neefjes, 2001).

6.5. Cross-linking to investigate the role of lateral mobility

FRAP occurs because unbleached fluorescently labeled molecules move into the bleached region. This movement is usually inferred to occur by lateral diffusion of molecules in the plane of the membrane, but other possible mechanisms include vesicular transport within the cell and, if SEP-tagged proteins are used, exocytosis within the ROI.

To define the contribution of lateral motion in FRAP, specific inhibition of lateral diffusion via antibody cross-linking can be used to form large aggregates in the plane of the plasma membrane (Fig. 6.4B). This cross-linking of multiple individual target molecules by saturating concentrations of antibody effectively prevents free lateral diffusion. Therefore, FRAP analysis under such cross-linking conditions can reveal the extent of fluorescence recovery caused by lateral diffusion (Dragsten et al., 1979). For example, through experiments in which antibody cross-linking effectively blocked FRAP completely, it was inferred that lateral diffusion of AMPA receptors in dendritic spines is responsible for the vast majority of rapid receptor exchange (Ashby et al., 2006). For quantitative analysis of the effects of lateral diffusion, FRAP should be compared under control (noncross-linked), cross-linked, and chemically fixed (minimum FRAP possible) specimens. Even in the presence of saturating antibody, if the density of plasma membrane target molecules is low, the efficiency of cross-linking is decreased because of their reduced spatiotemporal proximity. In some cases, this can be circumvented by the addition of a secondary ligand (often a polyvalent anti-IgG antibody) to form a second aggregated layer (Fig. 6.4C). Using cross-linking to analyze lateral diffusion assumes that there are no effects on other modes of molecule movement. To validate this, the effect of antibody binding and cross-linking should be assessed on membrane trafficking events such as endocytosis and recycling in independent assays (Ashby et al., 2006).

6.6. Combined FRAP and FLIP

Another method for distinguishing between lateral diffusion and vesicular trafficking of neuronal membrane proteins is combining fluorescence loss in photobleach (FLIP) with FRAP (Jaskolski et al., 2009). Here the contribution

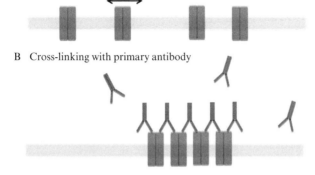
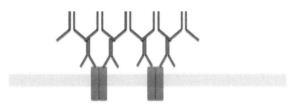
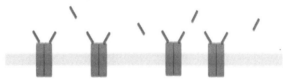

Figure 6.4 Antibody cross-linking of surface proteins. (A) Schematic view of the plasma membrane (yellow) with transmembrane proteins (red) able to move lateral in the plane of the membrane. (B) Addition of excess antibody causes cross-linking of proteins due to the divalent binding sites of the antibody, thus restricting lateral motion. (C) Cross-linking can be enhanced by the addition of a secondary antibody. (D) Antibody Fab fragments are monovalent and therefore can cause cross-linking. Therefore, Fab fragments can be used as a control to check that antibody binding itself does not influence lateral movement. (See Color Insert.)

of lateral diffusion in the FRAP region of a dendrite is removed by the continual and specific photobleaching (FLIP) of SEP-tagged membrane molecules in the regions flanking the bleached area of interest (Fig. 6.5). Using this approach, the contribution of vesicular trafficking to FRAP can be measured directly and the contribution of lateral diffusion can be inferred. Further, in the same experiment, FLIP can be used for qualitative assessment of diffusion in regions outside the photobleached ROIs.

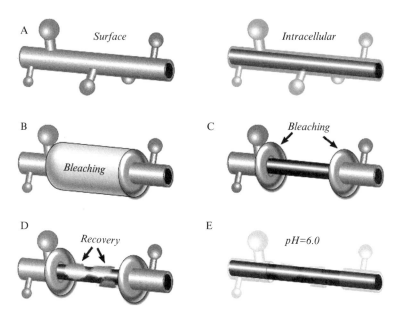

Figure 6.5 Assessing membrane insertion using FRAP and FLIP. (A) SEP-tagged proteins only emits fluorescence where inserted in the neuronal plasma membrane (left) but remains quenched in intracellular compartments (right). (B) High power excitation light causes a bleach of surface expressed tagged proteins. (C) To prevent recovery from diffusion, flanking region is continuously bleached (FLIP). (D) If recovery appears it can only be due to plasma membrane insertion from inner compartments. (E) pH 6.0 quenches the surface fluorescence. (For color version of this figure, the reader is referred to the Web version of this chapter.)

This is a self-contained, convenient, and powerful approach, but there are a number of technical issues that must be considered. The ROI has to be a section of dendritic shaft that is thin enough to be bleached by a rapid scan with the illumination volume. It is also important that the ROI is a linear section of dendrite with no ramifications. As discussed, it is also necessary to avoid illumination conditions for imaging that themselves cause slow bleaching of SEP. For example, it is expedient to image only once for every 10–20 bleaching scans. This is particularly important because the FRAP mediated by exocytosis is slow and weak compared to diffusion-mediated FRAP.

As for FRAP, this approach produces image stacks with one or more ROIs, and the measure of fluorescence intensity can be recorded using ImageJ software. If the SEP-tagged protein diffuses rapidly within the FRAP ROI, then the fluorescence increment is diluted in the noise. Fortunately, most membrane proteins undergo various diffusion modes and alternate between clustering and constrained diffusion, so newly exocytosed proteins tend to accumulate resulting in detectable fluorescence increases. These can be visualized by highlighting thin sub-ROIs along the shaft to record fluorescence variations in small domains. To confirm that fluorescence recovery is attributable to surface expression of the SEP-tagged protein, it is necessary to perform a pH 6.0 wash at the end of the experiment to ensure that the fluorescent signal is quenched (Fig. 6.5E).

6.7. Perspective

Although only a small subset of surface proteins in neurons has been investigated, FRAP of fluorophore-tagged proteins has already provided a wealth of information about the processes underlying membrane protein trafficking and localization. The application of FRAP in combination with other dynamic microscopy approaches such as single particle tracking and the newly emerging superresolution optical microscopy techniques will continue to provide insight and deepen understanding of the fundamental mechanisms that regulate neuronal function and dysfunction.

ACKNOWLEDGMENTS

We are grateful to the Wellcome Trust, the MRC, the BBSRC, and the ERC for funding. YG was a visiting scholar and IMGG and FJ were EMBO Fellows. We thank Dr. Andrew Doherty and Philip Rubin for technical expertise and support.

REFERENCES

Ashby, M. C., De La Rue, S. A., Ralph, G. S., Uney, J., Collingridge, G. L., and Henley, J. M. (2004a). Removal of AMPA receptors (AMPARs) from synapses is preceded by transient endocytosis of extrasynaptic AMPARs. *J. Neurosci.* **24,** 5172–5176.

Ashby, M. C., Ibaraki, K., and Henley, J. M. (2004b). It's green outside: Tracking cell surface proteins with pH-sensitive GFP. *Trends Neurosci.* **27,** 257–261.

Ashby, M. C., Maier, S. R., Nishimune, A., and Henley, J. M. (2006). Lateral diffusion drives constitutive exchange of AMPA receptors at dendritic spines and is regulated by spine morphology. *J. Neurosci.* **26,** 7046–7055.

Axelrod, D., Koppel, D. E., Schlessinger, J., Elson, E., and Webb, W. W. (1976). Mobility measurement by analysis of fluorescence photobleaching recovery kinetics. *Biophys. J.* **16,** 1055–1069.

Bouschet, T., Martin, S., and Henley, J. M. (2005). Receptor-activity-modifying proteins are required for forward trafficking of the calcium-sensing receptor to the plasma membrane. *J. Cell Sci.* **118,** 4709–4720.

Dragsten, P., Henkart, P., Blumenthal, R., Weinstein, J., and Schlessinger, J. (1979). Lateral diffusion of surface immunoglobulin, Thy-1 antigen, and a lipid probe in lymphocyte plasma membranes. *Proc. Natl. Acad. Sci. USA* **76,** 5163–5167.

Feder, T. J., Brust-Mascher, I., Slattery, J. P., Baird, B., and Webb, W. W. (1996). Constrained diffusion or immobile fraction on cell surfaces: A new interpretation. *Biophys. J.* **70,** 2767–2773.

Jaskolski, F., and Henley, J. M. (2009). Synaptic receptor trafficking: The lateral point of view. *Neuroscience* **158,** 19–24.

Jaskolski, F., Mayo-Martin, B., Jane, D., and Henley, J. M. (2009). Dynamin-dependent membrane drift recruits AMPA receptors to dendritic spines. *J. Biol. Chem.* **284,** 12491–12503.

Klonis, N., Rug, M., Harper, I., Wickham, M., Cowman, A., and Tilley, L. (2002). Fluorescence photobleaching analysis for the study of cellular dynamics. *Eur. Biophys. J.* **31,** 36–51.

Makino, H., and Malinow, R. (2009). AMPA receptor incorporation into synapses during LTP: The role of lateral movement and exocytosis. *Neuron* **64,** 381–390.

Martin, S., Bouschet, T., Jenkins, E. L., Nishimune, A., and Henley, J. M. (2008). Bidirectional regulation of kainate receptor surface expression in hippocampal neurons. *J. Biol. Chem.* **283,** 36435–36440.

Meier, J., Vannier, C., Serge, A., Triller, A., and Choquet, D. (2001). Fast and reversible trapping of surface glycine receptors by gephyrin. *Nat. Neurosci.* **4,** 253–260.

Miesenbock, G., De Angelis, D. A., and Rothman, J. E. (1998). Visualizing secretion and synaptic transmission with pH-sensitive green fluorescent proteins. *Nature* **394,** 192–195.

Pakhomov, A. A., and Martynov, V. I. (2008). GFP family: Structural insights into spectral tuning. *Chem. Biol.* **15,** 755–764.

Patterson, G. H., Knobel, S. M., Sharif, W. D., Kain, S. R., and Piston, D. W. (1997). Use of the green fluorescent protein and its mutants in quantitative fluorescence microscopy. *Biophys. J.* **73,** 2782–2790.

Reits, E. A., and Neefjes, J. J. (2001). From fixed to FRAP: Measuring protein mobility and activity in living cells. *Nat. Cell Biol.* **3,** E145–E147.

Sabirov, R. Z., and Okada, Y. (2009). The maxi-anion channel: A classical channel playing novel roles through an unidentified molecular entity. *J. Physiol. Sci.* **59,** 3–21.

Samhan-Arias, A. K., Garcia-Bereguiain, M. A., Martin-Romero, F. J., and Gutierrez-Merino, C. (2009). Clustering of plasma membrane-bound cytochrome b5 reductase within "lipid raft" microdomains of the neuronal plasma membrane. *Mol. Cell. Neurosci.* **40,** 14–26.

Sankaranarayanan, S., De Angelis, D., Rothman, J. E., and Ryan, T. A. (2000). The use of pHluorins for optical measurements of presynaptic activity. *Biophys. J.* **79,** 2199–2208.

Serge, A., Fourgeaud, L., Hemar, A., and Choquet, D. (2002). Receptor activation and homer differentially control the lateral mobility of metabotropic glutamate receptor 5 in the neuronal membrane. *J. Neurosci.* **22,** 3910–3920.

Swaminathan, R., Hoang, C. P., and Verkman, A. S. (1997). Photobleaching recovery and anisotropy decay of green fluorescent protein GFP-S65T in solution and cells: Cytoplasmic viscosity probed by green fluorescent protein translational and rotational diffusion. *Biophys. J.* **72,** 1900–1907.

Tsien, R. Y. (1998). The green fluorescent protein. *Annu. Rev. Biochem.* **67,** 509–544.

Weiss, M. (2004). Challenges and artifacts in quantitative photobleaching experiments. *Traffic* **5,** 662–671.

Wiedenmann, J., Oswald, F., and Nienhaus, G. U. (2009). Fluorescent proteins for live cell imaging: Opportunities, limitations, and challenges. *IUBMB Life* **61,** 1029–1042.

CHAPTER SEVEN

FLUORESCENT SPECKLE MICROSCOPY IN CULTURED CELLS

Marin Barisic,* António J. Pereira,* *and* Helder Maiato*,†

Contents

1. Introduction — 148
   1.1. Discovery and optimization of fluorescent speckle microscopy — 148
   1.2. Advantages of FSM — 149
   1.3. Applications of FSM—What have we learned with it? — 149
2. Technical Challenges of FSM — 151
   2.1. How to reach low expression of fluorescent proteins for FSM? — 151
   2.2. How to image fluorescent speckles? — 152
   2.3. How to analyze fluorescent speckles? — 152
3. Detailed Methodology for FSM in Cultured Cells — 153
   3.1. Choice of cells — 153
   3.2. Regulation of fluorescent protein expression — 153
   3.3. Live-cell imaging and microscopy setup — 153
   3.4. Comparative performance between wide-field and spinning-disk confocal for FSM — 154
   3.5. Image analysis — 156
4. Future Challenges — 159
Acknowledgments — 159
References — 159

Abstract

After slightly more than a decade since it was first established, fluorescent speckle microscopy (FSM) has been intensively used to investigate macromolecular dynamics, such as microtubule flux in mitosis and meiosis, microtubule translocation in neurons, microtubule-binding proteins, and focal adhesion proteins, as well as the assembly of actin filaments. This state-of-the-art technique is based on nonuniform distribution of fluorescently labeled subunits diluted in the endogenous, unlabeled ones, resulting in microscopy-detectable speckled patterns. In order to enable sufficient contrast between neighboring

---

* Chromosome Instability & Dynamics Laboratory, Instituto de Biologia Molecular e Celular, Universidade do Porto, Porto, Portugal
† Department of Experimental Biology, Faculdade de Medicina, Universidade do Porto, Porto, Portugal

diffraction-limited image regions, a low ratio between labeled and endogenous molecules is required, which can be achieved either by microinjection or by expression of limited amounts of fluorescently labeled subunits in cells. Over the years, the initial settings for FSM have been significantly improved by introduction of more sensitive cameras and spinning-disk confocal units, as well as by the development of specialized algorithms for image analysis. In this chapter, we describe our current FSM setup and detail on the necessary experimental approaches for its use in cultured cells, while discussing the present and future challenges of this powerful technique.

1. INTRODUCTION

1.1. Discovery and optimization of fluorescent speckle microscopy

Fluorescent speckle microscopy (FSM) is a technology developed in the late 1990s to investigate the dynamics of macromolecular assembly and disassembly *in vivo* and *in vitro* (reviewed by Danuser and Waterman-Storer, 2006). It derives from fluorescent analog cytochemistry and, as many other important findings in history, was discovered serendipitously. In fluorescent analog cytochemistry, fluorophore-labeled proteins are either expressed or microinjected in living cells, which gives them the ability to be tracked by fluorescence-based, time-lapse microscopy. Although with this technique fluorescent structures usually appear uniformly labeled, microscope images of X-rhodamine tubulin-injected cells showed variations in fluorescence intensity along microtubules, which in some cases resulted in the appearance of fluorescently labeled speckled microtubules (Waterman-Storer and Salmon, 1997). This pioneering observation enabled further development of FSM and its application to a wide range of biological problems.

It is now known that this speckled pattern was a result of unequal distribution of fluorescently labeled tubulin subunits diluted in the endogenous (unlabeled) ones along the microtubule lattice, defining a speckle as a diffraction-limited image region that is significantly higher in fluorophore concentration, and therefore also in fluorescence intensity, than its immediately neighboring regions (Waterman-Storer and Danuser, 2002). Speckles can be visualized only if the ratio between labeled and endogenous molecules is low enough, which can be achieved by microinjection or expression of very low amounts of fluorescently labeled subunits, enabling sufficient contrast between neighboring diffraction-limited image regions. Hence, speckle contrast increases by decreasing the fraction of labeled molecules, which was shown by comparing lower (1.25%) and higher (20%) fractions of fluorescent tubulin, where only microtubules with lower levels demonstrated speckled patterns (Waterman-Storer and

Salmon, 1998). Binomial statistics shows that contrast increases without bounds with decreasing concentrations of labeled molecules, but in practice, noise contributions to the detection signal obscures speckle signals at exceedingly small concentrations. In fact, it was found that the optimal fraction of fluorophore-containing molecules is between 0.5% and 2%, with each speckle consisting of three to eight fluorophores (Danuser and Waterman-Storer, 2003).

1.2. Advantages of FSM

Aside from reducing potential negative effects on cell behavior due to protein overexpression, a low fraction of fluorescent subunits provides two additional advantages compared to fluorescent analog cytochemistry, from which it derived. First, low fluorophore levels strongly reduce background fluorescence from out-of-focus and cytoplasmic-dispersed unincorporated fluorescent molecules, and second, speckled fluorescence distribution allows detection of the movement and turnover of fluorescent proteins, which is impossible to observe when structures are uniformly labeled (reviewed in Danuser and Waterman-Storer, 2006).

FSM has also several advantages over photoactivation and photobleaching techniques, which have been used to measure the dynamics of macromolecular structures. Each of these two methods is based on generating fiduciary marks across a defined region in the structures of interest using either laser-mediated photoactivation or photobleaching (Lippincott-Schwartz and Patterson, 2003; Mitchison, 1989; Theriot and Mitchison, 1991; Wadsworth and Salmon, 1986; Wang, 1985). These marks could then be followed by time-lapse microscopy, enabling the analysis of movement and turnover of the marked structures. However, monitoring the dynamics of macromolecular subunits by FSM is not limited to photoactivated or bleached regions but can be performed throughout the whole cell, and unlike induced photomarked regions, which are limited in duration, the speckle regime is persistent and can be followed for a longer period (Keating and Borisy, 2000). Taken together, FSM has proven to be the most precise and least invasive approach to investigate macromolecular dynamics in living cells.

1.3. Applications of FSM—What have we learned with it?

Since its implementation, this method has been mostly used to study the behavior of actin and tubulin polymers in the cytoskeleton, with a main focus on mitotic spindle dynamics during cell division. The mitotic spindle consists of overlapping microtubules coming from opposite poles (interpolar microtubules), as well as microtubules interacting with chromosomes at their kinetochores. The first class is an important mechanical element in the spindle required for the generation of outward pushing forces responsible for the maintenance of spindle bipolarity and spindle elongation during anaphase, while the latter is required for chromosome biorientation and segregation to the poles (reviewed in Kline-Smith and Walczak, 2004). Kinetochore microtubules are shorter, more stable, and form bundles called K-fibers. Each microtubule is a polymer, composed of α- and β-tubulin heterodimers organized into 13 protofilaments that form a hollow cylindrical structure. This organization defines on one side a minus end and on the other side a plus end containing a so-called GTP cap, which stabilizes the structure. After loss of the GTP cap due to hydrolysis, microtubules become unstable and depolymerize in a process called catastrophe, while the resumption of growth is known as rescue (reviewed in Howard and Hyman, 2009). Thus, microtubules show dynamic instability (Mitchison and Kirschner, 1984) as they stochastically switch between growth and shrinkage phases, unless their plus ends become "captured" and stabilized, for example, by interaction with kinetochores. Kinetochore attachment, however, does not prevent stabilized microtubules to turnover, while exhibiting poleward flux of tubulin subunits (Mitchison, 1989; Zhai et al., 1995). These dynamic properties, in conjunction with the capacity of kinetochores to remain attached with microtubules, play an important role in chromosome movements and error correction in mitosis (Ganem et al., 2005; Hayden et al., 1990; Matos and Maiato, 2011; Matos et al., 2009; Wollman et al., 2005). Although poleward flux was initially detected and described using photoactivation of microtubules labeled with caged-fluorescent tubulin (Mitchison, 1989), the discovery of FSM boosted the investigation on this conspicuous phenomenon. Using this technique, important studies of microtubule dynamics were performed in several different model organisms, such as *Xenopus* egg extracts, *Drosophila* cells and embryos, crane-fly spermatocytes, yeast, as well as mammalian cell lines. Waterman-Storer and colleagues measured flux in living newt lung epithelial cells using FSM and showed different dynamics in different classes of microtubules (Waterman-Storer et al., 1998). While interpolar microtubules exhibited faster flux velocities than kinetochore microtubules, astral ones showed no flux at all. These findings were later confirmed in other systems and are likely to reflect mechanical properties behind spindle architecture (LaFountain et al., 2004; Maddox et al., 2003; Matos et al., 2009; Waterman-Storer et al., 1998).

It became obvious that this method would be very useful to investigate kinetochore-microtubule flux and its roles in chromosome segregation. Experiments performed in *Xenopus* egg extract spindles, which were focused on understanding the dynamic interface between kinetochores and microtubules, showed that both kinetochore-based depolymerization (Pacman model) and flux-associated depolymerization of microtubules at minus ends are important for anaphase A and regulated at the metaphase–anaphase transition (Maddox et al., 2003). In contrast, crane-fly spermatocytes depend exclusively on flux-associated minus-end depolymerization for anaphase A (LaFountain *et al.*, 2004). It was also shown that, unlike in *Drosophila* S2 cells and embryos (Brust-Mascher *et al.*, 2009; Goshima *et al.*, 2005), kinetochore-microtubule flux in Ptk1 cells is largely kinesin-5 independent (Cameron *et al.*, 2006). Overall, these studies emphasize that conserved mechanisms related to mitotic spindle dynamics may operate through distinct molecular players and be used distinctively in different species. In this regard, the biggest difference was detected in yeast, where microtubule flux is absent and microtubules grow and shrink only at their plus ends (Maddox *et al.*, 1999, 2000). Investigation in *Drosophila* embryos and S2 cells showed that spindle elongation in anaphase B depends on suppression of flux-associated microtubule minus-end depolymerization (Brust-Mascher and Scholey, 2002; de Lartigue *et al.*, 2011; Matos *et al.*, 2009). This technique further helped to reveal the role of flux-associated microtubule slippage at kinetochores, which helps to equally distribute spindle forces along the metaphase plate (Matos *et al.*, 2009).

FSM experiments in *Xenopus* egg extract spindles were also the basis of the "slide-and-cluster" model for spindle assembly in this system, in which it is proposed that overlapping interpolar microtubules in anastral spindles are continuously nucleated near the chromosomes, from where they slide outward and cluster toward the spindle poles, where they eventually depolymerize (Burbank *et al.*, 2006, 2007).

Finally, in addition to the study of microtubule dynamics during mitosis, FSM has also been used to monitor the translocation of microtubules in neurons (reviewed in Waterman-Storer and Danuser, 2002).

As mentioned above, FSM has not been exclusively used to investigate the dynamic properties of microtubules. In fact, the first speckled structures were observed on actin filaments in an *in vitro* study (Sase *et al.*, 1995). Since then, a notable number of experiments were further performed in studies with actin (reviewed in Danuser and Waterman-Storer 2006), microtubule-binding proteins (Bulinski *et al.*, 2001; Perez *et al.*, 1999), and focal adhesion proteins (Adams *et al.*, 2004). The use of FSM was even expanded from single spectral channel to multispectral FSM, providing the ability to monitor simultaneously two speckled structures, such as actin and microtubules (Salmon *et al.*, 2002) or actin and focal adhesion proteins (Danuser and Waterman-Storer, 2006).

2. TECHNICAL CHALLENGES OF FSM

2.1. How to reach low expression of fluorescent proteins for FSM?

In order to achieve a low fraction of fluorophore-labeled subunits of a given structure, which is essential for optimal contrast between speckles and neighboring regions, fluorescently tagged proteins have to be either injected or expressed at very low concentrations. Although it was shown *in vitro* that these optimal conditions could be achieved when a single fluorophore represents one speckle, *in vivo* data suggest that 0.1–0.5% fractions of fluorescently labeled tubulin, corresponding up to seven fluorophores per speckle are ideal for most purposes, since single fluorophores produce extremely dim signals in living cells (Waterman-Storer and Salmon, 1999). However, the use of single fluorophore FSM in the investigation of meiotic spindle architecture in *Xenopus* egg extracts helped to decrease incorrect readouts since several fluorophores per speckle could represent more than a single microtubule and led to the proposal of a "tiled-array" model, where microtubules of different lengths are dynamically cross-linked throughout the spindle (Yang *et al.*, 2007).

Achieving the optimally low levels of fluorescently labeled subunits of the investigated structure is the first technical challenge of performing a successful FSM experiment. This can be accomplished either by titrating the required amount of microinjected fluorophore-conjugated proteins (e.g., X-rhodamine-conjugated tubulin) or by generating stable cell lines expressing very low levels of the respective GFP-tagged subunits. A good way to achieve the latter is by using the leakiness of inducible promoters that control protein expression.

2.2. How to image fluorescent speckles?

The second important technical challenge that has to be overcome in FSM studies is the capacity to perform high-resolution imaging of diffraction-limited regions (~ 0.25 µm), containing a small number of fluorophores (2–10), while avoiding photobleaching and phototoxicity due to the low signal and consequent longer exposure time to excitation light. To achieve this, the use of a very sensitive imaging system, with efficient light collection and a low noise/high quantum efficiency camera, is required. Initial settings for FSM included conventional wide-field epifluorescence light microscopes coupled to cooled charge-coupled-device (CCD) camera for digital imaging (Waterman-Storer and Salmon, 1997). Nowadays, this technique has been further improved by the use of spinning-disk confocal microscopes, as well as total internal reflection fluorescence (TIRF) microscopy, combined or not with structured illumination (Danuser and Waterman-Storer, 2006; Kner *et al.*, 2009).

2.3. How to analyze fluorescent speckles?

Development of image analysis tools for FSM is a third technical challenge in order to retrieve maximal quantitative information for the same experimental settings. In the case of microtubule flux, a convenient way to study it is by kymograph analysis (Waterman-Storer *et al.*, 1998), or its recently developed variations, guided-kymographs and chromo-kymographs, which offer increased readability and measurement reliability over conventional kymographs (Pereira and Maiato, 2010) but with the same (minimal) computational effort. Specialized computational tools for automatic speckle tracking combined with statistical analysis of their dynamics are also significantly improving the quality of FSM experiments (Danuser and Waterman-Storer, 2003; Yang et al., 2007, 2008).

3. Detailed Methodology for FSM in Cultured Cells

3.1. Choice of cells

Because of their numerous advantages, we chose *Drosophila* S2 cultured cells as the experimental model for the investigation of dynamics, architecture, and function of the mitotic spindle in mitosis. These cells contain a low number of chromosomes, and therefore a low number of K-fibers, which makes a complex system, such as the mitotic spindle, much simpler for observation by light microscopy. An additional important advantage is that the *Drosophila* genome is completely sequenced and annotated and has a high degree of conservation (more than 60%) with humans (Adams et al., 2000). Finally, the ease of RNAi together with an increasing collection of stable cell lines expressing fluorescently labeled components of the mitotic apparatus further strengthen *Drosophila* S2 cells as an ideal subject for investigation of mitosis (Moutinho-Pereira et al., 2010).

3.2. Regulation of fluorescent protein expression

To specifically investigate spindle microtubule dynamics and to successfully perform FSM in our lab, we use a *Drosophila* S2 cell line stably expressing GFP-α-tubulin (to label microtubules) and CID-mCherry (to specifically distinguish between kinetochore and nonkinetochore microtubules). In order to achieve the desired low expression of GFP-α-tubulin essential for FSM, this construct was put under control of the copper-inducible metallothionein promoter (pMT vector, Invitrogen). The leakiness of the promoter without induction allows expression of very low levels of fluorescent tubulin. Similar inducible vectors for expression in mammalian systems (e.g., with tetracycline) are also commercially available.

3.3. Live-cell imaging and microscopy setup

Prior to live cell imaging, exponentially growing *Drosophila* S2 cells are plated either into 22 × 22 mm #1.5 (0.17 mm) thickness coverslips coated with 0.25 mg/ml concanavalin A and mounted into modified Rose chambers with

Schneider's medium (Sigma–Aldrich) supplemented with 10% FBS (Pereira et al., 2009) or into 35 mm #1.5 (0.17 mm) thickness glass-bottomed dishes (MatTek) with the same medium and coating conditions. After preparation, chambers are maintained for imaging within an environmental perspex cage with temperature adjusted to 25 °C by an Air-Therm ATX heating control system (World Precision Instruments). As a rule of thumb for optimal FSM of spindle microtubules, one should choose among those cells with very dim, barely detectable GFP-α-tubulin expression through the microscope eyepiece.

As mentioned before, high-quality FSM requires a very sensitive and efficient imaging system. In our lab, we currently use a Yokogawa CSU-X1 spinning-disk confocal head equipped with two laser lines (488 and 561 nm) used for excitation of GFP and mCherry (assembled by Solamere Technology) on a Nikon TE2000U inverted microscope chassis, which allows easy handling of live cell samples. Detection is achieved by an iXon^{EM} + Electron Multiplying CCD camera with 512 × 512 active pixels, 16 μm pixel size, and over 90% maximum quantum efficiency in the required spectral window. The system is controlled by NIS-elements software (Nikon, Japan) and a digital-to-analog converter board (National Instruments, NI PCI-6733). Time-lapse images are collected in a single plane with a time interval of 5 s, using a 100 × 1.4 NA plan apochromatic DIC objective. Exposure time is set to 100 ms for both laser lines, with a typical power of 100 μW for 488 nm and 70 μW for 561 nm laser line. Due to the large pixel size of the EM-CCD camera, we introduced two additional optical elements in our imaging system: a 1.5× optivar that images the sample onto the spinning-disk holes and a 1.41× plan apochromatic imaging lens before the CCD, for an effective pixel size of 0.076 μm. Taken together, these settings allow us to perform FSM of GFP-α-tubulin in S2 cells for more than 10 min, without any evident phototoxicity-induced mitotic arrest, cell death, and/or photobleaching of the sample.

3.4. Comparative performance between wide-field and spinning-disk confocal for FSM

Our previous FSM setup was based on a wide-field microscope instead of a spinning-disk confocal and contained a less sensitive, but smaller pixel Coolsnap HQ2 CCD camera (Ropper Scientific) (Matos et al., 2009; Moutinho-Pereira et al., 2010). With this setup, images from FSM experiments had to be deconvolved afterwards to achieve higher signal-to-noise ratio and exclude out-of-focus information (Matos et al., 2009). Switching to spinning-disk confocal system with an EM-CCD camera enabled us with more sensitive acquisition and allowed a reduction of light exposure, giving us the opportunity to significantly extend our time-lapse recordings (Fig. 7.1). Additionally, by using confocal technology, we are able to increase the imaging quality by focusing a very thin optical slice within a cell,

Fluorescent Speckle Microscopy in Cultured Cells

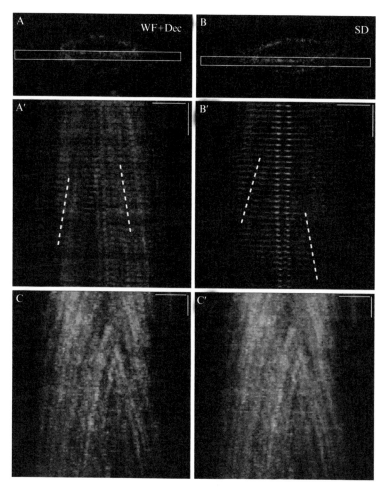

Figure 7.1 Comparative analysis of FSM by wide-field and spinning-disk confocal. (A–A' and B–B') FSM of K-fibers from bioriented chromosomes of *Drosophila* S2 cells stably expressing low levels of GFP-α-tubulin (green) and CID-mCherry (red) obtained, respectively, by wide-field followed by deconvolution (WF + Dec) and spinning-disk confocal (SD). Kymographs show two-dimensional (space × time) representation of selected regions in A and B. Note the poleward flux of tubulin subunits. Horizontal bars: 5 μm; vertical bars: 30 s. Figure A and A' were adapted from Matos et al. (2009). (C–C') Comparison between conventional kymograph (C) and chromokymograph (C') of a metaphase to early anaphase spindle in *Drosophila* S2 cells stably expressing low levels of GFP-α-tubulin. Horizontal bars: 5 μm; vertical bars: 1 min. (See Color Insert.)

which results in decreased phototoxicity and better and more detailed visualization of the processes of interest, due to a higher signal-to-noise ratio and less out-of-focus signals. Moreover, there is no need for time-consuming deconvolution of acquired images.

3.5. Image analysis

Focused on the characterization of speckle poleward motion in metaphase and anaphase spindles, we have put effort on developing algorithms that compensate for the major limitations encountered in conventional kymograph analysis (Pereira and Maiato, 2010). In general, the fact that in kymography a two-dimensional region of interest (ROI) is flattened onto a single dimension (x-axis) implies complete loss of the spatial information along the y-axis. This is the sole but essential limitation of kymography, which can show up in different forms, namely, velocity underestimation and particle degeneration, that is, the indistinguishability of those structures which coalesce in xx, albeit being separated in two-dimensional space (Fig. 7.2). A possible solution could involve the definition of very thin ROIs, a procedure which would naturally exclude structures with a y-axis velocity component and also avoid the structure degeneracy problem. These improvements would be as strong as the ROI is thin—in the limit all information is excluded—evidencing the need for alternatives.

We implement a two-step procedure to generate kymographs from which more reliable information can be extracted. First, instead of a static ROI we use a dynamic one, whose rotation and translation along time is defined by the two-dimensional coordinates of previously tracked references, the choice of these references being based on biological significance. We term this representation a *guided-kymograph*, after which we run a second routine that assigns chromatic information to subslices of the ROI

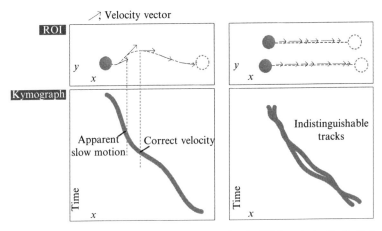

Figure 7.2 Limitations of conventional kymography for FSM analysis. (Left) Only the horizontal projection of the velocity vector contributes to the slope of the kymograph track, resulting in an underestimated velocity. In the case shown, a particle moves with constant velocity modulus (arrow length) but measured in the kymograph as having variable velocity. (Right) Separated particles will be regarded as coalescent in the kymograph if they occupy the same region in x.

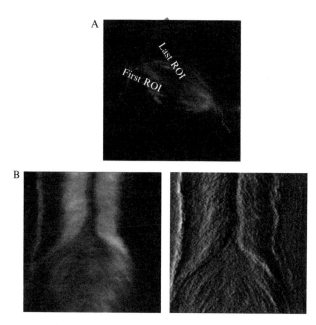

Figure 7.3 Improved kymography tools for FSM analysis. (A) Two-step procedure for guided- and chromo-kymograph generation. In this case, the spindle poles are used as guiding references for the ROI (guided-kymograph). Apart from dynamic, the ROI is structured in the color domain along the y-axis (chromo-kymograph) to allow discrimination of otherwise coalescent structures. (B) Chromo-kymograph of a metaphase–anaphase spindle. This particular example shows that, when tracks are not clearly visible (left), this can be improved by replacing the original images by x-gradient images for construction of the kymograph (right). (For color version of this figure, the reader is referred to the Web version of this chapter.)

in a smooth gradient running along the y-axis (Fig. 7.3). This assignment procedure, which we call a *chromo-kymograph*, results in a flattened (spatially one-dimensional) ROI which, however, retains some two-dimensional spatial information. As in a conventional kymograph, velocity measurements are done by inspection of tracks, with its slope being proportional to the instantaneous velocity. Visual inspection of kymographs to identify tracks is typically facilitated if images are preprocessed with a gradient operation applied in the x-axis (Fig. 7.4). This has two advantages, the first being the fact that speckle-unrelated morphological features of the spindle image, being concentrated along the y-axis, are therefore essentially excluded from the x-gradient processed image. A second advantage is entirely psychophysical and relates to the fact that light–shadow interfaces are more easily distinguishable from background.

Kymography represents a useful strategy to "print" motion. In any case, and even after the improvements of the guided- and chromo-kymograph, not all information is retained—kymography is still a data

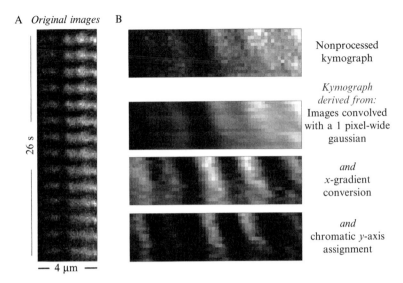

Figure 7.4 Effects of Gaussian convolution, x-gradient processing and chromokymography in FSM analysis. (A) Time-lapse series by FSM of GFP-α-tubulin on a single K-fiber. The respective kinetochore (not represented) is on the left side. (B) The original images in A are used to construct kymographs, where the example on the top is not processed. The first processing step is the convolution with a narrow Gaussian curve to remove information at the pixel scale, which may be assumed to be nonoptical since it lies below the diffraction limit (<100 nm). Then the x-gradient images are generated, which highlight intensity transition areas. Finally, chromatic assignment is performed, which in the case of thin slices is only marginally beneficial because a single structure is being selected. (For color version of this figure, the reader is referred to the Web version of this chapter.)

compression process. One might then ask why not to track individual speckles in sequential frames. Indeed, that is in principle the most natural and efficient strategy but requires sophisticated algorithms that can cope with obstacles such as low and time-variable speckle contrast, speckle coalescence and speckle "birth" and "death." Although difficult to deal with, some dedicated algorithms have been created and successfully used to measure poleward flux. However, the amount of data available to date is limited to *X. laevis* egg extracts and Ptk1 cells (Cameron *et al.*, 2006; Yang *et al.*, 2008), both systems having significantly larger spindles and therefore encompassing a very large mosaic of diffraction spots. The latter aspect also makes large spindles especially suited for the use of cross-correlation and, in general, statistical extraction processes to retrieve dynamic information (Miyamoto *et al.*, 2004). Although kymography is less sophisticated and more operator dependent, its judicious use allows valuable and accurate information to be retrieved.

4. FUTURE CHALLENGES

Since its introduction in 1997, FSM has been significantly improved over time. However, there are still several open directions for future advances to what concerns this powerful technique. Development of more sensitive CCD cameras with a smaller pixel size would enhance even more the detection of very dim signals at higher resolution that is optimal for FSM. At the same time, this would allow decrease of cell's exposure to light and result in increased cell viability, giving the possibility to monitor selected cells for longer time periods. An exciting and quite challenging future aspect would be to develop methods to increase the temporal resolution and increase speckle contrast. Temporal resolution would mean the capability to trigger the FSM regime without preventing the prior conventional imaging. Increased contrast could be achieved by introducing some deterministic variable that promotes clustering of fluorescent molecules along microtubule stubs. Finally, great potential for improvement lies in the design of new algorithms and software for more precise and simpler image analysis and evaluation of collected data.

ACKNOWLEDGMENTS

We would like to thank Gohta Goshima for providing the original pMT-GFP-α-tubulin S2 cell line. Work in the laboratory of H. M. is funded by grants PTDC/SAU-GMG/099704/2008 and PTDC/SAU-ONC/112917/2009 from Fundação para a Ciência e a Tecnologia of Portugal (COMPETE-FEDER), the Human Frontier Research Program, and the seventh framework program grant PRECISE from the European Research Council.

REFERENCES

Adams, M. D., Celniker, S. E., Holt, R. A., Evans, C. A., Gocayne, J. D., Amanatides, P. G., Scherer, S. E., Li, P. W., Hoskins, R. A., Galle, R. F., *et al.* (2000). The genome sequence of Drosophila melanogaster. *Science* **287**, 2185–2195.

Adams, M. C., Matov, A., Yarar, D., Gupton, S. L., Danuser, G., and Waterman-Storer, C. M. (2004). Signal analysis of total internal reflection fluorescent speckle microscopy (TIR-FSM) and wide-field epi-fluorescence FSM of the actin cytoskeleton and focal adhesions in living cells. *J. Microsc.* **216**, 138–152.

Brust-Mascher, I., and Scholey, J. M. (2002). Microtubule flux and sliding in mitotic spindles of Drosophila embryos. *Mol. Biol. Cell* **13**, 3967–3975.

Brust-Mascher, I., Sommi, P., Cheerambathur, D. K., and Scholey, J. M. (2009). Kinesin-5-dependent poleward flux and spindle length control in Drosophila embryo mitosis. *Mol. Biol. Cell* **20**, 1749–1762.

Bulinski, J. C., Odde, D. J., Howell, B. J., Salmon, T. D., and Waterman-Storer, C. M. (2001). Rapid dynamics of the microtubule binding of ensconsin in vivo. *J. Cell Sci.* **114**, 3885–3897.

Burbank, K. S., Groen, A. C., Perlman, Z. E., Fisher, D. S., and Mitchison, T. J. (2006). A new method reveals microtubule minus ends throughout the meiotic spindle. *J. Cell Biol.* **175**, 369–375.

Burbank, K. S., Mitchison, T. J., and Fisher, D. S. (2007). Slide-and-cluster models for spindle assembly. *Curr. Biol.* **17**, 1373–1383.

Cameron, L. A., Yang, G., Cimini, D., Canman, J. C., Kisurina-Evgenieva, O., Khodjakov, A., Danuser, G., and Salmon, E. D. (2006). Kinesin 5-independent poleward flux of kinetochore microtubules in PtK1 cells. *J. Cell Biol.* **173**, 173–179.

Danuser, G., and Waterman-Storer, C. M. (2003). Quantitative fluorescent speckle microscopy: Where it came from and where it is going. *J. Microsc.* **211**, 191–207.

Danuser, G., and Waterman-Storer, C. M. (2006). Quantitative fluorescent speckle microscopy of cytoskeleton dynamics. *Annu. Rev. Biophys. Biomol. Struct.* **35**, 361–387.

de Lartigue, J., Brust-Mascher, I., and Scholey, J. M. (2011). Anaphase B spindle dynamics in Drosophila S2 cells: Comparison with embryo spindles. *Cell Div.* **6**, 8.

Ganem, N. J., Upton, K., and Compton, D. A. (2005). Efficient mitosis in human cells lacking poleward microtubule flux. *Curr. Biol.* **15**, 1827–1832.

Goshima, G., Wollman, R., Stuurman, N., Scholey, J. M., and Vale, R. D. (2005). Length control of the metaphase spindle. *Curr. Biol.* **15**, 1979–1988.

Hayden, J. H., Bowser, S. S., and Rieder, C. L. (1990). Kinetochores capture astral microtubules during chromosome attachment to the mitotic spindle: Direct visualization in live newt lung cells. *J. Cell Biol.* **111**, 1039–1045.

Howard, J., and Hyman, A. A. (2009). Growth, fluctuation and switching at microtubule plus ends. *Nat. Rev. Mol. Cell Biol.* **10**, 569–574.

Keating, T. J., and Borisy, G. G. (2000). Speckle microscopy: When less is more. *Curr. Biol.* **10**, R22–R24.

Kline-Smith, S. L., and Walczak, C. E. (2004). Mitotic spindle assembly and chromosome segregation: Refocusing on microtubule dynamics. *Mol. Cell* **15**, 317–327.

Kner, P., Chhun, B. B., Griffis, E. R., Winoto, L., and Gustafsson, M. G. (2009). Super-resolution video microscopy of live cells by structured illumination. *Nat. Methods* **6**, 339–342.

LaFountain, J. R., Jr., Cohan, C. S., Siegel, A. J., and LaFountain, D. J. (2004). Direct visualization of microtubule flux during metaphase and anaphase in crane-fly spermatocytes. *Mol. Biol. Cell* **15**, 5724–5732.

Lippincott-Schwartz, J., and Patterson, G. H. (2003). Development and use of fluorescent protein markers in living cells. *Science* **300**, 87–91.

Maddox, P., Chin, E., Mallavarapu, A., Yeh, E., Salmon, E. D., and Bloom, K. (1999). Microtubule dynamics from mating through the first zygotic division in the budding yeast Saccharomyces cerevisiae. *J. Cell Biol.* **144**, 977–987.

Maddox, P. S., Bloom, K. S., and Salmon, E. D. (2000). The polarity and dynamics of microtubule assembly in the budding yeast Saccharomyces cerevisiae. *Nat. Cell Biol.* **2**, 36–41.

Maddox, P., Straight, A., Coughlin, P., Mitchison, T. J., and Salmon, E. D. (2003). Direct observation of microtubule dynamics at kinetochores in Xenopus extract spindles: Implications for spindle mechanics. *J. Cell Biol.* **162**, 377–382.

Matos, I., and Maiato, H. (2011). Prevention and correction mechanisms behind anaphase synchrony: Implications for the genesis of aneuploidy. *Cytogenet. Genome Res.* **133**, 243–253.

Matos, I., Pereira, A. J., Lince-Faria, M., Cameron, L. A., Salmon, E. D., and Maiato, H. (2009). Synchronizing chromosome segregation by flux-dependent force equalization at kinetochores. *J. Cell Biol.* **186**, 11–26.

Mitchison, T. J. (1989). Polewards microtubule flux in the mitotic spindle: Evidence from photoactivation of fluorescence. *J. Cell Biol.* **109**, 637–652.

Mitchison, T., and Kirschner, M. (1984). Dynamic instability of microtubule growth. *Nature* **312**, 237–242.

Miyamoto, D. T., Perlman, Z. E., Burbank, K. S., Groen, A. C., and Mitchison, T. J. (2004). The kinesin Eg5 drives poleward microtubule flux in Xenopus laevis egg extract spindles. *J. Cell Biol.* **167,** 813–818.

Moutinho-Pereira, S., Matos, I., and Maiato, H. (2010). Drosophila S2 cells as a model system to investigate mitotic spindle dynamics, architecture, and function. *Methods Cell Biol.* **97,** 243–257.

Pereira, A. J., and Maiato, H. (2010). Improved kymography tools and its applications to mitosis. *Methods* **51,** 214–219.

Pereira, A. J., Matos, I., Lince-Faria, M., and Maiato, H. (2009). Dissecting mitosis with laser microsurgery and RNAi in Drosophila cells. *Methods Mol. Biol.* **545,** 145–164.

Perez, F., Diamantopoulos, G. S., Stalder, R., and Kreis, T. E. (1999). CLIP-170 highlights growing microtubule ends in vivo. *Cell* **96,** 517–527.

Salmon, W. C., Adams, M. C., and Waterman-Storer, C. M. (2002). Dual-wavelength fluorescent speckle microscopy reveals coupling of microtubule and actin movements in migrating cells. *J. Cell Biol.* **158,** 31–37.

Sase, I., Miyata, H., Corrie, J. E., Craik, J. S., and Kinosita, K., Jr. (1995). Real time imaging of single fluorophores on moving actin with an epifluorescence microscope. *Biophys. J.* **69,** 323–328.

Theriot, J. A., and Mitchison, T. J. (1991). Actin microfilament dynamics in locomoting cells. *Nature* **352,** 126–131.

Wadsworth, P., and Salmon, E. D. (1986). Analysis of the treadmilling model during metaphase of mitosis using fluorescence redistribution after photobleaching. *J. Cell Biol.* **102,** 1032–1038.

Wang, Y. L. (1985). Exchange of actin subunits at the leading edge of living fibroblasts: Possible role of treadmilling. *J. Cell Biol.* **101,** 597–602.

Waterman-Storer, C. M., and Danuser, G. (2002). New directions for fluorescent speckle microscopy. *Curr. Biol.* **12,** R633–R640.

Waterman-Storer, C. M., and Salmon, E. D. (1997). Actomyosin-based retrograde flow of microtubules in the lamella of migrating epithelial cells influences microtubule dynamic instability and turnover and is associated with microtubule breakage and treadmilling. *J. Cell Biol.* **139,** 417–434.

Waterman-Storer, C. M., and Salmon, E. D. (1998). How microtubules get fluorescent speckles. *Biophys. J.* **75,** 2059–2069.

Waterman-Storer, C. M., and Salmon, E. D. (1999). Fluorescent speckle microscopy of microtubules: How low can you go? *Faseb J.* **13**(Suppl. 2), S225–S230.

Waterman-Storer, C. M., Desai, A., Bulinski, J. C., and Salmon, E. D. (1998). Fluorescent speckle microscopy, a method to visualize the dynamics of protein assemblies in living cells. *Curr. Biol.* **8,** 1227–1230.

Wollman, R., Cytrynbaum, E. N., Jones, J. T., Meyer, T., Scholey, J. M., and Mogilner, A. (2005). Efficient chromosome capture requires a bias in the 'search-and-capture' process during mitotic-spindle assembly. *Curr. Biol.* **15,** 828–832.

Yang, G., Houghtaling, B. R., Gaetz, J., Liu, J. Z., Danuser, G., and Kapoor, T. M. (2007). Architectural dynamics of the meiotic spindle revealed by single-fluorophore imaging. *Nat. Cell Biol.* **9,** 1233–1242.

Yang, G., Cameron, L. A., Maddox, P. S., Salmon, E. D., and Danuser, G. (2008). Regional variation of microtubule flux reveals microtubule organization in the metaphase meiotic spindle. *J. Cell Biol.* **182,** 631–639.

Zhai, Y., Kronebusch, P. J., and Borisy, G. G. (1995). Kinetochore microtubule dynamics and the metaphase-anaphase transition. *J. Cell Biol.* **131,** 721–734.

CHAPTER EIGHT

Green-to-Red Photoconvertible mEosFP-Aided Live Imaging in Plants

Jaideep Mathur, Sarah Griffiths, Kiah Barton, *and* Martin H. Schattat

Contents

1. Introduction	164
2. Expression of mEosFP Fusion Proteins in Plants	167
2.1. Transient expression	167
2.2. Creation of stable transgenic plants	169
3. Visualization of mEosFP Probes in Plants	170
3.1. Microscopy setup	170
3.2. General protocol for photoconversion	172
3.3. Caveats	172
4. Uses of mEosFP Probes in Plants	173
4.1. mEosFP for tracking organelles	173
4.2. Tracking proteins from one compartment to another	173
4.3. Using EosFP probes for understanding organelle fusion	175
4.4. Color recovery after photoconversion	175
5. Post Acquisition Image Processing and Data Creation	177
Acknowledgments	178
References	179

Abstract

Numerous subcellular-targeted probes have been created using a monomeric green-to-red photoconvertible Eos fluorescent protein for understanding the growth and development of plants. These probes can be used to create color-based differentiation between similar cells, differentially label organelle subpopulations, and track subcellular structures and their interactions. Both green and red fluorescent forms of mEosFP are stable and compatible with single colored FPs. Differential highlighting using mEosFP probes greatly increases spatiotemporal precision during live imaging.

Department of Molecular and Cellular Biology, Laboratory of Plant Development and Interactions, University of Guelph, Guelph, Canada

Methods in Enzymology, Volume 504   © 2012 Elsevier Inc.
ISSN 0076-6879, DOI: 10.1016/B978-0-12-391857-4.00008-2   All rights reserved.

1. INTRODUCTION

Fluorescent proteins (FPs) are essential tools for understanding gene activity, protein localization, and subcellular interactions. Numerous subcellular-targeted FP probes have been created for live imaging of plants at the organ, tissue, cell, subcellular, and suborganelle levels (Mano et al., 2008, 2011; Mathur, 2007; Mohanty et al., 2009; Nelson et al., 2007). The emission spectra of most commonly used FPs span discrete color bands (Shaner et al., 2007). Consequently, all targets highlighted by a particular FP fusion display one color only. Single color labeling becomes a limiting factor when the aim of an experiment is to understand interactions between similar organelles. Limitations of using single color FPs also become apparent when trying to visualize highly localized and transient changes in the organization of dynamic subcellular elements like the cytoskeleton and the endomembrane system. Finally, most live-imaging techniques suffer from the absence of built-in controls in the cells under observation. For most researchers, the decision of when to stop imaging of a cell and avoid artifacts due to photoinduced damage remains an empirical decision that is not based on clear imaging parameters. In most studies of living cells, internal controls for subcellular damage are missing, and it is generally assumed that such effects must be negligible (Mathur et al., 2010). However, as demonstrated recently (Schattat et al., 2011; Sinclair et al., 2009), plant cells respond rapidly and internal indicators are extremely important for minimizing artifacts while studying subcellular interactions.

"Optical highlighters" are recent additions to the FP toolbox. They are broadly categorized as photoactivable, photoswitchable, and photoconvertible (Ai et al., 2006; Shaner et al., 2007; Wiedenmann et al., 2009). These proteins undergo structural changes in response to specific wavelengths that result in their "switching on" to a bright fluorescent state (e.g., photoactivable FPs; Patterson and Lippincott-Schwartz, 2002) or cause a shift in their fluorescence emission wavelength (photoconvertible FPs; Wiedenmann et al., 2004; Gurskaya et al., 2006). Photoswitchable proteins can switch back and forth between two states (Adam et al., 2008). A number of photoconvertible proteins have been produced (Table 8.1). Among them EosFP, a homolog of Kaede derived from *Lobophyllia hemprichii* has been engineered to a monomeric form (mEosFP) without loss in fluorescence and photoconversion properties (Nienhaus et al., 2005; Wiedenmann et al., 2004). It has been used for demonstrating clathrin-dependent endocytosis during internalization of PIN auxin efflux carriers (Dhonukshe et al., 2007), for labeling F-actin (Schenkel et al., 2008) and peroxisomes (Sinclair et al., 2009) in plants. In its unconverted form, mEosFP displays bright green fluorescence, while upon illumination with approximately 390–405 nm light, it changes rapidly into an irreversible red fluorescent form

Table 8.1 Useful photoconvertible probes

Protein	Color	Chromophore	Excitation peak (nm)	Emission peak (nm)	Brightness (mM cm)$^{-1}$	State	Reference
tdEos	Green	HYG	506	516	55	Tandem dimer	Nienhaus et al. (2006)
	Red		569	581	20		
EosFP WT	Green	HYG	506	516	50	Tetramer	Wiedenmann et al. (2004)
	Red		571	581	23		
mEosFP	Green	HYG	505	516	43	Monomer	Wiedenmann et al. (2004)
	Red		569	581	23		
mEosFP2	Green	HYG	506	519	47	Monomer	McKinney et al. (2009)
	Red		573	584	30		
Dendra2	Green	HYG	490	507	23	Monomer	Gurskaya et al. (2006)
	Red		553	573	19		
Kaede	Green	HYG	508	518	87	Tetramer	Ando et al. (2002)
	Red		572	580	20		
mKikGR	Green	HYG	505	515	34	Monomer	Habuchi et al. (2008)
	Red		580	591	18		
KikGR	Green	HYG	507	517	20	Tetramer	Tsutsui et al. (2005)
	Red		583	593	18		
mClavGR2	Green	HYG	488	504	34	Monomer	Hoi et al. (2010)
	Red		566	583	18		
mIrisFP	Green	HYG	486	516	NA	Monomer	Fuchs et al. (2010)
	Red		546	578			
IrisFP	Green	HYG	488	516	NA	Tetramer	Adam et al. (2008)
	Red		551	580			
PS-CFP	Cyan	SYG	400	468	9	Monomer	Chudakov et al. (2004)
	Green		490	511	11		

Information provided in the table is primarily based on Shaner et al. (2007) and Wiedenmann et al. (2004)

Table 8.2 mEosFP-based probes targeted to different subcellular compartments in plants

Name of probe Target compartment	Sequence used for targeting/basic reference
mEosFP-cytosolic cytosol	Nontargeted monomeric EosFP/Wiedenmann et al. (2004)
mEosFP::PIP1 plasma membrane	At3g61430: CDS plasma membrane intrinsic protein 1, ATPIP1/Fetter et al. (2004)
mEosFP:: α−TIP1 vacuolar membrane	At1g73190: CDS alpha tonoplast intrinsic protein/Hunter et al. (2007)
mEosFP::ER mem ER membrane	At5g61790: membrane targeting sequence of calnexin 1/Runions et al. (2006)
Mito-mEosFP mitochondria	First 261 bp of the *N. plumbaginifolia* mitochondrial ATP2-1/Logan and Leaver (2000)
mEosFP-2xFYVE endosomes/PVC	2X-FYVE domain from mouse HGF-regulated tyrosine kinase substrate protein/Voigt et al. (2005)
mEosFP::GONST1 Golgi bodies	At2g13650: CDS GONST 1/Baldwin et al. (2001)
mEosFP::PTS1 peroxisome matrix	C-terminal tripeptide "SKL" (PTS1)/Mathur et al. (2002); Sinclair et al. (2009)
mEosFP::MBD-MAP4 microtubules	Microtubule-binding domain of mammalian MAP-4/Marc et al. (1998)
LIFEACT::mEosFP F-actin	17 aa peptide from yeast Abp140p/Riedl et al. (2008)
mEosFP::FABD-mTn F-actin	F-actin binding domain of mammalian Talin/Kost et al. (1998); Schenkel et al. (2008)

(Table 8.1). EosFP does not mature optimally at 37 °C (McKinney et al., 2009), but this limitation does not pose a major concern for its use in plants. Consequently, a number of mEosFP-based probes targeted to different components and compartments of the plant cell have been created (Mathur et al., 2010; Table 8.2).

2. Expression of mEosFP Fusion Proteins in Plants

2.1. Transient expression

Transient expression of gene constructs provides a fast alternative to study the gene of interest in plant cells without generating stable transgenic lines, which is often much more laborious and time consuming. Depending on the purpose, the plant species under investigation, and available resources, the different transient expression methods for plants include transformation of protoplasts using polyethylene glycol (PEG; Mathur and Koncz, 1997) or electroporation (Bates, 1999), direct microinjection (Miki et al., 1989), biolistic bombardment of gold or tungsten particles coated with DNA (Klein et al., 1987), and Agrobacterium (Kim et al., 2009; Wroblewski et al., 2005; Wydro et al., 2006) as well as virus-mediated gene expression (Scholthof et al., 1996). For studying the behavior of FP fusion constructs, infiltration of plant tissue with *Agrobacterium tumefaciens* and biolistic bombardment are probably the most commonly used techniques. The routinely used method for transiently expressing mEosFP constructs and pertinent notes are as follows.

2.1.1. Agro-infilteration
Infiltration of leaf tissue of tobacco with *A. tumefaciens* does not require special or expensive equipment and results in very efficient transient expression of the introduced transgene.

2.1.1.1. Materials required
2.1.1.1.1. Plants

- About 6-week-old *Nicotiana benthamiana* plants, grown on soil in a short-day light cycle (8 h light/16 h dark; light intensity 80–100 μ mol m^{-2} s^{-1}) at 21 °C during day and 16 °C during night. (*Note*: *N. benthamiana* can also be grown under long-day conditions with higher temperatures but leaf plastids accumulate more starch under these conditions. Although different developmental stages, from seedlings to mature flowering plants can be utilized for infiltration, leaves from young plants generally result in higher transformation rates.)

2.1.1.1.2. Reagents

- *Agrobacterium* Infiltration Media (AIM) consists of 10 mM MgCl$_2$, 5 mM MES, pH 5.6, and 150 µM acetosyringone; for creating 50 ml AIM, mix 0.5 ml MgCl$_2$ (stock 0.5 M), 0.5 ml MES (stock 1 M), pH 5.6, and 7.5 µl acetosyringone (stock 1 M in DMSO). Add H$_2$O to make up the final volume of 50 ml (*Note*: AIM should not be stored at 4 °C for longer than 2 weeks).
- YEB medium liquid or solidified with agar in a Petri dish.

2.1.1.1.3. Material and disposals

- Cork borer, 1-ml needleless syringe, 1-ml pipette, sterile tips, 2-ml reaction tubes, glass culture tubes, gloves, marker pen.

2.1.1.2. Protocol
2.1.1.2.1. Before infilteration

- Use fresh *Agrobacterium* cultures. Incubate over night at 28 °C. (*Note*: *A. tumefaciens* grows significantly slower than *Escherichia coli* and starting a liquid culture from a single colony takes usually 2–3 days. If a 28 °C incubator is not available, *A. tumefaciens* cultures can also be grown at lower temperature (e.g., 22 ° C) but will need longer to reach the same density.)

2.1.1.2.2. Infilteration

- Harvest bacteria by spinning 2 ml of a liquid culture in a table-top micro centrifuge at 10,000 rpm for a minute discarding the supernatant. (*Note*: Bacteria can be harvested directly from a YEB plate too.) Resuspend bacteria in 1.5 ml AIM and incubate for 1–2 h at room temperature.
- Obtain an optical density of 0.8 at 600 nm. (*Note*: The expression level can be influenced by the amount of infiltrated cells, and the optimal OD$_{600nm}$ might have to be evaluated and adjusted for each construct. Several different constructs can be coexpressed by mixing different *A. tumefaciens* cultures.)
- Perform the infiltration with a 1-ml needle-less syringe by gently pressing the syringe on the lower side of the leaf. Exert a counterpressure with a gloved finger on the other side of the leaf. Successful infiltration will be visible as a spreading dark green area. Mark its limits with a marker pen.

2.1.1.2.3. Postinfilteration

- Depending on the maturation time of the expressed protein and expression level, observations can usually be made 48–72 h after infiltration. Punch out a leaf disk by using a cork borer and observe the lower epidermis by epifluorescent or CLSM.

2.1.2. Biolistic bombardment

In comparison to the agro-infilteration method, the method involving coating of gold or tungsten particles with DNA is cumbersome, involves a proprietary biolistic particle delivery system (Bio-Rad PDS-1000/He; http://www.bio-rad.com/) and expensive consumables. The expression of mEosFP probes is usually assessed between 6 and 20 h after bombardment. This is a useful method if chlorophyll autofluorescence is a major impediment to observation since achlorophyllous cells such as those of the onion bulb epidermal layer can be used.

2.1.3. General notes

Great care must be taken in interpreting observations made using transient expression assays, since over a period of 6–90 h, the protein expression levels within a cell rise to a maximum and decline. Ideally, multiple observations spanning several hours should be taken since changes in protein expression levels invariably result in artifacts that may bias conclusion on subcellular localization and behavior. While not limited to mEosFP-based probes, certain endomembrane probes, such as CX-mEosFP (Table 8.2), tend to form aggregates or get sequestered into large brightly fluorescent vesicles. Overexpression of certain membrane-binding proteins such as the PI3P sensor mEosFP-2XFYVE appears to affect normal cellular functioning and generally increases the number of prevacuolar compartments within a cell (Mathur *et al.*, 2010; Vermeer *et al.*, 2006). Moreover, like monomeric GFP, nontargeted cytosolic mEosFP or its fusion with another small protein can diffuse freely in and out of the nucleus. For reasons that are unclear, over time the freely diffusing cytosolic mEosFP can become sequestered within the nucleus to suggest an artifactual nuclear localization pattern. Aggregates of mEosFP frequently appear as bright yellow-orange punctae and are easily visible in the 540–590 nm range even without photoconversion.

2.2. Creation of stable transgenic plants

A large number of plants can be transformed using *Agrobacterium* sps. to create stable transgenic lines. *Arabidopsis thaliana*, the model angiosperm, is easily transformed using the floral dip method (Clough and Bent, 1998). Although the creation of stable transgenics takes longer than transient assays, the availability of multiple lines provide the necessary range of observations that can point to phenotypic aberrations, specific tissue, cell, or subcellular pattern that might be associated with a probe. The availability of normally developing stable transgenic lines lends higher credibility to a particular probe for its use in cell biological observations. In our hands, the process (for *Arabidopsis*) usually requires up to 70 days.

2.2.1. Maintaining plants for experiments

Seeds from stably transformed Arabidopsis lines are grown on 1% agar-gelled Murashige and Skoog (1962) medium, supplemented with 3% sucrose, and with pH adjusted to 5.8. Plants are grown in Petri dishes in a growth chamber maintained at 21 ± 2 °C, and a 16/8 h light/dark regime using cool white light at approximately 80–100 μmol m^{-2} s^{-1}.

3. Visualization of mEosFP Probes in Plants

EosFP can be photoconverted from its green form (mEosFP-G) into a red fluorescent form (mEosFP-R) by a 405-nm wavelength centered violet-blue light. A diode 50 mW 405 nm violet laser is available and provides seamless functional coordination in both confocal and multiphoton microscopy systems under the control of pertinent software. However, for many laboratories, the addition of a 405-nm laser to an existing setup is not economically feasible. In such cases, a hybrid approach is advocated (see *Notes*).

3.1. Microscopy setup

- Any epifluorescent microscope equipped with a digital camera can be used for visualizing and gathering images from mEosFP-based probes. (*Note*: An epiflourescent microscopy setup with a digital camera is sufficient for visualizing and capturing images of EosFP probes. However, a confocal laser scanning microscope (CLSM) with 405, 488, 514, and 543 nm laser lines adds considerably to the clarity of images. CLSMs also provide a high degree of automation during photoconversion, image acquisition, and data processing.)
- Multipinhole iris diaphragm (*Note 2*: The hybrid approach involves carrying out the photoconversion step by using excitation wavelengths from glass filters (DAPI or D) followed by image acquisition using 488 and 543 nm laser scanning mode. In general, the broader bandwidth provided by glass filters is more efficient in carrying out the green-to-red photoconversion as compared to the 405-nm laser which is limited to the excitation peak for EosFP. The obvious limitation with the hybrid approach is the requirement for manually controlling the photoconversion process and thus missing the precision and automation that is possible through software-mediated control of the 405-nm laser. However, photoconversion via software control of laser positioning and scan time is best executed for large ROIs and for creating multiple ROIs in a sample. When dealing with motile approximately 1 µm diameter organelles such as mitochondria, peroxisomes, and Golgi bodies, the time between selection

of a target organelle and its photoconversion is generally long enough to allow the organelle to move away from the selected ROI. Thus, the limitation in creating a small ROI for photoconversion can be overcome partially by modifying the field iris diaphragm to obtain smaller pinholes. While diaphragms on standard microscopes contain aperture sizes distributed between 7 mm and 500 μm, most microscopy companies can create additional apertures upto 50 μm at relatively small cost).
- A high pressure mercury plasma arc-discharge lamp (e.g., Mercury Short ARC #HBO 103 W/2 (OSRAM GmbH, Steinerne Furt 62, 86167 Augsburg, Germany)).

3.1.1. Lens

- Standard lens (10×, 20×, 40×, 63×) for epifluorescent microscopy with the highest available numerical aperture are recommended.
- 40× and 63× water dipping lens recommended for live imaging. (*Note 3*: Ceramic-coated water-immersion lens with a long working distance are very convenient for visualizing living seedlings of *Arabidopsis*. Plants are mounted in deionized water on a depression slide and a coverslip (24 × 60 mm) placed on them while avoiding the air bubble formation. A drop of clean milli-Q water is placed on the coverslip and the lens dipped into it.)

3.1.2. Glass filter cubes

- DAPI/Hoechst/AMCA filter (Ex: 340–380; 400 DCLP; Em: 435–485 nm; Chroma Filter # 31000v2) or a "D" filter (excitation filter: 355–425 nm; dichromatic mirror 455 nm; suppression filter LP 470 nm). (*Note 4*: Photoconversion can be achieved using the DAPI filter commonly available on most epifluorescent microscopes). However, in living plant cells exposure to the UV wavelength from this filter causes a rapid increase in subcellular reactive oxygen species (ROS) and leads to photobleaching of both green and red forms of EosFP. We have found the narrower bandwidth "D" filter in combination with a high pressure mercury plasma arc-discharge lamp provides a violet-blue light optimal for use with plant cells. It minimizes exposure to harmful UV radiation while achieving maximum photoconversion. Depending upon the source and intensity of the light, an exposure time ranging from 2 to 10 s is sufficient for photoconversion.
- An Endow GFP bandpass emission filter (HQ 470/40X; Q 495/LP; HQ 525/50 m; Chroma Filter #41017)
- A TRITC (green) filter (HQ535/30x; Q570LP; HQ620/60 m; e.g., Chroma Filter # 41002c)

3.1.3. Material and disposals

- Glass slides (*Note*: Depression slides maintain plants in a moist condition and are very useful when using water-immersion lens.)
- Coverslips (premium glass 24 × 60 mm recommended for observing living plants)

3.2. General protocol for photoconversion

- Plant sample placed on microscope stage and area for visualization selected using the FITC filter. (*Note*: Plant cells should not be exposed to blue light for long as photobleaching occurs. If chloroplasts in a cell appear orange-yellow, choose another area for observing, as that region is already photodamaged. If a nonchlorophyllous cell looks yellow-green, the cell is displaying autofluorescence that might be due to drying or a mechanical damage. Roots of some plants exhibit green autofluorescence and should be checked beforehand.)
- A region of interest (ROI) brought into focus using appropriate iris aperture setting manually or by selecting appropriate software-mediated controls on CLSM.
- Image acquired using blue (FITC) and green (TRITC) filters on epifluorescent microscope or using 488 and 543 nm laser scanning on a CLSM. Only a green fluorescent image should be visible at this stage.
- Photoconversion for 3–10 s (or ROI scans in CLSM mode).
- Switch to TRITC filter for visualizing photoconverted ROI or on a CLSM switch to 488 and 543 nm laser-scanning mode for obtaining images in green and red channels simultaneously.
- Merge/overlay red and green images to observe photoconverted and non-photoconverted regions.
- Depending upon the experiment, build up image series in xyz, xyt, or $xyzt$ dimensions.

3.3. Caveats

Whereas our knowledge of mEosFP functioning in plants is recent, certain relevant lessons learnt from *Arabidopsis* plants deserve special attention.

3.3.1. Unintentional photoconversion

Transgenic *Arabidopsis* plants expressing cytosolic mEosFP probes under a strong promoter often exhibit an artifact whereby the nuclei in some cells appear bright red even when the plants have not been photoconverted specifically. The probable reason for this artifact might be the presence of a 405-nm peak within the spectrum of white fluorescent lamps commonly used in plant growth chambers (Source: Spectral Power Distributions of

SYLVANIA; Fluorescent Lamps: OSRAM SYLVANIA USA, www.sylvania.com). The red photoconversion observed in the nuclei of old plants exposed to white light clearly differs from the yellow-orange emission observed from vesicles that have sequestered high concentrations of mEosFP fusions.

3.3.2. Partial photoconversion of mEosFP probes

The green form of mEosFP does not undergo photoconversion without violet-blue excitation, while the red photoconverted form does not revert to the green one. Thus, it is possible to have a variable mixture of green and red forms in an ROI. Hues ranging from green to red are created depending upon the exposure time and can be quite confusing if the experiment involves colocalization of proteins. Care must be taken to completely photoconvert mEosFP and ensure that minimal green fluorescence is observed in the ROI.

4. USES OF MEOSFP PROBES IN PLANTS

The mEosFP probes (Table 8.2) are functional at different levels of plant organization and as shown in the following section can be used for a variety of purposes.

4.1. mEosFP for tracking organelles

Depending upon the size of the ROI selected for photoconversion, mEosFP probes can be used for differential color highlighting of tissues (e.g., for following lateral root development), a group of cells (e.g., for observing stomatal patterning in an expanding leaf), single cells (e.g., following changes in trichome or root hair position over time) (Fig. 8.1A), a subpopulation of organelles (e.g., dispersal of mitochondria from one subcellular region to another), a single organelle (e.g., tracking a single plastid during response to changing light conditions; Fig. 8.1B), and even suborganellar regions (e.g., observing changes in the location of nucleoli within the nucleus). The use of mEosFP probes for tracking is most effective when the probe is concentrated in a small organelle or vesicle and when it is not being constantly renewed through fresh protein turnover.

4.2. Tracking proteins from one compartment to another

Many proteins and lipids are moved between subcellular compartments for their modification or achieving a specific biochemical function. When fused to mEosFP, it is possible to photoconvert the proteins that are in the cytosol or sequestered in the ER-lumen and observe their progressive accumulation

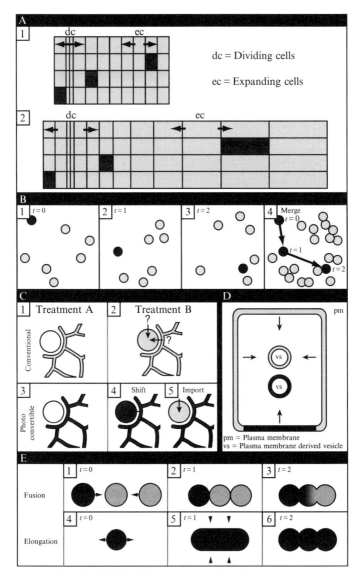

Figure 8.1 Diagrammatic depictions of use of mEosFP-based probes in plants. Fluorescent protein free areas are depicted in white; green fluorescent state is depicted in light gray, and the photoconverted red fluorescent state is shown as dark gray-black. (A) Tracking cells during development—cell division and cell expansion result in neighboring cells being shifted into relatively new positions. Photoconversion of mEosFP makes single cells easily recognizable within a population. The method works best for symplastically isolated cells and cell groups. (B) Tracking single organelles or a subpopulation of organelles—cytoplasmic streaming creates a complex mix of subcellular movements that makes it difficult to track a single organelle over time. Photoconversion using mEosFP (light gray dots in 1–4) creates color differentiation and allows long-term tracking using time-lapse imaging. The track of

in other compartments (Fig. 8.1C). Similarly, Eos-based probes have been used to track photoconverted vesicles from the plasma membrane into other regions of a cell (Dhonukshe *et al.*, 2007; Fig. 8.1D).

4.3. Using EosFP probes for understanding organelle fusion

As depicted in Fig. 8.1E, mEosFP probes can be used to demonstrate fusion between similar organelles (e.g., mitochondria; Arimura *et al.*, 2004) or provide compelling evidence for nonfusion of organelles like peroxisomes during their rapid elongation due to oxidative stress (Sinclair *et al.*, 2009).

4.4. Color recovery after photoconversion

A milder form of the FRAP (fluorescence recovery after photobleaching) technique is possible using mEosFP probes. FRAP is carried out by photobleaching an FP in an ROI with a strong pulse of the excitation wavelength. The recovery in fluorescent intensity increases over time as new fluorescent particles move into the bleached ROI until the prebleached state of fluorescence is reached (Fig. 8.2A). This information is used to calculate the diffusion coefficient for proteins under consideration (Axelrod *et al.*, 1976; Braga *et al.*, 2004). The FRAP procedure assumes that, although the light pulse is strong enough to bleach a fluorophore, it does not damage other molecules surrounding the ROI and in general does not interfere with cellular activity.

By contrast, the irreversible conversion of mEosFP from green to red allows observations on the dispersal of the red form and return of green fluorescence in an ROI after photoconversion. There is no dark

---

photoconverted organelles is recreated by simple maximum projection of a series of time-lapse images (dark gray dots in 1–4). (C) Tracking condition-dependent localization of proteins—many proteins exhibit condition-dependent localization to different subcellular compartments (depicted as treatment A (1, 3) and treatment B (2,4,5)). It is difficult to distinguish using conventional FPs (e.g., GFP; light gray in 1,2) whether the existing protein has shifted between compartments or if newly synthesized protein has been differentially targeted (2). Photoconversion of the protein before application of treatment B allows protein movement to be followed and the gradual localization of red fluorescence in another region. In case of newly synthesized and imported protein, the accumulation of unconverted green mEosFP rather than the red form would be observed. (D) Origin of membrane-derived vesicles—as demonstrated for Pin proteins (Dhonukshe *et al.*, 2007), mEosFP probes can be used to identify the origin of membrane-derived vesicles. Only vesicles derived from photoconverted areas exhibit the red fluorescence and can be tracked to other locations within the cell. (E) Morphological changes in organelles—fusion of organelles (1–3) and rapid organelle elongation (4–6) are frequently observed phenomena in living plant cells. Photoconversion-induced differential coloring of organelles is able to distinguish between fusion and membrane elongation events. Fusion results in organelles acquiring an intermediate color between the green and the red forms of mEosFP.

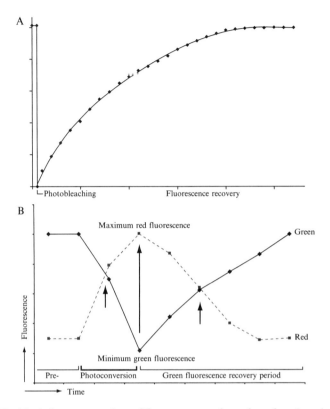

Figure 8.2 Typical representation of fluorescence values plotted against time for (A) fluorescence recovery after photobleaching (FRAP) and (B) color recovery after photoconversion (CRAP). Whereas photobleaching (A) results in a dark state in a region of interest photoconversion (B) brought about through a relatively mild excitation wavelength maintains a visible fluorescent state (green or red) throughout the experiment. The recovery curve for green fluorescence is dependent on the rate of movement of non-photobleached/non-photoconverted proteins from surrounding areas and is similar in both cases.

photobleached state and the red form provides visible proof of normal cellular functioning. Since after photoconversion, the intensity of the green fluorescence drops significantly, while the red form becomes predominant, the intensity of green as it recovers can be used in an identical manner to the way it would be used in a FRAP module. In many cases, FRAP is used in a qualitative manner, to determine whether diffusion is faster or slower compared to a control (Sprague et al., 2004). For this purpose, color recovery of mEosFP after photoconversion can provide comparable recovery graphs with the difference that there is no steep drop to a photobleached dark state (Fig. 8.2A vs. B). Instead, the green fluorescent converts to a red fluorescent state. Both the green and the red form are visible, and the recovery graph from red back to the green state is comparable to the graph obtained through FRAP.

5. POST ACQUISITION IMAGE PROCESSING AND DATA CREATION

The utility of mEosFP-based probes lies in their ability to become differentially colored. A wide variety of software tools are available for discriminating colors in an image. Many of the software that can be used for interpreting an image are available freely (e.g., ImageJ program, <http://rsb.info.nih.gov/ij> and its slightly enhanced distribution package version "Fiji" <http://mac.softpedia.com/get/Graphics/Fiji.shtml>). Most available software programs use the ICC compliant RGB triplet code for true colors as well as HTML-based Web applications code (Cowlishaw, 1985). Alternatively, the color picking tool and RGB value tables form integral components of commonly used digital coloring software such as Canvas and Adobe Photoshop. Table 8.3 lists a few useful plug-ins from the NIH public

Table 8.3 Some useful ImageJ plug-ins for image analysis of mEosFP probes

Plug-in	Description	URL
Color comparison	Color comparison of two 8-bit identically dimensioned gray scale images.	http://rsbweb.nih.gov/ij/plugins/color-comparison.html
RGB profiler	Draws the red, green, and blue profile plot of an image on the same plot, for each type of line selection (profile is refreshed).	http://rsbweb.nih.gov/ij/plugins/rgb-profiler.html
Color histogram	Generates a color histogram of RGB images.	http://rsbweb.nih.gov/ij/plugins/color-histogram.html
RGB measure	Separately measures the red, green, and blue channels of an RGB image.	http://rsbweb.nih.gov/ij/plugins/rgb-measure.html
RGB measure plus	Separately measures the red, green, and blue channels of an RGB image between user-defined threshold levels per channel. Should be combined with the threshold color plug-in.	http://rsbweb.nih.gov/ij/plugins/rgb-measure-plus.html

(Continued)

Table 8.3 (Continued)

Plug-in	Description	URL
Threshold color	Allows thresholding of color RGB images space.	http://www.dentistry.bham.ac.uk/landinig/software/software.html
Color profiler	Provides the same functionality as the Analyze/Plot Profile command but for RGB images.	http://rsbweb.nih.gov/ij/plugins/color-profiler.html
Interactive 3D surface plot	Creates interactive surface plots from all image types. Nonrectangular selections are supported.	http://rsbweb.nih.gov/ij/plugins/surface-plot-3d.html
RGB profile plot	Draws the red, green, and blue profile plot of an RGB image on the same Plot.	Comes with ImageJ
Color inspector	This plug-in shows the color distribution within a 3D-color space.	http://rsbweb.nih.gov/ij/plugins/color-inspector.html

domain funded ImageJ program. These tools allow breakdown of an image into red–green–blue values, creation of line traces, histograms, and 3D renditions that are useful for data presentation. Since protein levels are not really estimated, the programs mainly provide a qualitative comparison of ROIs in an image. However, with proper controls and through fluorescence comparisons with absolute green and red values on a 0- to 255-scale, ratiometric quantification can be achieved.

ACKNOWLEDGMENTS

We thank Joerg Wiedenmann for providing us the Eos FP. J. M. gratefully acknowledges funding from the Natural Sciences and Engineering Research Council of Canada (NSERC), the Canada Foundation for Innovation (CFI), the Ministry of Research and Innovation, Ontario, and the Keefer Trust, University of Guelph.

REFERENCES

Adam, V., Lelimousin, M., Boehme, S., Desfonds, G., Nienhaus, K., Field, M. J., Wiedenmaan, J., McSweeney, S., Nienhaus, G. U., and Bourgeois, D. (2008). Structural characterization of IrisFP, an optical highlighter undergoing multiple photo-induced transformations. *Proc. Natl. Acad. Sci. USA* **105**(47), 18343–18348.

Ai, H., Henderson, J. N., Remington, S. J., and Campbell, R. E. (2006). Directed evolution of a monomeric, bright and photostable version of *Clavularia* cyan fluorescent protein: Structural characterization and applications in fluorescence imaging. *Biochem. J.* **400**, 531–540.

Ando, R., Hama, H., Yamamoto-Hino, M., Mizuno, H., and Miyawaki, A. (2002). An optical marker based on the UV-induced green-to-red photoconversion of a fluorescent protein. *Proc. Natl. Acad. Sci. USA* **99**, 12651–12656.

Arimura, S., Yamamoto, J., Aida, G. P., Nakazono, M., and Tsutsumi, N. (2004). Frequent fusion and fission of plant mitochondria with unequal nucleoid distribution. *Proc. Natl. Acad. Sci. USA* **101**, 7805–7808.

Axelrod, D., Koppel, D. E., Schlessinger, J., Elson, E., and Webb, W. W. (1976). Mobility measurement by analysis of fluorescence photobleaching recovery kinetics. *Biophys. J.* **16**, 1055–1069.

Baldwin, T. C., Handford, M. G., Yuseff, M. I., Orellana, A., and Dupree, P. (2001). Identification and characterization of *GONST1*, a golgi-localized GDP-mannose transporter in Arabidopsis. *Plant Cell* **13**, 2283–2295.

Bates, G. W. (1999). Plant transformation via protoplast electroporation. *Plant Cell Cult. Protoc.* **111**, 359–366.

Braga, J., Desteroo, J. M. P., and Carmo-Fonseca, M. (2004). Intracellular macromolecular mobility measured by fluorescence recovery after photobleaching with confocal laser scanning microscopes. *Mol. Biol. Cell* **15**, 4749–4760.

Chudakov, D. M., Verkhusha, V. V., Staroverov, D. B., Souslova, E. A., Lukyanov, S., and Lukyanov, K. A. (2004). Photoswitchable cyan fluorescent protein for protein tracking. *Nat. Biotechnol.* **22**(11), 1435–1439.

Clough, S. J., and Bent, A. F. (1998). Floral dip: A simplified method for *Agrobacterium*-mediated transformation of *Arabidopsis thaliana*. *Plant J.* **16**, 735–743.

Cowlishaw, M. F. (1985). Fundamental requirements for picture presentation. *Proc. Soc. Inform. Display* **26**, 101–107.

Dhonukshe, P., Aniento, F., Hwang, I., Robinson, D. G., Mravec, J., Stierhof, Y., and Friml, J. (2007). Clathrin-mediated constitutive endocytosis of PIN auxin efflux carriers in *Arabidopsis*. *Curr. Biol.* **17**, 520–527.

Fetter, K., Van Wilder, V., Moshelion, M., and Chaumont, F. (2004). Interactions between plasma membrane aquaporins modulate their water channel activity. *Plant Cell* **16**, 215–228.

Fuchs, J., Böhme, S., Oswald, F., Hedde, P. N., Krause, M., Wiedenmaan, J., and Nienhaus, G. U. (2010). A photoactivatable marker protein for pulse-chase imaging with superresolution. *Nat. Methods* **7**(8), 627–630.

Gurskaya, N. G., Verkhusha, V. V., Shcheglov, A. S., Staroverov, D. B., Chepurnykh, T. V., Fradkov, A. F., Lukyanov, S., and Lukyanov, K. A. (2006). Engineering of a monomeric green-to-red photoactivatable fluorescent protein induced by blue light. *Nat. Biotechnol.* **24**, 461–465.

Habuchi, S., Tsutsui, H., Kochaniak, A. B., Miyawaki, A., and Van Oijen, A. M. (2008). mKikGR, a monomeric photoswitchable florescent protein. *PLoS One* **3**(12), e3394.

Hoi, H., Shaner, N. C., Davidson, M. W., Cairo, C. W., Wang, J., and Campbell, R. E. (2010). A monomeric photoconvertible fluorescent protein for imaging of dynamic protein localization. *J. Mol. Biol.* **410**, 776–791.

Hunter, P. R., Craddock, C. P., Di Benedetto, S., Roberts, L. M., and Frigerio, L. (2007). Fluorescent reporter proteins for the tonoplast and the vacuolar lumen identify a single vacuolar compartment in Arabidopsis cells. *Plant Physiol.* **145,** 1371–1382.

Kim, M. J., Baek, K., and Park, C. M. (2009). Optimization of conditions for transient Agrobacterium-mediated gene expression assays in Arabidopsis. *Plant Cell Rep.* **28,** 1159–1167.

Klein, T. M., Wolf, E. D., Wu, R., and Sanford, J. C. (1987). High-velocity microprojectiles for delivering nucleic-acids into living cells. *Nature* **327,** 70–73.

Kost, B., Spielhofer, P., and Chua, N. H. (1998). A GFP-mouse talin fusion protein labels plant actin filaments in vivo and visualizes the actin cytoskeleton in growing pollen tubes. *Plant J.* **16,** 393–401.

Logan, D. C., and Leaver, C. J. (2000). Mitochondria-targeted GFP highlights the heterogeneity of mitochondrial shape, size and movement within living plant cells. *J. Exp. Bot.* **51,** 865–871.

Mano, S., Miwa, T., Nishikawa, S., Mimura, T., and Nishimura, M. (2008). The plant organelles database (PODB): A collection of visualized plant organelles and protocols for plant organelle research. *Nucleic Acids Res.* **36,** D929–D937.

Mano, S., Miwa, T., Nishikawa, S., Mimura, T., and Nishimura, M. (2011). The Plant Organelles Database 2 (PODB2): An updated resource containing movie data of plant organelle dynamics. *Plant Cell Physiol.* **52,** 244–253.

Marc, J., Granger, C. L., Brincat, J., Fisher, D. D., Kao, T. H., McCubbin, A. G., and Cyr, R. J. (1998). A GFP-MAP4 reporter gene for visualizing cortical microtubule rearrangements in living epidermal cells. *Plant Cell* **10,** 1927–1940.

Mathur, J. (2007). The illuminated plant cell. *Trends Plant Sci.* **12,** 506–513.

Mathur, J., and Koncz, C. (1997). PEG-mediated protoplast transformation with naked DNA. *Methods Mol. Biol.* **82,** 267–276.

Mathur, J., Mathur, N., and Hülskamp, M. (2002). Simultaneous visualization of peroxisomes and cytoskeletal elements reveals actin and not microtubule-based peroxisome motility in plants. *Plant Physiol.* **128,** 1031–1045.

Mathur, J., Radhamony, R., Sinclair, A. M., Donoso, A., Dunn, N., Roach, E., Radford, D., Mohaghegh, P. S., Logan, D. C., Kokolic, K., and Mathur, N. (2010). mEosFP-based green-to-red photoconvertible subcellular probes for plants. *Plant Physiol.* **154**(4), 1573–1587.

McKinney, S. A., Murphy, C. S., Hazelwood, K. L., Davidson, M. W., and Looger, L. L. (2009). A bright and photostable photoconvertible florescent protein for fusion tags. *Nat. Methods* **6**(2), 131–133.

Miki, B., Huang, B., Bird, S., Kemble, R., Simmonds, D., and Keller, W. (1989). A procedure for the microinjection of plant cells and protoplasts. *Methods Cell Sci.* **12,** 139–144.

Mohanty, A., Luo, A., DeBlasio, S., Ling, X., Yang, Y., Tuthill, D. E., Williams, K. E., Hill, D., Zadrozny, T., Chan, A., Sylvester, A. W., and Jackson, D. (2009). Advancing cell biology and functional genomics in maize using fluorescent protein-tagged lines. *Plant Physiol.* **149,** 601–605.

Murashige, T., and Skoog, F. (1962). A revised medium for rapid growth and bio assays with tobacco tissue cultures. *Physiol. Plant.* **15,** 473–497.

Nelson, B. K., Cai, X., and Nebenführ, A. (2007). A multi-color set of in vivo organelle markers for colocalization studies in Arabidopsis and other plants. *Plant J.* **51,** 1126–1136.

Nienhaus, K., Nienhaus, G. U., Wiedenmann, J., and Nar, H. (2005). Structural basis for photo-induced protein cleavage and green-to-red conversion of fluorescent protein EosFP. *Proc. Natl. Acad. Sci. USA* **102,** 9156–9159.

Nienhaus, G. U., Nienhaus, K., Holzle, A., Ivanchenko, S., Red, R., Oswald, F., Wolff, M., Schmitt, F., Rocker, C., Vallone, B., Weidemann, W., Heilker, R., et al. (2006). Photoconvertible fluorescent protein EosFP: Biophysical properties and cell biology applications. *Photochem. Photobiol.* **82,** 351–358.

Patterson, G. H., and Lippincott-Schwartz, J. (2002). A photo-activatable GFP for selective photolabeling of proteins and cells. *Science* **297,** 1873–1877.

Riedl, J., Crevenna, A. H., Kessenbrock, K., Yu, J. H., Neukirchen, D., Bista, M., Bradke, F., Jenne, D., Holak, T. A., Werb, Z., Sixt, M., and Wedlich-Soldner, R. (2008). Lifeact: A versatile marker to visualize F-actin. *Nat. Methods* **5,** 605–607.

Runions, J., Brach, T., Kühner, S., and Hawes, C. (2006). Photoactivation of GFP reveals protein dynamics within the endoplasmic reticulum membrane. *J. Exp. Bot.* **57,** 43–50.

Schattat, M., Barton, K., Baudisch, B., Klösgen, R. B., and Mathur, J. (2011). Plastid stromule branching coincides with contiguous endoplasmic reticulum dynamics. *Plant Physiol.* **155,** 1667–1677.

Schenkel, M., Sinclair, A. M., Johnstone, D., Bewley, J. D., and Mathur, J. (2008). Visualizing the actin cytoskeleton in living plant cells using a photo-convertible mEos:: FABD-mTn fluorescent fusion protein. *Plant Methods* **4,** 21.

Scholthof, H. B., Scholthof, K. B. G., and Jackson, A. O. (1996). Plant virus gene vectors for transient expression of foreign proteins in plants. *Annu. Rev. Phytopathol.* **34,** 299–323.

Shaner, N. C., Patterson, G. H., and Davidson, M. W. (2007). Advances in fluorescent protein technology. *J. Cell Sci.* **120,** 4247–4260.

Sinclair, A. M., Trobacher, C. P., Mathur, N., Greenwood, J. S., and Mathur, J. (2009). Peroxule extension over ER-defined paths constitutes a rapid subcellular response to hydroxyl stress. *Plant J.* **59,** 231–242.

Sprague, B. L., Pego, R. L., Stavreva, D. A., and McNally, J. G. (2004). Analysis of binding reactions by fluorescence recovery after photobleaching. *Biophys. J.* **86,** 3473–3495.

Tsutsui, H., Karasawa, S., Shimizu, H., Nukina, N., and Miyawaki, A. (2005). Semi-rational engineering of a coral fluorescent protein into an efficient highlighter. *EMBO Rep.* **6,** 233–238.

Vermeer, J. E., van Leeuwen, W., Tobeña-Santamaria, R., Laxalt, A. M., Jones, D. R., Divecha, N., Gadella, T. W., Jr., and Munnik, T. (2006). Visualization of PtdIns3P dynamics in living plant cells. *Plant J.* **47,** 687–700.

Voigt, B., Timmers, A. C., Samaj, J., Hlavacka, A., Ueda, T., Preuss, M., Nielsen, E., Mathur, J., Emans, N., Stenmark, H., Nakano, A., Baluska, F., et al. (2005). Actin-based motility of endosomes is linked to the polar tip growth of root hairs. *Eur. J. Cell Biol.* **84,** 609–621.

Wiedenmann, J., Ivanchenko, S., Oswald, F., Schmitt, F., Röcker, C., Salih, A., Spindler, K., and Nienhaus, G. U. (2004). EosFP, a fluorescent marker protein with UV-inducible green-to-red fluorescence conversion. *Proc. Natl. Acad. Sci. USA* **101,** 15905–15910.

Wiedenmann, J., Oswald, F., and Nienhaus, G. U. (2009). Fluorescent proteins for live cell imaging: Opportunities, limitations, and challenges. *IUBMB Life* **61,** 1029–1042.

Wroblewski, T., Tomczak, A., and Michelmore, R. (2005). Optimization of Agrobacterium-mediated transient assays of gene expression in lettuce, tomato and Arabidopsis. *Plant Biotechnol. J.* **3,** 259–273.

Wydro, M., Kozubek, E., and Lehmann, P. (2006). Optimization of transient Agrobacterium-mediated gene expression system in leaves of *Nicotiana benthamiana*. *Acta Biochim. Pol.* **53,** 289–298.

CHAPTER NINE

METHODS FOR CELL AND PARTICLE TRACKING

Erik Meijering, Oleh Dzyubachyk, *and* Ihor Smal

Contents

1. Introduction	184
2. Tracking Approaches	185
2.1. Cell tracking approaches	185
2.2. Particle tracking approaches	186
3. Tracking Tools	187
3.1. Cell tracking tools	188
3.2. Particle tracking tools	192
4. Tracking Measures	192
4.1. Motility measures	192
4.2. Diffusivity measures	193
4.3. Velocity measures	194
4.4. Morphology measures	194
5. Tips and Tricks	195
5.1. Imaging	195
5.2. Tracking	196
5.3. Analysis	197
Acknowledgments	197
References	197

Abstract

Achieving complete understanding of any living thing inevitably requires thorough analysis of both its anatomic and dynamic properties. Live-cell imaging experiments carried out to this end often produce massive amounts of time-lapse image data containing far more information than can be digested by a human observer. Computerized image analysis offers the potential to take full advantage of available data in an efficient and reproducible manner. A recurring task in many experiments is the tracking of large numbers of cells or particles and the analysis of their (morpho)dynamic behavior. In the past decade, many methods have been developed for this purpose, and software tools based on

Biomedical Imaging Group Rotterdam, Departments of Medical Informatics and Radiology, Erasmus MC—University Medical Center Rotterdam, Rotterdam, The Netherlands these are increasingly becoming available. Here, we survey the latest developments in this area and discuss the various computational approaches, software tools, and quantitative measures for tracking and motion analysis of cells and particles in time-lapse microscopy images.

1. INTRODUCTION

A fundamental property of any real-world object is that it extends in both space and time. This is particularly true for living organisms, which, by definition, require the passage of time for their metabolism, growth, reaction to stimuli, and reproduction. Full understanding of any animate entity therefore necessitates studying not only its spatial (anatomic) but also its temporal (dynamic) properties (Tsien, 2003). It is therefore no surprise that research in medicine and biology has come to rely increasingly on time-lapse imaging and longitudinal examinations. In both the health sciences and the life sciences, the technologically deficient times when researchers had to draw conclusions based on static two-dimensional (2D) images are long gone, and it is now commonplace to image and study subjects in three dimensions over time (denoted $3D + t$ or 4D).

Live imaging of dynamic processes at the cellular and molecular levels has been made possible by the development of a vast spectrum of fluorescent proteins and nanocrystals and groundbreaking advances in optical microscopy technology. The resulting increase in the amount, size, dimensionality, and complexity of the image data has brought about new challenges for automated data analysis and management (Peng, 2008; Rittscher, 2010; Swedlow et al., 2009; Vonesch et al., 2006). A topic for which interest has increased exponentially in recent years (Fig. 9.1) is object tracking (Dorn et al., 2008; Jaqaman and Danuser, 2009; Meijering et al., 2006, 2009; Rohr et al., 2010; Zimmer et al., 2006). Indeed, it is practically impossible to manually follow hundreds to thousands of cells or particles through many hundreds to thousands of image frames, and sophisticated computerized methods are very much needed for these tasks.

Although first attempts to automate the tracking of cells or particles by digital image processing date back at least 30 years, the development of more advanced tracking methods really took off in the past decade, and it is only since a couple of years that biology at large is able to reap the fruits of these efforts through the increased availability of software implementations of such methods. The purpose of this chapter is to summarize these developments and to provide hands-on suggestions for practitioners in the field. After a brief description of the main tracking approaches, we highlight freely

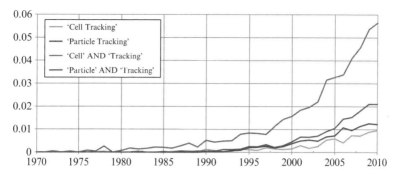

Figure 9.1 Percentage of publications in the PubMed database (National Library of Medicine, National Institutes of Health, Bethesda, MD, USA) as a function of publication year for the indicated combinations of words in the title and/or abstract. The plot shows the exponentially increasing interest in cell and particle tracking in the biomedical (and related) literature. Notice that by plotting percentages (of the total body of literature published in any given year), we have corrected for the intrinsic growth of the number of publications. In other words, the curves indicate a rising "market share" of tracking-related research. (For color version of this figure, the reader is referred to the Web version of this chapter.)

available software tools for cell and particle tracking, discuss frequently used measures to quantify dynamics, and conclude with concrete tips and tricks on various practical aspects.

2. TRACKING APPROACHES

Before discussing tracking tools, it is useful to survey the different methodological approaches on which these may be based. Since the appearance and behavior of cells can be quite different from particles, the image processing techniques developed to track them are usually also quite different and are therefore discussed separately here. In either case, there are generally two sides to the tracking problem: (1) the recognition of relevant objects and their separation from the background in every frame (the segmentation step) and (2) the association of segmented objects from frame to frame and making connections (the linking step).

2.1. Cell tracking approaches

In images where the cells have sufficiently and consistently different intensities than their surroundings, they are most easily segmented by thresholding, which labels pixels above the intensity threshold as "object" and the remainder as "background", after which disconnected regions can be automatically labeled as different cells. In the case of severe noise, autofluorescence, photobleaching (in fluorescence microscopy), very poor contrast, gradients, or halo artifacts (in phase-contrast or differential interference contrast microscopy), thresholding will fail, and more sophisticated segmentation approaches are needed. Popular examples (see Meijering *et al.*, 2008, 2009 for a more elaborate discussion) are template matching (which fits predetermined patches or models to the image data but is robust only if cells have very similar shape), watershed transformation (which completely separates images into regions and delimiting contours but may easily lead to oversegmentation), and deformable models (which exploit both image information and prior shape information).

The simplest approach to solving the subsequent association problem is to link every segmented cell in any given frame to the nearest cell in the next frame, where "nearest" may refer not only to spatial distance but also to difference in intensity, volume, orientation, and other features. This nearest-neighbor solution works well as long as the cells are well separated in at least one of the dimensions of the feature space. Essentially, this criterion also applies to so-called online cell tracking approaches, which alternate between segmentation and linking on a per-frame basis. For instance, template matching, mean-shift processing, or deformable model fitting is applied to one frame, and the found positions or contours are used to initialize the segmentation process in the next frame, and so on, which implicitly solves the linking problem (see Fig. 9.2 for an example result of applying such a scheme to a challenging tracking problem).

2.2. Particle tracking approaches

Individual proteins or other (macro)molecular complexes within cells (collectively referred to as particles) are hardly (if at all) visible in bright field or phase-contrast microscopy and require fluorescent labeling and imaging. Since fluorescent proteins are two orders of magnitude smaller (nanometer range) than the optical resolution of typical microscopes (100 nm or worse), they appear as diffraction-limited spots (foci) in the images. If their contrast to the background is sufficiently large throughout the image, they can be localized to nanometer resolution by intensity thresholding and computing the centroid position of each segmented spot or by fitting a theoretical or experimentally acquired model of the point spread function (Carter *et al.*, 2005; Cheezum *et al.*, 2001). However, in live-cell imaging, the contrast is often poor, and more sophisticated approaches are needed. The results of a recent comparison study (Smal *et al.*, 2010) suggest that better results can be obtained by specialized algorithms from mathematical morphology and supervised (machine-learning) approaches.

Similar to cell tracking, the most straightforward strategy to solve the association problem is to apply local nearest-neighbor linking. However,

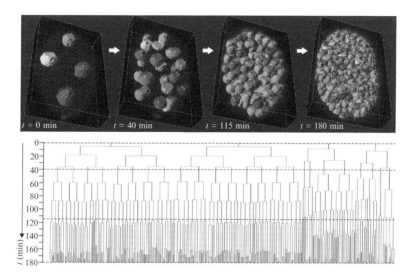

Figure 9.2 Cell tracking and lineage reconstruction for studying embryogenesis. The top row shows four time points of a 3D + t fluorescence microscopy image data set of a developing *Caenorhabditis elegans* embryo, starting from the 4-cell stage until approximately the 350-cell stage, with the segmentation and tracking results (surface renderings with arbitrary colors) overlaid on the raw image data (volume renderings). In this case, a level-set based model evolution approach was used for segmentation and tracking, modified from Dzyubachyk *et al.* (2010a). The bottom graph shows the lineage tree automatically derived from the tracking results, with the horizontal guidelines (red, dashed) corresponding to the four time points. (See Color Insert.)

in the case of particle tracking, the available information to resolve potential ambiguities in the matching process is much more limited (often the particles all have similar appearance). In addition, particles may disappear, (re)appear, split, and merge. More consistent results can be achieved by using global rather than local linking strategies. Examples include spatiotemporal tracing (Bonneau *et al.*, 2005) and graph-based optimization approaches (Jaqaman *et al.*, 2008; Sbalzarini and Koumoutsakos, 2005). Alternatively, various Bayesian estimation approaches have been explored in recent years (Genovesio *et al.*, 2006; Godinez *et al.*, 2009; Smal *et al.*, 2008), with promising results (see Fig. 9.3 for an example).

3. Tracking Tools

Computational approaches to cell and particle tracking as described in the previous section are interesting in their own right but have no value to practitioners in the field unless they are implemented and released in the

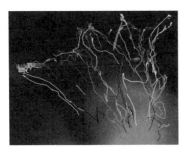 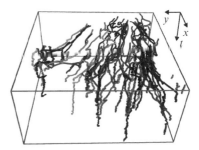

Figure 9.3 Particle tracking for studying vesicle dynamics. The left image shows the last frame of a 2D + t fluorescence microscopy image sequence of vesicles moving in the cytoplasm, with the detection and tracking results overlaid (arbitrarily colored trajectories). The results, adapted from Smal et al. (2008), were obtained using a tracking algorithm based on a Bayesian estimation framework, implemented by particle filtering. On the right, the trajectories are alternatively presented in a spatiotemporal rendering. (For color version of this figure, the reader is referred to the Web version of this chapter.)

form of user-friendly software tools. Fortunately, there is an increasing tendency among computer scientists, spurred by various open source and reproducible research movements, to go the extra mile and develop such tools. Table 9.1 lists 30 currently available tools for cell and/or particle tracking, their main features, and where to find more detailed information about them. Here, we briefly comment on common aspects.

3.1. Cell tracking tools

The general assumption made by most cell tracking tools is that the cells can be modeled as bright regions against a darker background (the fluorescence microscopy scenario). If this is not the case, or if the images are too noisy, it is necessary to apply suitable filters to match this assumption. Most commercial tracking tools (such as Volocity, ImarisTrack, MetaMorph, Image-Pro Plus), as well as more general purpose open-source software packages (CellProfiler, FARSIGHT, ICY, and ImageJ/Fiji), offer ample functionality for image preprocessing.

Virtually all cell tracking tools are capable of tracking multiple cells and allow the user to compute basic dynamics parameters from the resulting trajectories. Few tools (such as StarryNite) are designed specifically for the study of embryogenesis, which requires not only segmentation and tracking of individual cells but also accurate handling of all cell divisions and the reconstruction of the complete cell lineage tree. Several freeware tools (such as AceTree (Boyle et al., 2006; Murray et al., 2006), and ALES (Braun et al., 2003), not listed in the table) are available for the visualization, curation, and analysis of the cell lineages.

Table 9.1 Available tracking tools

Name	Available	Platform	Source	Cell	Particle	Multiple	Dimensions	Automation	Author of reference	Website
Braincells	Free	Win		√			2D	Manual	Gabor Ivancsy	http://pearl.elte.hu/~kyd
CellProfiler	Free	Win/Lin/Mac	√	√		√	2D	Auto	Carpenter et al. (2006)	http://www.cellprofiler.org/
CellTrack	Free	Win	√	√		√	2D	Auto	Sacan et al. (2008)	http://db.cse.ohio-state.edu/CellTrack/
CellTracker	Free	Win		√		√	2D	Semi	Shen et al. (2006)	http://go.warwick.ac.uk/bretschneider/celltracker/
ClusterTrack	Free	Matlab	√		√	√	2D	Auto	Matov et al. (2010)	http://lccb.hms.harvard.edu/software.html
DcellIQ	Free	Matlab	√	√		√	2D	Auto	Li et al. (2010)	http://www.cbi-tmhs.org/Dcelliq/
DIAS	Paid	Win/Mac		√		√	3D	Auto	Wessels et al. (2006)	http://keck.biology.uiowa.edu/
DiaTrack	Paid	Win			√	√	3D	Auto	Semasopht, Switzerland	http://ww.semasopht.com/
DYNAMIK	Free	Matlab	√	√	√	√	2D	Auto	Mosig et al. (2009)	http://www.picb.ac.cn/sysbio/DYNAMIK/
FARSIGHT	Free	Win/Lin/Mac	√	√	√	√	3D	Auto	Bjornsson et al. (2008)	http://www.farsight-toolkit.org/
ICY	Free	Java	√	√	√	√	3D	Auto	de Chaumont et al. (2011)	http://icy.bioimageanalysis.org/
Image-Pro Plus	Paid	Win		√	√	√	3D	Auto	Media Cybernetics, USA	http://www.mediacy.com/index.aspx?page=IPP
ImarisTrack	Paid	Win/Mac		√		√	3D	Auto	Bitplane, Switzerland	http://www.bitplane.com/go/products/imaristrack

(*Continued*)

Table 9.1 (Continued)

Name	Available	Platform	Source	Cell	Particle	Multiple	Dimensions	Automation	Author of reference	Website
LevelSetTracker	Free	Matlab	✓			✓	3D	Auto	Dzyubachyk et al. (2010b)	http://celmia.bigr.nl/
LineageTracker	Free	ImageJ		✓		✓	2D	Auto	Till Bretschneider	http://go.warwick.ac.uk/bretschneider/lineagetracker/
ManualTracking	Free	ImageJ	✓		✓	✓	3D	Manual	Fabrice Cordelières	http://rsb.info.nih.gov/ij/plugins/track/track.thtml
MetaMorph	Paid	Win		✓	✓	✓	3D	Auto	Molecular Devices, USA	http://www.moleculardevices.com/Products/Software.html
Mtrack2	Free	ImageJ	✓		✓	✓	2D	Auto	Nico Stuurman	http://valelab.ucsf.edu/~nico/IJplugins/MTrack2.html
MTrackJ	Free	ImageJ	✓		✓	✓	3D	Manual	Erik Meijering	http://www.imagescience.org/meijering/software/mtrackj/
Octane	Free	ImageJ	✓		✓	✓	2D	Auto	Ji Yu lab	http://www.ccam.uchc.edu/yu/Software.html
Oko-Vision	Paid	Win		✓		✓	2D	Semi	Okolab, Italy	http://www.oko-lab.com/cell_tracking.page
ParticleTracker	Free	ImageJ	✓		✓	✓	3D	Auto	Sbalzarini and Koumoutsakos (2005)	http://weeman.inf.ethz.ch/ParticleTracker/
plusTipTracker	Free	Matlab	✓		✓	✓	2D	Auto	Danuser lab	http://lccb.hms.harvard.edu/software.html

Name	Availability	Platform	Java	ImageJ	Matlab	Win	Lin	Mac	Dimensionality	Automation	Reference	Website
QuimP	Free	ImageJ		√					2D	Auto	Bosgraaf et al. (2009)	http://go.warwick.ac.uk/bretschneider/quimp/
SpotTracker	Free	ImageJ		√					2D	Auto	Sage et al. (2005)	http://www.bigwww.epfl.ch/sage/soft/spottracker
StarryNite	Free	Win/Lin				√	√		3D	Auto	Murray et al. (2006)	http://westerston.gs.washington.edu/
TIKAL	Request	Win/Lin				√	√		3D	Auto	Bacher et al. (2004)	http://ibios.dkfz.de/tbi/
TLA	Free	Matlab			√				2D	Auto	Kestler lab	http://www.informatik.uni-ulm.de/ni/staff/HKestler/tla/
u-track	Free	Matlab			√				2D	Auto	Jaqaman et al. (2008)	http://lccb.hms.harvard.edu/software.html
Volocity	Paid	Win/Mac				√		√	3D	Auto	Perkin Elmer, USA	http://cellularimaging.perkinelmer.com/products/volocity/demo/

The columns indicate (from left to right) the name of the tool, availability (Free = freeware, Paid = paid license code required or available as a paid service only, Request = freely available from the developers on request), the platform on which the tool runs (Java = runs on all platforms with Java installed, ImageJ = plugin for ImageJ and runs on all platforms with Java installed, Lin = distribution for Linux, Mac = distribution for Mac OS X, Matlab = runs on all platforms with Matlab installed, Win = distribution for Microsoft Windows), whether source code is available, whether it was developed primarily for cell tracking or for particle tracking, whether it can track multiple objects, the maximum spatial dimensionality per frame it can handle (2D image or 3D stack), the level of automation (Auto = automatic after initial parameter setting, Manual = requires continuous user input, Semi = partly automatic but requires user input), the author of the tool or a literature reference (with year) describing the tool, and finally the website where to find the tool.

3.2. Particle tracking tools

Similar to cell tracking, most particle tracking tools, too, assume the target objects (foci) to be significantly brighter than the local background, and prefiltering (noise reduction and deconvolution) of the images generally has a positive impact on their performance. In contrast with the mentioned commercial tools, many of which contain functionality for both cell tracking and particle tracking and are stand-alone applications, most freeware particle tracking tools are available either as a plugin of the widely used ImageJ/Fiji image analysis platform (MTrack2, Octane, ParticleTracker, SpotTracker) or as a Matlab module (plusTipTracker, u-track).

While all tracking tools generally perform well if the image data satisfy certain conditions (see the tips and tricks at the end of this chapter), experimental constraints often force these conditions to be violated, as a result of which automated tracking falls short or fails completely. Fixated on full automation, most tools offer very little functionality for manual trajectory inspection, curation, or creation. A tool specifically designed for this purpose is MTrackJ, an ImageJ/Fiji plugin, which at the time of writing has already been used in over a 100 scientific journal publications, testifying that fully automated tracking is still utopian in many situations.

4. TRACKING MEASURES

The direct result of applying tracking tools is a sequence of coordinates indicating the position of each tracked object at each time point. While this is an essential step and a tremendous data reduction, from millions to billions of (mostly irrelevant) pixels to a few (or perhaps a few tens or hundreds of) thousands of coordinate values, by itself this does not lead to new insights. The final step to knowledge is the computation of biologically meaningful quantitative measures from these coordinates. Here, we distinguish four categories of measures, characterizing the motility, diffusivity, velocity, and morphology of the moving objects, respectively.

4.1. Motility measures

The first step toward quantitative analysis is to reconstruct the trajectories of the tracked objects from the measured coordinates. This problem of "connecting the dots" is practically always solved by linear interpolation, resulting in piecewise-linear trajectories, although higher-order interpolation schemes (in particular cubic splines) can be expected to yield (bio)physically more accurate representations. Given the trajectories, a variety of measures can be straightforwardly computed. The most obvious measures of motility

Table 9.2 Quantitative tracking measures commonly found in the literature

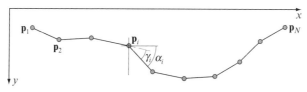

Measure	Definition
Total distance traveled	$d_{tot} = \sum_{i=1}^{N-1} d(\mathbf{p}_i, \mathbf{p}_{i+1})$
Net distance traveled	$d_{net} = d(\mathbf{p}_1, \mathbf{p}_N)$
Maximum distance traveled	$d_{max} = \max_i d(\mathbf{p}_1, \mathbf{p}_i)$
Total trajectory time	$t_{tot} = (N-1)\Delta t$
Confinement ratio	$r_{con} = d_{net}/d_{tot}$
Instantaneous angle	$\alpha_i = \arctan(y_{i+1} - y_i)/(x_{i+1} - x_i)$
Directional change	$\gamma_i = \alpha_i - \alpha_{i-1}$
Instantaneous speed	$v_i = d(\mathbf{p}_i, \mathbf{p}_{i+1})/\Delta t$
Mean curvilinear speed	$\bar{v} = \frac{1}{N-1}\sum_{i=1}^{N-1} v_i$
Mean straight-line speed	$v_{lin} = d_{net}/t_{tot}$
Linearity of forward progression	$r_{lin} = v_{lin}/\bar{v}$
Mean squared displacement	$MSD(n) = \frac{1}{N-n}\sum_{i=1}^{N-n} d^2(\mathbf{p}_i, \mathbf{p}_{i+n})$

The drawing (top) shows a sample trajectory consisting of N points $\mathbf{p}_i = (x_i, y_i)$ and the table (bottom) defines the measures. The example is given for the 2D + t case but the measures can be extended straightforwardly to 3D + t. A constant frame rate is assumed with a time interval of Δt seconds between successive frames. The distance $d(\mathbf{p}_i, \mathbf{p}_j)$ between any two points $\mathbf{p}_i$ and $\mathbf{p}_j$ is usually taken to be the Euclidean norm $||\mathbf{p}_i - \mathbf{p}_j||$.

include (see Table 9.2 for definitions) the total trajectory length (the total distance traveled by the corresponding object), the distance between start and end point (the net distance traveled), the maximum distance to the start (or any other reference) point, and the confinement ratio (also referred to as the meandering index or the straightness index; Beltman et al., 2009). Related measures, but for which varying definitions are found in the literature, include the chemotactic index and the McCutcheon index (Meijering et al., 2008). Other obvious measures to compute are local orientations (with respect to the coordinate system or a reference point), directional change (turning angle; Beltman et al., 2009; Soll, 1995), and the autocorrelation of the latter, which is indicative of directional persistence and process memory.

4.2. Diffusivity measures

A more sophisticated measure computable from a trajectory is the mean squared displacement (MSD). It is a function of time lag (see Table 9.2) and enables one to characterize the mode of motion of the corresponding object by inspection of the resulting MSD-time curve (Qian et al., 1991; Saxton and Jacobson, 1997). In the case of a pure random walk (such as Brownian motion of particles), the curve will be a straight line, given by MSD$(t) = cDt$, with $c = 4$ in 2D and $c = 6$ in 3D, and where D denotes the so-called diffusion coefficient. If the motion is impeded by obstacles, the diffusion is anomalous, and characterized by MSD$(t) = cDt^\alpha$ with $\alpha < 1$. An object whose motion is confined to some region will yield a curve that may be modeled as MSD$(t) = R[1 - a_1 \exp(-a_2 cDt/R)]$, which converges to a maximum value R proportional to the size of the region, where a_1 and a_2 are positive constants related to the geometry of that region. The ultimate case of $R = 0$, corresponding to immobile objects, results in a curve that is zero everywhere. On the other hand, if there is directed motion (flow) in addition to diffusion, the curve behaves as MSD$(t) = cDt + (vt)^2$, where v is the speed. Notice that the MSD is just the second-order moment of displacement. It may be helpful to compute other moment orders as well (Sbalzarini and Koumoutsakos, 2005).

4.3. Velocity measures

Other measures that can be easily derived from a trajectory are those concerning the rate of displacement. Instantaneous velocity, for example, is computed as the displacement from one frame to the next, divided by the time interval (Table 9.2). Notice that this quantity is a vector and its magnitude value is called speed, although in the literature the latter is also often referred to as velocity. A useful measure derived from instantaneous speed is the arrest coefficient (Beltman et al., 2009), defined as the fraction of time that the object is pausing (having a speed less than some minimum value). The mean curvilinear speed is computed as the arithmetic mean of the instantaneous speeds. If the frame rate is constant, that is, if the time elapsed between any two successive frames in the image sequence is the same, this is equal to computing the ratio between the total distance traveled and the total trajectory time. Alternatively, if we use the net distance traveled, the ratio yields the mean straight-line speed. The ratio between the latter and the former speeds is a measure expressing the linearity of forward progression. Rather than taking grand averages, it is often useful to make speed histograms (Bahnson et al., 2005; Qian et al., 1991), as they give more insight into the statistics of the dynamics.

4.4. Morphology measures

In contrast with particle tracking, cell tracking algorithms usually record the entire cell shape at each time point, from which a position estimate is commonly derived by computing the centroid to which, in turn, the above mentioned measures can be applied. Having the full shape as a function of time also allows for the computation of a host of measures characterizing the cell morphology (Bakal *et al.*, 2007; Soll, 1995). A distinction can be made here between measures of size and orientation versus measures of geometric complexity. Examples of the former in the case of 2D + t (and 3D + t) include the perimeter (surface area), area (volume), and the major and minor axes. Measures of (size and orientation invariant) complexity include circularity (sphericity), eccentricity (ellipticity), and convexity or concavity. More sophisticated analysis of morphology is possible by decomposing the shape in terms of Zernike polynomials or based on Fourier analysis, independent component analysis, or principal component analysis, the latter of which appears to be most suitable (see Pincus and Theriot (2007) for a thorough discussion).

5. Tips and Tricks

Concluding this chapter, we complement our discussion of tracking methods, tools, and measures with tips and tricks (including some serious warnings) concerning the imaging, tracking, and analysis. Since research goals, available equipment, and experimental conditions may vary widely, it is impossible to provide detailed protocols here. Nevertheless, the following general remarks should serve as a good basis for further consideration in designing cell and particle tracking experiments.

5.1. Imaging

- *Dimensionality*: The first thing to consider in preparing a time-lapse imaging experiment is whether to study the processes of interest in 2D or in 3D over time. If photobleaching and photodamage are to be minimized, or when dealing with very rapid motion of cells or particles, 3D + t imaging may simply be no option in view of the excess exposure and recording times required. However, biological processes naturally do take place in 3D + t, and it has been shown in various studies that 2D + t imaging and analysis may lead to different results (Meijering *et al.*, 2008). It is therefore important to verify one's assumptions.
- *Image quality*: One of the most critical factors affecting the performance of tracking tools is the signal-to-noise ratio (SNR) of the image data. Several studies (Carter *et al.*, 2005; Cheezum *et al.*, 2001; Smal *et al.*, 2010) have indicated that the accuracy, precision, and robustness of most particle detection and tracking methods drop rapidly for SNR <4. Thus, in order to minimize tracking errors, the illumination settings should be such that at least this SNR is reached. Even though for cell tracking the SNR may be somewhat less critical than for particle tracking (as cells are

much larger and therefore more clearly visible), it is advisable to use a similar minimum level.
- *Frame rate*: Another critical imaging parameter to be tuned carefully is temporal resolution. It is obvious that in the case of directed motion, the lower the frame rate, the larger the distances traveled by the objects between frames, and thus the higher the chance of ambiguities in reconstructing trajectories. For cell tracking, as a rule of thumb, the frame rate should be chosen such that cells move less than their diameter from frame to frame (Zimmer *et al.*, 2006). For particle tracking, it has been suggested (Jaqaman and Danuser, 2009) that in order for the nearest-neighbor linking scheme (used in many tools) to perform well, the ratio (ρ) of the average frame-to-frame displacement and the average nearest-neighbor distance within frames should be $\rho \ll 0.5$.

5.2. Tracking

- *Preprocessing*: In live-cell imaging, the SNR (directly related to light exposure) is often deliberately minimized to avoid photobleaching and photodamage, while it ought to be maximized to ensure high tracking performance. This conflict of requirements may be resolved to some extent by processing the data prior to tracking. Noise reduction filters are widely available and in some tracking tools they are an integral part of the processing pipeline. Sophisticated filters have also been developed to transform transmitted light contrast images into pseudofluorescence images (Xiong and Iglesias, 2010), making them suitable for processing by tracking tools designed for fluorescence microscopy.
- *Tool selection*: As shown in this chapter, quite a number of tools already exist for cell and particle tracking, and it seems likely that more tools will become available in the near future. There is no single criterion to decide which of these is best for a given purpose, but Table 9.1 provides hints where to start looking. Commercial tools usually offer the most user-friendly interfaces and extensive functionality but may be prohibitively expensive. For many tracking and motion analysis tasks, freeware tools are often sufficient, and if source code is available, it is usually not difficult to tailor a tool to one's needs.
- *Verification*: The results of automated tracking are rarely perfect. The lower the SNR or the larger the number of objects, their density, motility, or similarity, the larger the risk of tracking errors. It is therefore advisable to visually inspect (a representative part of) the trajectories prior to analysis and, where necessary, to fix erroneous track initiation, termination, duplication, switching, splitting, and merging events (Beltman *et al.*, 2009). Unfortunately, most tools (especially freeware) lack flexible track-editing functionalities, and it may be helpful to use tools (such as MTrackJ) designed specifically for this purpose.

5.3. Analysis

- *Diffusivity*: Several warnings are in order when estimating diffusion coefficients from MSD-time curves (Meijering *et al.*, 2008). First, the 2D diffusion coefficient (computed from 2D tracking) is equal to the 3D coefficient only in isotropic media, where displacements in the three spatial dimensions are uncorrelated. Second, the shorter the trajectories, the higher the inaccuracy of diffusion estimates. Third, even for long trajectories, the inherent localization uncertainty may cause apparent subdiffusion patterns at short time scales. Finally, a trajectory may contain both diffusive and nondiffusive parts, which are obscured if the MSD is computed over the entire trajectory.
- *Velocity*: The estimation of velocities based on finite differencing of subsequent position estimates implicitly assumes linear motion from frame to frame. It is important to realize that this minimalist approach yields the lowest possible estimate and results in underestimation of the true velocities in cases where the dynamics is more complex. Another warning concerns velocity estimation of migrating cells. Commonly, this is based on the cell centroid position. However, in the case of considerable shape changes, the centroid position may show a much larger fluctuation and is no longer representative.
- *Aggregation*: A final issue to consider is how to aggregate the estimates of a parameter when tracking multiple objects. In principle, there are two approaches (Beltman *et al.*, 2009): object based or frame based, which may lead to different results, depending on the statistic. For example, when tracking objects consisting of two subpopulations (slow and fast moving), a histogram of the per-object mean speeds will reveal this, whereas the per-frame mean speed histogram does not. Conversely, frame-based analyses would allow the detection of different modes of motion (for a single object or a population of synchronized objects), which may go unnoticed with an object-based approach.

ACKNOWLEDGMENTS

The authors gratefully acknowledge financial support from the European Commission in the Seventh-Framework Programme (FP7 grant 201842) and from the Dutch Technology Foundation (STW) in the Smart Optics Systems Programme (SOS grant 10443).

REFERENCES

Bacher, C. P., Reichenzeller, M., Athale, C., Herrmann, H., and Eils, R. (2004). 4-D single particle tracking of synthetic and proteinaceous microspheres reveals preferential movement of nuclear particles along chromatin-poor tracks. *BMC Cell Biol.* **5,** 1–14.

Bahnson, A., Athanassiou, C., Koebler, D., Qian, L., Shun, T., Shields, D., Yu, H., Wang, H., Goff, J., Cheng, T., Houck, R., and Cowsert, L. (2005). Automated measurement of cell motility and proliferation. *BMC Cell Biol.* **6**, 19.

Bakal, C., Aach, J., Church, G., and Perrimon, N. (2007). Quantitative morphological signatures define local signaling networks regulating cell morphology. *Science* **316**, 1753–1756.

Beltman, J. B., Marée, A. F. M., and de Boer, R. J. (2009). Analysing immune cell migration. *Nat. Rev. Immunol.* **9**, 789–798.

Bjornsson, C. S., Lin, G., Al-Kofahi, Y., Narayanaswamy, A., Smith, K. L., Shain, W., and Roysam, B. (2008). Associative image analysis: A method for automated quantification of 3D multi-parameter images of brain tissue. *J. Neurosci. Methods* **170**, 165–178.

Bonneau, S., Dahan, M., and Cohen, L. D. (2005). Single quantum dot tracking based on perceptual grouping using minimal paths in a spatiotemporal volume. *IEEE Trans. Image Process.* **14**, 1384–1395.

Bosgraaf, L., van Haastert, P. J. M., and Bretschneider, T. (2009). Analysis of cell movement by simultaneous quantification of local membrane displacement and fluorescent intensities using Quimp2. *Cell Motil. Cytoskeleton* **66**, 156–165.

Boyle, T., Bao, Z., Murray, J. I., Araya, C. L., and Waterston, R. H. (2006). AceTree: A tool for visual analysis of Caenorhabditis elegans embryogenesis. *BMC Bioinformatics* **7**, 275.

Braun, V., Azevedo, R. B. R., Gumbel, M., Agapow, P. M., Leroi, A. M., and Meinzer, H. P. (2003). ALES: Cell lineage analysis and mapping of developmental events. *Bioinformatics* **19**, 851–858.

Carpenter, A. E., Jones, T. R., Lamprecht, M. R., Clarke, C., Kang, I. H., Friman, O., Guertin, D. A., Chang, J. H., Lindquist, R. A., Moffat, J., Golland, P., and Sabatini, D. M. (2006). CellProfiler: Image analysis software for identifying and quantifying cell phenotypes. *Genome Biol.* **7**, R100.

Carter, B. C., Shubeita, G. T., and Gross, S. P. (2005). Tracking single particles: A user-friendly quantitative evaluation. *Phys. Biol.* **2**, 60–72.

Cheezum, M. K., Walker, W. F., and Guilford, W. H. (2001). Quantitative comparison of algorithms for tracking single fluorescent particles. *Biophys. J.* **81**, 2378–2388.

de Chaumont, F., Dallongeville, S., and Olivo-Marin, J. C. (2011). ICY: A new open-source community image processing software. Proceedings of the IEEE International Symposium on Biomedical Imaging, pp. 234–237.

Dorn, J. F., Danuser, G., and Yang, G. (2008). Computational processing and analysis of dynamic fluorescence image data. *Methods Cell Biol.* **85**, 497–538.

Dzyubachyk, O., van Cappellen, W. A., Essers, J., Niessen, W. J., and Meijering, E. (2010a). Advanced level-set-based cell tracking in time-lapse fluorescence microscopy. *IEEE Trans. Med. Imaging* **29**, 852–867.

Dzyubachyk, O., Essers, J., Baldeyron, C., van Cappellen, W. A., Inagaki, A., Niessen, W. J., and Meijering, E. (2010b). Automated analysis of time-lapse fluorescence microscopy images: From live cell images to intracellular foci. *Bioinformatics* **26**, 2424–2430.

Genovesio, A., Liedl, T., Emiliani, V., Parak, W. J., Coppey-Moisan, M., and Olivo-Marin, J. C. (2006). Multiple particle tracking in 3-D+t microscopy: Method and application to the tracking of endocytosed quantum dots. *IEEE Trans. Image Process.* **15**, 1062–1070.

Godinez, W. J., Lampe, M., Wörz, S., Müller, B., Eils, R., and Rohr, K. (2009). Deterministic and probabilistic approaches for tracking virus particles in time-lapse fluorescence microscopy image sequences. *Med. Image Anal.* **13**, 325–342.

Jaqaman, K., and Danuser, G. (2009). Computational image analysis of cellular dynamics: A case study based on particle tracking. *Cold Spring Harb. Protoc.* **2009**, pdb.top65.

Jaqaman, K., Loerke, D., Mettlen, M., Kuwata, H., Grinstein, S., Schmid, S. L., and Danuser, G. (2008). Robust single-particle tracking in live-cell time-lapse sequences. *Nat. Methods* **5**, 695–702.

Li, F., Zhou, X., Ma, J., and Wong, S. T. C. (2010). Multiple nuclei tracking using integer programming for quantitative cancer cell cycle analysis. *IEEE Trans. Med. Imaging* **29**, 96–105.

Matov, A., Applegate, K., Kumar, P., Thoma, C., Krek, W., Danuser, G., and Wittmann, T. (2010). Analysis of microtubule dynamic instability using a plus-end growth marker. *Nat. Methods* **7**, 761–768.

Meijering, E., Smal, I., and Danuser, G. (2006). Tracking in molecular bioimaging. *IEEE Signal Process. Mag.* **23**, 46–53.

Meijering, E., Smal, I., Dzyubachyk, O., and Olivo-Marin, J. C. (2008). Time-lapse imaging. In "Microscope Image Processing," (Q. Wu, F. A. Merchant, and K. R. Castleman, eds.), pp. 401–440. Academic Press, Burlington, MA. pp. 401–440, Chapter 15.

Meijering, E., Dzyubachyk, O., Smal, I., and van Cappellen, W. A. (2009). Tracking in cell and developmental biology. *Semin. Cell Dev. Biol.* **20**, 894–902.

Mosig, A., Jäger, S., Wang, C., Nath, S., Ersoy, I., Palaniappan, K. P., and Chen, S. S. (2009). Tracking cells in life cell imaging videos using topological alignments. *Algorithms Mol. Biol.* **4**, 10.

Murray, J. I., Bao, Z., Boyle, T. J., and Waterston, R. H. (2006). The lineaging of fluorescently-labeled Caenorhabditis elegans embryos with StarryNite and AceTree. *Nat. Protoc.* **1**, 1468–1476.

Peng, H. (2008). Bioimage informatics: A new area of engineering biology. *Bioinformatics* **24**, 1827–1836.

Pincus, Z., and Theriot, J. A. (2007). Comparison of quantitative methods for cell-shape analysis. *J. Microsc.* **227**, 140–156.

Qian, H., Sheetz, M. P., and Elson, E. L. (1991). Single particle tracking: Analysis of diffusion and flow in two-dimensional systems. *Biophys. J.* **60**, 910–921.

Rittscher, J. (2010). Characterization of biological processes through automated image analysis. *Annu. Rev. Biomed. Eng.* **12**, 315–344.

Rohr, K., Godinez, W. J., Harder, N., Wörz, S., Mattes, J., Tvaruskó, W., and Eils, R. (2010). Tracking and quantitative analysis of dynamic movements of cells and particles. *Cold Spring Harb. Protoc.* **2010**, pdb.top80.

Sacan, A., Ferhatosmanoglu, H., and Coskun, H. (2008). Cell track: An open-source software for cell tracking and motility analysis. *Bioinformatics* **24**, 1647–1649.

Sage, D., Neumann, F. R., Hediger, F., Gasser, S. M., and Unser, M. (2005). Automatic tracking of individual fluorescence particles: Application to the study of chromosome dynamics. *IEEE Trans. Image Process.* **14**, 1372–1383.

Saxton, M. J., and Jacobson, K. (1997). Single-particle tracking: Applications to membrane dynamics. *Annu. Rev. Biophys. Biomol. Struct.* **26**, 373–399.

Sbalzarini, I. F., and Koumoutsakos, P. (2005). Feature point tracking and trajectory analysis for video imaging in cell biology. *J. Struct. Biol.* **151**, 182–195.

Shen, H., Nelson, G., Kennedy, S., Nelson, D., Johnson, J., Spiller, D., White, M. R. H., and Kell, D. B. (2006). Automatic tracking of biological cells and compartments using particle filters and active contours. *Chemometr. Intell. Lab. Syst.* **82**, 276–282.

Smal, I., Meijering, E., Draegestein, K., Galjart, N., Grigoriev, I., Akhmanova, A., van Royen, M. E., Houtsmuller, A. B., and Niessen, W. (2008). Multiple object tracking in molecular bioimaging by Rao-Blackwellized marginal particle filtering. *Med. Image Anal.* **12**, 764–777.

Smal, I., Loog, M., Niessen, W., and Meijering, E. (2010). Quantitative comparison of spot detection methods in fluorescence microscopy. *IEEE Trans. Med. Imaging* **29**, 282–301.

Soll, D. R. (1995). The use of computers in understanding how animal cells crawl. *Int. Rev. Cytol.* **163**, 43–104.

Swedlow, J. R., Goldberg, I. G., Eliceiri, K. W., and OME Consortium (2009). Bioimage informatics for experimental biology. *Annu. Rev. Biophys.* **38,** 327–346.

Tsien, R. Y. (2003). Imagining imaging's future. *Nat. Rev. Mol. Cell Biol.* **4,** S16–S21.

Vonesch, C., Aguet, F., Vonesch, J. L., and Unser, M. (2006). The colored revolution of bioimaging. *IEEE Signal Process. Mag.* **23,** 20–31.

Wessels, D., Kuhl, S., and Soll, D. R. (2006). Application of 2D and 3D DIAS to motion analysis of live cells in transmission and confocal microscopy imaging. *Methods Mol. Biol.* **346,** 261–279.

Xiong, Y., and Iglesias, P. A. (2010). Tools for analyzing cell shape changes during chemotaxis. *Integr. Biol.* **2,** 561–567.

Zimmer, C., Zhang, B., Dufour, A., Thébaud, A., Berlemont, S., Meas-Yedid, V., and Olivo-Marin, J. C. (2006). On the digital trail of mobile cells. *IEEE Signal Process. Mag.* **23,** 54–62.

CHAPTER TEN

Correlative Light-Electron Microscopy: A Potent Tool for the Imaging of Rare or Unique Cellular and Tissue Events and Structures

Alexander A. Mironov *and* Galina V. Beznoussenko

Contents

1. Introduction	202
2. Observation of Living Cells and Fixation	203
2.1. Required materials	203
2.2. Procedure	204
3. Immunolabeling with NANOGOLD	207
3.1. Required materials	207
3.2. Procedure	207
4. Enhancement of Sample Contrast, Sample Locating, and Embedding	208
4.1. Required materials	208
4.2. Procedure	209
5. Identification of the Cell of Interest on EPON Blocks	211
5.1. Required materials	211
5.2. Procedure	212
6. Sample Orientation and EM Sectioning from the Very First Section	212
6.1. Required materials	212
6.2. Procedure	212
7. Picking up Serial Sections with the Empty Slot Grid	214
7.1. Required materials	214
7.2. Procedure	215
8. EM Analysis	216
8.1. Required materials	216
8.2. Procedure	216
Acknowledgments	217
References	218

Istituto FIRC di Oncologia Molecolare, Milan, Italy

Methods in Enzymology, Volume 504　　　　　　　　　　　　© 2012 Elsevier Inc.
ISSN 0076-6879, DOI: 10.1016/B978-0-12-391857-4.00010-0　　　　All rights reserved.

Abstract

In biology, light microscopy (LM) is usually used to study phenomena at a global scale and to look for unique or rare events, and it also provides an opportunity for live imaging, while the forte of electron microscopy (EM) is the high resolution. Observation of living cells under EM is still impossible. Traditionally, LM and EM observations are carried out in different populations of cells/tissues. The advent of true correlative light-electron microscopy (CLEM) has allowed high-resolution imaging by EM of the very same structure observed by LM. This chapter describes imaging with the help of CLEM. The guidelines presented herein enable researchers to analyze structure of organelles and in particular rare events captured by low-resolution imaging of a population or transient events captured by live imaging can now also be studied at high resolution by EM.

1. Introduction

Although advanced optical microscopy techniques can push resolution to 50–100 nm and even below, it is still much less than the resolution of electron microscopy (EM) and far from the resolution needed for the study of, for instance, the organization of assemblies of proteins and lipids in biological specimens. Often, the analysis of immunofluorescently labeled structures needs a better-than-light-microscopy resolution. On the other hand, although the spatial resolution of EM is superior, its field of view is limited: a resolution of 1 nm can only be realized when small ($2 \times 2\ \mu m^2$) areas are imaged. Consequently, the study of rarely occurring events in cells or tissues is extremely tedious and time consuming. This limitation has motivated researchers to embark on the development of imaging methods that combine, for example, light microscopy (LM) and EM, correlative light-electron microscopy (CLEM). CM uses a combination of microscopy methods for the study of rare cellular events of unique samples (Mironov and Beznoussenko, 2009; Polishchuk et al., 2000).

Now CLEM is a rather complex procedure with the possibility to use several techniques for the identification of the organelle of interest subsequently under LM and then under EM. The CLEM procedure includes several stages: (1) observation of the structures labeled with fluorescent protein (FP, i.e., green FP) or other fluorescent markers in living cells; (2) immobilization (fixation or freezing); (3) immuno- or other type of labeling with gold or other markers suitable for LM or directly for EM; (4) embedding; (5) identification of the just examined cell in the resin block or within the frozen sample; (6) sectioning of thin or thick serial sections and identification of the cell on the resin block and cutting of thin or thick serial sections; and (7) EM analysis and structure identification. Each of these steps could be performed by different ways, and all techniques have their own

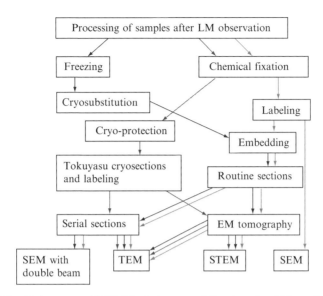

Figure 10.1 Main types of CLEM. Only steps following the LM or fluorescence microscopy (FM) observations are shown. More details see Fig. S1. Green arrows, CLEM based on chemical fixation, immunolabeling, and embedding or scanning EM analysis (Polishchuk et al., 2000). Red arrows, CLEM based on chemical fixation, immunolabeling, and Tokuyasu cryosections (van Rijnsoever et al., 2008). Blue arrows, CLEM based on immunolabeling, quick freezing, cryosubstitution, and epoxy resin embedding (Verkade, 2008). (See Color Insert.)

advantages and disadvantages (Fig. 10.1). Here, we are presenting an example of only one from many existing methods of CVLEM. We describe only those protocols that are indispensable for the presented type of CVLEM. The protocols of transfection, observation under a confocal microscope, and EM tomography can be found in corresponding protocol books.

To achieve success, the researcher should make all procedures with the reproducibility of about 99%. Therefore, we recommend performing all manipulations with grids under stereomicroscope. During manipulations with sections and samples, it is better to use self-closing tweezers and slot grids that are previously tested under the stereomicroscope.

2. Observation of Living Cells and Fixation

2.1. Required materials

2.1.1. Cells of interest, that is, HeLa cells (no. CCL 185; American Type Culture Collection, Rockville, MD), cDNA, and transfection reagents

2.1.2. Dulbecco's modified Eagle's medium (DMEM) supplemented with 10% fetal calf serum (FCS) and 2 mM L-glutamine (GIBCO BRL, Life Technologies)
2.1.3. MatTek Petri dish with CELLocate (MatTek, Ashland, MA)
2.1.4. A digitalized fluorescent-inverted microscope or laser scanning confocal microscope (i.e., from Leica Microsystems Spa, Milan, Italy, www.leica-microsystems.com)
2.1.5. HEPES buffer (0.2 M). Dissolve 4.77 g HEPES in 100 ml distilled water and add of 1N HCl droplet by droplet to provide a pH of about 7.2–7.3
2.1.6. Fixative 1. 0.05% glutaraldehyde plus 4% formaldehyde in 0.15 M HEPES (pH 7.2–7.3). Dissolve 8 g paraformaldehyde powder in 50 ml of 0.2 M HEPES buffer, stirring and heating the solution to 60 °C. Add drops of 1N NaOH to clarify the solution. Add 1.25 ml of 8% glutaraldehyde and 50 ml of 0.2 M HEPES buffer. Dilute twice before use
2.1.7. Fixative 2. 4% formaldehyde in 0.15 M HEPES (pH 7.2–7.3). Dissolve 4 g paraformaldehyde powder in 100 ml HEPES buffer, stirring and heating the solution to 60 °C. Add drops of 1N NaOH to clarify the solution

2.2. Procedure

2.2.1. Grow, that is, HeLa cells in DMEM supplemented with 10% FCS and 2 mM L-glutamine at 37 °C and 5% CO_2.
2.2.2. Suspend HeLa cells using standard procedures and plate cells for CVLEM on a MatTek Petri dish with the CELLocate coverslip attached to its bottom. The CELLocate coverslip contains an etched grid with coordinates that allow the localization of the cell of interest at any step in the preparation.
2.2.3. Transfect HeLa cells with cDNA of the GFP fusion protein using any standard method of transfection or microinjection of cDNA into the nucleus.
2.2.4. By 6–48 h (depending of the fusion protein and method of transfection used) after transfection, place the dish under an inverted fluorescence microscope or laser scanning confocal microscope.
2.2.5. Select the transfected cell of interest, and identify its position related to the coordinates of the CELLocate grid.
2.2.6. Draw (or photograph) the position of the cell on the map of the CELLocate grid. For instance, the cell could be near the cross of horizontal line A and vertical line 3 (see arrow in Figs. 10.2A and 10.3A). Then the figure will be visible after polymerization of EPON as the replica of the roof on the coverslip.

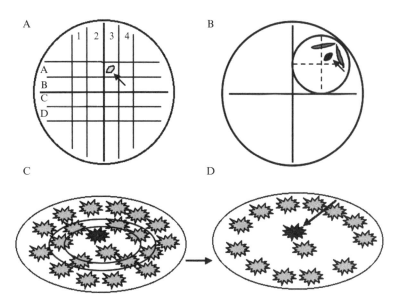

Figure 10.2 Methods of identification of the cell of interest. (A) Growing of cells on the coverslip with gridded coverslip. (B) Growing of cells on the coverslip without coordinated grid. It is possible to use the peculiar pattern of cell position for the identification of our cell. For instance in the upper-right-upper right position, there are two elongated cells (gray elongated profiles), and the cell of interest (dark profile) could be identified between them. (C, D) Growing of cells on the coverslip without coordinated grid. To label the cell of interest after fixation, one should take a wooden stick and make a ring (between two ovals) without cells by their scrapping around our cell (gray asterisk). After scrapping, the cell of interest (arrow in D) is easily identified.

Another way to map the cell of interest is to scrap cells around the cell of interest using wood stick and then cells remaining in the center of ring could be visible (Fig. 10.2C and D), Finally, the pattern of cell culture could be used for the labeling of the cell position (Fig. 10.2B).

2.2.7. Observe the dynamics of the GFP-labeled structures in the selected living cell using a multiphoton-, a laser scanning confocal-, or a digitalized fluorescent-inverted microscope, which allows the grabbing of a time-lapse series of images by a computer.

2.2.8. At the moment of interest, add fixative A to the cell culture medium while still grabbing images (fixative A: medium volume ratio is 1:1). Fixation usually induces the fast fading of GFP fluorescence and blocks the motion of labeled structures in the cell.

2.2.9. Stop grabbing time-lapse images and keep the cells in fixative for 5–10 min (during this time it is useful to grab a Z-series of images of the cell).

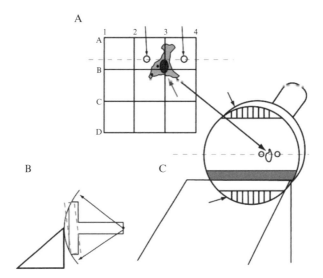

Figure 10.3 Orientation of samples for sectioning. (A) The sample map after embedding. The cell of interest (gray arrow) is located at the position B3. Two black arrows indicate the marks that should be done by the steel needle. (B) Orientation of the sample surface along the vertical direction. Arc shows the trajectory of the knife related to the sample. Arrows indicate the limits of movement of the holder. If the orientation of the sample is vertical, the ends of the blocks should be equally cut by the knife (triangle on the left) whereas the central zone of the sample where the cell is situated will not be trimmed. If the orientation of block surface is not vertical (showed by gray dashed parallelogram), the upper end of the sample will be trimmed first and more intensively. (C) The surface of the sample should be parallel to the glass knife that is used for trimming. In order to measure the verticality, one should measure the thickness of the shadow (gray area below the cell of interest: long arrow). This shadow should be wider when the cell of interest and marks of the needle are situated at the level of the edge of the knife and equally narrower when the marks and the cell are upper or lower than the edge of the knife. Hatching and arrows indicate the ends of the samples already cut by the glass knife. The position of the cell is shown by long black arrow. The distances between the position of the cell and the beginnings of the sectioned zones (horizontal lines bordering the hatching zones) should be equal.

2.2.10. Replace the mixture with the fixative 1 and keep the cells in the fixative 1 for 5 min.
2.2.11. Replace fixative 1 with fixative 2 and keep cells there for 30 min.
2.2.12. Wash with 0.2 M HEPES (pH 7.2–7.3) for 10 min. For subsequent immunolabeling, cells should be treated in the culture dish without removing the cover glass.

After step 2.2.12, one could apply immunolabeling for several antigens (see corresponding chapters in this book) and then go to the labeling for EM.

3. IMMUNOLABELING WITH NANOGOLD

3.1. Required materials

3.1.1. See item 2.1.4
3.1.2. Blocking solution. Dissolve 0.50 g BSA, 0.10 g saponin, 0.27 g NH_4Cl in 100 ml of 0.2 M HEPES (pH 7.2–7.3)
3.1.3. NANOGOLD-conjugated monovalent Fab fragments (Nanoprobes, Incorporated, Yaphank, NY)
3.1.4. Gold-enhance mixture. Use a gold-enhance kit (Nanoprobes, Incorporated, Yaphank, NY). Using equal amounts of the four components (Solutions A, B, C, and D), prepare about 200 μl of reagent per Petri dish (a convenient method is to use an equal number of drops from each bottle). First mix Solution A (enhancer; green cap) and Solution B (activator; yellow cap). Wait for 5 min, and then add Solution C (initiator; purple cap), and finally Solution D (buffer; white cap). Mix well
3.1.5. 1% Glutaraldehyde in 0.2 M HEPES buffer (pH 7.2–7.3). Mix 1 ml of 50% glutaraldehyde (EM grade; Electron Microscopy Sciences, Hatfield, PA) with 49 ml of 0.2 M HEPES (pH 7.2–7.3). Store at 0–4 °C

3.2. Procedure

3.2.1. Wash the fixed cells for 3 × 5 min with 0.2 M HEPES (pH 7.2–7.3).
3.2.2. Incubate the cells with the blocking solution for 30 min, and then with the primary antibodies diluted in blocking solution overnight.
3.2.3. Wash the cells for 6 × 2 min with 0.2 M HEPES (pH 7.2–7.3).
3.2.4. Dilute the NANOGOLD-conjugated Fab fragments of the secondary antibodies 50 times in the blocking solution and add this to the cells; incubate for 2 h.
3.2.5. Wash the cells again for 6 × 2 min with 0.2 M HEPES (pH 7.2–7.3).
3.2.6. Fix the cells with 1% glutaraldehyde in 0.2 M HEPES buffer (pH 7.2) for 5 min.
3.2.7. Wash the cells for 3 × 5 min with 0.2 M HEPES (pH 7.2–7.3), and then for 3 × 5 min in distilled water.
3.2.8. Incubate the cells with the gold-enhancement mixture for 6–10 min according to the manufacturer's instruction. The cells will become violet–gray in color if the gold enhancement is successful.
3.2.9. Wash the cells for 3 × 5 min with distilled water and subject to contrasting (see item 4).

4. ENHANCEMENT OF SAMPLE CONTRAST, SAMPLE LOCATING, AND EMBEDDING

If you plan to use EM tomography, it could be helpful to avoid the use of lead citrate staining (see corresponding chapters in EM book) that often gives unequal staining of the thick sections. For this purpose, one could use thiocarbohydrazide staining after fixation with aldehyde.

4.1. Required materials

4.1.1. 0.2 M Cacodylate buffer. Dissolve 2.12 g sodium cacodylate (Electron Microscopy Sciences, Hatfield, PA) in 100 ml of distilled water. Add 1N HCl to provide a pH of about 6.9

4.1.2. Reduced OsO_4. Mix 2% OsO_4 in water with 3% potassium ferrocyanide in 0.2 M cacodylate buffer (pH 7.2) 1:1

4.1.3. 0.3% Thiocarbohydrazide. Dissolve 0.1 g of thiocarbohydrazide in 33.3 ml of 0.2 M cacodylate buffer (pH 6.9). Use immediately after preparation

4.1.4. 1% OsO_4. Mix 2 ml of 4% OsO_4 (Electron Microscopy Sciences, Hatfield, PA) with 6 ml of 0.2 M cacodylate buffer (pH 6.9). Use immediately after preparation

4.1.5. 1% Tannic acid. Dissolve 0.1 g tannic acid (TAAB Laboratories Equipment Ltd) in 0.05 M sodium cacodylate buffer (pH 6.9) prepared by dilution of one volume of 0.2 M cacodylate buffer (pH 6.9) in three volumes of distilled or deionized water

4.1.6. Ethanol. To prepare N% ethanol, mix N ml of 100% ethanol (Electron Microscopy Sciences, Hatfield, PA) with 100-N ml of deionized water

4.1.7. EPON mixture (TAAB 812, DDSA, MNA, and DMP30 in a ratio of 24:9.5:16.5:1, respectively). Put 20.0 g EPON, 13.0 g dodecenyl succinic anhydride (DDSA), and 11.5 g methyl nadic anhydride (MNA) into the same test tube. Heat the tube in the oven for 2–3 min at 60 °C and then vortex it well. Add 0.9 g tri-dimethylaminomethyl phenol (DMP30; all from Electron Microscopy Sciences, Fort Washington, PA) and immediately vortex the tube again. It is possible to freeze the EPON in aliquots and to store it for a long time at 20 °C before use

4.1.8. EPON/100% ethanol mixture (1:1). Mix 10 ml of 100% ethanol with 10 ml of EPON mixture

4.2. Procedure

4.2.1. Wash the cells three times with 0.2 M cacodylate buffer (pH 6.9).
4.2.2. Treat the cells with the reduced osmium for 1 h on ice in the darkness.
4.2.3. Wash the cells with 0.2 M cacodylate buffer (pH 6.9) 3 × 5 min.
4.2.4. Treat the samples with 0.3% thiocarbohydrazide in 0.2 M cacodylate buffer (pH 6.9) for 10 min.
4.2.5. Wash the cells three times with 0.2 M cacodylate buffer (pH 6.9).
4.2.6. Treat the cells with 1% OsO_4 in 0.2 M cacodylate buffer (pH 6.9) for 30 min. (To further enhance contrast, one can add three additional steps between steps number 4.2.6 and 4.2.7).
4.2.7. Rinse the samples with 0.1 M sodium cacodylate (pH 6.9) buffer until all traces of the yellow osmium fixative have been removed.
4.2.8. Stain the sample with a 1% solution of tannic acid in 0.05 M sodium cacodylate buffer for 45 min.
4.2.9. Replace the tannic acid with a 1% solution of anhydrous sodium sulfate in 0.05 M sodium cacodylate buffer and leave for 5 min.
4.2.10. Rinse the sample in deionized water and then dehydrate stepwise with 70%, 90%, and 100% ethanol (3 × 5 min for each) for the dehydration of the specimens.
4.2.11. Prepare EPON solution as described in Section 4.1.7.
4.2.12. Prepare a 1:1 EPON/100% ethanol mixture (acetone and propylene oxide could dissolve plastic Petri dishes). Vortex until thoroughly mixed and store all solutions in the fume hood.
4.2.13. Treat the sample with a 1:1 EPON/100% ethanol mixture for 1 h.
4.2.14. Replace the 1:1 EPON/100% ethanol mixture with EPON alone for 4 h with one exchange of EPON after 2 h.
4.2.15. Keep the cells in EPON for 1–2 h at room temperature, and then leave them in an oven at 60 °C for 12 h.
4.2.16. After 12 h of polymerization of the EPON, place a small droplet of a fresh resin on the site where the examined cell is located, and insert a resin cylinder (prepared before by polymerization of the resin in a cylindrical mold) with a flat lower surface; leave the samples for an additional 18 h in the oven at 60 °C. The resin cylinder should be placed on the semi-polymerized EPON just over the cell of interest. The central position of the cylinder is important to ensure the possibility of using the transmission light for the examination of EPON samples. If the wall of the cylinder is situated near the cell of interest, the light will be scattered.
4.2.17. Polymerize for 24 h at 60 °C.
4.2.18. Carefully pick up the resin from the Petri dish and glass; this is easy to do by gentle bending of the resin cylinder to and fro. The resin

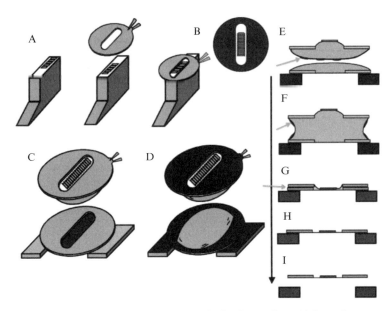

Figure 10.4 Picking up of serial sections with the donor slot grid from the water. (A) Left. Serial sections in the glass bath. Center. Approaching of the donor grid. Right. Orientation of the slot grid related to the position of serial sections within the surface of the bath. (B) Correct positioning of the grid and serial sections. (C) Touching of the water with the empty slot grid leads to the capturing of sections together with the droplet of water by the transfer grid. The acceptor slot grid with carbon-coated formvar film should be placed on the scotch holders (the parallel holders covered with scotch). (D) In order to avoid the hit of dirt on the sections, it is useful to place a very small droplet of distilled water on the acceptor grid that is covered with the formvar film. (E, F). After placement of the transfer grid with sections and droplet of water on the acceptor grid covered by a small droplet of water as well the dirt appeared on the lateral surface of the droplet and after elimination of water with a filter paper sections appeared being attached to the formvar (G). (H) The transfer grid should be eliminated. (I) The acceptor grid should detached from the scotch holder. Green arrows in E–G show dirt that is eliminated from the section during this manipulation being shifted to the lateral surface of grids. (See Color Insert.)

block and the empty MatTek Petri dish after block detachment could be separated easily. If the cover glass with a coordinated grid cannot be detached from the cells included into the resin, the latter should be placed into commercially available hydrofluoric acid (do not use glassware for this) for 30–60 min (Fig. 10.5A and B).

4.2.19. Remove the glass bottom of tissue culture dishes by hydrofluoric acid for 30–60 min.
4.2.20. Control the completeness of the glass dissolution under a stereomicroscope.
4.2.21. Wash the samples in water after the complete removal of the glass.

Correlative Light-Electron Microscopy

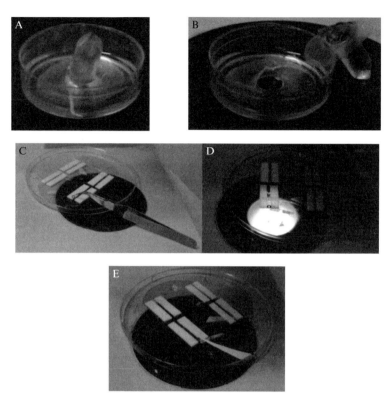

Figure 10.5 Processing of samples and grids. (A, B) Detachment of block from the Petri dish. (C, D) Self-made parallel holders for grids covered with scotch tape. (E) Removal of water from the space between the donor and acceptor grids (see Fig. 10.4F).

4.2.22. Leave the samples in 0.2 M HEPES buffer (pH 7.3) for 60 min to neutralize the hydrofluoric acid.

4.2.23. Wash the samples in water and allow them to dry. The round basement of the sample with diameter of about 1 cm or more should not be cut.

5. IDENTIFICATION OF THE CELL OF INTEREST ON EPON BLOCKS

5.1. Required materials

5.1.1. A stereomicroscope (Leica Microsystems Spa, Milan, Italy, www.leica-microsystems.com) that is equipped with a fiber optic illuminator

5.2. Procedure

5.2.1. Place the embedded sample under a stereomicroscope using a fiber optic illuminator.

5.2.2. If the gridded coverslip was used, the furrows on the coverslips are filled with EPON, and after polymerization and dissolution of the glass or its detachment, the furrows appeared as combs because EPON forms a replica from the gridded glass. The relief of grid is visible under a stereomicroscope. Using numbers and letters on the grid, it is easy to find the position on the surface of the sample where the cells are situated.

5.2.3. Use the composite of bright field DIC images or manual drawings of the region of interest to locate the region of interest on the EPON block.

6. SAMPLE ORIENTATION AND EM SECTIONING FROM THE VERY FIRST SECTION

This method is applicable for CLEM even if cells were looked at as CVLEM. If you use EM tomography, the thickness of EM sections should be 200–250 nm. If you use routine serial sections, the thickness could be from 40 to 80 nm. Thinner sections demand more grids necessary for the picking up of serial section. Thus, one should find a compromise between the difficulty, cost, and demands.

6.1. Required materials

6.1.1. Ultramicrotome Leica EM FC6 or 7 (Leica Microsystems Spa, Milan, Italy, www.leica-microsystems.com). Other ultramicrotomes are not convenient

6.2. Procedure

6.2.1. Place the sample under the transmission light microscope.

6.2.2. Find the cell of interest among the cells within the sample according to the coordinated grid (Fig. 10.2A) or pattern of the cell layer (Fig. 10.2B).

6.2.3. Put the resin block into the holder of an ultratome and examine it under a stereomicroscope.

6.2.4. Place the holder with the sample under the stereomicroscope, and using a steel needle and rotating the sample in the holder, make two small cavities in such a way that they form a horizontal line (broken line in Fig. 10.3A and C) with our cell appearing in the center of the sample.

6.2.5. Introduce the holder into the ultratome in such a way that the segment arc of the ultratome is in the vertical position and the two cavities form a horizontal line.

6.2.6. By rotating the glass knife stage, align the bottom edge of the pyramid parallel to the knife-edge. Using the segment arc, orient the plane of the sample vertically.

6.2.7. Bring the sample as close as possible toward the glass knife.

6.2.8. Adjust the gap (which is visible as a bright band if all three of the lamps of an ultratome are switched on) between the knife-edge and the surface of the sample. The gap has to be identical in width between the most upper and lower edges of the sample during the up and down movement of the resin block. This ascertains that every point of the sample surface containing the cell of interest is at the same distance from the knife-edge.

6.2.9. Slowly moving the sample up and down, continue its approach until the knife begins to cut one of the edges of the sample. The sectioning begins from either the upper or the lower part of the sample, the middle part of the sample where the cell of interest is situated will be unaffected because the length of the radius passing through the cell is shorter than the radii passing through the upper and the lower edges of the sample.

6.2.10. If the sectioning is to begin from the upper part of the sample, tilt the segment arc to approach the lower edge toward the knife. If the sectioning is to begin from the lower edge of the sample, tilt the segment arc and approach the upper edge of the sample toward the knife. A vertically oriented sample should produce equal sections from both the upper and lower edges of the samples (Fig. 10.3C).

6.2.11. Note down precisely all of parameters relating to the position of the sample in the ultratome, that is, the degree of rotation of the sample in the holder, the degree of tilting of the segment arc, and the degree of rotation of the knife in its stage. Do not take the sample from the holder and do not rotate the sample inside the holder.

6.2.12. Take the sample and trim it to provide a narrow horizontal pyramid. The pyramid should be as narrow as possible, and the cell of interest should be at the center of the pyramid. The length of the pyramid should be smaller than 0.9 mm and its height smaller than 0.1 mm. In this case, you could place it on the slot 18 serial sections. An experienced person can trim a pyramid directly with a razor blade.

6.2.13. Introduce the sample back into the ultratome, and lock it in exactly the same position as before (preserving all of the parameters of sample positioning; this is very important).

6.2.14. Replace the glass knife with the diamond one, and position the latter toward the pyramid. If the sample is not parallel to the knife, adjust the angle of the diamond knife by rotating the knife stage to make its edge parallel to the plane of the pyramid. Do not change any other parameters of the sample position.

6.2.15. Approach the sample toward the edge of the knife until the gap is extremely narrow. Using a 200-nm approaching step, begin the sectioning. Take serial sections according to the instructions with the ultratome. It is enough to take only 10 sections to pass 2 μm from the bottom of the cell.

6.2.16. Identify the position of the organelle of interest within the EM sample according to the Z-stacking, and select those thick sections that should correspond to this position. For instance, if the organelle of interest is situated at 500 nm from the surface of the sample (bottom of the cell), it is enough to collect only the first three 200-nm serial sections (or ten 50-nm sections). If the position is at 1 μm in height, it will be necessary to collect from the fourth to the eighth serial 200-nm sections (or twenty 50-nm sections).

7. Picking up Serial Sections with the Empty Slot Grid

Standard method to pick up the serial sections for routine EM is not sufficient for CVLEM due to significant rate of possible mistakes during manipulations with sections. For CVLEM the demands for the reproducibility of this step are much higher. The method should give almost 100% of success. Here, we described a specific method that after some training gives such a result.

7.1. Required materials

7.1.1. Anti-capillary self-closing tweezers (Ted Pella Inc., Redding, CA)
7.1.2. Diamond knife (Electron Microscopy Sciences, Hatfield, PA)

7.1.3. Eyelash with handle (Ted Pella Inc., Redding, CA)
7.1.4. Chloroform
7.1.5. 1 × 2 mm slot grids (Electron Microscopy Sciences, Hatfield, PA)
7.1.6. 1 × 2 mm slot grids covered with formvar and carbon (Electron Microscopy Sciences, Hatfield, PA)
7.1.7. Parallel holders for grids covered with scotch tape are not commercially available. Take a sheet of thick paper and cover it from both sides with a double-covered scotch tape. Then cut with scissor two narrow (3 mm in width) pieces and attach them to the glass with the distance of 2 mm between them (Fig. 10.5C–E)
7.1.8. Whatman No. 1 filter paper (Fisher Scientific, Pittsburgh, PA)

7.2. Procedure

7.2.1. Clean the empty (not covered) slot grid (the donor transfer grid) and one slot grid covered with formvar–carbon film (the acceptor grid).
7.2.2. When the normal (not bent) slot grid is taken by the tweezer, it is not convenient to take sections from the water because the axes of grid cannot be oriented parallel to the plane of water surface. One needs a holder on the grid. The holder on the slot grid could be done by bending just the small part of the grid from horizontal position. Using a tweezer, bend just the small part of the edge of the empty slot grid from horizontal position. This bent rim will serve as a holder for the tweezer.
7.2.3. Take the slot grid with supporting film from a container by the self-closing tweezer and place it film-down on parallel holders covered with scotch tape (Fig. 10.4C, below).
7.2.4. Take the empty slot grid using the just made holder and place it over the band of serial sections not touching the water and in such a way that the slot is projected on the band of serial section.
7.2.5. Touch the water with the donor grid in such a way that sections should be inside the slot and move the slot grid away. The droplet of water and sections will be inside the slot (Fig. 10.4A–C).
7.2.6. To avoid dirt on the sections, it is better to place a small droplet of glass-handled distilled water on the acceptor slot grid. Using stereomicroscope, put a small droplet of water on the supporting film of the acceptor grid (Fig. 10.4D).
7.2.7. Put the donor grid with serial sections on acceptor grid as it is shown in Fig. 10.4E and F.

7.2.8. Using sharp and very narrow filter paper, orient slot in parallel to each other and eliminate the excess of water from the space between slot grids (Fig. 10.4G).
7.2.9. Dry the grids for at least 20 min.
7.2.10. Carefully eliminate the donor grid (Fig. 10.4H).
7.2.11. Take the dried acceptor grid and check whether sections are in correct position (Fig. 10.4I).

8. EM Analysis

8.1. Required materials

8.1.1. A stereomicroscope equipped with a fiber optic illuminator (i.e., Leica Microsystems Spa, Milan, Italy, www.leica-microsystems.com)
8.1.2. An electron microscope with a motorized goniometer (i.e., FEI, Eindhoven, The Netherlands)

8.2. Procedure

8.2.1. Using stereomicroscope (to avoid mistakes during manipulations), place the slot grid into the holder of EM and introduce it into EM.
8.2.2. Find the very first serial EM sections using the traces (usually letters) of the coordinated grid filled with the resin. These are visible mostly on the first two sections. If the central position of the cell of interest within the pyramid is used as a marker, identify the central cell within the sections.
8.2.3. Take consecutive photographs (or grab the images with a computer using a video camera) of the serial sections beginning from the very first section since the organelle of interest (just observed under the LSCM) appears and until the organelle of interest is no longer seen. Having images from LSCM is useful to identify the cell of interest.
8.2.4. If EM and FM images are reflective, it is necessary to change the reflection position of FM images.
8.2.5. Prepare the figure for the paper showing different steps of sample preparation and proving that this is CLEM (Figs. 10.6 and 10.7).

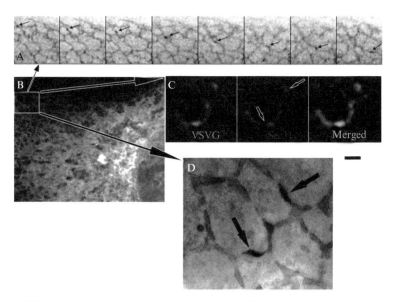

Figure 10.6 *In vivo* dynamics and ultrastructure of individual ER carriers studied using CLEM. Cells transfected with the temperature-sensitive G protein of vesicular stomatitis virus that is tagged with green fluorescence protein were placed at 40 °C for 16 h to block folding of this protein and, thus, to prevent its exit from the ER. Then, cells were shifted to the permissive temperature 32 °C and examined under the confocal microscope in 10 min after the shift. Video recording was performed during 3 min, and then cells were fixed and prepared for immuno EM. The G protein was labeled with antibody against folded protein and then with monovalent Fab fragments of antibody against the primary antibody that were conjugated with peroxidase/DAB. (A) The consecutive frames of the video-film examining the behavior of the ER-to-Golgi carrier (arrows). (B) Cells were fixed. (C) Cells were labeled for VSVG (green) and Sec31 (red). (D) Identification of the very same area in EM images. VSVG was labeled with DAB (thick arrows). Bars: 2 μm (A), 4 μm (B), 1 μm (C), and 0.5 μm (D). (See Color Insert.)

ACKNOWLEDGMENTS

The research relevant to this paper was supported by Telethon Italia and Consorzio Mario Negri Sud.

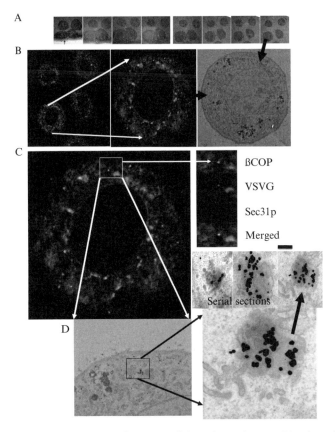

Figure 10.7 Multiple labeling of ER-to-Golgi carriers using combination of LM and CLEM. Cells were prepared as in Fig. 10.6. The G protein was labeled with antibody against folded protein and then with monovalent Fab fragments of antibody against the primary antibody that were conjugated with horseradish peroxidase/DAB. (A) Serial EM sections. (B) Pictures immunofluorescence (left and middle) and EM (right) images of the same cell. (C) Triple labeling for VSVG (red), Sec31 (blue), and ßCOP (blue). (D) Identification of the carrier after nanogold labeling for VSVG on serial EM sections. Bars: 17 μm (A); 10 μm (B, left), 2 μm (B, middle and right); 1 μm (C); and 0.1 μm (D). (For interpretation of the references to color in this figure legend, the reader is referred to the Web version of this chapter.)

REFERENCES

Mironov, A. A., and Beznoussenko, G. V. (2009). Correlative microscopy: A potent tool for the study of rare or unique cellular and tissue events. *J. Microsc.* **235**(3), 308–321.

Polishchuk, R. S., Polishchuk, E. V., Marra, P., Alberti, S., Buccione, R., Luini, A., and Mironov, A. A. (2000). Correlative light-electron microscopy reveals the tubular-saccular ultrastructure of carriers operating between Golgi apparatus and plasma membrane. *J. Cell Biol.* **148**, 45–58.

van Rijnsoever, C., Oorschot, V., and Klumperman, J. (2008). Correlative light-electron microscopy (CLEM) combining live-cell imaging and immunolabeling of ultrathin cryosections. *Nat. Methods* **5**(11), 973–980.

Verkade, P. (2008). Moving EM: The rapid transfer system as a new tool for correlative light and electron microscopy and high throughput for high-pressure freezing. *J. Microsc.* **230**, 317–328.

CHAPTER ELEVEN

Optical Techniques for Imaging Membrane Domains in Live Cells (Live-Cell PALM of Protein Clustering)

Dylan M. Owen, David Williamson, Astrid Magenau, *and* Katharina Gaus

Contents

1. Introduction	222
2. Sample Preparation and Data Acquisition	224
2.1. Cell culture and transfection procedure	225
2.2. Synapse formation and data acquisition	226
3. Data Analysis	228
3.1. PALM data analysis	228
3.2. Ripley's K-function and Getis and Franklin's local point pattern analysis	231
3.3. Extracting numbers and final conclusions	233
Acknowledgments	234
References	234

Abstract

It is now recognized that the plasma membrane is not homogeneous but instead contains a variety of membrane microdomains. These include lipid microdomains (lipid rafts) and protein clusters which exist on a range of size and time scales but are often small and short-lived. The small size and dynamic nature of membrane domains has made them difficult to study by conventional fluorescence microscopy approaches. Photoactivated localization microscopy (PALM) is a super-resolution technique capable of localizing the positions of individual molecules with tens of nanometers precision. Here, we describe a method for imaging membrane proteins using PALM, including live-cell PALM, to detect the molecular clustering of plasma membrane proteins using a statistical cluster-analysis method based on Ripley's K-function. While the method is applicable to a wide variety of proteins in various biological systems, to illustrate the technique, we will image and analyze the clustering behavior of the adaptor protein Linker for activation of T cells (LAT) at the T cell immunological

Centre for Vascular Research, University of New South Wales, Sydney, Australia synapse [Williamson, D. J., Owen, D. M., Rossy, J., Magenau, A., Wehrmann, M., Gooding, J. J. , and Gaus, K. (2011). Pre-existing clusters of the adaptor Lat do not participate in early T cell signaling events. *Nat. Immunol.* **12**, 655–662.].

1. Introduction

Conventional fluorescence microscopy is subject to limited resolution imposed by diffraction. Depending on the nature of the microscope and the wavelengths involved, this resolution limit is typically in the range of 200–400 nm. Two fluorescent objects, such as individual fluorophores, which are closer than this distance cannot be resolved as being two separate objects. In recent years, several techniques have emerged to break the diffraction limit and enhance the resolution of fluorescence microscopes. These techniques can be broadly divided into three main classes: stimulated emission depletion (STED, which uses a doughnut shaped depletion beam to narrow the excitation volume) (Hein *et al.*, 2008; Hell and Wichmann, 1994), structured illumination microscopy (SIM, which uses a shifting grid pattern of excitation light to allow high-spatial-frequency components to be imaged) (Gustafsson, 2000), and localization microscopy approaches including photoactivated localization microscopy (PALM) (Betzig *et al.*, 2006; Hess *et al.*, 2006; Shroff *et al.*, 2008) and (direct) stochastic optical reconstruction microscopy ((d)STORM) (Heilemann *et al.*, 2008; Rust *et al.*, 2006). Of these, localization microscopy is especially well suited to studying the cell plasma membrane as it is typically performed using total internal reflection fluorescence (TIRF) illumination to limit the excitation volume to within 100 nm of the cell–coverslip interface. This illumination system is required to reduce background light to the point where fluorescence from individual molecules can be detected.

Fundamentally, PALM and (d)STORM work by temporally separating closely spaced molecules such that only a small fraction are available for imaging at any given time. In PALM, this is done by means of photoactivatable or photoconvertible fluorescent proteins. The protein mEos2, for example, fluoresces in the green region of the spectrum, however, after illumination with 405 nm light, it is converted to a red-fluorescing form (Wiedenmann *et al.*, 2004) (Fig. 11.1A). If very low intensities of 405 nm light are used, the probability of switching is also very low, meaning only a few of the many mEos2 molecules in a sample will be red at any one time. When imaged on a sensitive camera, a sparse set of point spread functions (PSFs) is observed—each originating from an individual molecule (example shown in Fig. 11.1B). It is possible to mathematically calculate the center of these (approximately Gaussian) PSFs to an accuracy of tens of nanometers and store those coordinates in computer memory. Since the red fluorescent

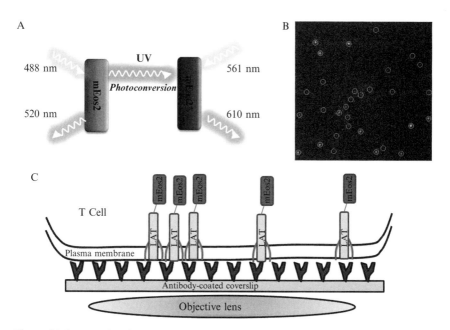

Figure 11.1 (A) The photoconvertible fluorescent protein mEos2 is green in its native form; being excited at 488 nm and emitting fluorescence around 520 nm. Upon irradiation with UV light at 405 nm, the protein converts to a red-fluorescent form. In this state, it is excited at 561 nm and emits around 610 nm. (B) When imaged under TIRF illumination through a red filter, individual PSFs originating from single mEos2 molecules can be observed and localized. (C) The mEos2 fluorescent protein is fused to LAT and the construct is transfected into Jurkat T cells where it localizes to the plasma membrane. The T cells can then be made to form immunological synapses by placing them on activating antibody-coated glass coverslips. (See Color Insert.)

molecules are imaged under intense laser illumination, they bleach rapidly and a new set of molecules can be converted to the red form and imaged, and the process repeated. This way, after many cycles, all the mEos2 molecules in the sample can be imaged and localized and the coordinate data used to generate a single super-resolution image.

PALM requires activated molecules to be sparsely distributed, so many (thousands) of frames need to be recorded in order to localize all molecules in the field of view. This means that a typical PALM image, composed of data from thousands of raw frames, can take several minutes (or even tens of minutes) to acquire. Because of this slow acquisition time, PALM has typically been thought of as being only applicable to imaging fixed cells. However, with advances in camera technology allowing higher rates of imaging and new data analysis methods, PALM can now be applied to live cells (Shroff *et al.*, 2008).

It is now recognized that the plasma membrane of cells is not homogeneous but contains a variety of domains such as protein clusters, ordered-phase lipid microdomains, caveolae, etc. (Binder *et al.*, 2003; Simons and

Ikonen, 1997). Membrane lipid microdomains, enriched in cholesterol, sphingolipids, and saturated phospholipids are thought to be involved in a number of cell signaling and membrane trafficking processes. This is because different membrane proteins have different affinities for either the liquid-ordered or liquid-disordered phases. Ordered domains therefore concentrate and cluster some molecules while excluding others. Protein clusters may also form in the plasma membrane through direct protein–protein interactions (Brown, 2006). PALM microscopy is applicable to protein clusters generated by these mechanisms.

Here, our example protein is the T cell adaptor protein *linker for activation of T cells* (*LAT*). LAT is a transmembrane protein that also contains two post-translational lipid modifications (palmitoylation) which are thought to target LAT to high-order membrane regions (Hartgroves *et al.*, 2003; Tanimura *et al.*, 2006; Zhang *et al.*, 1998). It is now known that LAT is clustered in the plasma membrane of resting T cells and that this clustering is enhanced upon T cell activation during formation of an immunological synapse (Lillemeier *et al.*, 2010). Membrane order has also been observed to increase during immune synapse formation (Owen *et al.*, 2010a,b). While immunological synapses usually form between T cells and antigen-presenting cells, it is possible to form synapses between T cells and coverslips coated with antibodies against the T cell receptor protein CD3, thereby making them accessible to TIRF (and hence PALM) imaging. A schematic of this biological situation is shown in Fig. 11.1C.

Here, we describe a protocol to image and analyze clusters of LAT at the T cell immunological synapse at super-resolution using PALM in both fixed and live cells where molecular clusters can be observed to be dynamic over time. While here we focus on one example—LAT-mEos2 at T cell synapses—the methodology of data acquisition and data analysis is generally applicable to other membrane proteins of interest, other fluorescent fusion constructs, and other biological systems.

2. Sample Preparation and Data Acquisition

In this section, we will provide details on generating T cell immunological synapses with fluorescent fusion constructs suitable for imaging by live-cell PALM microscopy. We use the immortal T cell line *Jurkat E6.1*, which can be transfected using an Invitrogen *NEON* microporator device. The user will also require plasmid DNA of the protein of interest (in this example LAT) fused to a photoactivatable or photoconvertible fluorescent protein. There are many such proteins available such as PS-CFP2 (which converts from blue to green upon UV illumination) and mEos2 (which converts from green to red). mEos2 is the protein that will be used in this example (Fig. 11.1A). For a more comprehensive list of applicable fluorescent proteins, the reader is referred to Gould *et al.* (2009). After transfection of the cells and generation of immunological synapses on antibody-coated glass coverslips (Fig. 11.1C), the samples are placed in live-cell imaging chambers with immobile gold beads to allow sample drift correction.

We then outline the method of data acquisition using a commercial PALM microscope (Zeiss ELYRA PS-1) and associated software. This includes the steps needed to achieve high-quality TIRF illumination and recommended imaging parameters to maximize the quality of single-molecule imaging for later analysis.

2.1. Cell culture and transfection procedure

2.1.1. Materials

Required equipment

- Invitrogen *NEON* Microporator (invitrogen.com, Cat. No. MPK5000S) and associated consumables for transfection of Jurkat T cells.
- TIRF-suitable coverslips are mounted for TIRF/PALM imaging in 18 mm diameter Chamlide CMB magnetic sample holders (chamlide.com, Cat. No. CM-B-30).
- Zeiss ELYRA PS-1 PALM/TIRF microscope with a 100×, 1.46 NA oil-immersion objective, and an Andor iXon DU-897D EM-CCD camera. Photoconversion and imaging of mEos2 require lasers at 405, 488, and 561 nm.

Required materials

- Jurkat E6.1 T cells (or other cells of interest) which can be obtained from the American Type Culture Collection (ATCC, atcc.org, Cat. No. TIB-152). These should be maintained between 0.2 and 1.0×10^6 cells/ml and below passage 6.
- High-quality plasmid DNA encoding a fluorescent fusion construct containing a photoactivatable or photoconvertible fluorescent protein. In this example, LAT-mEos2 will be used. Plasmid DNA must be endotoxin free and at a concentration of at least 1 μg/μl.
- *Growth medium*: RPMI1640 media supplemented with 10% FCS and 2% L-glutamine (e.g., Invitrogen Cat. No. 72400-047).
- Phosphate buffered saline (PBS) without magnesium or calcium (e.g., Invitrogen Cat. No. 10010-023).
- CD3 antibody (e.g., clone OKT3, eBioscience Cat. No. 16-0037).
- CD28 antibody (e.g., clone CD28.2, eBioscience Cat. No. 16-0289).
- 200 nm gold colloid beads (BBInternational Cat. No. GC200).

2.1.2. Cell transfection

The day before imaging: Gently resuspend the cells to disrupt cell clumps. Centrifuge the cells in a 15 ml tube at $200 \times g$ for 5 min at room temperature and remove the supernatant. Resuspend the cell pellet in 15 ml of PBS. Repeat the centrifugation and resuspend again in PBS such that the final cell concentration is 20×10^6 cells/ml. Centrifuge 50 μl of the PBS-cell suspension at $200 \times g$ for 5 min in a benchtop microcentrifuge. This should be just enough cells to form a visible pellet. Remove the supernatant and replace with 50 μl of the Invitrogen *NEON* Buffer R and mix very gently to resuspend the pellet. *Note*: the cells at this point have to be used within 15 min. Gently mix 10 μl of the cell-Buffer R suspension with 1 μg of plasmid DNA. It is recommended that the volume of plasmid DNA be kept to less than 10% of the total transfection volume, that is, for 10 μl transfections DNA concentration should be at least 1 μg/μl.

Assemble the Invitrogen NEON transfection station with 3 ml Buffer E and fit the electroporation pipette with a 10 μl tip. Aspirate 10 μl of the cell–DNA–buffer suspension taking care to avoid introducing bubbles into the tip. Microporate the cells using two 30 ms pulses at 1150 V. Immediately transfer the suspension into 1 ml fresh, prewarmed RPMI1640 media containing 10% FCS and 2% L-glutamine. Incubate the cells overnight at 37 °C in a humidified 5% CO_2 atm.

Tip: Neon electroporation tips can be used at least twice (for identical plasmids) without affecting transfection efficiency or cell viability and transfected cells can be pooled to recover in the same well. Identical transfections can be performed with the remaining cells in Buffer R to avoid wasting reagents and to have more cells on hand for subsequent imaging.

Rinse high-quality, #1.5 (0.17 mm) thickness, 18 mm diameter round glass coverslips with sterile water or 0.22 μm filtered PBS and place at the bottom of a 12-well tissue-culture plate. For T cell activation, dilute mouse monoclonal anti-human CD3 and CD28 antibody to 10–20 μg/ml each in filter-sterilized PBS. Pipette 50–70 μl of diluted antibody solution into the center of each coverslip and incubate overnight at 37 °C in a humidified 5% CO_2 atm. Wash the coverslips three times with sterile, filtered PBS.

2.2. Synapse formation and data acquisition

The day of imaging: Turn on all imaging equipment, including heaters (we recommend a fully enclosed incubation chamber set to 25 °C for fixed cell imaging or 37 °C for live-cell imaging), and allow the system to equilibrate for at least 2 h. This will help reduce drift during the acquisition. Set the camera cooling temperature to -63 °C and the Optivar lens to $\times 1.6$ (this will mean that the camera pixel size at the sample plane is 100 nm).

Check transfected Jurkat cells for plasmid expression using a standard epi-fluorescence microscope with 488 nm excitation. Healthy cells should appear round and with minimal debris. Gently resuspend the cells to disrupt clumps and pipette 400 µl of the cell suspension onto an antibody-coated coverslip in a 12-well tissue-culture plate and incubate for 10 min at 37 °C in a humidified 5% CO_2 atm. *If required*: fix the cells by adding 133 µl of 16% paraformaldehyde (PFA) to each well (for a final PFA concentration of 4%) and incubate at 37 °C in a humidified 5% CO_2 atm for 30 min. Gently wash the coverslips with filtered PBS. *Note*: After fixation, *do not* mount the cells in mounting media (such as Mowiol) as TIRF excitation requires the cells to be in an aqueous environment.

For fixed cell imaging: Mount the sample coverslips into an 18-mm Chamlide chamber and add 20 µl of 100 nm gold beads. Allow the beads to settle for 2 min and then cover the sample with sterile filtered PBS. Place a drop of immersion oil on the objective and place the Chamlide chamber on the stage ensuring the chamber is flat and secure. Using the eye-pieces and the brightfield lamp, focus upon the cells and ensure they are well spread (i.e., have formed synapses) on the antibody-coated coverslips. Set the camera EM gain to 200, camera integration time to 30 ms, 405 nm laser power to ~ 8 µW, and 561 nm laser power to ~ 12 mW (at the sample).

For live-cell imaging: Set the camera integration time to 5 ms, 405 nm laser power to 8 µW, and 561 nm laser power to 18 mW or higher (at the sample). Mount an antibody-coated coverslip in the Chamlide chamber, pipette 20 µl of 100 nm gold beads and allow them to settle. Secure the chamber to the microscope stage and use the eyepieces and brightfield lamp to find and focus on the fiducial beads. Supplement the transfected cell media with 50 mM HEPES (unless using an incubated stage with 5% CO_2 supply) and ensure the stage and chamber are at the correct temperature. Gently add the cell suspension to the coverslip, ensuring even coverage. Wait for 1 min to allow the cells to settle and begin spreading on the antibody-coated coverslip. This time can be used to find and focus on cells which are beginning to interact with the coated glass prior to starting the image acquisition. Immunological synapses will form over a period of 10 min.

Using the camera, locate a fluorescent cell that has several fiducial beads in close proximity, place it in the center of the field-of-view and focus on the basal membrane. Crop the camera field-of-view to a 256 × 256 pixel region centered on the cell. Using the beam angle slider, set the excitation beam to TIRF mode. Good quality TIRF illumination is critical to maximizing the signal-to-noise ratio in single-molecule detection.

Tip: TIRF illumination can be tested by changing the focus: when in TIRF mode, no deeper structures can be observed when focusing up through the cell.

Tip: It is easier to observe single-molecule signals by changing the look-up-table from a linear table to a min–max version to enhance fluorescence image contrast.

Adjust the 405 nm laser power such that PSFs are sparse and not overlapping. Set the software to acquire 20,000 frames and click "Start." For live-cell imaging, set the total frame count much higher (e.g., 80,000 frames) to compensate for the reduced camera integration time and to ensure the acquisition does not end prematurely. It is important to monitor the acquisition to ensure the sample remains in focus over the time series. As the density of viable molecules declines over the course of the acquisition, the 405 nm laser power should be increased to ensure a steady rate of switched molecules. Once the acquisition has finished, save the image series as a Zeiss .lsm file.

3. Data Analysis

In this section, we will describe the steps required to extract quantitative super-resolution data from the saved raw image series. This procedure can be broadly split into two parts: PALM processing in which single-molecule fluorescent events are detected and their positions determined, and quantitative cluster analysis of the resulting spatial point pattern. We recommend the first stage be performed using Zeiss Zen 2010 software although other alternatives such as the QuickPALM (Henriques *et al.*, 2010) plugin for ImageJ (http://rsbweb.nih.gov/ij/; National Institute of Health, NIH) are available. The second stage can be performed in a variety of programming languages but we recommend Matlab (Mathworks) in conjunction with ImageJ. Plugins for other software (such as Microsoft Excel), for example, SpPack (Perry, 2004), can also be used. The final outcome is a system where the number of molecular clusters, the number of molecules per cluster, cluster size and shape, and a variety of other parameters can be obtained. Note that the cluster analysis is statistical and therefore requires a certain number of points in each region to generate reliable results; regions with fewer than 100 molecules/μm^2 should not generally be analyzed by this method.

3.1. PALM data analysis

The first stage in the analysis is to detect single-molecule events above background noise. Begin by loading the PALM image sequence into Zen and select the PALM processing option. There are a number of user-definable parameters which must be entered to optimally detect single molecules.

The software first applies a Gaussian (to reduce noise) and a Laplace (to enhance single-molecule events) filter to each frame of the raw image. The image mean (M) and standard deviation (S) are then computed. Fluorescent events are then defined to originate from single-molecules (rather than noise) when the peak intensity (I) satisfies:

$$I - M > S \times \text{SNR} \tag{11.1}$$

where SNR is a user-definable signal-to-noise ratio (called "Peak intensity to noise" in Zen). In our experience, the optimal SNR to select when using mEos2 is 5. Below this, many background noise "events" are counted as single molecules. Above this, many single fluorophores do not emit enough photons to be detected and are lost from later analysis. The "peak mask size" slider determines the area analyzed around each event and should typically be set to 9 pixels. Select "discard overlapping" which prevents the software analyzing events whose PSF overlap. Note that if there are many overlapping PSFs, the density of single-molecule activation during acquisition was too high. In this case, the intensity of the 405 nm activation laser should be reduced in the future.

There are two methods for calculating the center of detected PSFs—the mask fit (Thompson et al., 2002) and the full 2D Gaussian fit. The mask fit reduces computational processing time, whereas the Gaussian fit increases accuracy, and we therefore generally use the Gaussian fit approach. Once these parameters are set, select "Apply."

The software generates a PALM image together with a table containing the $x-y$ coordinates (and other details, notably the frame the molecule was first detected and the precision of localization) of all the single molecules that have been localized. A typical PALM acquisition of LAT at an immunological synapse processed with the above parameters might contain of the order of 50,000–200,000 total detected events.

Next, select the PAL-drift tab and select "fiducials table." The sliders for "Min on Time," "Off Gap," and "Capture Radius" should be selected appropriately such that all the fiducials in the image are detected. The software will then automatically correct for image drift during the acquisition. This is essential to maximize image resolution. Click on the "PAL-statistics" tab, select "Table," and save the data table as a text file (example shown in Fig. 11.2A).

The next step is to crop the data table into analyzable regions. Regions selected for analysis should contain no cell boundaries or other such features as these will distort the cluster analysis. Regions can be selected to have any size (or aspect ratio) but in general should be chosen to have between 1000 and 10,000 molecules and no cell edges. Square regions deliver the best results as these minimize edge effects in later processing. We therefore recommend using either 3×3 or 4×4 μm square regions. Save the

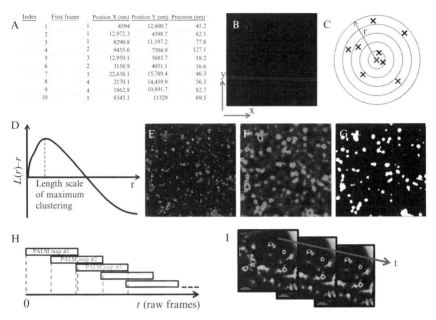

Figure 11.2 (A) After single-molecule detection and localization, the PALM data is in the form of a table containing the x–y coordinates of each localized molecule. (B) The x–y coordinates can be plotted against each other to create a map of the molecular positions. (V) For Ripley's K-function, concentric circles are drawn around each molecule and the number of other points encircled is counted. (D) After normalization to the total particle density and linearization, the height of the $L(r) - r$ function gives a measure of the degree of clustering at that length scale. (E) By calculating the value of $L(r)$ at $r = 50$ nm for each molecule, each point can be pseudo-colored with its clustering value at $r = 50$ nm. (F) These values can be used to interpolate a 2D pseudo-colored cluster "heat map." (G) The heat map can then be thresholded to create a binary map from which clustering parameters can easily be extracted. (H) For live-cell PALM, a "moving window" can be used to create cluster maps from different time periods of the raw data acquisition. (I) These maps can then be assembled into a live-cell PALM movie.

cropped regions as separate sheets in a Microsoft Excel file. We also recommend cropping the data by precision—thereby removing poorly localized molecules which might otherwise mask the presence of clusters. We recommend discarding events localized worse than 50 nm. Plot the x–y coordinates as a scatter plot and check for data anomalies, edges, and appropriate particle densities. An example 2D scatter plot is shown in Fig. 11.2B.

If the user is performing live-cell PALM, the data must also be cropped temporally using the "first frame" column—the value of the frame in which that molecule was first detected. It is important to note that a cluster map requires at least approximately 1000 molecules to be reliable, therefore, for a 3×3 μm region, around 1000–2000 raw frames may be required to deliver this many molecules. A live-cell PALM movie is generated by a "moving window" approach: crop in x–y and then select frames 0–1000, 200–1200, 400–1400, etc. (Fig. 11.2H). Copy each temporal window and save as separate sheets in an MS Excel file.

This concludes the first stage of the analysis—generating analyzable 2D spatial point pattern data from detected single-molecule fluorescent events.

3.2. Ripley's K-function and Getis and Franklin's local point pattern analysis

3.2.1. Ripley's K-function

Univariate Ripley's K-function and univariate Getis and Franklin's analysis are related techniques for detecting clustering in sets of spatial point pattern data (Getis and Franklin, 1987; Ripley, 1977). Both have previously been demonstrated for analyzing 2D point patterns generated from PALM microscopy (Owen et al., 2010a,b; Williamson et al., 2011). Ripley's K-function is calculated as:

$$K(r) = A \sum_{i=1}^{n} \sum_{j=1}^{n} \left(\frac{\delta_{ij}}{n^2}\right) \text{ where } \delta_{ij} = 1 \quad \text{if } \delta_{ij} < r, \quad \text{otherwise } 0 \quad (11.2)$$

where $K(r)$ is the K-function, A is the area being analyzed, n is the total number of points in that region, δ_{ij} is the distance between points i and j, and r is the spatial scale. Essentially, this draws concentric circles of radius r around each point and counts how many other points have been encircled (Fig. 11.2C). However, as the K-value scales with the area of the circles, it is usually linearized to scale with radius, represented by the L-function:

$$L(r) = \sqrt{\frac{K(r)}{\pi}} \quad (11.3)$$

For a completely spatially random (CSR) set of molecules, $L(r) = r$. That is, the value of $L(r)$ simply increases linearly with the radius of the circles as more points are encircled. We therefore plot a graph of $L(r) - r$ versus r (Fig. 11.2D). For the CSR case, the value of $L(r) - r$ will be zero for all r. If at some radius the value of $L(r) - r$ is positive, it means that more points have been encircled at that radius than would have been expected for a random distribution—that is, at that spatial scale, the molecules are clustered. If $L(r) - r$ is negative, fewer molecules have been encircled as would have been if they were randomly distributed.

Note: it is important to take account of edge effects. Encircling molecules at the edge of the distribution will capture fewer molecules simply because a fraction of the circle lies outside the area being analyzed. This can be overcome by weighting the edge points to remove this bias. For example, if at some radius half the circle lies outside the analyzed region, the $K(r)$ value of that point should be doubled. Confidence intervals should also be calculated which can be done by plotting the same graph for CSR simulated data points which can then be used to indicate what level of clustering $(L(r) - r)$ is required to be, say, 95% or 99% certain that value is not the result of random chance.

3.2.2. Getis and Franklin's local point pattern analysis

Ripley's K-function averages over all molecules in a region and produces a single plot for each analyzed area. This plot contains simple information on the level of clustering $(L(r) - r)$ at each spatial scale (r) and thus answers the question "on average, how clustered are these points?" and "at what spatial scale are the points most clustered?"

By dispensing with the averaging and calculating the value of $L(r)$ for each molecule individually, more rich information can be extracted from the spatial point pattern. The value of $L(r)$ for each molecule can be calculated for any length scale; however, generating a quantitative cluster map requires the user to select a spatial scale to be studied. We recommend either selecting the value of the peak clustering scale in the Ripley's K-function plot or in general selecting a value of the scale of the clusters that are expected in the sample. In this example, we will search for clustering on a 50 nm spatial scale. In this case, the value of $L(r)$ for each molecule is calculated as:

$$K_i(50) = A \sum_{j=1}^{n} \left(\frac{\delta_{ij}}{n^2}\right) \quad \text{where } \delta_{ij} = 1 \quad \text{if } \delta_{ij} < 50, \text{otherwise } 0 \quad (11.4)$$

and

$$L_i(50) = \sqrt{\frac{K_i(50)}{\pi}} \quad (11.5)$$

Once each molecule in the region has been assigned a value of $L(50)$, the points can be pseudo-colored and displayed on a graph (Fig. 11.2E). Here, points with high $L(r)$ values are colored red and low values colored blue.

To extract quantitative data this plot needs to be converted to an "image" by interpolating cluster values between the points. This can be done using Matlab, and we recommend the "$v4$" interpolation algorithm applied to a grid with 10×10 nm resolution (10 nm is typically higher resolution than the localization precision of a PALM acquisition). Note that a linear interpolation algorithm can be used in the case of limited computational time, with a resulting decrease in accuracy. Once the interpolated map has been generated it can be pseudo-colored to highlight regions of high clustering (Fig. 11.2F). *Note*:

- To correctly compare maps between data sets, the minimum and maximum values of the color scale should be set to fixed values (e.g., Min, blue, $L(r) = 0$. Max, red, $L(r) = 300$).
- The color gradient should be set to give a smooth transition from blue-red. 100 color levels are typically sufficient.
- Be careful the image display aspect ratio is set to 1:1.

In the example in Fig. 11.2F, the 2D spatial point pattern (Fig. 11.2B) has been overlaid with the interpolated color map. This is purely for illustrative purposes and can be achieved with the Matlab image "hold" and "hold off" commands. The interpolated map should be saved as a TIFF (.tif) image file and a .fig Matlab file which allows future modification of the display parameters.

Next, the color map should be thresholded to generate a binary map. The value of $L(r)$ to threshold (i.e., the value above which an area is defined as being "a cluster") is user-definable but of course should be kept consistent between experiments. A simple way to generate the thresholded map is simply to change the color scale from (in this case) "Jet" to a binary map and select the value of $L(r)$ of the black-white transition (Fig. 11.2G). Save the binary map as a TIFF file. For live-cell PALM, the entire cluster mapping procedure is repeated for each temporal image window in the data series (Fig. 11.2I).

3.3. Extracting numbers and final conclusions

Once the binary map (Fig. 11.2G) has been generated, it is trivial to extract quantitative numbers (from either fixed-cell individual maps or live-cell map series). A simple way is to use ImageJ (NIH) but this can also be achieved in Matlab or other image analysis software packages. Load the binary map into ImageJ and select "Process" → "Binary" → "Make Binary."

Note: You may have to crop the image to remove the border and leaving only the square cluster map if this was not done automatically in Matlab.

Next select Analyze → Analyze Particles, check the boxes "Display results" and "Summarize" and click "OK." The range of parameters calculated by ImageJ is controlled in the Analyze → Set Measurements menu. ImageJ now displays the selected parameters, potentially including the number of clusters, cluster size (in pixels by default. Use the "Set Scale" function to return measurements in nanometers), fraction of area covered by clusters, and so on. The number of molecules in clusters is calculated from the $L(r)$ values for each molecule calculated earlier. It is simply the number of molecules that have $L(r)$ values above the level of the threshold that was set to produce the binary image. All these parameters too can be calculated for each "window" of a live-cell time series.

In conclusion, the above procedure can be used to: (a) transfect Jurkat T cells with fluorescent fusion constructs. (b) Generate immunological synapses between these cells and activating, antibody-coated glass coverslips. (c) Image the resulting molecular distributions by TIRF and PALM in fixed or live cells. (d) Generate quantitative super-resolution cluster maps based on the acquired data, and (e) extract quantitative data on various parameters that describe the degree of molecular clustering. It should be noted that the analysis method is, in general, also extensible to 3D molecular clustering. It is hoped that these techniques will deliver a better understanding of the role of molecular clustering and cluster dynamics in the activation of T cells and other cell systems.

ACKNOWLEDGMENTS

We acknowledge funding from the National Health and Medical Research Council (NHMRC), the Australian Research Council (ARC), and the Human Frontier science Program (HFSP).

REFERENCES

Betzig, E., Patterson, G. H., Sougrat, R., Lindwasser, O. W., Olenych, S., Bonifacino, J. S., Davidson, M. W., Lippincott-Schwartz, J., and Hess, H. F. (2006). Imaging intracellular fluorescent proteins at nanometer resolution. *Science* **313,** 1642–1645.
Binder, W. H., Barragan, V., and Menger, F. M. (2003). Domains and rafts in lipid membranes. *Angew. Chem. Int. Ed.* **42,** 5802–5827.
Brown, D. A. (2006). Lipid rafts, detergent-resistant membranes, and raft targeting signals. *Physiology* **21,** 430–439.
Getis, A., and Franklin, J. (1987). Second-order neigborhood analysis of mapped point patterns. *Ecology* **68,** 473–477.
Gould, T. J., Verkhusha, V. V., and Hess, S. T. (2009). Imaging biological structures with fluorescence photoactivation localization microscopy. *Nat. Protoc.* **4,** 291–308.
Gustafsson, M. G. L. (2000). Surpassing the lateral resolution limit by a factor of two using structured illumination microscopy. *J. Microsc.* **198,** 82–87.
Hartgroves, L. C., Lin, J., Langen, H., Zech, T., Weiss, A., and Harder, T. (2003). Synergistic assembly of linker for activation of T cells signaling protein complexes in T cell plasma membrane domains. *J. Biol. Chem.* **278,** 20389–20394.
Heilemann, M., van de Linde, S., Schüttpelz, M., Kasper, R., Seefeldt, B., Mukherjee, A., Tinnefeld, P., and Sauer, M. (2008). Subdiffraction-resolution fluorescence imaging with conventional fluorescent probes. *Angew. Chem. Int. Ed.* **47,** 6172–6176.
Hein, B., Willig, K. I., and Hell, S. W. (2008). Stimulated emission depletion (STED) nanoscopy of a fluorescent protein-labeled organelle inside a living cell. *Proc. Natl. Acad. Sci. USA* **105,** 14271–14276.

Hell, S. W., and Wichmann, J. (1994). Breaking the diffraction resolution limit by stimulated emission: Stimulated-emission-depletion fluorescence microscopy. *Opt. Lett.* **19,** 780–782.

Henriques, R., Lelek, M., Fornasiero, E. F., Valtorta, F., Zimmer, C., and Mhlanga, M. M. (2010). QuickPALM: 3D real-time photoactivation nanoscopy image processing in ImageJ. *Nat. Methods* **7,** 339–340.

Hess, S. T., Girirajan, T. P. K., and Mason, M. D. (2006). Ultra-high resolution imaging by fluorescence photoactivation localization microscopy. *Biophys. J.* **91,** 4258–4272.

Lillemeier, B. F., Mortelmaier, M. A., Forstner, M. B., Huppa, J. B., Groves, J. T., and Davis, M. M. (2010). TCR and Lat are expressed on separate protein islands on T cell membranes and concatenate during activation. *Nat. Immunol.* **11,** 90–96.

Owen, D. M., Oddos, S., Kumar, Sunil, Davis, D. M., Neil, M. A., French, P. M., Dustin, M. L., Magee, A. I., and Cebecauer, M. (2010a). High plasma membrane lipid order imaged at the immunological synapse periphery in live T cells. *Mol. Membr. Biol.* **27,** 178–189.

Owen, D. M., Rentero, C., Rossy, J., Magenau, A., Williamson, D., Rodriguez, M., and Gaus, K. (2010b). PALM imaging and cluster analysis of protein heterogeneity at the cell surface. *J. Biophoton.* **3,** 446–454.

Perry, G. L. W. (2004). SpPack: Spatial point pattern analysis in Excel using Visual Basic for applications (VBA). *Environ. Modell. Softw.* **19,** 559–569.

Ripley, B. D. (1977). Modelling spatial patterns. *J. R. Stat. Soc. Ser. B Stat. Methodol.* **39,** 172–192.

Rust, M. J., Bates, M., and Zhuang, X. (2006). Sub-diffraction-limit imaging by stochastic optical reconstruction microscopy (STORM). *Nat. Methods* **3,** 793–796.

Shroff, H., Galbraith, C. G., Galbraith, J. A., and Betzig, E. (2008). Live-cell photoactivated localization microscopy of nanoscale adhesion dynamics. *Nat. Methods* **5,** 417–423.

Simons, K., and Ikonen, E. (1997). Functional rafts in cell membranes. *Nature* **387,** 569–572.

Tanimura, N., Saitoh, S. I., Kawano, S., Kosugi, A., and Miyake, K. (2006). Palmitoylation of LAT contributes to its subcellular localization and stability. *Biochem. Biophys. Res. Commun.* **341,** 1177–1183.

Thompson, R. E., Larson, D. R., and Webb, W. W. (2002). Precise nanometer localization analysis for individual fluorescent probes. *Biophys. J.* **82,** 2775–2783.

Wiedenmann, J., Ivanchenko, S., Oswald, F., Schmitt, F., Röcker, C., Salih, A., Spindler, K.-D., and Nienhaus, G. U. (2004). EosFP, a fluorescent marker protein with UV-inducible green-to-red fluorescence conversion. *Proc. Natl. Acad. Sci. USA* **101,** 15905–15910.

Williamson, D. J., Owen, D. M., Rossy, J., Magenau, A., Wehrmann, M., Gooding, J. J., and Gaus, K. (2011). Pre-existing clusters of the adaptor Lat do not participate in early T cell signaling events. *Nat. Immunol.* **12,** 655–662.

Zhang, W., Trible, R. P., and Samelson, L. E. (1998). LAT palmitoylation: Its essential role in membrane microdomain targeting and tyrosine phosphorylation during T cell activation. *Immunity* **9,** 239–246.

CHAPTER TWELVE

Single Live Cell Topography and Activity Imaging with the Shear-Force-Based Constant-Distance Scanning Electrochemical Microscope

Albert Schulte,* Michaela Nebel,[†] *and* Wolfgang Schuhmann[†]

Contents

1. Introduction	238
2. Shear-Force-Based Constant-Distance Scanning Electrochemical Microscopy for Live Cell Studies: Apparatus, Probes, and Operation	240
2.1. Basic system requirements	240
2.2. Tailored vibrating scanning probes for catecholamine, oxygen, and nitric oxide detections with isolated live cells	242
2.3. Instructions for the execution of SECM cell topography and activity imaging	246
3. Selected Applications of SF-CD-SECM Live Cell Studies	248
Acknowledgments	253
References	254

Abstract

In recent years, scanning electrochemical microscopy (SECM) has become an important tool in topography and activity studies on single live cells. The used analytical probes ("SECM tips") are voltammetric micro- or nanoelectrodes. The tips may be tracked across a live cell in constant-height or constant-distance mode, while kept at potentials that enable tracing of the spatiotemporal dynamics of functional chemical species in the immediate environment. Depending on the type of single live cells studied, cellular processes addressable by SECM range from the membrane transport of metabolites to the stimulated release of hormones and neurotransmitters and processes such as cell respiration or cell death and differentiation. In this chapter, we provide the key practical details of the constant-distance mode of SECM, explaining the

---

* Biochemistry-Electrochemistry Research Unit, Schools of Chemistry and Biochemistry, Institute of Science, Suranaree University of Technology, Nakhon Ratchasima, Thailand
[†] Analytische Chemie, Elektroanalytik & Sensorik, Ruhr-Universität Bochum, Bochum, Germany

establishment, and operation of the tailored distance control unit that maintains a stable tip-to-cell separation during scanning. The continuously maintained tip positioning of the system takes advantage of the decreasing impact of very short-range hydrodynamic tip-to-surface shear-forces on the vibrational amplitude of an oscillating SECM tip, as the input for a computer-controlled feedback loop regulation. Suitable microelectrode probes that are nondestructive to soft cells are a prerequisite for the success of this methodology and their fabrication and successful application are the other topics covered.

1. INTRODUCTION

Simultaneous acquisition of the morphological fine structure and local biochemical and physiological activity of living cells has become an important goal of live and medical sciences, since information from such a scaled-down experimental approach is required for expanding our knowledge of both functioning and of disease-affected cellular networks. To enable bioanalysis at the level of the single cells, scanning near-field (Dickensen et al., 2010; Mooren et al., 2006), laser scanning confocal (Halbhuber and König, 2003), total internal reflection (Axelrod, 2008) fluorescence microscopy (FM) schemes, and various nonoptical scanning probe microscopy (SPM) techniques have been established for microscopic observations of isolated cells of various kinds under normal or pathological conditions. The nonoptical SPM options for structural and functional live cell studies include atomic force microscopy (AFM) (Allison et al., 2010; Francis et al., 2010), scanning ion conductance microscopy (SICM) (Liu et al., 2009), and scanning electrochemical microscopy (SECM) (Schulte and Schuhmann, 2007; Schulte et al., 2010). As is evident from the above citations, use of these techniques for assessments of micromechanical and microchemical cellular properties and the immediate cell environment have reached a sophisticated level. Imaging by SECM, described here, uses the tips of rastered electrochemical (EC) microsensors that work with a constant or ramped electrode potential for spatially resolved amperometric or voltammetric current acquisition, respectively (Amemiya et al., 2008). The so-called SECM tip is moved horizontally in the imaging plane across the cell(s) under inspection while the current (I_{tip}) produced by a potential-induced oxidation or reduction of redox species is measured, stored, and displayed as function of the x and y tip coordinates (Fig. 12.1A). During scanning, the tip potential is adjusted relative to the redox potential of cell-consumed or -released chemical species. If no other compound is electroactive under the chosen condition, I_{tip} variations in the SECM current plots indicate spatially confined uptake or secretion activity of the cell.

Live Cell Scanning Electrochemical Microscopy

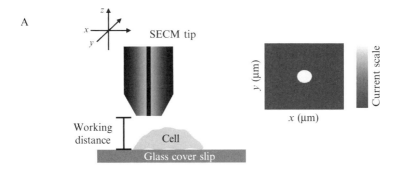

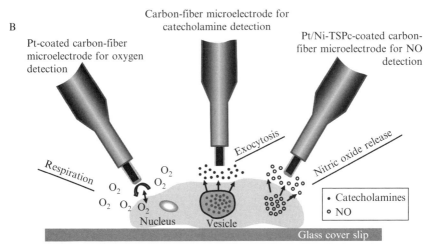

Figure 12.1 Introduction to cellular scanning electrochemical microscopy. (A) The local probe for measurements on biological objects is usually a disk-shaped voltammetric micro- or nanoelectrode that is movable in the x, y, and z directions. The so-called SECM tip is scanned horizontally at a nondestructive working distance above a selected cell while its current response is measured, stored, and displayed as function of the x, y coordinates. SECM images are plots of the tip current which, in the constant-height mode, will reflect cell topography and/or activity. In the constant-distance mode, the convolution of topography and activity effects are eliminated. (B) Well-known target processes of live cell SECM studies are the spatiotemporal detection of respiration (metabolic oxygen consumption or release), vesicular hormone, or neurotransmitter release (exocytosis) and the stimulated liberation of the messenger nitric oxide (NO).

Applied to well-adhering live cells, SECM tips may either be safely guided at a preadjusted constant height (z) that, based on approach curve data, avoids contact with the cells, or moved at a smaller, contactless, constant distance (CD) from the cell. CD mode scanning is best for live cell SECM imaging, as (i) there is less risk of cell damage through sporadic contact with the tip, (ii) the tip is always at a optimal sensing distance from the cell, and (iii) the interpretation of probe currents is not complicated by topographic convolutions. A valuable by-product of the CD mode is that in addition to local activity data, topographic details are collected through storage of the z coordinates of the micropositioning device that executes scanning. A CD-SECM setup has the components of a "normal" (= constant-height) mode biological SECM. The essential components are a low-noise potentiostat, a good micropositioning system with sub-µm precision, and an inverted stereomicroscope platform. An important additional feature is a distance control and scanning unit that employs a regulated feedback loop for continuous probe repositioning and maintenance of constant vertical tip-to-sample separation during line scans. The input signal for the feedback loop must be a distance-dependent property of the SECM tip and, among others, the vibrational amplitude of an oscillating scanning probe has been reported to be suitable (Ballesteros Katemann *et al.*, 2003; Hengstenberg *et al.*, 2000). Here, technical aspects of CD-SECM will be explained for an instrument with an optical readout of the impact of surface-proximal hydrodynamic shear-forces (SFs) on the actual vibration amplitude of resonating SECM tips. We also comment on device optimization and probe fabrication and the application of the methodology is illustrated with representative single-cell SF-CD-SECM experiments in which cellular chemical release and respiration are measured (Fig. 12.1B).

2. Shear-Force-Based Constant-Distance Scanning Electrochemical Microscopy for Live Cell Studies: Apparatus, Probes, and Operation

Detailed information and instructions on the design, construction, and use of a device suitable for performing SF-CD-SECM at single live cells are available in previous reports (Hengstenberg *et al.*, 2000; Pitta Bauermann *et al.*, 2004). It is assumed that the required hardware and software are available, enabling electroanalysis, feedback-controlled micropositioning, and visual inspection of cells and SECM tips. The following guidance is on the issues of system fine-tuning and tip preparation; establishment of the shear-force distance between a live cell and SECM tip. Its maintenance in the scanning mode and use for topography acquisition and probe placement at desired locations are additionally described.

2.1. Basic system requirements

Figure 12.2 shows the SF-CD-SECM setup with a laser-assisted readout of the tip-to-sample proximity and indicates the assembly of the components and their wiring. With the system in place, the following precautions are required for successful work on cultured cells:

Live Cell Scanning Electrochemical Microscopy

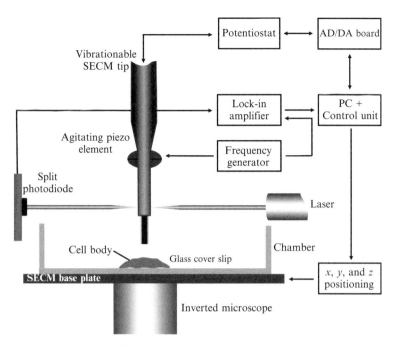

Figure 12.2 A diagram of a set-up for shear-force based constant-distance scanning electrochemical microscopy on live cells with a laser-assisted readout of the tip-to-sample separation. Computer-controlled entities are the potentiostat for the execution of electroanalysis schemes, the joint stepper motor and piezoelectric positioning elements for coarse and fine x, y, and z movements of the sample stage, the SECM tip with special holder, the inverted microscope for optical cell inspection, and finally the distance control unit with an agitation piezo and frequency generator for triggering SECM tip vibration, a laser and split photodiode for optical detection of tip-to-sample shear-force interaction, and lock-in amplifier for the phase-sensitive acquisition of the alternating difference current from the split photodiode.

- Currents produced by chemical activities of single cells are very small and often transient. The SECM for cell studies thus has to be able to record current with picoampere sensitivity and low millisecond time resolution. For amperometric experiments, for instance, a peak-to-peak noise of 5 pA or less is desirable at a 3 kHz filter setting. Hence, all SECM components have to be assembled within a Faraday cage and proper grounding, use of shielded cables and good cable placement ensured.
- During CD cell experiments, the flexible oscillating tip is operated at sub-micrometer distance from the cell surface. Accordingly, protection from external vibration is crucial to avoid cell damage. Efficient active vibration dampening for the equipment base plate is thus essential and even then operation in a quiet laboratory area is recommended.

- It is useful to attach a digital camera to the inverted microscope. This enables video observation of the cells and the SECM tip or at least the shadow of the approaching SECM tip. Their relative position can be followed at any time without the tip-to-sample distance regulation being affected by physical contact of the operator with the eyepieces.
- Two types of chambers have been adapted for SF-CD-SECM: (1) a cuvette in which the shear-force-detecting laser beam passes through glass windows and solution to the vibrating SECM tip, then to a photodiode for sensing of the Fresnel diffraction pattern (Hengstenberg et al., 2000) and (2) a flat-wall alternative in which the laser travels above cell walls through air from the source to the detector (e.g., Pitta Bauermann et al., 2004). Both versions work, but the flat-wall chamber is easier to use, with better signal intensities from the photodiode, while beam disturbances caused by scattering by the cuvette wall or solution are excluded.

2.2. Tailored vibrating scanning probes for catecholamine, oxygen, and nitric oxide detections with isolated live cells

The quality of the vibrating probes, and their mechanical and analytical attributes, are important for SF-CD-SECM on cultured cells. SECM tips must, for example, not be too stiff as to cause damage at shear-force contact with the soft objects they are scanning; but they must also not be too compliant, as they would then fail to properly report shear-forces to the distance control unit, through changes in their vibrational characteristics. For optical shear-force recognition a laser is focused on an upper part of the swinging tip and shear-force appearance is detected through changes in the difference current on the split photodiode. Suitable SECM tips for this scheme can be carbon fiber disk electrodes, with thinly polymer-insulated graphitic filaments sealed into long flexible glass pipettes (Fig. 12.3A). The laser spot is focused on the tapered glass capillary, though as close as possible to the protruding carbon fiber (Fig. 12.2). The stiffness of a pipette/fiber assembly can be adjusted through the shape of the glass taper and the length of the insulated carbon fiber that remains after the trimming step that exposes the electrode disc. Bare carbon micro disks work well for catecholamine detection while for oxygen (O_2) and nitric oxide (NO) measurements chemical surface modification (Fig. 12.1B) with platinum (O_2) or, for example, nickel (Ni) phthalocyanine (NO) (Bedioui et al., 1997) is required, to achieve the necessary sensitivity and selectivity. Details of the fabrication of the different variants of scanning tips (refer also to Fig. 12.3B) and advice on testing their quality is given below.

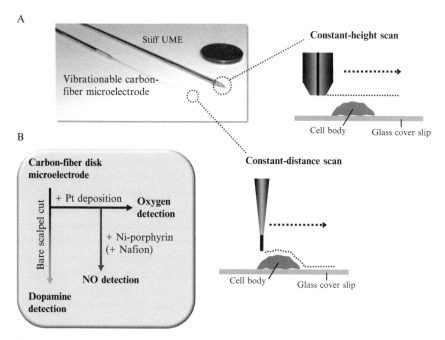

Figure 12.3 The geometry and surface modification of probes for shear-force based constant-distance scanning electrochemical microscopy on live cells. (A) Comparison of a flexible and thus vibrating tapered carbon fiber microdisk SECM tip suitable for shear-force based constant-distance SECM scans with a Pt microdisk electrode with a rigid body, as routinely used for the conventional constant-height mode of scanning. (B) A diagram showing the pathways of surface modifications that are required to activate carbon fiber microdisk SECM tips for catecholamine, oxygen, or nitric oxide live cell analysis.

2.2.1. Flexible carbon fiber SECM tips for measurements of chemical release from secretory cells

Materials and equipment. Carbon fibers (marketed with 5–10 μm diameters; boil fibers in acetone to remove surface sizing); glass capillaries (e.g., 100 mm long borosilicate glass capillaries with an outer and inner diameter of 1.5 and 0.75 mm); a two-component epoxy resin; an electrodeposition paint (EDP, available as anodic or cathodic systems; both work); straight pieces of copper wire of ∼0.5 mm diameter and 15 cm length; conductive carbon or silver paste; a platinum wire ring electrode; a glue stick; a dissection needle; scalpels; forceps; a wire cutter; a stereo (dissection) microscope; a homemade or commercial pipette puller; a DC power supply; an oven; $K_3[Fe(CN)_6]$; $[Ru(NH_3)_6]Cl_3$; dopamine; KCl, ultrapure water.

Instructions for probe preparation

1. Dip the end of the needle into the glue stick and use the sticky point to extract one carbon fiber from the bundle. Transfer the fiber (a few cm long) to the end of a Cu wire, using a drop of the conductive paste for attachment.

2. Feed the Cu wire from the fiber-free end into a glass capillary and push it through. Pull the end with the attached carbon fiber into the tube as far as to the middle. Fix the protruding Cu wire to the capillary wall with a drop of epoxy resin, and cut the outer part of the Cu wire to a convenient length.
3. Place a carbon fiber filled capillary in the holder of the pipette puller, taking care that the internal wire/fiber junction is just above the heating coil. Initiate pulling with parameters that produce pipettes with long and flexible tapers (Fig. 12.3A). The end of the fine glass tip should grip and seal the projecting fiber.
4. Setup an electrochemical cell with a Pt ring and the fiber/pipette assembly as electrodes. Fill the cell with EDP solution and let the carbon fiber enter the EDP suspension in the center of the Pt ring. Connect the carbon fiber electrode to the positive terminal of the power supply if anodic EDP is in use, or the negative pole for cathodic paints, respectively. Deposit the organic paint film at constant potential. Adjust the deposition voltage and time so as to obtain a thin and uniform EDP film on the carbon fiber. Cure the EDP coat by heat treatment according to the manufacturer's instructions.
5. Using the microscope and a scalpel blade, trim the EDP-insulated carbon fiber to a length of 0.3 mm or shorter, as then it is able to act as nondestructive transducer of shear-forces. Trimming also exposes a polymer-insulated graphitic disk as the SECM tip surface.
6. Check the prepared tips by cyclic voltammetry (CV) in electrolytes containing $K_3[Fe(CN)_6]$, $[Ru(NH_3)_6]Cl_3$, or dopamine. In all cases, characteristic steady-state microelectrode current–voltage curves should be obtained, with a steep increase to a plateau at potentials below and well above the formal potential of the compounds, respectively (see Fig. 12.4A and B). Identify good tips, rinse them with water, and store until use. Just prior to an SECM experiment at live cells expose a fresh carbon disk and check voltammetrically again the electrode quality to ensure the best analytical performance.

2.2.2. Flexible Pt-modified carbon fiber SECM tips for O_2 measurements close to single metabolically active cells

Materials and equipment. All items mentioned in Section 2.2.1 and also hexachloroplatinic acid (IV) hexahydrate, sulfuric acid, a potentiostat for platinum electroplating (e.g., the one in the SECM setup), a silver/silver chloride reference electrode, a platinum counter electrode, a nitrogen gas supply for oxygen removal from the plating solution.

Instruction for probe preparation

1. As described in Section 2.2.1, prepare some carbon fiber-based SECM tips.

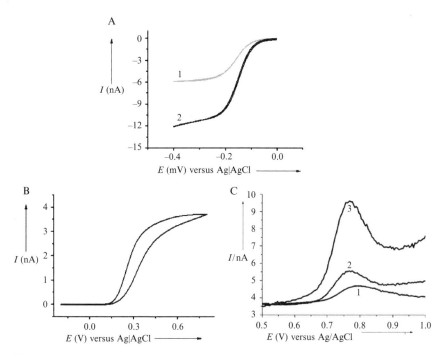

Figure 12.4 Voltammetric quality checks of carbon fiber (CF) microdisk tips for constant-distance scanning electrochemical microscopy. (A) Cyclic voltammograms obtained in the presence of about 5 mM [Ru(NH$_3$)$_6$]Cl$_3$ in 100 mM KCl with (1) a bare carbon fiber microdisk surface and (2) a Pt-coated carbon fiber microdisk surface. The superior diffusion-limited current of the Pt-modified tip indicates the gain in active surface area through the porosity of the finely dispersed Pt deposit. (B) Cyclic voltammogram of 1 mM dopamine in 100 mM phosphate buffer, pH 7, at a bare carbon fiber (CF) tip as prepared for constant-distance electrochemical microscopy (C). Differential pulse voltammograms obtained in the presence of about 10 μM NO with (1) a bare CF microdisk surface, (2) a platinum-coated CF microdisk surface, and (3) a platinum-coated CF microdisk surface that was additionally covered with a layer of a polymerized nickel phthalocyanine (Isik and Schuhmann, 2006, *Angew. Chem. Int. Ed.* 45, p. 7452—Reproduced with the permission of Wiley-VCH Verlag GmbH & Co. KGaA).

2. Ensure that the plating solution has been freed of oxygen by purging with nitrogen for several minutes. In three-electrode configuration deposit finely dispersed Pt particles onto the surface of freshly trimmed carbon fiber SECM tips using multiple cyclic sweeps between +300 and −500 mV versus reference at 100 mV s^1 in 5 mM H$_2$PtCl$_6$ at pH 0. Use a smaller (e.g., 3) or larger (e.g., 10) number of cycles to vary the amount of deposited Pt.
3. Rinse Pt-modified SECM tips with water and inspect the electrode modification through CV in a low mM K$_3$[Fe(CN)$_6$] or [Ru(NH$_{3)6}$]Cl$_3$ solution. A platinum electrodeposit is highly porous and increased values of the plateau currents are the sign of a successful surface modification (refer to Fig. 12.4A).

4. For live cell SECM, perform Pt-modifications just before use, or carry out further modification as described in Section 2.2.3.

2.2.3. Flexible nickel phthalocyanine-modified carbon fiber SECM tips for NO release measurements, for example, from endothelial cells

Materials and equipment. All items mentioned in Sections 2.2.1 and 2.2.2 and also nickel (II) phthalocyanine-tetrasulfonic acid tetrasodium salt (NiPcTS), sodium hydroxide, a 5% ethanolic solution of Nafion®, a NO standard of known concentration as described in published procedures (e.g., Lantoine *et al.*, 1995).

Instruction for the tip preparation

1. Using the steps described in Section 2.2.1, and 2.2.2 prepare a set of Pt-modified carbon fiber-based SECM tips.
2. Fill an electrochemical cell with a 1 mM solution of NiPcTS in 0.1 M NaOH and purge with nitrogen. In three-electrode configuration, electrochemically deposit NiPcTS onto the platinized SECM tips by 20 repetitive potential cycles between 0 and 1200 mV versus reference.
3. Rinse freshly Pt/NiPcTS-modified tips with water and inspect the quality of the catalytic deposits with differential pulse voltammetry (DPV) in the presence of NO for calibration of the NO concentration proportional current response. For good sensors, the DPV peak currents should correlate with the NO concentration (refer to Fig. 12.4C). If necessary, a semipermeable Nafion film may be placed on top of the Pt/NiPcTS by dip-coating in a solution of the material to create a top layer that protects against anionic interferences and thus enhances selectivity.
4. As they work best when freshly prepared, the immediate use of NO-sensitive CD-SECM tips is recommended.

2.3. Instructions for the execution of SECM cell topography and activity imaging

With healthy, well-adhering target cells and appropriate SECM tips fabricated as described above, the following sequence of steps has to be undertaken to assess the cell topography and/or local chemical profile with CD-SECM:

1. Put the measuring chamber containing the cultured cells in a minimal medium preferentially phosphate buffer (PBS) on the SECM sample stage. Add a chloridized silver wire as the second electrode of a two-electrode configuration.

2. Mount the vibrating carbon fiber-based SECM tip as working electrode and bring the agitation piezo into gentle physical contact with the glass stem of the probe. Adjust the vertical position of the tip electrode. It should be immersed and the tip should be positioned, based on microscope observation, a few tenths of micrometers above the cell-covered bottom of the measuring container.
3. With aid of the optical microscope choose a suitable cell and, using back- and forth focusing between the chosen cell and the SECM tip, place the tip at a horizontal (x, y) location next to the cell of choice that will bring it during a later sample approach into gentle shear-force contact with the chamber bottom.
4. Start the laser and center the light beam on the thin glass wall of the tapered CD-SECM tip at a point about 1 mm above the rim of the electrolyte chamber. Fine-tune the position of the split photodiode relative to the laser source and tip so that the resulting diffraction pattern is optimally placed on the detecting panel.
5. To establish shear-force-regulated tip-to-sample distance control, the carbon fiber SECM probe has to oscillate at a suitable resonance frequency. Frequency selection is based on recordings of the magnitude, R, of the AC photodiode signal on the log-in amplifier as a function of the frequency (f) of the AC voltage that stimulates the agitation piezo. R peaks appear at f values that produce resonance vibration. Typical resonance frequencies for flexible carbon fiber CD-SECM probes, as displayed in Fig. 12.3, are in the range 1–5 kHz. A resonance frequency is appropriate for use if (i) the corresponding R value with a freely swinging probe in bulk solution is stable, (ii) during a tip approach a rapid R drop toward zero is observed, indicating the onset of short-range tip/sample shear-force contact and consequent vibration hindrance, and (iii) the R versus tip-to-sample distance curves from multiple approaches and retractions are superimposable and reproducible.
6. At the predetermined optimal frequency for tip agitation, use the R value of the log-in amplifier as regulative input signal for the computer-controlled feedback loop that maintains constant tip-to-sample separation. Allow the positioning device to bring the tip slowly to shear-force distance from the sample, by smooth piezoelectric fine positioning (here the bottom of the chamber just next to a selected cell) and stop movement automatically when R decreases to, say, 75% of the value observed in bulk solution with free probe vibration (see Fig. 12.5A). Based on an approach curve analysis, the user-defined set point for R corresponds to a particular distance between the probe and the chamber surface and any change in R relates to a corresponding change in tip-to-sample separation. During a line scan, a decreasing R value may be observed, indicating an area with stronger vibrational dampening and thus a smaller tip-to-sample distance. The distance control software module acts to

keep R constant at the chosen set point level, causing the z-piezo positioning element to separate the tip and sample smoothly by a distance calculated from the slope dR/dz of the approach curve. The set point value for R, dR/dz, and an empirical dampening factor affecting the speed and strength of the responsiveness of the feedback loop are parameters that have to be fine-tuned with each CD-SECM experiment, based on the observations made for the used SECM tip.

Feedback distance control guides the SECM tip over a selected individual live cell in continuous shear-force distance. Plots of the operating voltage of the z-piezo positioning element as function of the position (Fig. 12.5B) provide information on the topography of the biological object. An obvious exploitation of linescan topography information during cell activity measurements is the accurate positioning of the tip on, for instance, the sloping side or flat top of a cell. A "sliced" chemical profiling of the environment of live cells is also possible, through a series of voltammetric measurements, the first acquired at shear-force distance and the succeeding ones recorded at known larger separations, after multiple retractions to defined tip-to-sample distances (Nebel et al., 2010).

3. SELECTED APPLICATIONS OF SF-CD-SECM LIVE CELL STUDIES

In vitro cell systems involved in specific SECM imaging experiments could be primary cell cultures prepared from fresh tissue, secondary cultures with cells coming from a reproduction of originally primary cells, or established continuous cell lines with members conveniently renewed for a reasonable number of passages by continuous cell division. In any case, it is important for optimal results in a SECM experiment that the studied cells grew well on their seeding plates prior to experiments with no sign of abnormalities. Additionally, their preferred artificial physiological buffer solution (e.g., PBS), with adapted composition and osmolarity, has to be used as electrolyte during the measurements.

Common model cells for studies concerning the mechanisms of the controlled release of hormones and neurotransmitters are neuroendocrine medullary chromaffin cells from bovine, rat or mouse adrenal glands, pheochromocytoma (PC) 12 cells derived from the cancerous rat adrenal medulla, mast cells, pancreatic β-cells, and human or animal endothelial cells. Their mechanisms of regulated release of catecholamines (chromaffin, PC12, and mast cells), insulin (β-cells), or NO (endothelial cells) can currently be studied with a number of optical, molecular biology, electrophysiological, and electrochemical techniques. These involve, for instance, the detection of FM-stained membrane components, protein mutant or knockout generation

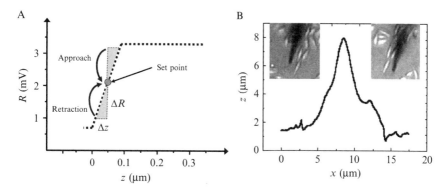

Figure 12.5 Live cell topography imaging with constant-distance mode scanning electrochemical microscopy. (A) The establishment of the constant-distance mode of scanning using approach curve measurements. A vibrating carbon fiber SECM tip is brought from bulk solution (shear-force-free vibration = maximal photodiode output R as provided by the log-in amplifier) into the region of shear-force (vibration dampening = decrease of photodiode output R). The approach to the sample is stopped at a user-defined set point and the distance control unit is instructed to keep the related R/z parameters constant during line scans. R/z variations due to up and downs on the scanned sample surface are corrected by balancing z tip displacements. Plots of the observed z displacements as a function of the x position are representative of the sample topography. (B). A single topographic line scan of a fibroblast cell as acquired in the constant-distance mode by scanning a vibrating carbon fiber SECM tip from one to the other side of the living object and displaying the z-piezo displacement required for maintenance of the same tip-to-sample separation as a function of the tip position in scan direction. The insets are optical images from the digital camera of the SECM device, confirming the location of the probe tip prior and subsequent to a full scan. (Pitta Bauermann et al., 2004—Reproduced with the permission of the PCCP Owner Societies; http://dx.doi.org/10.1039/B405233A.)

and analysis, patch clamp membrane capacitance measurements, and carbon fiber amperometry. Traditionally, the latter approach is performed in a classical electrophysiology setup with the polarized disk of the carbon fiber microelectrode brought manually into the necessary sub-μm detection distance from the cell membrane of the secretory cell, using a micromanipulator and stereo microscope. Achieving moderate cell contact, the observation of slight plastic plasma membrane deformation and a subsequent controlled retraction to the point of membrane leveling are the prerequisites for tip positioning, prior to cell stimulation and concomitant detection of compound release from the cell. Direct physical contact between the cell membrane and the microelectrode tip is, however, not optimal, since contamination of the active carbon disc of the microelectrode with cell debris is likely and an adverse impact on local membrane properties is also possible. When, on the other hand, the SF-CD-SECM is used for live cell studies, the active distance control unit prevents physical contact of the electroactive disk of the carbon fiber tip with the cell. Instead, the tip is continuously maintained at a shear-force

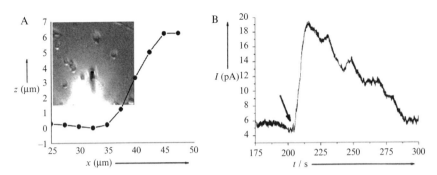

Figure 12.6 Local nitric oxide (NO) release measurements from a single transformed human umbilical vein endothelial cell (T-HUVEC) with the constant-distance mode scanning electrochemical microscope. (A) Utilization of a topography line scan for positioning a Ni phthalocyanine-modified NO-sensitive SECM probe in shear-force distance of about 300 nm immediately above the center of the cell. (Isik and Schuhmann, 2006, *Angew. Chem. Int. Ed.* 45, p. 7543—Reproduced with the permission of Wiley-VCH Verlag GmbH & Co. KGaA.) (B) Amperometric current response of a NO-sensitive SECM tip in a location as described in (A) to cell stimulation with 1 μM bradykinin. The arrow shows the time of bradykinin application (Isik and Schuhmann, 2006, *Angew. Chem. Int. Ed.* 45, p. 7543—Reproduced with the permission of Wiley-VCH Verlag GmbH & Co. KGaA).

distance of a few hundred nanometers during both cell approach and scanning. Accordingly, the problems associated with membrane contact are avoided, and the chances of unaffected measurements are hence correspondingly higher.

The placement of the SECM tip at a desired location above an individual single cell for a release activity assessment works by repetition of a line scan that was originally acquired for determining cell topography stopping the scan movement at a predefined favored place. At the end of such a procedure, the SECM tip is positioned in a gap-like configuration above the live cell membrane and able to electrochemically oxidize (or reduce) secreted molecules. The real-time electrochemical signal generation corresponds to the release of the compounds from the cell. Figure 12.6A illustrates the procedure for the placement of a Pt/NiPcTS-modified SECM tip with NO sensing ability on top of a single transformed human umbilical vein endothelial cell (T-HUVEC) cell. With the tailored sensor in place and poised at the working potential (e.g., +0.75 V vs. Ag/AgCl) while continuously recording the NO oxidation current amperometrically, temporary exposure of the cell to the stimulating agent bradykinin causes NO release and, after a short delay, a transient increase in the current response with a maximum reached within a few seconds is observed (see Fig. 12.6B). Similarly, a bare carbon fiber SECM tip positioned on top of a single bovine chromaffin cell and polarized to a potential of +0.75 V versus Ag/AgCl detects the

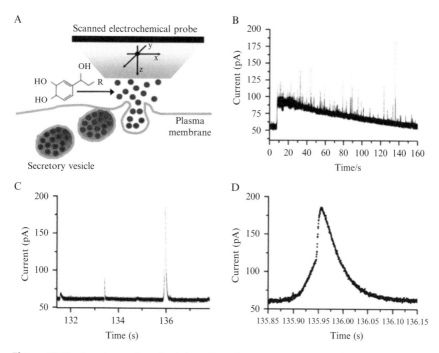

Figure 12.7 Local noradrenalin/adrenalin release measurements from single bovine chromaffin cells with the constant-distance mode scanning electrochemical microscope. (A, B) During Ca^{2+}-dependent single vesicle exocytosis a storage vesicle is moved through priming and docking processes into physical contact with the intracellular side of the plasma membrane. Membrane depolarization by high KCl application and consequent opening of Ca^{2+} channels leads to a rise in membrane-proximal intracellular [Ca^{2+}] and then to the fusion of the storage vesicle with the plasma membrane, opening of fusion pores, and finally burst release of intravesicular catecholamines. Information from a shear-force controlled topography line scan brought the bare sensing disk of a carbon fiber SECM on top of the live cell into proximity to the release sites in the plasma membrane. The carbon fiber SECM tip adjacent to release sites detects the secreted catecholamine by their oxidation at $+800$ mV as a shower of current spikes, each representing liberation of the contents of a single vesicle. (C, D) Individual amperometric current spikes displayed at higher temporal resolution. (Schulte and Schuhmann, 2006, In: *Electrochemical Methods for Neuroscience* (eds. A.C. Michael, L.M. Borland), chapter 17, p. 366—Reproduced with the permission of Taylor & Francis group LLC via the Copyright Clearance Center CCC.)

Ca^{2+}-dependent vesicular release of adrenaline/noradrenaline upon cell stimulation via, for example, membrane depolarization by application of KCl solution (100 mM; Fig. 12.7). Many mature vesicles from the so-called readily releasable pool can fuse with the membrane and release their content into the gap between the positioned SECM tip and the cell membrane with each membrane depolarization. The current versus time plots thus show the expected appearance of a shower of amperometric

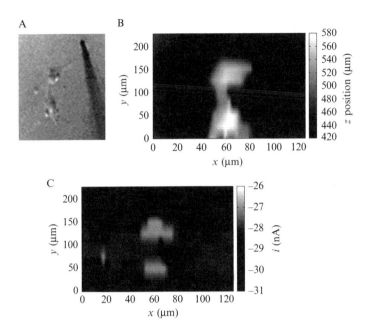

Figure 12.8 Topography and local oxygen (O_2) measurements on a single transformed human umbilical vein endothelial (T-HUVEC) cell with the constant-distance mode scanning electrochemical microscope. (A) Two adjacent T-HUVEC cells as seen in the experiment with the digital camera of the inverted microscope of the SECM device. (B) A bird's-eye view presentation of the topography of the two target cells as obtained by constant-distance SECM imaging and shear-force controlled guidance of the SECM probe, (a Pt/NiPcTS catalyst-modified vibrational carbon fiber microelectrode of about 7 μm diameter), across the sampled area. The plot shows the change in the z-piezo elongation or contraction needed to maintain constant distance throughout line scans as a function of x and y probe position. (C) O_2 reduction SECM tip current as function of x, y tip position. The image is a map of the local distribution of oxygen in the imaging plane above the sampled area. Lower analyte levels are observed around the two living cells due to their metabolic activity and aerobic respiration. The oxygen measurement at the SECM tip was carried out in the redox competition (RC) mode by applying repetitively the following potential pulse profile to the sensing disk: 1 s at 0.5 V, 0.2 s at 1.4 V, and 0.5 s at −0.65 V (all vs. reference electrode). The last parameter of the sequence is the cathodic detection potential and current values for SECM image creation were sampled at 20 ms after applying the potential of −0.65 V.

spikes with fast rise and decay times and amplitudes of up to a few hundreds of pA (see Fig. 12.7B). The observation of a prespike "foot" signals, which reflects leakage of catecholamine molecules through partially open fusion pores ahead of complete fusion (Fig. 12.7D), is an evidence of the excellent quality of SF-CD-SECM live cell release measurements.

Cellular respiration, the continuous exchange of dissolved O_2 or CO_2 through the plasma membrane is essential for the function of most cells, of tissue and ultimately of the whole body. An inadequate intracellular O_2

partial pressure occurs in the pathological conditions of hypoxia or anoxia, deriving from a lowered concentration of O_2 within the arterial blood. Moreover, an abnormally high rate of cellular O_2 uptake is characteristic of cancer cell agglomerations with elevated metabolic demands. This involvement in the mentioned two disease states is a good reason for *in vitro* studies of live cell respiration and many examples are available in the literature. Here, we present the outcome of a novel SF-CD-SECM single-cell respiration experiment that takes advantage of the distance regulation and topography imaging capability of the technique, together with a recently proposed type of electrochemical analyte detection: amperometric O_2 reduction in the redox competition (RC) mode (Eckhard *et al.*, 2006). In the RC mode, the electrochemical assessment of oxygen works with a pulsed potential, instead of the conventional constant potential profile. The advantage of a pulsed O_2 detection is that the SECM tip is only intermittently, and on a millisecond time scale, reducing the analyte species, so in terms of oxygen consumption it is not permanently in competition with the respiring cell. Figure 12.8A shows two adjacent T-HUVEC cells as seen in course of the experiment with the aid of the inverted microscope and digital camera of the SECM setup. Figure 12.8B illustrates in a bird's-eye view the topography of the two target cells, obtained by constant-distance SECM imaging and a shear-force controlled guidance of the SECM probe (a Pt-modified vibrational carbon fiber microelectrode of about 7 μm diameter) across the sampled area. Figure 12.8C shows the simultaneously acquired RC oxygen reduction SECM tip current as function of the x, y tip position, which is a representative measure for the local distribution of oxygen in the sampled area. Clearly, lower oxygen levels are observed at the location of the two living cells, confirming their aerobic respiration.

In summary, we have shown with three examples of biological applications that SF-CD-SECM is a promising complementary analytical scheme for real-time, *in vitro* topography, and activity studies of live cells. The methodology requires a skilled and dedicated operator with cross-disciplinary knowledge of instrumentation, computation, and electroanalysis and cell biology. However, efforts with this approach are worthwhile because of the valuable information that can be obtained about the local properties of the microenvironment of single live cells.

ACKNOWLEDGMENTS

The authors thank Dr. David Apps, Biochemistry Reader (retired), Centre for Integrative Physiology, Edinburgh University, Scotland for his critical manuscript reading and language improvements. A. S. is grateful for support from the Suranaree University of Technology through funds to the Biochemistry—Electrochemistry Research Unit. The authors are grateful to Dr. Kathrin Eckhard for the images of the O_2 consumption at a pair of living cells.

REFERENCES

Allison, P. A., Mortensen, N. P., Sullivan, C. J., and Doktycz, M. J. (2010). Atomic force microscopy of biological samples. *Nanomed. Nanobiotechnol.* **2**, 618–634.

Amemiya, S., Bard, A. J., Fan, F. R. F., Mirkin, M. V., and Unwin, P. R. (2008). Scanning electrochemical microscopy. *Annu. Rev. Anal. Chem.* **1**, 95–131.

Axelrod, D. (2008). Total internal reflection fluorescence microscopy. *Methods Cell Biol.* **89**, 169–221.

Ballesteros Katemann, B., Schulte, A., and Schuhmann, W. (2003). Constant-distance mode scanning electrochemical microscopy (SECM). Part I: Adaptation of a non-optical shear-force-based positioning mode for SECM tips. *Chem. Eur. J.* **9**, 2025–2033.

Bedioui, F., Trěvin, S., Devynck, J., Lantoine, F., Brunet, A., and Devynck, M.-A. (1997). Elaboration and use of nickel planar macrocyclic complex-based sensors for the direct electrochemical measurement of nitric oxide in biological media. *Biosens. Bioelectron.* **12**, 205–212.

Dickensen, N. E., Armendariz, K. P., Huckabay, H. A., Livanec, P. W., and Dunn, R. C. (2010). Near-field scanning optical microscopy: A tool for nanometric exploration of biological membranes. *Anal. Bioanal. Chem.* **396**, 31–43.

Eckhard, K., Chen, X., Turcu, F., and Schuhmann, W. (2006). Redox-competition mode of scanning electrochemical microscopy for visualisation of local catalytic activity. *Phys. Chem. Chem. Phys.* **8**, 5359–5365.

Francis, L. W., Lewis, P. D., Wright, C. J., and Conlan, R. S. (2010). Atomic force microscopy comes of age. *Biol. Cell* **102**, 133–143.

Halbhuber, K.-J., and König, K. (2003). Modern laser scanning microscopy in biology, biotechnology and medicine. *Ann. Anat.* **185**, 1–20.

Hengstenberg, A., Kranz, C., and Schuhmann, W. (2000). Facilitated tip-positioning and applications of nonelectrode tips in scanning electrochemical microscopy using a shear force based constant-distance mode. *Chem. Eur. J.* **6**, 1547–1554.

Isik, S., and Schuhmann, W. (2006). Detection of nitric oxide release from single cells by using constant-distancemode scanning electrochemical microscopy. *Angew. Chem. Int. Ed.* **44**, 7451–7454.

Lantoine, F., Trěvin, S., Bedioui, F., and Devynck, J. (1995). Selective and sensitive electrochemical measurement of nitric oxide in aqueous solution: Discussion and new results. *J. Electroanal. Chem.* **392**, 85–89.

Liu, L., Wang, Y., and Zhang, Y. (2009). Scanning ion conductance microscopy and its application in nanobiology and nanomedicine. *Adv. Mater. Res.* **60–61**, 27–30.

Mooren, O. L., Erickson, E. S., Dickenson, N. E., and Dunn, R. C. (2006). Extending near-field scanning microscopy for biological studies. *J. Assoc. Lab. Autom.* **11**, 268–272.

Nebel, M., Eckard, K., Erichsen, T., Schulte, A., and Schuhmann, W. (2010). 4D shearforce-based constant-distance mode scanning electrochemical microscopy. *Anal. Chem.* **82**, 7842–7848.

Pitta Bauermann, L., Schuhmann, W., and Schulte, A. (2004). An advanced biological scanning electrochemical microscope (BIO-SECM) for studying individual living cells. *Phys. Chem. Chem. Phys.* **6**, 4003–4008.

Schulte, A., Nebel, M., and Schuhmann, W. (2010). Scanning electrochemical microscopy in neuroscience. *Annu. Rev. Anal. Chem.* **3**, 299–318.

Schulte, A., and Schuhmann, W. (2007). Single-cell microelectrochemistry. *Angew. Chem. Int. Ed.* **46**, 8760–8777.

Schulte, A. and Schuhmann, W. (2006). 17. Scanning electrochemical microscopy as a tool in neuroscience. In: *Electrochemical methods in neuroscience.* (eds. A.C. Michael, L.M. Borland), Taylor & Francis - CRC press, Boca Raton, USA, pp. 353–372.

CHAPTER THIRTEEN

Visualization of TGN-Endosome Trafficking in Mammalian and Drosophila Cells

Satoshi Kametaka *and* Satoshi Waguri

Contents

1. Introduction 256
2. Molecular Tools 258
3. Live-Cell Imaging in Mammalian Cells 260
    3.1. Transfection in HeLa cells 260
    3.2. Setup of microscope system 261
    3.3. Simple time-lapse observation of the TGN-derived transport carriers 262
    3.4. Tracking transport carriers after photobleaching 262
    3.5. Dual-color imaging of the TGN-derived transport carriers and clathrin adaptor molecules 263
    3.6. Visualization of the interaction between TGN-derived transport carriers and transferrin-containing endosomes 264
    3.7. FRAP analysis for TGN-endosome transport kinetics 265
    3.8. FRAP analysis for exchange kinetics of AP1 and GGAs between the TGN membrane and cytosolic pools 265
4. Live-Cell Imaging in Drosophila Cells 266
    4.1. Generation of S2 clones stably expressing mCherry-LERP 267
    4.2. Visualization of LERP-positive transport carriers in S2 cells. 268
5. Conclusion Remarks 269
References 269

Abstract

Mannose 6-phosphate receptors (MPRs) are known to be shuttled between the trans-Golgi network (TGN) and endosomes, thereby several lysosomal hydrolases are delivered through the endocytic pathway into lysosomes. This interorganellar transport is mediated by transport intermediates, now called transport carriers. Previous studies employing green fluorescent protein (GFP)-based live-cell imaging demonstrated that these transport carriers are

Department of Anatomy and Histology, Fukushima Medical University School of Medicine, Fukushima, Japan pleiomorphic structures composed of tubular and vesicular elements. Introducing a time-axis into light microscopic observations enabled us to identify transport carriers that are derived from or targeted at a distinct organelle. In this study, we describe several methods for the observation of GFP-tagged MPRs. Photobleaching the peripheral region of a cell before a time-lapse observation allows us to monitor TGN-derived transport carriers for longer periods (more than 4 min). Events of their targeting into endosomes can be visualized by dual-color imaging of both GFP-MPRs and fluorescently tagged transferrin that is internalized by cells. By using a technique of fluorescence recovery after photobleaching (FRAP), we can analyze overall cycling kinetics of MPRs in a single cell. Transport of MPRs is regulated by several cytosolic factors like clathrin adaptors, AP1, and GGAs. The adaptors on the TGN membranes are exchanging with their cytosolic pool, which can also be analyzed by FRAP. In addition, the relationships of the MPR-containing transport carriers that left the TGN and the adaptors can be visualized by dual-color imaging. A similar system of membrane transport and its regulation is well documented in drosophila cells. As *Drosophila melanogaster* has only a single MPR (LERP), AP1, or GGA, it is an ideal model system for the understanding of specific functions of each cytosolic factor. To visualize these molecules in drosophila cells, however, we need to consider that multiple Golgi dots exist scattered in the cytoplasm. Thus, the Golgi dots or endosomes should be identified before live-cell imaging.

1. INTRODUCTION

Targeting of cellular components to their own destination is crucial for cells to maintain homeostasis and to adopt various environmental conditions. Integral membrane proteins and secretory proteins are initially synthesized in the endoplasmic reticulum and delivered through the Golgi compartments to the trans–Golgi network (TGN), where they are sorted for the post-Golgi compartments including plasma membrane, endosomes, and lysosomes. In previous studies, each transport pathway between organelles had been described as being mediated by transport vesicles (Rothman and Wieland, 1996). The introduction of live-cell imaging using GFP (green fluorescent protein)-technology clearly demonstrated that these organelles, in fact, look like pleiomorphic structures composed of vesicular or tubular elements (Bonifacino and Lippincott-Schwartz, 2003; Ghosh *et al.*, 2003). Importantly, observations of such transport carriers along the temporal axis have provided us with the directional information of the moving structures; we can therefore describe them as "x-derived" or "x-targeting" transport carriers.

The following section mainly deals with the trafficking of the mannose 6-phosphate receptors (MPRs). MPRs capture newly synthesized lysosomal enzymes at the TGN and transport them to endosomes for the degradation

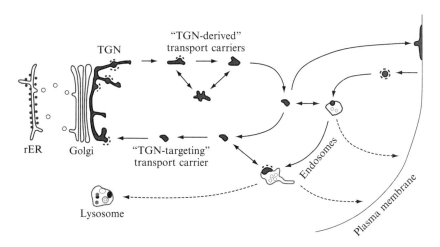

Figure 13.1 A model for the post-Golgi trafficking of MPR proposed by visualization studies. Intracellular compartments or membrane domains containing high concentrations of MPRs are depicted as gray with a black limiting line. The dotted lines indicate clathrin coats. For more details please see the text.

of endocytic or autophagic materials (Fig. 13.1). After the delivery of enzymes, MPRs can return to the TGN for the next round of transportation, thus cycling between the TGN and endosomes. Moreover, they are known to appear on the cell surface contributing extracellular secretion of lysosomal enzymes. Further, one of the receptors, cation-independent MPR (CIMPR: another is cation-dependent MPR), functions on the cell surface as a scavenger receptor for insulin-like growth factor II (Ghosh et al., 2003). Thus, they cycle between at least three post-Golgi organelles, making the live-cell observations complicated. By fluorescence recovery after photobleaching (FRAP) technique, however, we can investigate the overall transport dynamics of MPRs, which easily tell us that MPRs are in rapid equilibrium between the post-Golgi compartments (Waguri et al., 2006). Another important aspect in MPR trafficking is the regulatory mechanisms by the clathrin adaptor molecules, adaptor protein complex 1 (AP1) and Golgi-localized, gamma ear-containing, Arf-bindig proteins 1–3 (GGA1–3). At the TGN these adaptors connect the MPRs or other cargo proteins to the clathrin triskelions contributing to the incorporation of MPRs into the clathrin-coated vesicles. This segregation and packaging events are thought to increase the efficiency in the correct sorting of MPRs. FRAP analysis also revealed that the association of the adaptors to the TGN membrane is not stable but rapidly exchanging with the cytosolic pool (Kametaka et al., 2005; Puertollano et al., 2001; Wu et al., 2003). Moreover, live-cell imaging of these proteins revealed that they are not only localized on the TGN and endosomes but also on the transport carriers (Puertollano et al., 2001, 2003; Waguri et al., 2003). It appears that both adaptors are involved in anterograde- and retrograde-transport, but the sites and modes of function are still under controversy.

Recently, Dennes and coworkers identified LERP (*l*ysosomal *e*nzyme *r*eceptor *p*rotein), as an ortholog of mammalian CIMPR in the genomic database of fruit fly, *Drosophila melanogaster* (Dennes *et al*., 2005). Many of the mammalian proteins that are involved in cargo sorting and transport carrier formation are also conserved in the fly genome, most often as a single protein for each (Boehm and Bonifacino, 2001), indicating that this organism possesses similar mechanisms to those used in mammalian cells. More recently, we (Kametaka *et al*., 2010) and Hirst *et al*. (Hirst *et al*., 2009) showed that transport of LERP involves the function of drosophila AP1 and GGA. As a result, we now know that the sorting system of CIMPR is conserved from the fly to mammals. Mammalian cells contain multiple isoforms for each subunit of AP1 and GGA (Robinson and Bonifacino, 2001), which often caused difficulties in understanding specific functions of each molecule. On the other hand, drosophila cells have only a single set of AP1 subunits or single GGA, being an ideal model system for the analysis of TGN-endosome transport.

In this section, we describe methods for live-cell imaging of TGN-endosome transport by using GFP-tagged MPRs, AP1, and GGA. We especially focus on the detection of TGN-derived or -targeting transport carriers, and some application of FRAP analysis for the overall post-Golgi transport kinetics and exchange kinetics of AP1 and GGAs. We also mention how to observe drosophila cells for live-cell imaging in the final section.

2. Molecular Tools

There have been several fluorescent proteins (FPs)-tagged tools in previous studies that have described the behavior of molecules involved in the TGN-endosome transport (summarized in Table 13.1). MPRs are type-I transmembrane proteins whose cytoplasmic domains are implicated in the interaction with several cytoplasmic factors such as AP1, GGAs, PACS1, TIP47/rab9, and retromer complex. Thus, FP fused to the cytoplasmic domain of MPRs might be sufficient for mimicking the MPR transport, making it applicable as an indicator for the TGN-endosome transport in a model experimental system (Puertollano *et al*., 2003; Waguri *et al*., 2003). However, it should be noted that the luminal domain of the CIMPR is involved in the tight interaction with the endocytic compartment by unknown mechanisms (Waguri *et al*., 2006). As GGAs are also known to interact with the free C-terminal end of the MPR tail, FP-tagging at the C-terminus may cause aberrant behavior of the molecule (Cramer *et al*., 2010). Therefore, we recommend the insertion of FP after the first processing site in the N-terminal region of MPRs.

Table 13.1 FP-tagged molecular tools for visualizing the TGN-endosome transport

Name of the construct	Description	Literature
Mammals		
CDMPR-CFP	CFP fused to the C-terminus of human CDMPR	Puertollano et al. (2001)
GFP-CI-MPR	GFP fused to the TM^a and tail region of mouse CIMPR.	Waguri et al. (2003)
G-CIMPRtail	GFP fused to the TM^a and tail region of human CIMPR	Waguri et al. (2006)
G-CIMPRfull	GFP fused to the N-terminus of the full length human CIMPR	Waguri et al. (2006)
YFP-µ1	YFP inserted in the middle of µ1 subunit of human AP1	Huang et al. (2001)
YFP-γ1	YFP fused to the C-terminus of γ1 subunit of human AP1	Waguri et al. (2003)
YFP-GGA1	YFP fused to the N-terminus of human GGA1	Puertollano et al. (2001)
PAGFP-GGA3	PAGFP^b fused to the N-terminus of human GGA3	Kametaka et al. (2005)
GFP-rab9	GFP fused to the N-terminus of human RAB9	Barbero et al. (2002)
YFP-VPS35	YFP fused to the N-terminus of human VPS35	Arighi et al. (2004)
Drosophila		
mCherry-LERPfull	mCherry inserted in the site right after the signal sequence of LERP	Kametaka et al. (2010)
GFP-LERP	GFP fused to the TM^a and tail region of LERP	Hirst et al. (2009)
GFP-dGGA	GFP fused to the N-terminus of dGGA	Kametaka et al. (2010)
GFP-CLC	GFP-fused to the N-terminus of dClc	Kochubey et al. (2006)

^{a} TM, transmembrane.
^{b} PAGFP, photoactivatable GFP.

Using FP-tagged AP1 complex is more difficult due to the heterotetrameric complex of AP1, which is composed of β-, γ-, σ-, and µ-subunits. Moreover, it functions as an adaptor between membrane cargo proteins and clathrin triskelion. Therefore, addition of FP of the size 30 × 40 Å in the case of GFP to any subunit would impede the complex formation and the adaptor function. So far, the insertion of yellow FP (YFP) in the middle region of μ-subunit and the tagging with FP at the C-terminus of γ-adaptin have been successful; they are incorporated into the AP1 complex (Huang *et al.*, 2001; Waguri *et al.*, 2003). As for GGA, only tagging at N-terminus has worked well (Kametaka *et al.*, 2005; Puertollano *et al.*, 2001, 2003).

3. LIVE-CELL IMAGING IN MAMMALIAN CELLS

3.1. Transfection in HeLa cells

Expression vectors containing FP-fusion constructs are transiently expressed in HeLa cells using a lipofection reagent as follows:

1. HeLa cells are grown in α-MEM (Invitrogen) supplemented with 10% FBS (Hyclone Laboratories, Inc.), 2 mM glutamine, 100 U/ml penicillin, and 100 μg/ml streptomycin at 37 °C in a humidified atmosphere of 5% CO_2. As a passage procedure, ~10% of cells are transferred to a new dish when they reach a confluency of approximately 80%.
2. Two days before the live-cell imaging, HeLa cells are seeded onto the Delta T dish at a confluency of approximately 50% in 1 ml of culture medium and culture overnight. The Delta T dish is coated with thin underside film surface and has electrodes for heating, constituting a part of the Delta T® Open Dish System (Bioptechs, Inc., Butler, PA; Fig. 13.2).
3. One day before the imaging, a half microgram of the expression vector is diluted in 100 μl of Opti-MEM (Invitrogen) and 1.0 μl of FuGene6 transfection reagent (Roche) is added into the diluted DNA solution directly, and immediately mixed vigorously for a few seconds at room temperature.

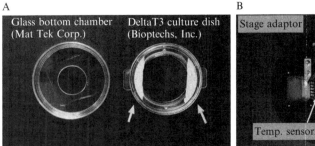
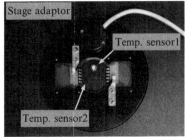

Figure 13.2 Cell culture units for live-cell imaging. (A) Glass-bottom dishes with (Delta T dish, right) or without (Mat Tek glass-bottom dish, left) electrodes (arrows) for heating. (B) Microscope stage unit for Delta T dish with two temperature sensors.

4. After incubation for 10 min at room temperature, the DNA/FuGene6 complex solution is added to the cells prepared in step 2. The cells are cultured for 18–24 h.

   Note 1:. To avoid generation of autofluorescent materials in the cells, cells must be passaged before overgrowing.

   Note 2:. The best window of the transfection time is relatively narrow. Although longer transfection time usually causes higher levels of protein expression, they should be kept as low as possible. Otherwise perturbation of intracellular trafficking might occur. Overexpression of integral membrane protein, MPRs, often causes retention at the rER. Also, high expression of AP1 subunits or GGA might cause dominant negative effects. Thus, the optimum transfection conditions should be determined in each construct.

   Note 3:. Using cell lines stably expressing FP-tagged proteins are preferable because this at least indicates that the expression of the FP-tagged protein is not so harmful as it causes cell death.

3.2. Setup of microscope system

To observe live-cells, the Delta T® Open Dish System (Bioptechs, Inc.; http://www.bioptechs.com/index.html) that allows more accurate controlling of the microenvironment temperature is used (Fig. 13.2). The Delta T dish can be set on most of the contemporary inverted microscopic systems by specific stage adaptors. If an observation period exceeds 2 h, we use another stage top incubation system equipped with a CO_2 gas mixer (TOKAI HIT, Japan). As for the microscope system, we have used two types of laser scanning confocal microscope, Carl Zeiss system (LSM510: Axiovert 100M equipped with a Plan-Apochromat 63× 1.4 NA objective lens) and Olympus system (FV1000: IX81 equipped with UPlanSApo 60× 1.35 NA objective lens). A video-microscope system with cooled CCD can also be used for simple time-lapse observation of a single molecule. However, a laser scanning microscope system would be preferable in the case of dual-color observation of transport carriers and the applications of the photobleaching technique.

1. One hour prior to observation, the culture medium in the Delta T dish is replaced with 1.0 ml of fresh complete medium supplemented with 10 mM HEPES (pH 7.4).
2. The completion of the whole microscope system including the mercury lamp, laser emission, time-lapse parameter setting in the PC program, and the Delta T dish controller device is confirmed.
3. The Delta T dish with transfected cells is set on the heat controlling stage adaptor. Meanwhile, a drop of immersion oil is laid on the lens and the temperature sensor 1 (Fig. 13.2) of the stage adaptor.

4. Approximately 800 µl of mineral oil (SIGMA, M8410) is slowly overlaid on the medium to prevent evaporation and pH change of the medium. The medium temperature is then set at 37 °C by the controller device.
   Note:. Switching on the microscope system causes a temporal increase in the temperature around optical pathway that might lead a shift of focal plane during recording periods. Thus, it is better to wait for about 1 h before live-cell observation, especially if each time-lapse session takes more than 30 min.

3.3. Simple time-lapse observation of the TGN-derived transport carriers

1. The first step is selecting cells or part of cells to be observed. When a FP-tagged molecule is transiently expressed, cells expressing low or modest levels of FP-tagged proteins should be chosen. Cells with diffuse high background signal or several aggregated signals for FP-MPR or FP-AP1 should be avoided. After capturing the whole cell image, one-third to half of the cell area containing the TGN region is selected.
2. The pinhole size of the confocal parameters is set to near maximum to obtain more signals at the expense of lowering confocal effects. Laser power should be as low as possible to prevent photobleaching during scanning. Gain and offset should be arbitrarily adjusted by comparing the final images. These parameter settings should be performed in a preliminary experiment.
3. The level of focal plane should be determined so that both the TGN and peripheral areas are included in the same focal plane. Selected cells are often excluded from the analysis by this process.
4. Under normal conditions, the TGN-derived transport carriers containing FP-MPR or FP-AP1 move at a maximum speed of ~ 1 µm/s (Puertollano et al., 2003; Waguri et al., 2003). Therefore, we routinely capture images with 1 or 2 s of scanning without interval. Three to 5 min of observation periods would be enough to track some TGN-derived transport carriers.
5. Captured images can be further analyzed by software such as Meta-Morph Imaging System (Universal Imaging Corporation, West Chester, PA) or Image J with appropriate plug-ins (http://rsbweb.nih.gov/ij/).

3.4. Tracking transport carriers after photobleaching

TGN-derived transport carriers can fuse or interact with preexisting transport carriers or endosomes that also contain FP-MPRs, making a peripheral network (Fig. 13.1; Waguri et al., 2003). This characteristic behavior often makes it difficult to follow each transport carrier for longer periods of time. However, this situation can be overcome by combining the photobleaching procedure before time-lapse observations.

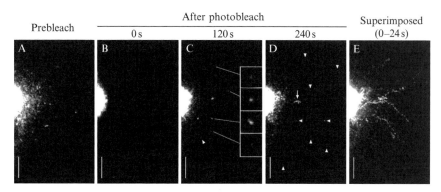

Figure 13.3 Time-lapse observations of the TGN-derived transport carriers containing GFP fused to the cytoplasmic domain of CIMPR (G-CIMPR-tail). HeLa cells expressing G-CIMPR-tail (A) were processed for selective photobleaching in their peripheral signal (B), followed by image acquisition every 2 s. Images taken before photobleaching, and at 0 (B), 120 (C), and 240 s (D) after photobleaching are shown. TGN-derived transport carriers are indicated by white arrowheads or shown in insets in (C). A white arrow indicates a long tubular carrier that left the TGN. Tracks followed by each transport carrier can be traced by superimposing time-lapse images (E). Scale bar: 10 μm. Modified from Waguri et al., 2006.

1. After setting a region of interest within a cell (as mentioned in Section 3.3, step 1), the peripheral region was selectively photobleached with a strong laser light, followed by time-lapse image acquisition with attenuated laser light as mentioned in Section 3.3. (Fig. 13.3). The program for this photobleaching operation should be found in latest laser scanning confocal microscope systems.
2. Likewise, transport carriers that come from the peripheral region and target to the TGN can also be observed (Waguri et al., 2006).
   Note:. Conditions for photobleaching should be carefully determined so that the total amount of light exposure is within a minimal requirement. Scanning with strong laser light for longer times in large areas are toxic, often causing detachment of cells.

3.5. Dual-color imaging of the TGN-derived transport carriers and clathrin adaptor molecules

Transport of MPRs is well known to be regulated by clathrin adaptors, AP1 and GGA1-3 (Ghosh et al., 2003; Robinson and Bonifacino, 2001). Therefore, it is interesting to observe the relationship between TGN-derived transport carriers and clathrin adaptors. Basic methods for transfection and microscopic observations are described as above. Here, we describe some notes that are specific for dual-color imaging of CFP-MPR and YFP-AP1.

1. *Transfection*: vector plasmids containing CFP-MPR (CFP is fused to the cytoplasmic domain of CIMPR) and YFP-γ adaptin are cotransfected

into cells. The relative amount of each plasmid to be mixed can be adjusted considering that the expression of CFP-MPR appears to be much easier than YFP-γ adaptin.
2. *Filter set*: CFP is excited by a laser beam of 458-nm and detected through a 475–515-nm band pass filter, while YFP is excited by a laser beam of 514-nm and detected through a 530–600-nm band pass filter (Zeiss LSM510).
3. *Time-lapse observation*: We may need to detect coating or uncoating events of AP1 on the carriers, which is much faster (less than a few seconds) than the movement of transport carriers. Therefore, Y-axis length of a scanning frame should be as narrow as possible so that each scanning time is less than 1 s. During the scanning, detection modes for both signals should be interchanged line-by-line, not frame-by-frame. This operation enables near simultaneous acquisition of both signals and reduces signal cross talk. The latest laser scanning confocal microscopes should include this program.

3.6. Visualization of the interaction between TGN-derived transport carriers and transferrin-containing endosomes

To see the complete transport pathway from the TGN to endosomes, we need to observe TGN-derived transport carriers that target the endosomal compartments or plasma membrane. For this purpose, we labeled the early/recycling endosomes with Alexa594-conjugated transferrin (Tfn), which is continuously internalized by receptor-mediated endocytosis. In this case, dual-color imaging should be combined with photobleaching procedures as mentioned in Sections 3.4. and 3.5.

1. Alexa594-conjugated Tfn (Invitrogen, Cat. T-13342) is reconstituted by dissolving in 1 ml of dH_2O. The stock solution (concentration of 5 mg/ml) was kept at 4 °C and away from light.
2. A Delta T dish containing cells expressing GFP-MPR is set on the microscope as mentioned in Section 3.2.
3. 5 µl of the stock solution is added to the Delta T dish with 1 ml culture medium (final concentration of 25 µg/ml). The medium is gently mixed with a 200 µl pipette tip.
4. A cell is chosen as mentioned in Section 3.3. In this step, you cannot see the Alexa594-Tfn-containing endosomes through the eyepieces as the medium still contains a high concentration of Alexa594-Tfn. The endosomal signal can only be checked by operating laser scanning confocal microscope.
5. Several GFP-MPR-positive dots in the peripheral region are eliminated by photobleaching and dual-color time-lapse imaging is carried out with 1–2 s of scanning without intervals.

3.7. FRAP analysis for TGN-endosome transport kinetics

Although it is difficult to follow all the transport carriers in a single cell, overall cycling kinetics of FP-MPR can be assessed by FRAP technique. This application was first described while analyzing the cycling of lipid raft between the Golgi and plasma membrane (Nichols et al., 2001, JCB). GFP-MPR has been shown to cycle between the TGN and peripheral region including endosomes and plasma membrane within an hour (Waguri et al., 2006).

1. Transfected cells are incubated with cycloheximide (10 µg/ml) for 1 h to inhibit *de novo* synthesis of proteins including GFP-MPR. Thus, we do not have to consider an additional pool that should otherwise be included in a compartment modeling as described below in step 6.
2. A microscope system is set up as described in Section 3.2 and the detector gain is adjusted so that there is no saturated region in the signal intensity.
3. The entire TGN region is selectively photobleached with strong laser light and the fluorescence recovery is monitored with an attenuated laser light at minute intervals for 30 min (Fig. 13.4A).
4. Changes in the signal intensities in the TGN, whole cell, and extracellular space as the background are quantified using the MetaMorph® software.
5. Fractional values for TGN- and peripheral-pools are calculated. (Fig. 13.4B and C).
6. The values are fitted to a two-compartment model where both pools exchange their contents with first order kinetics (Fig. 13.4D), and two rate constants ($K_{TGN \to peri}$ and $K_{peri \to TGN}$) are optimized by computer-based simulation analysis using the SAAMII software (SAAM institute, Seattle, WA). Resident time can be expressed as the reciprocal of the rate constant.

Note 1:. Fractional TGN values may not be recovered completely after longer observation. This may be due to the presence of immobile fraction of GFP-MPR or some changes in cellular activities during the recording session.

Note 2:. This method does not necessarily give us the absolute values for the kinetics, because the GFP signal may fluctuate according to micromilieu. As the GFP signal is reduced in acidic conditions, careful consideration should be taken while designing the experiment. Alternatively, other FPs such as mCherry that are more resistant to the acidic environment can be used.

3.8. FRAP analysis for exchange kinetics of AP1 and GGAs between the TGN membrane and cytosolic pools

FRAP analysis is also applicable for detecting exchange kinetics of clathrin adaptors between the TGN membrane and cytosolic pools. Methods for photobleaching and time-lapse observations are the same as in Section 3.6

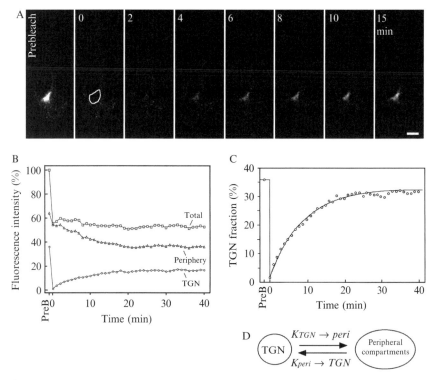

Figure 13.4 FRAP analysis for overall cycling kinetics of MPRs. HeLa cells expressing G-CIMPR-tail were processed for selective photobleaching in the TGN region (enclosed by a white line; see at 0 min). The recovery in the TGN fluorescence was then monitored at the indicated interval times. Scale bar: 10 μm. (B) Total (square), TGN region (circle), and peripheral (triangle) signals were measured and their relative values were calculated as the total fluorescence at prebleaching state is 100. (C) Fractional values of the TGN region were plotted and a fitting line was drawn according to the SAAMII software. (D) A two-compartment model for MPR cycling kinetics (see the text for more detail). Modified from the literature of Waguri et al., 2006.

except image acquisition should be as rapid as possible, as TGN signal recovers within a minute. For more information, please refer to previous studies (Kametaka et al., 2005; Liu et al., 2005; Puertollano et al., 2003).

4. Live-Cell Imaging in Drosophila Cells

In drosophila Schneider S2 cells (hereafter referred to as S2 cells), Golgi compartments are identified as punctate structures scattering throughout the cytoplasm (Kametaka et al., 2010; Kondylis et al., 2001). This characteristic distribution pattern (Fig. 13.5) makes it hard to tell which dots are Golgi and which are endosomes at a light microscopic level.

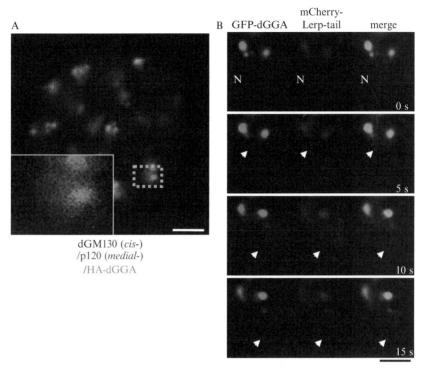

Figure 13.5 (A) dGGA localization in the punctate Golgi structures in the S2 cells. S2 cells transiently expressing HA-dGGA were immunostained with anti-HA (green), anti-p120 Golgi protein (*medial*-Golgi, red), and anti-dGM130 (*cis*-Golgi, blue). (B) Dual-color imaging of GFP-dGGA and mCherry-LERP-tail (Table 13.1). Transport carriers emerging from the Golgi structure (arrowheads) were subjected to the time-lapse imaging with 5 s intervals. N, nucleus. Modified from Kametaka *et al.* (2010), bars: 2 μm. (See Color Insert.)

However, it has recently been shown that clathrin and clathrin adaptor dGGA are mainly localized at the Golgi compartments in S2 cells where dGGA binds to the cytoplasmic tail of LERP, an ortholog of mammalian CIMPR, for its efficient sorting into CCVs (Hirst *et al.*, 2009; Kametaka *et al.*, 2010). Thus, to monitor LERP emerging from the Golgi compartments, FP-dGGA can be introduced as a marker of the Golgi. Here, we describe methods for generation of S2 clones stably expressing mCherry-LERP and live-cell observation using S2 cells (Fig. 13.5).

4.1. Generation of S2 clones stably expressing mCherry-LERP

1. S2 cells (Invitrogen, Cat. R69007) are maintained in Schneider medium (Invitrogen, Cat. 11720–034) supplemented with 10% FBS and 100 U/ml penicillin, and 100 μg/ml streptomycin at 25 °C.

2. S2 cells (1×10^6) cultured in a well of a 12-well plate are subjected to cotransfection with 1 μg of mCherry-LEPP in pAc5.1/V5-His-A vector and 0.1 μg of pCoBlast (Invitrogen) using Fugene6 as described in Section 4.2.
3. Forty eight hours after transfection, the cells are transferred to 10 cm culture dishes and cultured in the presence of 25 μg/ml blasticidin-HCl (Invitrogen, Cat. R21001) for 10–14 days. Cell colonies are picked up in a clean bench and further grown in 24 well-plates in the presence of 5 μg/ml blasticidin-HCl. When each clone becomes confluent, the expression level and distribution pattern of mCherry-LERP are assessed under a microscope.

Note 1:. pCoBlast harbors a gene that confers the resistance to the antibiotic blasticidin, thereby used for the selection of stable transfectants. During the selection period, do not shake the dish frequently, otherwise the growing colonies detach from the dish and cannot form monoclonal colonies. We usually keep them untouched for 2 weeks in the incubator after starting the antibiotic selection.

Note 2:. The Actin5 promoter on the pAc5.1 vector is a strong promoter and the expression level of the FP-tagged protein could be too high for live-cell imaging, even in the stable clones with the lowest expression level. In such case, the gene of interest can be subcloned into a pMT/V5-His vector (Invitrogen) that harbors the copper-inducible metallothionein promoter, and use the stable clone without induction (Hirst et al., 2009). Low expression at a basal level would be expected with this method.

4.2. Visualization of LERP-positive transport carriers in S2 cells.

1. S2 cells expressing mCherry-LERP are seeded in a well of a 24-well plate to a confluency of 80% in 0.5 ml medium.
2. A half μg of GFP-dGGA expression vector is diluted with 25 μl of Opti-MEM (Invitrogen), then 3 μl of FuGene6 (Roche) is added to the DNA/Opti-MEM mixture. The solution is mixed well by tapping the tube, and incubated at room temperature for 10 min.
3. The mixture is added to the cells, which are further cultured for 24 h.
4. The setup of the microscope system does not require special dishes for temperature control unless a specific condition is required, because the optimum temperature for the growth of S2 cells is between 18 and 25 °C.
5. S2 cells are floating cells, therefore should be attached to the bottom of dishes for microscopic observation. To promote this attachment,

glass-bottom dish (Mat Tek Corp. Cat. P35G-1.0-14-C) is coated with poly-L-lysine: 100 µl of 0.01% poly-L-lysine solution (Sigma, Cat. P4707) is spotted on the center of the bottom, incubated for 5 min, and aspirated.

6. The transfected cells are suspended in the well, and 50 µl of the cell suspension is spotted onto the coated area and let to sink for 5 min in order to attach to the bottom.
7. One ml of culture medium is added to the dish for live-cell imaging.

Note 1:. Since two types of FPs are introduced into S2 cells, dual-color imaging as described above (Sections 3.5. and 3.6) are necessary.

Note 2:. S2 cells are originally derived from phagocyte-like cells called hemocytes (Schneider, 1972) and have phogocytic activity. Most of the cells seeded on the glass-bottom dish look spherical in shape, but some show ameba-like flat cell shape. These flat cells are better for live-cell imaging.

5. Conclusion Remarks

We described practical live-cell imaging techniques for observation of the TGN-endosome transport from several aspects. Using these methods, we can successfully visualize transport carriers for MPRs that move between the TGN and endosomes, together with clathrin adaptors. We can also analyze the overall transport kinetics of MPR in a single cell, or exchange kinetics of clathrin coat-associated molecules between the TGN membrane and cytosolic pool. Moreover, the TGN-endosome transport can be visualized in drosophila S2 cells, which would be a good model to elucidate the complex molecular mechanisms involving clathrin coat formation and other transport events. We have to stress, however, that most of these observations have been made in system where aberrant FP-fusions are expressed in addition to the normal amount of original proteins. Therefore, the observed phenomenon could be more or less deviated from a natural occurrence. Nevertheless, the live-cell imaging has been and will provide us with exciting and valuable understanding in the field of biological science.

REFERENCES

Arighi, C. N., Hartnell, L. M., Aguilar, R. C., Haft, C. R., and Bonifacino, J. S. (2004). Role of the mammalian retromer in sorting of the cation-independent mannose 6-phosphate receptor. *J. Cell Biol.* **165**, 123–133.

Barbero, P., Bittova, L., and Pfeffer, S. R. (2002). Visualization of Rab9-mediated vesicle transport from endosomes to the trans-Golgi in living cells. *J. Cell Biol.* **156**, 511–518.

Boehm, M., and Bonifacino, J. S. (2001). Adaptins: The final recount. *Mol. Biol. Cell* **12**, 2907–2920.

Bonifacino, J. S., and Lippincott-Schwartz, J. (2003). Coat proteins: Shaping membrane transport. *Nat. Rev. Mol. Cell Biol.* **4**, 409–414.

Cramer, J. F., Gustafsen, C., Behrens, M. A., Oliveira, C. L., Pedersen, J. S., Madsen, P., Petersen, C. M., and Thirup, S. S. (2010). GGA autoinhibition revisited. *Traffic* **11**, 259–273.

Dennes, A., Cromme, C., Suresh, K., Kumar, N. S., Eble, J. A., Hahnenkamp, A., and Pohlmann, R. (2005). The novel Drosophila lysosomal enzyme receptor protein mediates lysosomal sorting in mammalian cells and binds mammalian and Drosophila GGA adaptors. *J. Biol. Chem.* **280**, 12849–12857.

Ghosh, P., Dahms, N. M., and Kornfeld, S. (2003). Mannose 6-phosphate receptors: New twists in the tale. *Nat. Rev. Mol. Cell Biol.* **4**, 202–212.

Hirst, J., Sahlender, D. A., Choma, M., Sinka, R., Harbour, M. E., Parkinson, M., and Robinson, M. S. (2009). Spatial and functional relationship of GGAs and AP-1 in Drosophila and HeLa cells. *Traffic* **10**, 1696–1710.

Huang, F., Nesterov, A., Carter, R. E., and Sorkin, A. (2001). Trafficking of yellow-fluorescent-protein-tagged mu1 subunit of clathrin adaptor AP-1 complex in living cells. *Traffic* **2**, 345–357.

Kametaka, S., Mattera, R., and Bonifacino, J. S. (2005). Epidermal growth factor-dependent phosphorylation of the GGA3 adaptor protein regulates its recruitment to membranes. *Mol. Cell. Biol.* **25**, 7988–8000.

Kametaka, S., Sawada, N., Bonifacino, J. S., and Waguri, S. (2010). Functional characterization of protein-sorting machineries at the trans-Golgi network in Drosophila melanogaster. *J. Cell Sci.* **123**, 460–471.

Kochubey, O., Majumdar, A., and Klingauf, J. (2006). Imaging clathrin dynamics in *Drosophila melanogaster* hemocytes reveals a role for actin in vesicle fission. *Traffic* **7**, 1614–1627.

Kondylis, V., Goulding, S. E., Dunne, J. C., and Rabouille, C. (2001). Biogenesis of Golgi stacks in imaginal discs of Drosophila melanogaster. *Mol. Biol. Cell* **12**, 2308–2327.

Liu, W., Duden, R., Phair, R. D., and Lippincott-Schwartz, J. (2005). ArfGAP1 dynamics and its role in COPI coat assembly on Golgi membranes of living cells. *J. Cell Biol.* **168**, 1053–1063.

Nichols, B. J., Kenworthy, A. K., Polishchuk, R. S., Lodge, R., Roberts, T. H., Hirschberg, K., Phair, R. D., and Lippincott-Schwartz, J. (2001). Rapid cycling of lipid raft markers between the cell surface and Golgi complex. *J. Cell Biol.* **153**, 529–541.

Puertollano, R., Aguilar, R. C., Gorshkova, I., Crouch, R. J., and Bonifacino, J. S. (2001). Sorting of mannose 6-phosphate receptors mediated by the GGAs. *Science* **292**, 1712–1716.

Puertollano, R., van der Wel, N. N., Greene, L. E., Eisenberg, E., Peters, P. J., and Bonifacino, J. S. (2003). Morphology and dynamics of clathrin/GGA1-coated carriers budding from the trans-Golgi network. *Mol. Biol. Cell* **14**, 1545–1557.

Robinson, M. S., and Bonifacino, J. S. (2001). Adaptor-related proteins. *Curr. Opin. Cell Biol.* **13**, 444–453.

Rothman, J. E., and Wieland, F. T. (1996). Protein sorting by transport vesicles. *Science* **272**, 227–234.

Schneider, I. (1972). Cell lines derived from late embryonic stages of Drosophila melanogaster. *J. Embryol. Exp. Morphol.* **27**, 353–365.

Waguri, S., Dewitte, F., Le Borgne, R., Rouille, Y., Uchiyama, Y., Dubremetz, J. F., and Hoflack, B. (2003). Visualization of TGN to endosome trafficking through fluorescently labeled MPR and AP-1 in living cells. *Mol. Biol. Cell* **14**, 142–155.

Waguri, S., Tomiyama, Y., Ikeda, H., Hida, T., Sakai, N., Taniike, M., Ebisu, S., and Uchiyama, Y. (2006). The luminal domain participates in the endosomal trafficking of the cation-independent mannose 6-phosphate receptor. *Exp. Cell Res.* **312,** 4090–4107.

Wu, X., Zhao, X., Puertollano, R., Bonifacino, J. S., Eisenberg, E., and Greene, L. E. (2003). Adaptor and clathrin exchange at the plasma membrane and trans-Golgi network. *Mol. Biol. Cell* **14,** 516–528.

CHAPTER FOURTEEN

Live Cell Imaging with Chemical Specificity Using Dual Frequency CARS Microscopy

Iestyn Pope,[*] Wolfgang Langbein,[†] Paola Borri,[*] and Peter Watson[*]

Contents

1. Introduction	274
2. "Noninvasive" Live Cell Imaging	275
2.1. Raman scattering	276
2.2. Coherent anti-Stokes Raman scattering	277
2.3. CARS microscopy	279
3. Experimental Setup	279
3.1. Picosecond laser sources	279
3.2. Dual femtosecond laser sources	280
3.3. Tuning and spectral focusing of femtosecond sources	282
3.4. Dual frequency CARS	285
3.5. Correlative CARS for live cell imaging	287
4. Maximizing Collection Efficiency for Live Cell Imaging	287
Acknowledgments	289
References	290

Abstract

Live cell microscopy using fluorescent proteins and small fluorescent probes is a well-established and essential tool for cell biology; however, there is a considerable need for noninvasive techniques able to study tissue and cell dynamics without the need to introduce chemical or genetically encoded probes. Coherent anti-Stokes Raman scattering (CARS) microscopy is an emerging tool for cell biologists to examine live cell dynamics with chemical specificity in a label-free, noninvasive way. CARS is a multiphoton process offering intrinsic three-dimensional submicron resolution, where the image contrast is obtained from light inelastically scattered by the vibrations of endogenous chemical bonds. CARS is particularly well suited to study lipid

[*] School of Biosciences, Cardiff University, Cardiff, United Kingdom
[†] School of Physics and Astronomy, Cardiff University, Cardiff, United Kingdom

biology, since the CARS signal of localized lipids (exhibiting a large amount of identical bonds in the focal volume) is very strong. Conversely, photostable, lipid-specific markers for fluorescence microscopy are difficult to produce and the process of labeling often affects lipid localization and function, making imaging lipids in live cells challenging, and accurate quantification often impossible. Here, we describe in detail the principles behind our experimental setup for performing CARS microscopy of lipid droplets on live cells. Since typical vibrational resonances in liquid have coherence times in the picosecond range, CARS is preferably implemented with picosecond lasers which are however expensive and less efficient than femtosecond lasers, which could also be used for other multiphoton techniques such as two-photon fluorescence. In our setup, we show that femtosecond lasers can be spectrally focused in a simple, alignment insensitive, and cost-effective way to achieve a vibrational excitation similar to picosecond lasers. This opens the way to integrate CARS and two-photon fluorescence in a single multimodal instrument for its widespread application. We also describe our dual frequency CARS system which eliminates the nonresonant CARS background offering superior sensitivity and image contrast.

1. INTRODUCTION

The ideal cell imaging system would be able to deliver contrast of cellular structures with excellent spatial resolution, but at no "cost" to the biological processes occurring within the cell. In conventional fluorescence microscopy, an item of interest must be fluorescently tagged in order for it to be visualized (Watson *et al.*, 2005), but the process of labeling this structure (be it a small molecule, nucleic acid, or protein) introduces artifacts. Chemical modification in order to introduce a fluorescent dye into a small molecule can alter the properties of that molecule. Expressing proteins engineered with a fluorescent tag from a plasmid can introduce problems controlling expression level (therefore, affecting the biological balance within the cell) as well as affecting the efficiency and specificity of the protein in question via steric interference by the fluorescent protein tag. Introduction of expression plasmids requires some kind of disruptive event (electroporation, chemical/lipid based transformation, microinjection) and care must be taken to ensure cell survivability and retention of normal cell function. For cell imaging, correct fluorophore selection could be considered as critical as microscope selection and is dependent on a number of factors including properties that affect the relative brightness (such as quantum yield and molar absorption), through to its ease in biological manipulation (chemical or genetic modification) and chemical and photo stability. Although considerable effort has been, and still is (Mehta and Zhang, 2010; Miyawaki, 2010) being, made to discover and engineer novel fluorescent biomarkers for imaging, photostability and photobleaching remains a major limiting factor for long-term image acquisition of live biological samples. Fluorophores also suffer from saturation due to their excited state lifetime (the fluorophore cannot accept another photon until it has returned back to its ground state); hence there is a limit to the number of photons they can emit within a given time. Phototoxicity, caused by the photochemical modification of biomolecules such as the creation of reactive oxygen species (Hoebe et al., 2007), is also a process that can effect biological events within the cell, mask, or confuse phenotypes, and ultimately lead to cell death. The factors that need to be considered for efficient and sustained detection of a biological entity become increasingly critical as one moves away from "model" cell types, into three-dimensional cell and tissue cultures, or move away from "commonly used" microscopy techniques. Because of this, there is considerable interest in noninvasive techniques, able to study tissue and cell biology without the need for introduced probes. These techniques are critical in areas of biology where the biological specimen is to be utilized following assessment. One example is in embryo assessment where live, noninvasive imaging can predict which embryos will reach the blastula stage (Wong et al., 2010), which in the long term should result in improved IVF outcomes.

2. "Noninvasive" Live Cell Imaging

Traditional brightfield modes include phase contrast, differential interference contrast (DIC) and Hoffman modulation contrast (HMC), which are sensitive to differences in the optical path length (thickness, refractive index) but not inherently sensitive to chemical changes in the sample. Second harmonic imaging microscopy (Campagnola and Loew, 2003) of biological specimens exploits a nonlinear optical effect known as second harmonic generation (SHG). As intense laser light passes through a polarisable material with a noncentrosymmetric (lack of inversion symmetry) molecular organization, nonlinear mixing of the excitation light results in the generation of a wave at twice the optical frequency (i.e., half the wavelength). This frequency doubled SHG signal can be relatively easily separated from the incident light through the use of frequency filters and provides high axial and lateral resolution without the need for a detection pinhole, similar to multiphoton fluorescence microscopy. As a label-free technique for cell biology SHG is typically used to detect collagen in the extracellular matrix (Schenke-Layland, 2008), since SHG of cellular components is generally weak. To extend this technique, noncentrosymmetric nanoprobes are also under development. These have a number of advantages over fluorophores as the generated signal does not saturate with increasing illumination intensity, and they display excellent photostability (Pantazis et al., 2010), however, this converts SHG into an invasive technique. Optical coherence tomography (OCT) utilizes an optical beam directed at the sample and measures magnitude, phase, frequency shift, and polarization of the back reflected light from the sample (Huang et al., 1991). Depth information is calculated via interference with a reference beam, enabling the acquisition of three-dimensional datasets from live samples with video rate temporal resolution. Predominantly used for ophthalmic studies to examine the structures within the eye, it also has uses in cardiovascular imaging, gastroenterology and dermatology (Fercher, 2010). However, in essence OCT is sensitive only to the optical path length, hence lacks chemical specificity similar to DIC and phase contrast. Another limitation with OCT is its relatively poor spatial resolution (>1 µm), below that required for individual cell analysis. Although whole animal biophotonic imaging (BPI) is often classed as a noninvasive technique (Andreu et al., 2011), there is still the requirement for a fluorescent or luminescent signal, usually a genetically modified microorganism, whose location and spread of infection can be recorded using a highly sensitive imaging system. BPI is showing great potential in identifying novel insights into microorganism behavior, and the niche's they exploit within their host, whilst minimizing the numbers of animals needed during this process.

2.1. Raman scattering

Light scattering can be either elastic, in which the frequency of light (hence photon energy) remains the same but its direction of propagation is changed (Rayleigh and Mie scattering), or inelastic, where the optical frequency is changed (Raman scattering). More specifically, in Raman scattering, light interacts with vibrations between atoms within the system, resulting in an optical frequency change, either up (anti-Stokes shift) or down (Stokes shift) by an amount equal to the bonds vibrational frequency. In a single molecule, the majority of light scattering that occurs is Rayleigh scattering, whilst the probability of obtaining inelastic scattering is low (one event per 10^{18} photons; Evans and Xie, 2008) due to a typical Raman scattering cross section per molecule of $\sim 10^{-29}$ cm^2, 10^{14} times lower than the typical absorption cross section used in fluorescence. Hence, Raman Spectroscopy suffers from low sensitivity and requires high-excitation intensities and long integration times, neither of which are compatible with live cell imaging. Further, the Raman signal may be lost in the inherent autofluorescence of some biological molecules. Sensitivity can be dramatically increased via the local field enhancement occurring near a metal nanostructured substrate (surface enhanced Raman spectroscopy (SERS) reviewed in Stiles et al., 2008). This enhancement could be as much as 10^{11}, leading to the possibility of single molecule detection. However, the need of a nanostructured substrate complicates sample preparation and limits the applicability of the method to thin regions near the substrate.

2.2. Coherent anti-Stokes Raman scattering

To enhance the Raman signal, it is possible to drive the molecular vibration with coherent light and generate coherent anti-Stokes Raman scattering (CARS) (Fig. 14.1). CARS is a third-order nonlinear process (four-wave mixing) in which two laser fields of frequencies ω_P (Pump) and ω_S (Stokes) coherently drive a molecular vibration which is resonant at the frequency difference between the two fields ($\omega_{vib} = \omega_P - \omega_S$). A third beam (the Probe ω_{PB}) is then inelastically scattered by the driven vibration, and frequency upshifted into $\omega_{as} = \omega_{PB} + \omega_{vib}$. Experimentally the Probe is usually derived from the Pump thus $\omega_{as} = 2\omega_p - \omega_s$. Typically, the anti-Stokes signal is detected as it is easier to filter the blue shifted emission from the excitation wavelength and any sample autofluorescence. Importantly, the beat frequency ($\omega_p - \omega_s$) coherently drives all identical vibrations within the focal volume hence the anti-Stokes Raman fields constructively interfere. The coherent superposition of all the dipoles (vibrating bonds)

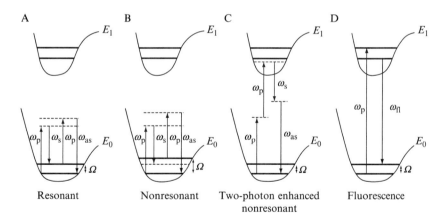

Figure 14.1 Energy level diagrams for CARS and fluorescence transitions. Energy level diagrams for CARS (A–C) and fluorescence (D) transitions. E_1 and E_0 are electronic levels, with $E_1 - E_0$ typically UV for molecules studied with CARS and UV to visible for fluorophores. Ω is the energy separation between vibtational modes, typically mid-infrared. Dashed lines represent virtual energy states. (A) Resonant CARS contribution. (B) Nonresonant CARS contribution from other nonresonant molecular vibrations. (C) Two-photon enhanced electronic nonresonant contribution. ω_p, ω_s, ω_{as}, ω_{fl} refer to the Pump, Stokes, anti-Stokes, and fluorescence frequencies, respectively. (For color version of this figure, the reader is referred to the Web version of this chapter.)

induced by the beat frequency generates a third-order polarization ($P^{(3)}$) which is proportional to the third-order susceptibility ($\chi^{(3)}$).

$$P^{(3)} \propto \chi^{(3)} E_P^2 E_S^* \tag{14.1}$$

where E_P and E_S are the Pump and Stokes fields, respectively. $\chi^{(3)}$ is the third-order susceptibility which near a vibrational resonance ($\omega_p - \omega_s \approx \Omega$) can be approximated as (Cheng and Xie, 2004)

$$\chi^{(3)} = \frac{A_R}{\Omega - (\omega_p - \omega_s) - i\Gamma_R} + \chi^{(3)}_{nr} + \frac{A_t}{\omega_t - 2\omega_p - i\Gamma_t} \tag{14.2}$$

(first given by Lotem et al., 1976) where A_R is a constant proportional to the Raman scattering cross sections and to the number of identically vibrating bonds probed within the focal volume. Γ_R is the half-width at half-maximum of the Raman line. χ_{nr} is the nonresonant susceptibility and Ω is the vibration frequency. ω_t represents an electronic resonance with two-photon absorption cross section A_t and linewidth Γ_t. The first term in Eq. (14.2) is a contribution from vibrationally resonant states. The second term is a nonresonant contribution (known as the nonresonant background) independent of the Raman shift and thus constant as the beat frequency is tuned. The last term in Eq. (14.2) is the enhanced nonresonant contribution due to two-photon electronic resonance which occurs if $2\omega_p \approx \omega_t$. The anti-Stokes intensity (I_{AS}) may be obtained by solving the wave equation, assuming plane Pump and Stokes waves (Evans and Xie, 2008).

$$I_{AS} \propto |\chi^{(3)}|^2 I_p^2 I_s \left(\frac{\sin(\Delta k z/2)}{\Delta k/2}\right)^2 \tag{14.3}$$

where I_p and I_s are the Pump and Stoke intensities, z is the sample thickness, and $\Delta k = k_{as} - (2k_p - k_s)$ is the wavevector mismatch. From Eq. (14.3), we can see that although $\chi^{(3)}$ is linearly dependent on the number of vibrating bonds the CARS intensity scales as $|\chi^{(3)}|^2$. This relationship is the reason for the improved signal strength as compared to the Raman intensity when a large number of identically vibrating bonds are probed. Rewriting Eq. (14.3) in terms of resonant and nonresonant parts of $\chi^{(3)}$ gives

$$I_{AS}(\Delta) \propto \left|\chi^{(3)}_{nr}\right|^2 + \left|\chi^{(3)}_R(\Delta)\right|^2 + 2\chi^{(3)}_{nr} \mathrm{Re}\chi^{(3)}_R(\Delta) \tag{14.4}$$

where $\Delta = \omega_p - \omega_s - \Omega$ is the detuning (Raman shift). The third term is generated by mixing between the resonant and nonresonant parts.

2.3. CARS microscopy

By applying these principles of CARS field/signal generation, CARS microscopy has emerged as a new technique for imaging in cell biology, combining noninvasive, chemical specificity with high-resolution imaging for biological samples. For microscopy applications, it is advantageous that the nonlinear CARS process only takes place in the focal volume where high photon densities are reached, allowing for intrinsic 3D spatial resolution without the need of a detector pinhole, similar to multiphoton techniques such as TPF and SHG microscopy. Moreover, as in the case of these multiphoton techniques, CARS utilizes laser pulses such that photons are concentrated in time whilst low average powers are maintained to minimize thermal damage. However, the complexity and cost of these laser systems can be a major obstacle for the widespread application of CARS microscopy. The majority of research utilizing CARS microscopy for cell biology has examined lipid dynamics (reviewed extensively in Pezacki *et al.*, 2011), due to the strong CARS signal that arises from lipid C-H bond stretches. Whilst fluorescence microscopy of proteins is well-established, stable, lipid-specific markers are difficult to produce as the process of labeling affects lipid localization and function. Hence, fast timelapse and long timecourse experiments on lipids and lipid droplets in live cells is challenging, and accurate quantification is often impossible. CARS microscopy is developing as an excellent tool for cell biologists to examine live cell lipid dynamics with chemical specificity. Here we describe our experimental setup, and the rationale behind the choices made, for performing dual frequency CARS microscopy utilizing femtosecond lasers as an excitation source (Fig. 14.2).

3. EXPERIMENTAL SETUP

3.1. Picosecond laser sources

Typical CARS resonances in liquids have coherence times in the picosecond (10^{-12} s) range, hence in order to obtain optimal spectral selectivity, picosecond pulses (Fig. 14.3A) are the obvious choice to generate a CARS signal as their bandwidth is matched to the vibrational resonance. Whilst conceptually simple to have two picoseconds sources for Pump and Stokes; for example, two electronically synchronized Ti:sapphire oscillators (Cheng *et al.*, 2001) or an optical parametric oscillator pumped by a picoseconds frequency-doubled Nd:vanadate laser (Evans and Xie, 2008), this is technically challenging and expensive.

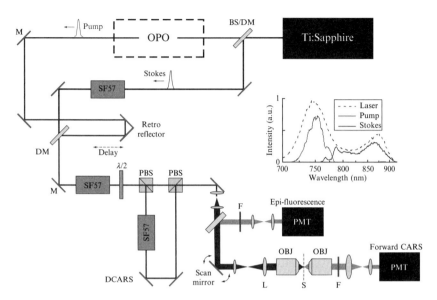

Figure 14.2 Sketch of the optical layout of our CARS microscope. Two different layouts are represented. Option 1: (2300–3400 cm^{-1}) 100 fs Ti:sapphire laser provides the Stokes beam and pumps the OPO used to generate the Pump beam. A beam splitter (BS) is used to split the Ti:sapphire output into the two arms. Option 2: (800–1800 cm^{-1}) 8 fs Ti:sapphire laser provides both the Stokes and Pump beams (the OPO is not present), a dichroic mirror (DM) is used to split the laser output into the two arms. The rest of the optical setup is the same for both options. Graph: typical spectra of laser, Pump and Stokes for option 2. SF57, glass blocks used to chirp pulses; M, mirror; PBS, polarizing beam splitter; $\lambda/2$, half-wave plate; F, filter; L, lens; S, sample; OBJ, objective. (For color version of this figure, the reader is referred to the Web version of this chapter.)

3.2. Dual femtosecond laser sources

Tunable femtosecond (10^{-15} s) laser systems are more common than their picoseconds counterparts and are also better suited for quasi-instantaneous multiphoton processes such as TPF and SHG; producing the opportunity that they could therefore be implemented in a multimodal CARS/TPF/SHG microscope for live cell imaging. However, femtosecond pulses are spectrally broad (compare the broad range of driving wavelengths in Fig. 14.3B to the narrow range delivered from a picoseconds laser in Fig. 14.3A), leading to a decrease in spectral selectivity of Raman resonances that can be probed and an increased contribution from the spectrally constant nonresonant background. For our dual femtosecond source we use a 100 fs Ti:Sa laser system (Coherent Verdi/Mira) to synchronously pump an optical parametric oscillator (OPO) (APE PP2), see option 1 in Fig. 14.2.

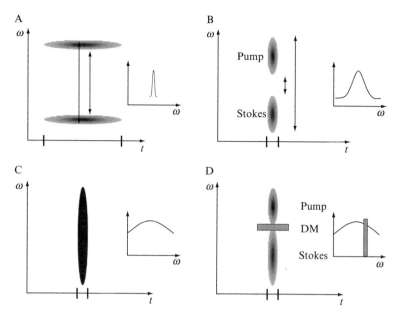

Figure 14.3 Comparisons of pico- and femtosecond sources for CARS microscopy. (A) Dual picoseconds lasers can be used to produce Pump and Stokes pulses that are spectrally narrow and temporally long. This generates a narrow bandwidth for the IFD represented on the insert (~ 10 cm^{-1}). (B) 100 fs lasers produce pulses shorter in time, but broader in wavelength, generating broad IFD (~ 100 cm^{-1}). (C) Broadband sub-10 fs lasers produce a broad range of wavelengths, that can be separated (D) into Pump and Stokes beams through the use of a dichroic mirror (DM). t, time; ω, frequency. (For color version of this figure, the reader is referred to the Web version of this chapter.)

The Ti:Sa delivers 100 fs pulses centered at 832 nm at 76 MHz repetition rate, which is used as the Stokes beam. The OPO supplies the Pump beam (via intracavity frequency doubling of the OPO signal). Pulses ranging from 700 to 650 nm can be used to excite vibrational resonances from 2300 to 3400 cm^{-1}. Downstream pulse shaping is used to improve the spectral resolution.

3.2.1. Single femtosecond laser source

The use of a *single* laser source to generate the Pump and Stokes beams would provide a particularly attractive (and cost-effective for commercial exploitation) way to generate CARS. A few groups have proposed the use of a 100 fs Ti:Sapphire laser and a photonic crystal fiber as a nonlinear medium (such that as the 100 fs pulse passes through the fiber it is dispersed into a broadband spectrum), in order to achieve the correct Pump and Stokes beams (Motzkus and von Vacano, 2006). This is particularly suitable for upgrades to multiphoton systems which have a 100 fs laser source already in place. Olympus (www.olympusamerica.com) supplies an add-on to their multiphoton laser scanning microscope (FV1000MPE FemtoCARS) that allows simultaneous visualization of lipid (CH_2), TPF and SHG signals. Newport (www.newport.com) provides a "wavelength extension unit" (WEU-02) that uses a photonic crystal fiber to produce a Stokes beam for CARS microscopy. Both of these systems will effectively upgrade an existing two-photon microscope to enable CARS imaging. However, continuum generation in fibers is sensitive to laser fluctuations and alignment and also suffers from aging. Further, since the bandwidths of the 100 fs laser and the broadband continuum are different by an order of magnitude, the scheme is not efficient for single-frequency CARS since only a small part of the continuum is creating CARS, whilst the rest is both wasting laser power and additionally exposing the sample.

An alternative approach overcoming these limitations has been developed by our group (Langbein et al., 2009b) based on a single broadband sub-10 fs laser (Figs. 14.2 and 14.3C). To create Pump and Stokes pulses from a single laser, we use a broadband sub-10 fs Ti:Sapphire laser (KMLabs MTS) with 200 nm bandwidth covering the wavelength range from 700 to 900 nm with 300–500 mW of power at 95 MHz repetition rate. The laser bandwidth contains Pump and Stokes wavelengths capable of exciting vibrational frequencies in the characteristic region of 800–1800 cm^{-1}. As shown in option 2 Figs. 14.2 and 14.3D, the broadband sub-10 fs laser is split into two frequency bands corresponding to Pump (higher frequency band) and Stokes (lower frequency band) using a dichroic mirror (DM) (Chroma 770DCXRU). This is a simple and effective way to generate Pump and Stokes pulses without temporal jitter between them (since they originate from the same source), as opposed to using two separate lasers which require synchronization. As with dual femtosecond sources, downstream pulse shaping is used to improve the spectral resolution.

3.3. Tuning and spectral focusing of femtosecond sources

Each femtosecond band generated is still much broader than the typical Raman linewidths (i.e., the duration of Pump and Stokes pulses are much shorter than the vibrational coherence time of ~ 1 ps). However, the vibrational excitation in CARS is governed by the interference of the Pump and Stokes fields; therefore, the spectral resolution is not determined by these individual spectrum, but by the spectrum of their temporal interference. Hence by carefully shaping the Pump and Stokes pulses in time, it is possible to drive a narrow vibrational frequency range, even though the individual pulses are spectrally broad. Although a number of ways to achieve this have been proposed (Dudovich et al., 2002; Hellerer et al., 2004), we recently demonstrated a simple, highly efficient, alignment insensitive, and cost-effective method which utilizes glass elements of known dispersion (Langbein *et al.*, 2009a; Rocha-Mendoza *et al.*, 2008) through a process known as Chirp.

3.3.1. Chirp (spectral focusing)

The refractive index (n) of a dispersive medium is not constant but varies with wavelength, $n(\lambda)$ (and thus frequency). This means that when a short pulse travels through a dispersive medium, its different wavelength (or frequency) components travel at different speeds (Fig. 14.4A). Typically, the longer

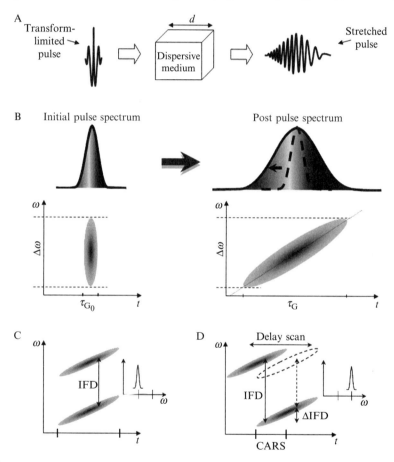

Figure 14.4 Spectral focusing and tuning. (A) Cartoon illustrating how a transform limited pulse is stretched out after passing through a dispersive medium, note how longer wavelengths emerge first. (B) Spectral representation illustrating that the spectral range of the pulse $\Delta\omega$ remains constant as the pulse is stretched out in time due to chirp. (C) Illustration demonstrating that the IFD between two equally chirped pulses is constant with time. (D) Illustration demonstrating how varying the overlap of the two pulses changes the IFD, thus allowing spectral tuning via simple delay scanning. t, time; ω, frequency; τ_{G_0}, initial pulse width; τ_G, final pulse width. (For color version of this figure, the reader is referred to the Web version of this chapter.)

wavelengths travel faster and thus emerge from the medium first. This stretches out the pulse from its initial width τ_{G_0} to a width τ_G (Fig. 14.4B).

For a Gaussian pulse, this may be expressed as

$$E(t) = E_0 \exp\left(-(t/\tau_G)^2 + it(\omega_0 + \beta t)\right) \quad (14.5)$$

$$\beta = \frac{2k''d}{\tau_{G_0}^2 \tau_G^2} \quad \text{and} \quad k'' = \frac{\lambda^3}{2\pi c^2}\frac{d^2 n}{d\lambda^2}$$

where ω_0 is the center frequency, t is time, β is the chirp parameter, d is the thickness of the dispersive medium, c is the speed of light, λ is the wavelength and k'' is the group velocity dispersion (GVD). From Eq. (14.5) the instantaneous frequency ($\omega_{\text{inst}}(t)$) is then defined as

$$\omega_{\text{inst}}(t) = \omega_0 + 2\beta t. \quad (14.6)$$

In our setup (see Fig. 14.2), we use SF57 glass (Changchun Fortune Optronics Inc.) as dispersive medium as it provides the required chirp from reasonable lengths of material. The key point for spectrally selective CARS is to introduce the *same* chirp parameter in Pump and Stokes pulses via the appropriate lengths of SF57 glass blocks such that the instantaneous frequency difference (IFD) $\omega_D = \omega_p - \omega_s$ between the two pulses remains constant (Fig. 14.4C). In practice, in our setup the Stokes beam first propagates through a block of SF57 glass before recombination with the pump and propagation through a common SF57 block, this is to account for the different GVDs of Pump and Stokes. The molecular vibration is driven at the beat frequency ($\omega_p - \omega_s$) which will be centered at the IFD $\omega_D = \omega_{p_0} - \omega_{s_0}$ and has a spectral width given by the Fourier limit of the temporal envelope of the pulses, which can be elongated by the applied linear chirp to a few picoseconds (τ_G). This is a highly efficient (>70% transmission), alignment insensitive, and cost-effective way to spectrally focus broadband lasers, as compared to spatial light modulators or grating-lens based systems.

3.3.2. Tuning to a Raman resonance

We are able to tune the vibrational excitation by simply adjusting the arrival time of the pump (Fig. 14.4D) using a mechanical delay stage (Physik Instrumente M-403.6DG, 150 mm travel, 0.2 μm minimum incremental motion), (Fig. 14.2) allowing vibrational spectroscopy to be performed via simple delay scanning, without the need of multiplex detection or laser tuning. One should also bear in mind that the time ordering between the creation of the vibration and its readout by the Pump matters. The molecular vibration of interest cannot be probed until it has first been driven at resonance. Since experimentally a single beam is used for both the Pump and the Probe, this implies that a small temporal shift between Pump and Stokes t_0 (negative for Stokes leading) can be used to optimize time ordering and hence maximize the CARS signal (see Langbein et al., 2009a for a detailed description). In our setup, Pump and Stokes are focused onto the sample using a water immersion 1.2 NA microscope objective (Leica HCX PL APO 63× W Corr CS). CARS is collected by an identical microscope objective in forward direction, transmitted through a bandpass filter to reject Pump and Stokes excitation wavelengths (e.g., Semrock FF01-562/40 for 100 fs Stokes at 832.5 nm and excitation/detection at the 2845 cm^{-1} CH$_2$ symmetric stretch vibration), and detected by a photomultiplier (Hamamatsu H7422-40). Beam scanning is achieved using a single-mirror tip-tilt system (Physik Instrumente PI S-334), see Fig. 14.2.

3.4. Dual frequency CARS

One of the major problems in CARS is the presence of the nonresonant background (as discussed previously, see Eq. (14.2)) that normally severely limits both sensitivity and image contrast. Several methods have been employed in an attempt to reduce or remove this background (reviewed in Arora et al., 2011; Day et al., 2011). In order to enhance the image contrast against this nonresonant CARS background, and to improve the chemical specificity of the technique, we have recently invented and demonstrated (Rocha-Mendoza et al., 2009) a dual frequency differential CARS (D-CARS) method. The Pump and Stokes pulse pair is divided into two pairs using a $\lambda/2$ waveplate (CASIX Achromatic waveplate) rotating the linear polarization direction, and a polarizing beamsplitter (PBS, Lambda Research Optics) to separate the two orthogonally polarized components. By adjusting the traveled distance, the deflected second pair is delayed by half the separation $T_{\rm rep}$ between two subsequent pulses from the laser source (i.e., half of the laser repetition rate). This second pair travels through a third SF57 glass block of appropriate length so that the IFD of the second pair differs from the IFD in the first pair by an amount controlled via the length of the glass block. Importantly, the thickness of this third block is sufficiently small (a few mm) so that we can neglect the effect of the change in chirp parameter β. The main point of this third block is to introduce a delay between Pump and Stokes due to their different group velocities. The resulting CARS intensity is a periodic function of period $T_{\rm rep}$, and within one period $[0, T_{\rm rep})$ can be written as the sum of the intensities from each Pump–Stokes pair, that is, $I(t)=A_1\delta(t) + A_2\delta(t - T_{\rm rep}/2)$. The duration of the CARS intensity created by each pair is below the time-resolution of the detector and is therefore represented as "instantaneous," that is, by a delta-function. By developing the periodic intensity $I(t)$ in a Fourier series $I(t) = a_0/2 + a_1\cos(2\pi t/T_{\rm rep}) + \ldots$ one can deduce that the dc Fourier coefficient is $a_0 = 2(A_1 + A_2)/T_{\rm rep}$ and the coefficient associated with the first harmonic is $a_1 = (A_1 - A_2)/T_{\text{rep}}$. Therefore, by measuring not only the dc component of the CARS signal but also the ac component using appropriate phase-sensitive frequency filtering (e.g., a Lock-in amplifier with reference signal at frequency $2\pi/T_{\text{rep}}$) one can simultaneously record the total CARS signal (Fig. 14.5C) as well as the differential effect from the two exciting Pump–Stokes pairs (Fig. 14.5D). In this way, the spectrally constant nonresonant CARS background can be completely eliminated in the differential detection and the chemical contrast enhanced in relation to the specific CARS spectral lineshape (see Fig. 14.5D). As the Stokes-Pump pulse pairs are originally derived from the same pulses, they do not suffer from classical relative intensity fluctuations. Further, the high modulation frequency of $1/T_{\text{rep}} \sim 80$ MHz eliminates noise due to mechanical fluctuations. Hence D-CARS is shot noise-limited. Importantly, the D-CARS concept can be extended to more pairs by cascading the splitting and

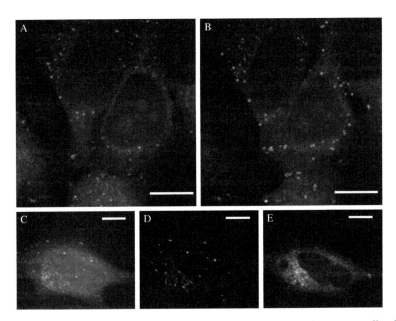

Figure 14.5 CARS images of live cells. (A,B) CARS images of live HeLa cells after (A) 2 h and (B) 3 h incubation in a lipid rich media (DMEM supplemented with BSA: oleic acid). CARS is centered at 2845 cm^{-1} corresponding to the C–H$_2$ symmetric stretching vibration resonance. Note the increase in size and number of structures, corresponding to cytosolic lipid droplets. (C–E) CARS, D-CARS, and TPF multimodal imaging of live HeLa cell expressing eCFP-tagged proteins in the cytosol. (C) CARS and (D) D-CARS images at the sum and difference between 2845 cm^{-1} (C–H$_2$ symmetric stretching vibration resonance) and 2920 cm^{-1}, respectively. (E) TPF acquired in epi-geometry simultaneously with forward CARS. All Images were acquired using the setup option 1. Bar = 10 μm. Excitation power<10 mW. 100 nm pixel size, 0.1 ms/pixel dwell time.

delaying of subsequent Stokes–Pump pairs. Moreover, by adding a $\lambda/4$ wave plate after the second PBS the pairs can be converted into a cross-circularly polarized configuration to minimize polarization artifacts. Alternatively, one can use the orthogonal polarization of the pairs to implement polarization D-CARS. Since D-CARS measures at two frequencies, it allows one to distinguish the concentration of two different materials. We have previously imaged a mixture of polystyrene and polymethyl methacrylate (PMMA) beads, the latter lacking CH aromatic vibrations and thus exhibiting a different CARS spectrum (Rocha-Mendoza et al., 2009). We have also demonstrated D-CARS for live cell microscopy, with HeLa cells grown overnight in a lipid rich medium, showing variations in the relative CH_2–CH_3 content between droplets (hence their chemical composition differences) revealed throughout the cell in the form of bright and dark droplets when imaged by D-CARS (Rocha-Mendoza et al., 2009). Importantly, in our D-CARS method the difference between two frequencies is acquired *simultaneously* in a single scan (Fig. 14.5C and D), hence the comparison is not affected by motion artifacts, a key point for live cell imaging. This process removes the nonresonant background (as can be seen in Fig. 14.5D) and therefore increases sensitivity and image contrast.

3.5. Correlative CARS for live cell imaging

Besides extracting Pump and Stokes pulses from a single source, the use of a femtosecond laser is also ideal to perform TPF. We have already demonstrated the ability to use the Stokes pulse to drive two-photon fluorescence in live cells, thus performing correlative TPF/D-CARS (Fig. 14.5C–E). We are currently developing a system in which the long wavelength section of the laser pulse can be picked out from the Stokes beam using a second DM, without traveling over the SF57 glass blocks in order to generate a beamline suitable for TPF and SHG. A recent study has characterized the two-photon excitation profiles of a wide number of commonly used fluorophores (Drobizhev et al., 2011), and this information will be critical in deciding which fluorophores will be suitable for correlative TPF/CARS to ensure no cross talk between CARS and fluorescence excitation and emission.

4. MAXIMIZING COLLECTION EFFICIENCY FOR LIVE CELL IMAGING

Similar to techniques utilizing fluorescence, in order to maintain optimal cell health on the microscope, it is important to minimize excitation intensity and exposure times in CARS microscopy. Unlike fluorescence emission which is emitted in all directions, CARS emission has a strong

Table 14.1 Theoretical collection efficiencies of objective-condenser combinations suitable for live cell imaging

Objective	Condenser	Collection efficiency (%)
Water (1.27 NA)	Oil (1.4 NA)	100.0
Water (1.27 NA)	Air (0.72 NA)	11.8
Air (0.75 NA)	Oil (1.4 NA)	100.0
Air (0.75 NA)	Air (0.72 NA)	83.7
Air (0.75 NA)	Air (0.72 NA)	9.7 (96-well plate)
Air (0.75 NA)	Air (0.72 NA)	74.7 (microwell slide)

Due to the CARS signal emitting predominantly in the forward direction, careful matching of excitation and collection optics is required to ensure maximum collection efficiency for live cell imaging.

directionality depending on the size and shape of the scatterer. If the object emitting CARS has an axial size larger than the wavelength of the excitation source CARS is emitted predominantly in the forward direction (Cheng et al., 2002) within the transmitted light cone of the excitation source. Assuming the entire emitted CARS signal falls within the transmitted light cone of the excitation source one can calculate the collection efficiencies for objective-condenser combinations (Table 14.1; Fig. 14.6). The calculated collection efficiencies are the maximum theoretical values based on the numerical aperture (NA) of the objective and condenser lens, additional losses due to reflections at interfaces are not taken into account (refractive indices of air (1), water (1.33), glass and oil (1.5)). Whilst such collection efficiencies may be realized when working with slide or microfluidic based samples, careful consideration needs to be taken when preparing samples for other applications. High-throughput fluorescence microscopy is generally based around 96- or 384-well plates on inverted epi-detection systems, where the well parameters have little or no influence on collection efficiencies. Conversely, for CARS forward-detection, well parameters have a significant impact on the overall collection efficiency as the plate walls limit the collection angle in transmission geometry. For an average 96-well glass bottomed plate, the overall collection efficiency could be reduced to as little as 9.7% for a 0.75 NA air objective and 0.72 NA condenser combination suitable for high throughput. Clearly for high-throughput CARS to be effective, shallow well depths will need to be used, since simply halving the well depth leads to a fourfold increase in collection efficiency. We are currently investigating the compatibility of "nonstandard" live cell imaging chambers for use in CARS microscopy, such as Hamamatsu's microwell slides (A10657-01, Hamamatsu Photonics) that have a small well depth (3.1 mm) compared to their diameter. Through a process of optimizing collection efficiency, and implementation of the

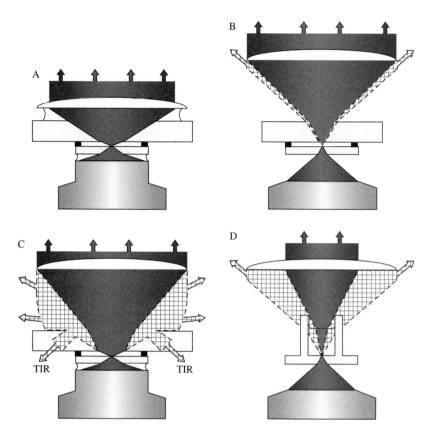

Figure 14.6 CARS collection efficiency in typical live cell imaging setups. Schematic representation of some of the CARS collection efficiency listed in Table 14.1 for (A) 1.27 NA water objective with a 1.4 NA oil condenser; (B) 0.75 NA air objective with a 0.72 NA condenser; (C) 1.27 NA water objective with a 0.72 NA air condenser; (D) same as (B) but with the sample in a typical well of a 96-well plate, as opposed to the slide and cover slip geometries of (A)–(C). The solid cone and hatched areas represent the collected and lost portions of the CARS signal. TIR refers to total internal reflection. (For color version of this figure, the reader is referred to the Web version of this chapter.)

D-CARS modality in order to remove nonresonant background it should be possible to obtain high signal-to-noise ratios for CARS, whilst implementing correlative fluorescent and SHG techniques to study live cells with chemical specificity.

ACKNOWLEDGMENTS

P. W. is an RCUK Fellow. P. B. is an EPSRC Leadership Fellow funded under the grant agreement EP/I005072/1. I. P., W. L., P. B., and P. W. are funded by BBSRC Project Grant BB/H006575/1.

REFERENCES

Andreu, N., Zelmer, A., and Wiles, S. (2011). Noninvasive biophotonic imaging for studies of infectious disease. *FEMS Microbiol. Rev.* **35,** 360–394.

Arora, R., Petrov, G. I., Liu, J., and Yakovlev, V. V. (2011). Improving sensitivity in nonlinear Raman microspectroscopy imaging and sensing. *J. Biomed. Opt.* **16,** 021114.

Campagnola, P. J., and Loew, L. M. (2003). Second-harmonic imaging microscopy for visualizing biomolecular arrays in cells, tissues and organisms. *Nat. Biotechnol.* **21,** 1356–1360.

Cheng, J. X., Jia, Y. K., Zheng, G., and Xie, X. S. (2002). Laser-scanning coherent anti-Stokes Raman scattering microscopy and applications to cell biology. *Biophys. J.* **83,** 502–509.

Cheng, J. X., Volkmer, A., Book, L. D., and Xie, X. S. (2001). An epi-detected coherent anti-stokes raman scattering (E-CARS) microscope with high spectral resolution and high sensitivity. *J. Phys. Chem. B* **105,** 1277–1280.

Cheng, J. X., and Xie, X. S. (2004). Coherent anti-Stokes Raman scattering microscopy: Instrumentation, theory, and applications. *J. Phys. Chem. B* **108,** 827–840.

Day, J. P., Domke, K. F., Rago, G., Kano, H., Hamaguchi, H. O., Vartiainen, E. M., and Bonn, M. (2011). Quantitative coherent anti-Stokes Raman scattering (CARS) microscopy. *J. Phys. Chem. B* **115,** 7713–7725.

Drobizhev, M., Makarov, N. S., Tillo, S. E., Hughes, T. E., and Rebane, A. (2011). Two-photon absorption properties of fluorescent proteins. *Nat. Methods* **8,** 393–399.

Dudovich, N., Oron, D., and Silberberg, Y. (2002). Single-pulse coherently controlled nonlinear Raman spectroscopy and microscopy. *Nature* **418,** 512–514.

Evans, C. L., and Xie, X. S. (2008). Coherent anti-Stokes Raman scattering microscopy: Chemical imaging for biology and medicine. *Annu. Rev. Anal. Chem. (Palo Alto Calif)* **1,** 883–909.

Fercher, A. F. (2010). Optical coherence tomography—development, principles, applications. *Z. Med. Phys.* **20,** 251–276.

Hellerer, T., Enejder, A. M. K., and Zumbusch, A. (2004). Spectral focusing: High spectral resolution spectroscopy with broad-bandwidth laser pulses. *Appl. Phys. Lett.* **85,** 25–27.

Hoebe, R. A., Van Oven, C. H., Gadella, T. W., Jr., Dhonukshe, P. B., Van Noorden, C. J., and Manders, E. M. (2007). Controlled light-exposure microscopy reduces photobleaching and phototoxicity in fluorescence live-cell imaging. *Nat. Biotechnol.* **25,** 249–253.

Huang, D., Swanson, E. A., Lin, C. P., Schuman, J. S., Stinson, W. G., Chang, W., Hee, M. R., Flotte, T., Gregory, K., Puliafito, C. A., et al. (1991). Optical coherence tomography. *Science* **254,** 1178–1181.

Langbein, W., Rocha-Mendoza, I., and Borri, P. (2009a). Coherent anti-Stokes Raman micro-spectroscopy using spectral focusing: Theory and experiment. *J. Raman Spectrosc.* **40,** 800–808.

Langbein, W., Rocha-Mendoza, I., and Borri, P. (2009b). Single source coherent anti-Stokes Raman microspectroscopy using spectral focusing. *Appl. Phys. Lett.* **95**.

Lotem, H., Lynch, R., and Bloembergen, N. (1976). Interference between Raman resonances in four-wave difference mixing. *Phys. Rev. A* **14,** 1748–1755.

Mehta, S., and Zhang, J. (2011). Reporting from the field: Genetically encoded fluorescent reporters uncover signaling dynamics in living biological systems. *Annu. Rev. Biochem.* **80,** 375–401.

Miyawaki, A. (2011). Development of probes for cellular functions using fluorescent proteins and fluorescence resonance energy transfer. *Annu. Rev. Biochem.* **80,** 357–73.

Motzkus, M., and von Vacano, B. (2006). Time-resolved two color single-beam CARS employing supercontinuum and femtosecond pulse shaping. *Opt. Commun.* **264,** 488–493.

Pantazis, P., Maloney, J., Wu, D., and Fraser, S. E. (2010). Second harmonic generating (SHG) nanoprobes for in vivo imaging. *Proc. Natl. Acad. Sci. USA* **107,** 14535–14540.

Pezacki, J. P., Blake, J. A., Danielson, D. C., Kennedy, D. C., Lyn, R. K., and Singaravelu, R. (2011). Chemical contrast for imaging living systems: Molecular vibrations drive CARS microscopy. *Nat. Chem. Biol.* **7,** 137–145.

Rocha-Mendoza, I., Langbein, W., and Borri, P. (2008). Coherent anti-Stokes Raman microspectroscopy using spectral focusing with glass dispersion. *Appl. Phys. Lett.* **93,** 201103-1–201103-3.

Rocha-Mendoza, I., Langbein, W., Watson, P., and Borri, P. (2009). Differential coherent anti-Stokes Raman scattering microscopy with linearly chirped femtosecond laser pulses. *Opt. Lett.* **34,** 2258–2260.

Schenke-Layland, K. (2008). Non-invasive multiphoton imaging of extracellular matrix structures. *J. Biophotonics* **1,** 451–462.

Stiles, P. L., Dieringer, J. A., Shah, N. C., and Van Duyne, R. P. (2008). Surface-enhanced Raman spectroscopy. *Annu. Rev. Anal. Chem. (Palo Alto Calif)* **1,** 601–626.

Watson, P., Jones, A. T., and Stephens, D. J. (2005). Intracellular trafficking pathways and drug delivery: Fluorescence imaging of living and fixed cells. *Adv. Drug Deliv. Rev.* **57,** 43–61.

Wong, C. C., Loewke, K. E., Bossert, N. L., Behr, B., De Jonge, C. J., Baer, T. M., and Reijo Pera, R. A. (2010). Non-invasive imaging of human embryos before embryonic genome activation predicts development to the blastocyst stage. *Nat. Biotechnol.* **28,** 1115–1121.

CHAPTER FIFTEEN

IMAGING INTRACELLULAR PROTEIN DYNAMICS BY SPINNING DISK CONFOCAL MICROSCOPY

Samantha Stehbens,* Hayley Pemble,* Lyndsay Murrow,[†] *and* Torsten Wittmann*

Contents

1. Introduction	294
2. Instrument Design	298
2.1. Microscope, stage, and environmental control	298
2.2. SDC scanner and illumination	299
2.3. Image acquisition	303
3. Combination with Other Imaging Techniques	304
3.1. SDC and TIRF	305
3.2. SDC and fast photoactivation	306
4. Specimen Preparation	307
4.1. Lentivirus-mediated stable expression	307
4.2. Epithelial sheet migration assay	310
Acknowledgments	312
References	312

Abstract

The palette of fluorescent proteins (FPs) has grown exponentially over the past decade, and as a result, live imaging of cells expressing fluorescently tagged proteins is becoming more and more mainstream. Spinning disk confocal (SDC) microscopy is a high-speed optical sectioning technique and a method of choice to observe and analyze intracellular FP dynamics at high spatial and temporal resolution. In an SDC system, a rapidly rotating pinhole disk generates thousands of points of light that scan the specimen simultaneously, which allows direct capture of the confocal image with low-noise scientific grade-cooled charge-coupled device cameras, and can achieve frame rates of up to 1000 frames per second. In this chapter, we describe important components of

---

* Department of Cell & Tissue Biology, University of California, San Francisco, San Francisco, California, USA
[†] Department of Pathology, University of California, San Francisco, San Francisco, California, USA

a state-of-the-art spinning disk system optimized for live cell microscopy and provide a rationale for specific design choices. We also give guidelines of how other imaging techniques such as total internal reflection microscopy or spatially controlled photoactivation can be coupled with SDC imaging and provide a short protocol on how to generate cell lines stably expressing fluorescently tagged proteins by lentivirus-mediated transduction.

1. INTRODUCTION

The discovery of the original green fluorescent protein (GFP) from the jellyfish *Aequorea victoria*, the realization that GFP could be used to label cells and proteins, and subsequent development of an entire color palette of fluorescent proteins (FPs) from different organisms have revolutionized cell biology (Chalfie et al., 1994; Shaner et al., 2005; Shimomura et al., 1962; Tsien, 1998; Zimmer, 2009). Over the past 10 years, thousands of papers have been published employing FPs to label cells or intracellular components. Such experiments were substantially more difficult before the advent of GFP and required *in vitro* fluorescent labeling of purified proteins and cumbersome microinjection into tissue culture cells (Sammak and Borisy, 1988; Wittmann et al., 2003, 2004). The availability of genetically encoded FP tags has allowed specific labeling of virtually any protein component in living cells and has been a major driving force of discovery in cell biology and innovation in fluorescence light microscopy.

Out-of-focus light obscures detail in conventional, widefield epifluorescence microscopy, which is particularly problematic with FP-tagged proteins that often have a high cytoplasmic background. Hence, the demand has risen for confocal techniques that allow rapid, multidimensional imaging of FP-tagged protein dynamics in live cells. All confocal microscopes rely on the same concept that was initially invented by Marvin Minsky in the 1950s. Pinhole apertures placed in the emission light path allow light to pass that originates from the point of focus, but largely block light from out-of-focus planes. Confocal laser scanning (CLS) microscopes utilize a single excitation laser beam that is rapidly scanned over the specimen. The light returning from each point through the pinhole is typically quantified by a photomultiplier tube and the image is reconstructed by a computer (Conchello and Lichtman, 2005). In contrast, spinning disk confocal (SDC) microscopes use a rapidly rotating disk with thousands of pinholes such that thousands of points of light scan the specimen simultaneously (Fig. 15.1). The spinning disk design is sometimes referred to as Nipkow disk because of its similarity to rotating pinhole disks used for early experimental TV transmissions in the late 1800s. However, spinning disk systems were not used for fluorescence microscopy until

Spinning Disk Confocal Live Cell Microscopy

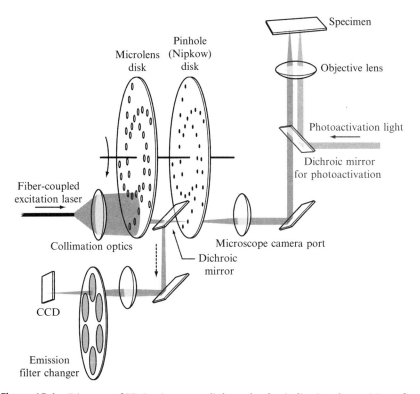

Figure 15.1 Diagram of SDC microscope light path, also indicating the position of an optional dichroic mirror for coupling of a photoactivation light source. Dashed arrow indicates the emission light path.

recently because it was not possible to transmit enough excitation light through the pinhole disk to the specimen to obtain a sufficiently bright confocal signal. In combination with high-intensity laser light sources and sensitive digital cameras, this problem was overcome by introduction of a dual disk design introduced by the Yokogawa Electric Corporation ~10 years ago in which incoming excitation light is focused through the pinholes by a second disk containing microlenses that are aligned with the pinholes on the Nipkow disk (Tanaami *et al.*, 2002). In the Yokogawa design, the pinholes are arranged in a spiral array of equal pitch to generate uniform illumination of the specimen. Longer wavelength emission light returning from the specimen is reflected out of the excitation light path by a dichromatic mirror located between the two disks (Fig. 15.1).

SDC microscopy has a number of important advantages for fluorescence live cell imaging compared with single-point CLS systems. Because the specimen is scanned by thousands of points of light in parallel, the rate of frame acquisition is high. SDC microscopes can achieve theoretical frame rates of up to 2000 frames per second, and clearly this is not the limiting factor for most cell biology live imaging experiments. Fast frame rates have the advantage that scientific grade-cooled charge-coupled device (CCD) cameras can be used for direct confocal image acquisition rather than photomultiplier tubes that are still the light detectors most often employed in CLS microscopes. Although photomultiplier tubes can achieve extremely high gain, they are typically limited by low quantum efficiency and high noise as a result of the charge multiplication process. Thus, in a real-life setting, modern cooled CCD cameras will always produce a better, less noisy image, and the development of large format, high-resolution, and low-noise scientific grade complementary metal-oxide semiconductor (CMOS) cameras holds additional promise for the future.

SDC microscopy is also less prone to fluorescence saturation, or ground state depletion, in which high illumination intensity results in the majority of fluorophores populating the excited state. In this situation, additional excitation light only contributes to photodamage and out-of-focus fluorescence, and does not yield additional signal. During a typical camera exposure of a few hundred milliseconds on an SDC system, each point in the specimen is illuminated several hundred times. The image formed results from many short exposures at comparably low peak illumination intensity. In contrast, each point is illuminated only once during the acquisition of one frame on a CLS microscope resulting in longer dwell times per pixel and several thousand fold higher peak illumination intensities. Thus, fluorescence saturation is reached quite easily on a CLS microscope (Wang et al., 2005) and is most likely the reason for anecdotal reports of increased photobleaching compared with SDC microscopy.

One disadvantage of spinning disk and other multipoint confocal systems is that the theoretical confocality is somewhat reduced compared with single-point laser scanning systems (Conchello and Lichtman, 1994). The reason for this is that some out-of-focus emission light will be transmitted to the camera through adjacent pinholes. In addition, spinning disk systems have a fixed pinhole size that is optimized for 100× magnification. In theory, these pinholes are too large at lower magnification resulting in transmission of more out-of-focus light and reduced confocality. However, in live cell imaging experiments, this slight loss of confocality is more than compensated by a better signal-to-noise ratio and reduced photobleaching.

Different variations of fast multipoint confocal scanning systems are now available, but the original dual disk design by Yokogawa is still the most successful and most widely used. It is utilized by different companies offering turn-key SDC microscopes including the Andor Revolution and the PerkinElmer UltraView systems. The objective of this chapter is to give an overview of an SDC microscope system that we have recently built in our lab (Fig. 15.2). We highlight aspects of instrument design that we think are essential for optimal live cell imaging performance, and provide a rationale for specific design choices. This is important because SDC imaging

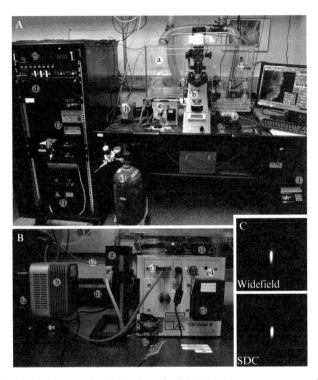

Figure 15.2 (A) Photograph of the described SDC microscope system: (a) In Vivo Scientific environmental chamber, (a1) CO_2 supply for environmental control, (b) Nikon TI inverted microscope stand, (c1) Applied Scientific Instrumentation MS-2000 linear encoded stage, (c2) joystick, and (c3) rack-mounted controller, (d) rack-mounted 64-bit Windows 7 PC, (e) Yokogawa CSU-X1 spinning disk confocal head, (f1) Spectral Applied Research LMM5 laser merge module, (f2) LMM5 controller, and (f3) individual DPSS laser control units, (g) Sutter Instrument emission filter changer, (h) Andor Clara Interline CCD camera, (i) Andor iXon EMCCD camera for TIRF, (j) Sutter Instrument Lambda XL Xenon lamp for epifluorescence illumination, (k) camera port for widefield epifluorescence or diascopic imaging. (B) Close-up of Borealis-modified CSU-X1 SDC scan head: (e1) manual dichroic mirror changer, (e2) multimode optical fiber (from LMM5), (e3) connector for disk rotation speed control (from LMM5 controller), (e4) built-in laser safety shutter button (cannot be software-controlled), (g) Sutter Instrument emission filter changer, (h) Andor Clara Interline CCD camera, (h1) focus tube with lock screws to adjust distance between camera and Nipkow disk, (h2) TTL-out connector ("fire" from Andor cameras to AOTF controller), (i1) Nikon TIRF attachment is visible behind the environmental chamber, (i2) optical fiber to TIRF attachment (from LMM5). (C) Comparison of measured XZ point spread functions in widefield epifluorescence (through port k), and SDC mode with 100×1.49 N.A. objective lens.

systems always combine components from many different manufacturers, and the combined effect of multiple small design mistakes can add up to overall suboptimal performance of the system. It is thus essential for anybody who wishes to set up their own SDC microscope to understand and think about what is expected from the instrument and how different components will affect the capabilities of the system.

2. INSTRUMENT DESIGN

2.1. Microscope, stage, and environmental control

An SDC microscope does not only consist of the confocal head, but careful consideration should be given to other optical and mechanical components of the system. An inverted microscope stand is necessary for the observation of live specimens in open tissue culture chambers. Mammalian cells will also require temperature and may require CO_2 control. We have found that the best way to provide a stable environment is an acrylic box that encloses the entire microscope stage, and that should be sufficiently large to allow easy manipulation of the specimen. Such incubators are available from different manufacturers. We are using an environmental chamber made by In Vivo Scientific in which warm air enters the incubator from below and is recycled to the heater at the top, a design that minimizes temperature gradients in the incubator chamber. We always leave the temperature control at 37 °C and also store objective lenses and stage adaptors for different cell culture chambers inside the incubator.

Repeated cooling and warming of objective lenses is not recommended and will negatively affect the lens over time. It takes several hours for the microscope body to reach temperature equilibrium during warm-up, and temperature fluctuations will cause focus drift. In general, focus drift can be a major problem in any high magnification imaging experiment. Particularly if quantitative analysis is to be performed, shifts in focus over time can invalidate otherwise perfectly good datasets. For these reasons, we are using a Nikon TI inverted microscope stand that includes Nikon's proprietary Perfect Focus technology although other companies are selling similar autofocus systems. The Perfect Focus system utilizes a low-power infrared laser beam that is partially reflected on the interface between the cover glass and the aqueous media above. The reflected beam is a sensitive readout for the position of this interface and directly controls a focus feedback mechanism. It is important to note that this works well with live specimens in which there is a large difference in refractive index between cover glass and tissue culture media, but often fails with fixed samples in which the refractive index of the mounting media is too similar to that of glass. For imaging with the highest possible optical resolution, we typically use Nikon CFI Apo TIRF N.A. 1.49 100× or

60× oil immersion objective lenses. For thicker specimens, we empirically adjust the coverslip thickness correction collar on these lenses to minimize spherical aberration, but we have also used a Nikon CFI Plan Apo N.A. 1.2 60× water immersion lens for 3D specimens. An MS-2000 motorized stage from Applied Scientific Instrumentation with linear encoders is used to be able to record multiple positions during one experiment. At high magnification, positional repeatability is extremely important, and in our hands, this stage will return to positions within 300 nm accuracy. Positional accuracy can often be improved by reducing stage speed. Nikon NIS Elements software running on a Windows 7 64-bit PC is used to control all components of the microscope.

2.2. SDC scanner and illumination

We are using both Yokogawa CSU-10 and CSU-X1 SDC scan heads on different microscope systems. One advantage of the newer CSU-X1 is a higher and adjustable disk rotation speed, which can be important in combination with short camera exposure times. The pinhole pattern on the Nipkow disk is arranged such that each rotation of the disk scans the image field 12 times. The CSU-10 fixed rotation speed is 1800 rpm resulting in a rate of 360 scans per second. Thus, the image formed on the camera chip represents the integration of multiple individual scans by the Nipkow disk, and short exposure times can therefore result in streaking artifacts. For example, in a 5-ms exposure different image areas are scanned either once or twice resulting in a 50% signal variation (Fig. 15.3). This can be resolved by synchronization of disk speed and camera exposure time. However, exact timing can be difficult to achieve and an alternative, simpler solution is to increase disk speed so that variations in image coverage do not matter as much (the CSU-X1 can reach up to 10,000 rpm, i.e., 10 scans in 5 ms). In live cell imaging experiments, the minimum exposure time is most often

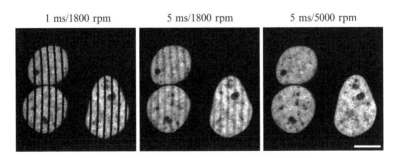

Figure 15.3 Streaking artifacts resulting from unequal coverage of the specimen at low disk rotation speeds and short camera exposure times. The streaking pattern becomes negligible if disk rotation speed is increased. The specimen shown are HeLa cells stably expressing mCherry-histone acquired with a 100 × 1.49 N.A. objective lens. Scale bar, 10 μm.

dictated by the fluorescence signal that can be obtained from cells expressing FP-tagged proteins at low, physiological levels. We find that high-resolution imaging with a nonintensified interline CCD camera (see below) generally requires exposure times of 100 ms or longer at 100% laser power. Thus, streaking artifacts do not present a problem in most experiments.

A weak point in the design of Yokogawa spinning disk heads is the optics by which the excitation laser light is coupled into the spinning disk head. Factory-shipped Yokogawa heads are typically not well aligned, display a relatively large variation of intensity across the field as a result of the Gaussian beam profile exiting the single-mode fiber, and throughput of excitation light to the sample is quite inefficient. For these reasons, we use a CSU-X1 custom-modified by Spectral Applied Research. Although the details of this Borealis modification are proprietary, it involves the use of a multimode optical fiber to couple excitation light into the confocal head, which significantly improves field flatness and increases excitation light throughput severalfold over a wide range of laser wavelengths.

The better excitation light efficiency of the Borealis CSU-X1 also allows the use of less powerful lasers. The described system employs four 100-mW diode-pumped solid-state (DPSS) lasers at 442 nm (Power Technology Inc.), 488 nm (Coherent Sapphire), 515 nm (Cobolt Fandango), and 561 nm (Cobolt Jive), in contrast to 200-mW DPSS lasers we use on an older CSU-10 system. Hundred milliwatts are significant laser power levels that can cause retina damage, and appropriate laser safety precautions and interlocks should be used. Solid state lasers now offer a large selection of available wavelengths at these power levels; have a longer lifetime and lower power consumption; develop less heat; and are significantly smaller than high-power Krypton–Argon gas lasers (Adams *et al.*, 2003). These lasers are mounted in an LMM5 laser merge module by Spectral Applied Research and wavelength and intensity are controlled by an Acousto-Optical Tunable Filter (AOTF). Laser combiners are available from other companies, but in our hands the LMM5 has performed exceptionally well and, on our older setup, has not required any alignment in over 4 years. Although the LMM5 is currently not directly supported by NIS Elements software, it is easy to control by serial commands and we implemented laser line and intensity selection as general filter changers in the software.

Because laser wavelengths are selected by the AOTF, excitation filters in the spinning disk head are not necessary and should be omitted. We use three dichroic mirrors from Semrock that we find optimal for imaging GFP alone, or different combinations of FPs (Table 15.1). Spinning disk dichroic mirrors only transmit narrow bands of excitation light, which is different from dichroic mirrors in epifluorescence filter sets. The dichroic mirrors are mounted in a manual filter changer in our Borealis-modified CSU-X1 because the motorized dichroic mirror changer offered by Yokogawa is expensive and not fast enough for switching during a live cell experiment.

Table 15.1 Dichroic mirrors and emission filters used in the light path of the described spinning disk confocal system

Spinning disk head dichroic mirrors		
	Semrock part number	>90% Transmission
GFP, EGFP	Di01-T488	405–488 nm
GFP, EGFP	Di01-T405/488/561	400–410 nm
RFP, tdTomato, mApple, mCherry		488 nm
		561 nm
ECFP, Cerulean	Di01-T445/515/561	441–449 nm
EYFP, Venus		513–517 nm
RFP, tdTomato, mApple, mCherry		559–563 nm
Emission filters		
	Semrock part number	>93% Transmission
GFP, EGFP (longpass)	BLP01-488R	above 505 nm
GFP, EGFP (bandpass)	FF02-525/50	500–550 nm
RFP, mCherry, etc. (longpass)	LP02-568RS	above 575 nm
ECFP, Cerulean (bandpass)	FF02-475/50	429–522 nm
EYFP, Venus (longpass)	BLP01-514R	529 nm
EYFP, Venus (bandpass)	FF01-542/27	528-555 nm

FPs listed are examples and many other FPs will fall into these bands. We recommend careful examination of excitation and emission spectra of a particular FP before choosing a filter set. Particularly, red FPs vary widely in their excitation and emission characteristics and 561 nm excitation is a compromise.

Photobleaching is the number one enemy of FP live cell imaging, and selection of optimized emission filters is extremely important for maximal signal, a fact that is often strangely overlooked in the design of a live cell imaging system. Emission filters are mounted in a fast Sutter Instrument filter changer between the spinning disk head and the camera. For shortest switching time of ~40 ms, combinations of filters that are used together in dual wavelength experiments should be mounted in adjacent positions. In any case, emission filter switching will result in a delay between image acquisitions at different wavelengths. If speed is of the essence, a multiband emission filter can be used, but one should be aware that longer wavelength FPs (e.g., red fluorescent protein (RFP)) will be excited by the shorter excitation wavelength (e.g., 488 nm for GFP) and will contribute to signal cross-talk into that channel. To optimize fluorescence signal, we use emission filters with cut-off wavelengths as close as possible to excitation laser lines (Table 15.1). Bandpass filters should be selected with a wide transmission band, and we use longpass emission filters for the longest wavelength in an experiment to collect as much emission light as possible. For example, imaging of enhanced GFP (EGFP) alone with the GFP/RFP dual wavelengths dichroic mirror and emission filter results in an ~20% loss of signal compared with the optimized GFP longpass set. Thus, acquiring images with similar signal will require more excitation light and result in increased photobleaching.

To further minimize specimen exposure to excitation light, and thus photobleaching, laser shutters should not be controlled through software, but instead be triggered directly by the camera. Most cameras output a transistor–transistor logic (TTL) signal when an image is acquired, which can be used to directly trigger a shutter. However, in multiwavelength

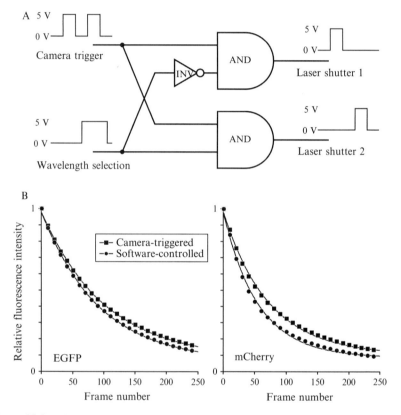

Figure 15.4 Shutter control by camera trigger decreases photobleaching. (A) Diagram of a simple TTL circuit to switch between two excitation laser wavelengths and trigger shutter opening through the camera. The laser wavelength can be selected by the microscope control software through a pin on the parallel port or better through an inexpensive USB DAQ-board (e.g., from National Instruments). For more than two wavelengths, more extensive TTL circuits can be designed. (B) Photobleaching curves of cells expressing EGFP- or mCherry-tagged proteins acquired at 50 ms exposure times and 100% laser power. Particularly, the less photostable mCherry bleaches significantly faster when the laser shutter is controlled by the imaging software and not directly triggered by the camera.

experiments, a simple logic circuit is necessary so that the computer can control which shutter is going to be opened (Fig. 15.4A). In many systems, the software-induced delay in opening and closing shutters can be several hundred milliseconds and thus significantly contribute to excessive photobleaching. In the described system that is controlled by a fast 64-bit PC, and in which camera control during time-lapse experiments takes advantage of enhanced communication between Andor cameras and Nikon NIS Elements software, the software-induced delay between shutter opening and camera exposure is much shorter. Nevertheless, at short exposure times, decreased photobleaching is still evident for both EGFP and mCherry in camera-triggered acquisition mode (Fig. 15.4B).

2.3. Image acquisition

Even in the Yokogawa dual disk design, transmission of excitation light to the sample is inefficient. Spinning disk systems were and are still sold with insufficient excitation laser power, and low-light electron multiplication CCD (EMCCD) cameras are used to compensate for low signal. In any epifluorescence microscope, confocal or not, the objective lens functions both as objective and condenser, and the smallest resolvable distance at which the maximum of one point spread function (PSF) coincides with the first minimum of an adjacent PSF is calculated by the Rayleigh criterion $r = 0.61\lambda/\text{N.A.}$ At 510 nm, the emission maximum of EGFP, and using a 1.49-N.A. objective lens, r is $\sim$210 nm. The Nyquist sampling criterion, which states that the sampling frequency must be two times larger than the highest frequency of a signal, is often inaccurately referred to in this context. In most cases, depending on where the center of the circular PSF is projected onto the square pixel grid of the camera CCD, Nyquist sampling is not sufficient to resolve two PSFs separated by r in the digital image (Wittmann *et al.*, 2004). Camera noise will make this situation even worse. To accurately reproduce the circular PSF image on a square pixel grid without excessive aliasing or signal clipping, the pixel size should be at least three times smaller than r, that is, $\sim$70 nm in the example above.

A typical, high-end EMCCD chip such as the one used in the Andor iXon camera has 16×16 μm pixels. At $100\times$ magnification, this translates into a 160-nm pixel size in the image formed on the camera sensor. Thus, although the high sensitivity of EMCCD cameras has the advantage of potentially very low photobleaching, because of the large pixel size, significant optical resolution is lost (Fig. 15.5). In contrast, most high-resolution cooled CCD cameras utilize a Sony Interline CCD chip with 6.45×6.45 μm pixels, which in theory yields 64.5-nm pixels on the image. It is worth noting that some intermediate magnification occurs in the spinning disk head, and the measured pixel size in the described system is $\sim$60 nm at $100\times$ magnification. The Sony Interline ICX285 chip has been

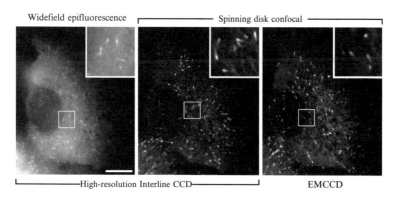

Figure 15.5 Comparison of images acquired by epifluorescence widefield and SDC microscopy acquired with either a Clara interline CCD camera (Andor) or an iXon EMCCD camera (Andor). Images are displayed at the same magnification, and insets show the indicated regions at higher magnification demonstrating aliasing with the large pixel EMCCD camera. The specimen shown are HaCaT keratinocytes stably expressing EB1-EGFP acquired with a $100 \times$ 1.49 N.A. objective lens. Scale bar, 10 μm.

around for over a decade, and is used in cameras from many manufacturers including the Hamatsu ORCA series, Photometrics HQ2, and Andor Clara. Cooling of the CCD chip and readout electronics are important to minimize dark current and read noise, and cameras from different manufacturers should be compared side-by-side.

The SDC image is formed in the plane of the Nipkow disk, and it is thus essential that the camera is well focused on the pinholes in the disk. This is done by switching off the spinning disk motor and imaging the pinholes with diascopic illumination from above. The optimal focus is determined by adjusting the distance between camera and the spinning disk head. Pinholes in the Yokogawa disk have a diameter of 50 μm and at $100 \times$ magnification should appear in the image as evenly illuminated disks with well-defined edges of ~500 nm in diameter (Fig. 15.6). This is slightly larger than the theoretical 420 nm diameter of the Airy disk for a $100 \times$ 1.49 N.A. objective calculated above.

3. COMBINATION WITH OTHER IMAGING TECHNIQUES

Because both illumination and imaging in an SDC system occur through the camera port, the widefield epifluorescence port remains available for other imaging modalities. In this section, we describe briefly how total internal reflection (TIRF) microscopy or photoactivation techniques can be integrated with an SDC microscope.

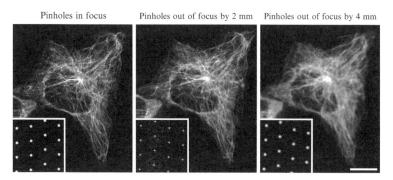

Figure 15.6 Wrong focal distance between camera CCD and Nipkow disk results in severe image deterioration. The specimen shown is an EGFP-tubulin expressing cell acquired with a 100 × 1.49 N.A. objective lens. The camera was moved by the indicated distance away from the optimal position at which pinholes are focused on the camera chip (left panel) and appear as evenly illuminated disks with well-defined edges. Insets show images of the stopped Nipkow disk in transmitted light illumination.

3.1. SDC and TIRF

In TIRF, the evanescent wave generated by a laser beam that is reflected at the interface between the cover glass and the aqueous medium above only excites fluorophores in very close proximity (100–500 nm) to the cover glass (Mattheyses et al., 2010). Thus, TIRF is particularly useful to observe events at the bottom of the cell with high contrast such as vesicle fusion with the plasma membrane or adhesion site dynamics (Hu et al., 2007; Toomre et al., 2000). In the described system, we are using a motorized Nikon TIRF illuminator in which the incident angle of the TIRF laser beam can be controlled by software. The TIRF image is captured by an iXon EMCCD camera on the right side camera port. The LMM5 laser merge module is outfitted with a software-controlled fiber switch that allows the usage of the same lasers for SDC imaging and TIRF illumination and rapid switching between the two modes. Switching between SDC and TIRF, however, also requires moving the epifluorescence filter turret in the microscope stand between an empty position for SDC imaging and an optimized TIRF filter set (Chroma Technology Corp., Cat. No. 91032), as well as switching between the left and the right camera port. It makes economic sense to integrate a TIRF illuminator on an SDC system because lasers are expensive. However, long switching times between the different modes on the microscope stand limit its usefulness for simultaneous confocal and TIRF imaging. Although switching between filter cube positions in the Nikon TI microscope stand is quite rapid, changing from the left to the right camera port is slow and takes about a second.

3.2. SDC and fast photoactivation

Photobleaching, photoactivation, or photoconversion of FPs are useful techniques to quantitatively investigate protein and organelle dynamics in live cells (Lippincott-Schwartz and Patterson, 2008; Lukyanov et al., 2005; Shaner et al., 2007). SDC microscopy is particularly amenable to fast photoconversion of FPs because 405-nm light required for photoconversion can be coupled into the widefield epifluorescence port with an appropriate dichroic mirror installed in one of the epifluorescence filter cubes (e.g., Semrock Di01-R405-25×36; Fig. 15.1). Longer wavelength light used for SDC imaging is transmitted through the dichroic mirror while the shorter wavelength photoactivation light entering through the epifluorescence port is reflected toward the specimen. Near complete photoconversion of mEos2, for example, can be readily achieved in this setup with relatively low intensity of photoactivation light (Fig. 15.7; McKinney et al., 2009). This setup has the important advantage that SDC imaging and photoactivation can be performed simultaneously. In CLS microscopes, fluorescence recovery after photobleaching experiments are often performed by photobleaching an area of interest with the scanning laser set to maximum intensity, but the delay between photobleaching and acquisition scanning is a severe limitation of this method. This is particularly important for the measurement of rapid diffusion-limited protein dynamics where recovery may already occur during the bleaching process. Such fast dynamics occur, for example, in proteins binding reversibly to cytoskeleton structures (Wittmann and Waterman-Storer, 2005). One limitation of utilizing 405-nm wavelength light for photobleaching is, however, that many FPs have been optimized to abolish the excitation peak in the near

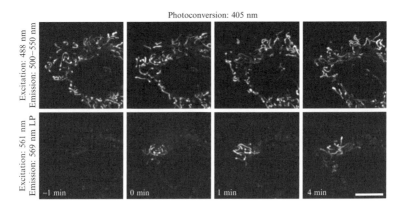

Figure 15.7 Photoconversion of mitochondria matrix-targeted mEosS2 in mouse fibroblasts imaged by SDC illustrating mixing of mitochondria content. Photoconversion was performed with a Mosaic Digital Illumination device (Andor) and a 405-nm DPSS laser coupled into the SDC light path as described. Scale bar, 10 μm.

UV, and EGFP, for example, is bleached rather inefficiently at 405 nm. Photobleaching of EGFP may be more easily achieved with longer wavelength lasers (e.g., 442 nm or even 470 nm) that could still be coupled into the SDC light path as described.

4. Specimen Preparation

In order to image intracellular FP dynamics, the cells that are to be imaged have to express the FP-tagged protein of interest. Transient expression can be achieved by a variety of different commercially available transfection reagents. However, many cell types are not easily transfected, and most, if not all, transfection reagents are cytotoxic and may adversely affect cell behavior. Some of these problems can be overcome by infection with recombinant adenovirus particles that yield high infection efficiency with most cell types (Gierke et al., 2010). However, protein expression levels in transiently expressing cells always suffer from high cell-to-cell variability and generally excessively high expression levels. Clearly, the gold standard is to generate stable cell lines that express FP-tagged proteins at low and relatively even levels across the population. Although this can be achieved by transfection, antibiotic selection, and subsequent fluorescence-activated cell sorting (FACS), we find that lentivirus-mediated transduction is the most reliable technology to rapidly generate stably expressing cell lines (Vigna and Naldini, 2000). Lentiviral transduction has many benefits over both transfection and adenovirus transduction. Lentivirus will stably integrate into the host cell genome resulting in more physiological expression level. This is extremely important in quantitative live cell imaging experiments, as overexpression of FP-tagged proteins will alter the dynamics of the process under investigation.

4.1. Lentivirus-mediated stable expression

Although reagents and protocols on how to generate stable cell lines by lentivirus transduction are available from many sources, we include a short detailed protocol that has worked well for us to generate many different cell lines stably expressing FP-tagged proteins. Different lentivirus systems use different packaging vectors that are not interchangeable, and we routinely use the ViraPower Lentiviral Expression System (Invitrogen) that is not inexpensive, but has proven to be highly reliable in our hands. It is important to ensure proper biosafety procedures and clearance to work with amphotrophic lentivirus. Although third generation lentivirus systems such as ViraPower produce replication-incompetent lentivirus particles, and have enhanced biosafety features, all local safety guidelines should be followed and American laboratories require BSL-2 clearance. Contaminated surfaces, containers, and pipette tips should be decontaminated with 10% bleach before disposal.

4.1.1. Required materials

Gateway entry vector (e.g., pENTR/D-TOPO; Invitrogen, Cat. No. K2400-20)
Gateway destination vector (e.g., pLenti6/V5 DEST; Invitrogen, Cat. No. V496-10)
Gateway LR Clonase II Enzyme Mix (Invitrogen, Cat. No. 11791-020)
One Shot Stbl3 *E. coli* (Invitrogen, Cat. No. C7373-03)
S.O.C. medium (Invitrogen, Cat. No. 15544-034)
Lipofectamine 2000 Transfection Reagent (Invitrogen, Cat. No. 11668-027)
ViraPower Lentiviral Packaging Mix (Invitrogen, Cat. No. K4975-00)
Opti-MEM I Reduced Serum Medium (Invitrogen, Cat. No. 31985-070)
293FT Cell Line (Invitrogen, Cat. No. R700-07)
Dulbecco's modified Eagle's medium (DMEM; Invitrogen, Cat. No. 10313) supplemented with 10% fetal bovine serum (Invitrogen, Cat. No. 26140), 2 mM L-glutamine (Invitrogen, Cat. No. 25030), 0.1 mM MEM nonessential amino acids (Invitrogen, Cat. No. 11140-050); 500 μg/ml Geneticin (Invitrogen, Cat. No. 10131-035) is added for 293FT maintenance, but not during lentivirus production
Sequabrene (Sigma, Cat. No. S-2667)
Blasticidin (Invitrogen, Cat. No. R210-01)

4.1.2. Generation of lentiviral destination vector

We routinely use Gateway Cloning Technology (Invitrogen) to create entry and destination clones for lentivirus production. The first step is the generation of an entry clone with your gene of interest. The exact cloning strategy will depend on the desired product, but we usually include the FP tag at this point as well as additional unique restriction sites in case we want to use the DNA construct for traditional cloning into another vector at a later date. A Kozak consensus sequence and appropriate stop codons should be included in the entry clone. Gateway cloning into the destination vector is essentially done following Invitrogen protocols with reduced reaction volumes to conserve reagents.[1] Briefly, mix 25 ng of the entry clone and 50 ng of pLenti6/V5 DEST in 4 μl water, add 1 μl LR Clonase II Enzyme Mix, mix by pipetting up and down. Incubate for at least 1 h at room temperature. We generally incubate overnight for inserts longer than 5 kb.

---

[1] Useful Web site for frugal Gateway Cloning Protocols: http://www.untergasser.de/lab/protocols/index.htm

Stop reaction by adding 0.5 μl Proteinase K solution and incubate at 37 °C for 10 min. Transform the entire reaction using Stbl3 competent cells following Invitrogen's instructions. Stbl3 cells are recommended as they reduce the frequency of unwanted homologous recombination of lentiviral long terminal repeats. The Gateway recombination reaction is very efficient and to ensure single colonies, plate several dilutions on ampicillin plates. We generally isolate plasmid DNA from 5 to 10 clones to screen by sequencing. Ensure that the Mini-prep protocol uses a resuspension buffer containing 10 mM EDTA required to inactivate the Stbl3 endonuclease. Contamination-free DNA used for 293FT transfection should be prepared with a commercial Midi- or Maxi-prep kit.

4.1.3. Lentivirus production

293FT cells produce the highest quality lentivirus when they are carefully handled and maintained. Unlike adenovirus, lentivirus is secreted into the medium and cell health is essential. It is important to ensure that the 293FT cells never overgrow and are passaged at least twice before lentivirus production. We do not use cells after passage 10. Take care when changing growth medium as 293FT cells become easily detached. When changing media, we recommend removing the old growth medium by decanting, and slowly adding fresh DMEM to the edge of the dish. Cells will detach from the dish if left without medium even for short periods of time, and if multiple transfections are carried out in parallel, growth medium changes should be carried out one dish at a time.

Day 1: Split 293FT cells into complete, but Geneticin-free DMEM growth medium, following Invitrogen's instructions. Cells should be 80–90% confluent the following day, and we generally plate 2–3 different dilutions in 6-cm tissue culture dishes to ensure cells are at the optimal confluency.

Day 2: Cotransfect 293FT cells with the ViraPower packaging mix and the destination clone according to Invitrogen's instructions. We commonly use Lipofectamine 2000. Other transfection reagents can be used but will require optimization. We divide all reagent quantities by 2.8 to account for the smaller surface area of 6-cm dishes. It is important that the transfection is performed in the absence of Geneticin. Add transfection mixture dropwise and very gently mix by rocking plate front to back, then side to side. Do not swirl.

Day 3: Carefully replace growth medium with fresh DMEM without Geneticin. Remember that you are now handling active, infectious virus particles. Do not be too concerned if cells at the edge of the dish become detached slightly. Expression of the vesicular stomatitis virus glycoprotein (VSV-G) causes 293FT cells to fuse resulting in large, multinucleated cells. This morphological change is normal and does not affect lentivirus production.

Day 4: Harvest lentivirus by pipetting cell supernatants into 15-ml conical centrifuge tubes. We use serological pipettes or filter tips when handling virus. If the 293FT cells remain attached to the tissue culture dishes during this process, gently add fresh DMEM for a second round of virus harvest the following day. We have found no difference in virus harvested at 48 or 72 h posttransfection. Centrifuge supernatants at 3000 rpm for 15 min at 4 °C to pellet cellular debris. Snap freeze lentivirus-containing supernatant in 1 ml aliquots in cryovials in liquid nitrogen and store at -80 °C. Lentivirus particles can be further concentrated and purified using the PEG-it virus precipitation system (Systems Biosciences), or filtration through 0.45 μm syringe filters, although this is not necessary for production of simple stable cell lines (Reiser, 2000).

4.1.4. Production of stable cell line

Split cells to be 70–80% confluent at the time of transduction. We generally use one 6-cm dish for each cell line we want to create. To each dish, add 1 ml of growth medium, 0.1–1 ml of virus, and 1–2 μg/ml Sequabrene, a cationic polymer that neutralizes the charge between virions and sialic acid on the cell surface and greatly increases infection efficiency (Davis *et al.*, 2004). The exact amount of virus will depend both on virus titer (Sastry *et al.*, 2002) and on the cell type used, but 1 ml should generally be more than enough. Rock dish side to side and back and forth; return to tissue culture incubator overnight. Change growth medium the next day remembering that the old medium still contains active virus. After 24 h, start Blasticidin selection, making sure that cells will have enough room to divide. The exact concentration of Blasticidin (between 2 and 10 μg/ml) will need to be determined for the cell type used. We always include a control dish to ensure that Blasticidin is efficiently killing nontransduced cells. Change the growth medium every 2–3 days to remove dead cells. We generally collect $\sim$6 million cells by FACS, selecting cells in the top 20% of fluorescence intensity.

4.2. Epithelial sheet migration assay

Cell migration is often analyzed by wounding of a cell monolayer. Traditional methods to generate such wounds in confluent fibroblast cultures involve scratching of the cell monolayer with a pipette tip, and subsequent imaging of the closure of this "wound." For a number of reasons, however, we find that epithelial cells are better suited to investigate cytoskeleton dynamics during cell migration. Epithelial cells that retain cell–cell contacts, for example, HaCaT keratinocytes, migrate as a cell sheet, and leading edge cells become highly polarized for long periods of time resulting in highly coordinated turnover of cytoskeleton structures (Kumar *et al.*, 2009). However, scratching of an epithelial cell monolayer with a pipette tip usually results in excessive tearing of the wound edge, detachment of leading edge cells from the cover glass, and disturbance of the deposited extracellular matrix. This disrupts coordinated cell migration and is particularly problematic for high-resolution SDC imaging. We therefore use a modified scratch assay in which half of the epithelial monolayer is removed from the cover glass by a razor blade swipe.

4.2.1. Required materials

15 mm Round #1.5 coverslips (CS-15R15; Warner Instruments, Cat. No. 64-0713). To clean coverslips for live cell imaging, immerse coverslips in a glass jar in deionized water with 1–2 drops of Versa-Clean detergent (Fisher Scientific, Cat. No. 04-343) and sonicate for 20 min. Rinse five to six times in deionized water until detergent bubbles disappear, sonicate for 20 min, rinse again five to six times in water, and finally rinse two to three times in ethanol. Store the coverslips in ethanol. Flame coverslips briefly to remove ethanol before placing in tissue culture dishes

Custom-made anodized aluminum imaging chambers: these are slide size pieces of aluminum with a central hole and a shallow counterbore on either side to hold coverslips (Wittmann *et al.*, 2004)

High-quality stainless steel razor blades (e.g., GEM brand, single edge industrial blade 62-0165)

Appropriate cell culture growth medium

0.25% Trypsin–EDTA (Invitrogen, Cat. No. 25200-056)

1 M 4-(2-hydroxyethyl)-1-piperazineethanesulfonic acid (HEPES) buffer solution (Invitrogen, Cat. No. 15630-080)

High vacuum grease (Dow Corning, Cat. No. 1597418)

4.2.2. Epithelial migration assay

Day 1: We generally plate cells 2 days prior to wounding at several different dilutions to ensure ~90% confluency on the day of wounding. This allows epithelial cells to secrete extracellular matrix, and to properly adhere and spread. Coating coverslips may also improve migration, and for HaCaT keratinocyte migration assays we usually coat coverslips with 10 μg/ml Fibronectin (Roche, Cat. No. 11051407001) in PBS for 30–40 min at 37 °C, 5% CO_2. Rinse coated coverslips with PBS before plating.

Day 3: Perform the wound assay in the morning following sterile tissue culture practice. Wash coverslips in 37 °C PBS and return to incubator for 5 min. This is important and allows cell–cell contacts to loosen, and prevents tearing of the cell sheet during wounding. Place coverslip on parafilm with cells facing up. Place the edge of a sterile, sharp razor blade in the middle of the coverslip, and while holding the coverslip on one side, scrape the blade across half of the coverslip to remove all cells from this half. Take care not to apply too much force, or the coverslip will break. Return coverslip to a tissue culture dish, and immediately cover with 0.25% Trypsin–EDTA for 10 s; then rinse five to six times by directly pipetting media across the wound to remove loosely attached and dead cells. Return to incubator for 2–3 h.

Prepare live cell imaging chamber by coating both sides of the rim with a thin layer of vacuum grease. This works best by squeezing grease out of a plastic syringe. Remove excess grease from the inside of the rim. Place a clean coverslip onto the greased rim, gently pushing down on the coverslip edge. Clean with ethanol to remove grease on the coverslip. Flip chamber over so that the coverslip-mounted side is facing down. Add 150 μl of media, supplemented with 20 mM HEPES, pH 7.5 to the chamber well. Mount coverslip with cells onto the greased rim, gently pressing down and aligning the wound to ensure it is vertical when on the microscope. Do not push on the center of the coverslip, as this will create air bubbles or break the coverslip. Clean the outside coverslip surface with ethanol. Make sure it is 100% clean as this will face the objective. Place in tissue culture incubator overnight and image the following day.

ACKNOWLEDGMENTS

S.S. and H.P. are recipients of American Heart Association postdoctoral and predoctoral research fellowships. T.W. is supported by National Institutes of Health grant R01 GM079139. This research was in part conducted in a facility constructed with support from the Research Facilities Improvement Program grant C06 RR16490 from the National Center for Research Resources of the National Institutes of Health, and the described instrument was funded by Shared Equipment Grant S10 RR26758.

REFERENCES

Adams, M. C., Salmon, W. C., Gupton, S. L., Cohan, C. S., Wittmann, T., Prigozhina, N., and Waterman-Storer, C. M. (2003). A high-speed multispectral spinning-disk confocal microscope system for fluorescent speckle microscopy of living cells. *Methods* **29,** 29–41.

Chalfie, M., Tu, Y., Euskirchen, G., Ward, W. W., and Prasher, D. C. (1994). Green fluorescent protein as a marker for gene expression. *Science* **263,** 802–805.

Conchello, J. A., and Lichtman, J. W. (1994). Theoretical analysis of a rotating-disk partially confocal scanning microscope. *Appl. Opt.* **33,** 585–596.

Conchello, J. A., and Lichtman, J. W. (2005). Optical sectioning microscopy. *Nat. Methods* **2,** 920–931.

Davis, H. E., Rosinski, M., Morgan, J. R., and Yarmush, M. L. (2004). Charged polymers modulate retrovirus transduction via membrane charge neutralization and virus aggregation. *Biophys. J.* **86,** 1234–1242.

Gierke, S., Kumar, P., and Wittmann, T. (2010). Analysis of microtubule polymerization dynamics in live cells. *Methods Cell Biol.* **97,** 15–33.

Hu, K., Ji, L., Applegate, K. T., Danuser, G., and Waterman-Storer, C. M. (2007). Differential transmission of actin motion within focal adhesions. *Science* **315,** 111–115.

Kumar, P., Lyle, K. S., Gierke, S., Matov, A., Danuser, G., and Wittmann, T. (2009). GSK3beta phosphorylation modulates CLASP-microtubule association and lamella microtubule attachment. *J. Cell Biol.* **184,** 895–908.

Lippincott-Schwartz, J., and Patterson, G. H. (2008). Fluorescent proteins for photoactivation experiments. *Methods Cell Biol.* **85,** 45–61.

Lukyanov, K. A., Chudakov, D. M., Lukyanov, S., and Verkhusha, V. V. (2005). Innovation: Photoactivatable fluorescent proteins. *Nat. Rev. Mol. Cell Biol.* **6,** 885–891.

Mattheyses, A. L., Simon, S. M., and Rappoport, J. Z. (2010). Imaging with total internal reflection fluorescence microscopy for the cell biologist. *J. Cell Sci.* **123,** 3621–3628.

McKinney, S. A., Murphy, C. S., Hazelwood, K. L., Davidson, M. W., and Looger, L. L. (2009). A bright and photostable photoconvertible fluorescent protein. *Nat. Methods* **6,** 131–133.

Reiser, J. (2000). Production and concentration of pseudotyped HIV-1-based gene transfer vectors. *Gene Ther.* **7,** 910–913.

Sammak, P. J., and Borisy, G. G. (1988). Direct observation of microtubule dynamics in living cells. *Nature* **332,** 724–726.

Sastry, L., Johnson, T., Hobson, M. J., Smucker, B., and Cornetta, K. (2002). Titering lentiviral vectors: Comparison of DNA, RNA and marker expression methods. *Gene Ther.* **9,** 1155–1162.

Shaner, N. C., Steinbach, P. A., and Tsien, R. Y. (2005). A guide to choosing fluorescent proteins. *Nat. Methods* **2,** 905–909.

Shaner, N. C., Patterson, G. H., and Davidson, M. W. (2007). Advances in fluorescent protein technology. *J. Cell Sci.* **120,** 4247–4260.

Shimomura, O., Johnson, F. H., and Saiga, Y. (1962). Extraction, purification and properties of aequorin, a bioluminescent protein from the luminous hydromedusan, Aequorea. *J. Cell. Comp. Physiol.* **59,** 223–239.

Tanaami, T., Otsuki, S., Tomosada, N., Kosugi, Y., Shimizu, M., and Ishida, H. (2002). High-speed 1-frame/ms scanning confocal microscope with a microlens and Nipkow disks. *Appl. Opt.* **41,** 4704–4708.

Toomre, D., Steyer, J. A., Keller, P., Almers, W., and Simons, K. (2000). Fusion of constitutive membrane traffic with the cell surface observed by evanescent wave microscopy. *J. Cell Biol.* **149,** 33–40.

Tsien, R. Y. (1998). The green fluorescent protein. *Annu. Rev. Biochem.* **67,** 509–544.

Vigna, E., and Naldini, L. (2000). Lentiviral vectors: Excellent tools for experimental gene transfer and promising candidates for gene therapy. *J. Gene Med.* **2,** 308–316.

Wang, E., Babbey, C. M., and Dunn, K. W. (2005). Performance comparison between the high-speed Yokogawa spinning disc confocal system and single-point scanning confocal systems. *J. Microsc.* **218,** 148–159.

Wittmann, T., and Waterman-Storer, C. M. (2005). Spatial regulation of CLASP affinity for microtubules by Rac1 and GSK3beta in migrating epithelial cells. *J. Cell Biol.* **169,** 929–939.

Wittmann, T., Bokoch, G. M., and Waterman-Storer, C. M. (2003). Regulation of leading edge microtubule and actin dynamics downstream of Rac1. *J. Cell Biol.* **161,** 845–851.

Wittmann, T., Littlefield, R., and Waterman-Storer, C. M. (2004). Fluorescent speckle microscopy of cytoskeletal dynamics in living cells. *In* "Live Cell Imaging: A Laboratory Manual," (D. L. Spector and R. D. Goldman, eds.), pp. 187–204. Cold Spring Harbor Press, New York.

Zimmer, M. (2009). GFP: From jellyfish to the Nobel prize and beyond. *Chem. Soc. Rev.* **38,** 2823–2832.

SECTION TWO

TOOLS

CHAPTER SIXTEEN

Visualizing Dynamic Activities of Signaling Enzymes Using Genetically Encodable FRET-Based Biosensors: From Designs to Applications

Xin Zhou,* Katie J. Herbst-Robinson,* and Jin Zhang*,†

Contents

1. Introduction	318
2. Generalizable Modular Designs	319
2.1. Fluorescent protein pair	319
2.2. Sensing unit	321
2.3. Unimolecular and bimolecular designs	323
3. FRET-Based Biosensors for Monitoring Signaling Enzymes	324
3.1. Protein kinases	325
3.2. Small GTPases	329
4. Example: A-kinase Activity Reporter (AKAR)	332
4.1. Development of AKAR	332
4.2. Microscope setup	334
4.3. Cellular expression of AKAR	334
4.4. Cell imaging	335
4.5. Controls	335
4.6. Data analysis and FRET quantification	336
5. Summary and Perspectives	336
Acknowledgments	337
References	337

Abstract

Living cells respond to various environmental cues and process them into a series of spatially and temporally regulated signaling events, which can be tracked in real time with an expanding repertoire of genetically encodable FRET-based biosensors. A series of these biosensors, designed to track

* Department of Pharmacology and Molecular Sciences, The Johns Hopkins University School of Medicine, Baltimore, Maryland, USA
† The Solomon H. Snyder Departments of Neuroscience and Oncology, The Johns Hopkins University School of Medicine, Baltimore, Maryland, USA

dynamic activities of signaling enzymes such as protein kinases and small GTPases, have yielded invaluable information regarding the spatiotemporal regulation of these enzymes, shedding light on the orchestration of signaling pathways within the native cellular context. In this chapter, we first review the generalizable modular designs of FRET-based biosensors, followed by a detailed discussion about biosensors for reporting protein kinase activities and GTPase activation. Two general designs, uni- and bimolecular reporters, will be discussed with an analysis of their strengths and limitations. Finally, an example of using both uni- and bimolecular kinase activity reporters to visualize PKA activity in living cells will be presented to provide practical tips for using these biosensors to explore specific biological systems.

1. INTRODUCTION

In an ever changing environment, a living cell relies on exquisite spatial and temporal regulation of signal transduction machinery to make vital decisions, such as differentiation, migration, and apoptosis. For some of these signal transduction events, such as changes in pH, Ca^{2+} level, and membrane potential, specific fluorescent biosensors (Cohen et al., 1974; Grynkiewicz et al., 1985; Ross et al., 1974) have existed for a few decades to allow tracking of these events in real time. However, with the discovery of green fluorescent protein (GFP) and the advances in imaging technologies, the past decade has brought an escalation in both the development and the application of genetically encoded fluorescent biosensors for imaging signal transduction in complex biological systems such as living cells and organisms. These genetically encodable biosensors can be introduced in living cells by standard molecular and cellular biology techniques and targeted to specific cells or subcellular locations to monitor local dynamic signaling processes, such as changes in protein expression, localization, turnover, posttranslational modification or interactions with other proteins in the cellular milieu.

Many types of fluorescent biosensors have been developed for visualizing a variety of cellular and molecular events, such as those based on probe translocation, direct sensitization of a fluorescent protein (FP), or fragment complementation of FPs (Newman et al., 2011). In this chapter, we focus on Förster Resonance Energy Transfer (FRET)-based biosensors for tracking activities of signaling enzymes in living cells. FRET is a quantum mechanical phenomenon in which an excited donor fluorophore transfers energy in a nonradiative fashion to an acceptor fluorophore in its close proximity (i.e., <10 nm apart). For FRET (Forster, 1948), the efficiency of energy transfer is inversely proportional to the sixth power of the distance between the donor and acceptor and is also dependent on the relative orientation of the fluorophores. As FRET is particularly sensitive to variations in distance in the range of macromolecular dimension (from 10 to 100 nm), this technique has been applied to analyze the molecular dynamics of biologically relevant processes in a variety of different ways. In the context of FRET-based biosensors for the characterization of signaling enzymes such as protein kinases (PKs) and GTPases, which will be discussed here in detail, changes in the activation state or activity of the signaling enzymes are translated into changes in FRET. Importantly, FRET-based biosensors provide ratiometric readout, which is desirable to eliminate variations in probe concentration and cell thickness. Further, these biosensors have an established generalizable modular design, greatly facilitating the process of generating customized FRET-based biosensors for signaling enzymes in the same families. Consequently, this approach has the potential to be readily adopted to track a large number of dynamic signaling events in space and time.

Here, we first describe strategies to design genetically encodable FRET-based biosensors for signaling molecules, with a brief discussion about some differences between the two general classes, uni- and bimolecular reporters. We then discuss the development of FRET-based biosensors for two classes of signaling enzymes: PKs and small GTPases, and representative applications for each. We end with a specific example of FRET-based biosensors for tracking the activity of cAMP-dependent protein kinases, highlighting the experimental design and practical tips for using such sensors.

2. Generalizable Modular Designs

The generalizable modular design of genetically encodable FRET-based biosensors consists of two units: a signal sensing unit to recognize biologically relevant signals and a reporting unit, that is, FP pair, to convert the relevant signaling event into a change in FRET readout. There are two general classes of FRET-based biosensors, unimolecular and bimolecular which are based on intra- and intermolecular FRET, respectively. Importantly, both types of biosensors can be used to track signaling dynamics in real time, and each has its respective strengths and limitations. In this section, we discuss the general designs of FRET-based biosensors, with a focus on the makeup of both functional units.

2.1. Fluorescent protein pair

After being translated in cells, GFP and its derivatives form fluorophores via an autocatalytic mechanism (Chalfie *et al.*, 1994; Cubitt *et al.*, 1995; Tsien, 1998). Thus, as a general requirement for using FPs in living cells, FPs should express readily, mature rapidly, and should be stable and nontoxic when expressed in cells. Because of recent and extensive protein engineering efforts, not only have these above requirements for FPs been met, but there is also an abundance of FPs with distinct biophysical properties which can facilitate the design of an optimal biosensor to best suit the need of a specific study (Newman et al., 2011). Below we highlight some recommendations for selecting a FRET pair, which is typically used as the reporting unit in a FRET-based biosensor.

First of all, it is important to select a FP pair that has sufficient spectral overlap between donor emission and acceptor excitation to allow FRET to occur, as well as clearly separated excitation and emission peaks for both the donor and the acceptor to avoid cross-excitation and signal bleedthrough. Currently, the most widely used FRET pair in biosensor design is a cyan fluorescence protein (CFP)–yellow fluorescent protein (YFP) pair. Examples of cyan FPs include Cerulean, mTFP1, and CyPet, and yellow FPs include Venus, Citrine, and YPet. Cerulean exhibits superior brightness and efficient folding and maturation at 37 °C but has lower photostability compared to some of the other CFP variants (Rizzo et al., 2004). A more recent variant, mCerulean3, shows greatly reduced fluorescence photoswitching behavior and enhanced photostability (Markwardt et al., 2011). A different cyan variant, monomeric teal fluorescent protein (mTFP), also exhibits improved brightness and photostability and has a narrow emission peak (Henderson et al., 2007). Both Cerulean and mTFP have proven to be efficient donors for monomeric yellow FP variants such as Citrine and Venus. Moreover, in an effort to generate an optimal FRET pair, another CFP, cyan fluorescent protein for energy transfer (CyPet), was coevolved with yellow fluorescent protein for energy transfer (YPet), and this CyPet/YPet FRET pair has proven a useful FRET pair (Nguyen and Daugherty, 2005). While YPet is one of the brightest YFP variants and displays desirable pH- and photostability, CyPet suffers from poor folding efficiency, and it is thus often advantageous to use another suitable CFP variant, such as eCFP, as the donor for YPet. Further, circularly permutated FPs (cpFPs), which have identical secondary structure but rearranged N- and C-termini relative to the WT protein, can be introduced in a biosensor as a means to optimize the energy transfer efficiency between donor and acceptor by tweaking the relative orientation of the fluorophores in the basal and/or stimulated state (Allen et al., 2006; Allen and Zhang, 2006; Gao and Zhang, 2008; Nagai et al., 2004). In addition to the Cyan/Yellow FRET pair, other FRET pairs which have been utilized include Green/Red (EGFP/mCherry) (Albertazzi et al., 2009; Ni et al., 2011), Cyan/Orange (MiCy/mKO) (Karasawa et al., 2004), and Yellow/Red (mAmetrine/tdTomato) (Ai et al., 2008), and some of these spectrally distinct FRET pairs have been advantageous for simultaneous imaging of two or more signaling events.

In addition to the spectral considerations, an ideal FP should be bright, pH- and photostable, and monomeric. Some bright FPs have high extinction coefficients near 1000,000 $M^{-1}cm^{-1}$ and high quantum yields approaching the theoretic limit of 1. In the cellular context, factors like maturation speed, folding efficiency, and susceptibility to photobleaching affect the performance of a FP. In addition, in most cases, monomeric FPs are used in a biosensor, because dimerization or oligomerization may interfere with the function and localization of biosensors, and this can sometimes lead to the formation of protein aggregate. Therefore, FP selections are based on a comprehensive set of parameters. Many comprehensive reviews are available to aid in the optimal selection of an FP pair (Day and Schaufele, 2008; Newman et al., 2011; Shaner et al., 2005).

2.2. Sensing unit

The sensing unit of a FRET-based biosensor determines which biological event the biosensor monitors. The sensing unit senses a signaling activity within the intracellular environment, such as ligand binding or enzyme activation, and converts these signals into conformational changes or association/dissociation events that can alter FRET between the FP pair, the reporting unit.

When the signaling activity to be monitored generates a conformational change in a particular protein, such as a conformational change induced by ligand binding or covalent modification, this protein itself could be used as a sensing unit (Fig. 16.1A). For example, as cAMP binding induces an intrinsic conformational change in the Exchange Protein Activated by cAMP (Epac1), ICUE1 (Indicator of cAMP using Epac), a sensitive reporter of cAMP levels, was developed by sandwiching Epac1 between a FRET pair (DiPilato et al., 2004). Moreover, as the cAMP-induced conformational change is reversible, ICUE1 can be used to detect both the accumulation and degradation of cAMP. This strategy has been applied to many proteins that undergo a large conformational change upon activation (Calleja et al., 2007; Fujioka et al., 2006; Schleifenbaum et al., 2004). For instance, in the specific case of signaling enzymes, a kinase activation sensor can be designed by fusing the kinase between a FRET pair, as long as the kinase undergoes an inherent conformational change upon activation (Ananthanarayanan et al., 2007; Calleja et al., 2007).

In the case that a protein does not undergo a sufficient conformational rearrangement, a "molecular switch" can be engineered by coupling a "receiving" segment to a "sensing" segment to take advantage of signal-induced association/dissociation of the two segments (Fig. 16.1B–D). For example, the molecular switch of a kinase activity reporter (KAR) is designed by linking a peptide substrate which can be recognized and phosphorylated by a kinase-of-interest to a phosphoamino acid binding domain (PAABD). When phosphorylated, the substrate falls into the binding pocket of PAABD, and this engineered conformational change of the molecular switch can be read out as a change in FRET. Importantly, this

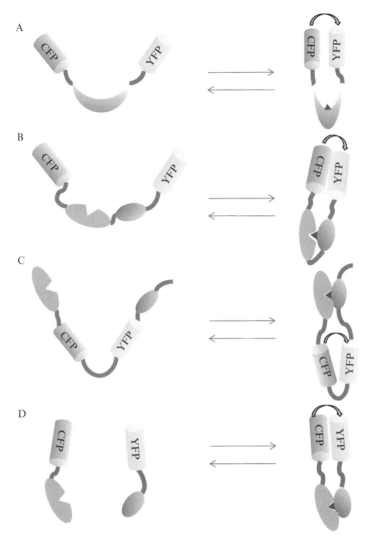

Figure 16.1 Modular designs of genetically encodable FRET-based unimolecular biosensors (A–C) and bimolecular biosensor (D). Two basic units include a reporting unit, consisting of CFP (blue cylinder) and YFP (yellow cylinder), and a sensing unit. The sensing unit can be a single conformationally responsive domain (orange U shape in a) or a "molecular switch" consisting of a sensing segment (purple oval in B–D) and a receiving segment (orange irregular shape in B–D) with different topology in (B) and (C); or in two different polypeptide chains in (D). The sensing unit recognizes the biologically relevant events, such as ligand binding and phosphorylation (depicted by adding the red triangles), and converts them into a change in FRET (orange arrows indicate the energy transfer from donor to acceptor). (For interpretation of the references to color in this figure legend, the reader is referred to the Web version of this chapter.)

design is also modular and can be applied to measure a variety of phosphorylation events by simply replacing the surrogate substrate in an existing biosensor with a substrate specific to the kinase under investigation (Ni et al., 2006). Moreover, this design strategy can be applied to study the activity of enzymes responsible for other posttranslational modifications, such as acetylation and O-GlcNAc glycosylation (Aye-Han et al., 2009). Similarly, this molecular switch concept can be applied to the engineering of FRET-based biosensors for small molecules. For example, a ligand-induced molecular switch can incorporate a pseudoligand, which binds to the sensing domain in the basal state and will be displaced when concentrations of the endogenous ligand are suitably increased, inducing the critical conformational change in the switch. This particular design is used in the biosensor, InPAkt (Indicator for phosphoinositides based on Akt), which probes the dynamics of membrane restricted lipid second messengers, phosphatidyl inositol (3,4,5) trisphosphate (PIP_3) and phosphatidyl inositol (3,4) bisphosphate ($PIP(3,4)_2$) (Ananthanarayanan et al., 2005). Specifically, a molecular switch was constructed by linking the pleckstrin homology (PH) domain from Akt, which specifically binds to PIP_3 and $PI(3,4)P_2$, to an acidic patch of amino acids from nucleolin 1 as the pseudoligand. When produced, 3' phosphoinositides bind to the designed binding site within the PH domain, displacing the pseudoligand and inducing a conformational change in the molecular switch and a FRET change. As such, InPAkt serves as a FRET-based indicator for intracellular levels of 3' phosphoinositides and, when combined with subcellular targeting sequences, can monitor 3' phosphoinositde levels at different subcellular locations.

2.3. Unimolecular and bimolecular designs

In a unimolecular FRET-based biosensor, the fluorescent donor and acceptor are attached to the same protein to form a single-chain reporter. By design, this class of reporters relies on an intramolecular conformational change to alter the distance and/or orientation between the FRET pair (Fig. 16.1A–C). In contrast, in bimolecular reporters, the donor and acceptor FP are fused to either portion of the molecular switch (Fig. 16.1D) and expressed separately. With this design, the two FPs are physically separated in the basal state and are brought into close proximity only upon activation of the molecular switch, or vice versa. Each design has its strengths and limitations and special care should be taken to choose probes which are best suited to address the specific biological questions under study.

The design characteristics of each class of biosensor confer their unique features. First, the FRET response generated from a unimolecular reporter originates from the repositioning of two fluorophores with respect to one another, and this change in distance and orientation can be small due to structural constraints of the single peptide chain. However, the FRET change generated from a bimolecular system has the potential to be much larger since there is physical separation of the two FPs in the basal or unstimulated state. Consequently, bimolecular biosensors typically exhibit a larger dynamic range than their unimolecular counterparts, presumably due to lower FRET in the dissociated state.

Second, while bimolecular reporters can achieve superior sensitivity compared to unimolecular counterparts, they lack the fixed stoichiometry of the donor and the acceptor which is inherent to the unimolecular design. Thus special care needs to be taken to match the expression levels of the two components of bimolecular reporters to ensure proper analysis of signals. Unmatched and variable ratios between the two components will not only compromise the sensitivity of the biosensors but also complicate the quantification (Zhang et al., 2002).

Another consideration when using bimolecular probes is that, because the two components of the molecular switch are expressed independently, they are more likely to interact with endogenous molecules (Miyawaki, 2003), which may affect the performance of the biosensor or interfere with the signaling machinery. For example, in the case of bimolecular KARs, the designed PAABD may have to compete with endogenous phosphoamino acid binding proteins for the phosphopeptide, although this does not appear to be a concern in the case of bimolecular PKA and PKC activity reporters (Herbst et al., 2011). A final caveat of bimolecular sensors is that they are potentially less temporally sensitive than their unimolecular counterparts as they are more sensitive to diffusion constraints.

In many cases, both unimolecular (DiPilato et al., 2004; Miyawaki et al., 1997) and bimolecular FRET-based reporters (Herbst et al., 2011; Miyawaki et al., 1997; Zaccolo et al., 2000) have been effectively applied to monitor the same signaling process, for example, to track kinase activities, monitor the activation of small GTPases, and probe the intracellular dynamics of second messengers, such as Ca^{2+} and cAMP. Below we highlight the development of both uni- and bimolecular biosensors to probe the dynamics of signaling enzymes, with a focus on their designs and application.

3. FRET-BASED BIOSENSORS FOR MONITORING SIGNALING ENZYMES

There are many biosensors available to track different enzymatic modifications, such as phosphorylation/dephosphorylation (Newman and Zhang, 2008; Ni et al., 2006), O-glycosylation (Carrillo et al., 2006), histone methylation (Lin et al., 2004) and acetylation (Sasaki et al., 2009), and protein ubiquitination (Perroy et al., 2004). However, kinase activation/activity biosensor and GTPase activation reporters represent two major classes where multiple examples exist (Aoki and Matsuda, 2009; Hodgson *et al.*, 2008; Ni *et al.*, 2006; Zhang and Allen, 2007). Here, we focus on these two families of biosensors, concentrating on their design and use in investigating specific signaling pathways.

3.1. Protein kinases

3.1.1. Introduction

Protein kinases are enzymes that catalyze the transfer of the γ-phosphate of ATP to the protein substrates, thus altering their functions. These signaling enzymes play a critical and complex role in regulating cellular signal transduction as they orchestrate multiple intracellular processes such as glycogen synthesis, hormone responses, and ion transport. Many signaling cascades involving protein kinases require dynamic control and spatial compartmentalization of kinase activity, which needs to be tracked continuously in different compartments and signaling microdomains in living cells; however, traditional methods to study protein kinases activity only provide static and limited snapshots of signaling events and fail to capture the dynamic changes of kinase activity. On the other hand, genetically encodable FRET-based biosensors offer a versatile and powerful approach to elucidate the spatiotemporal patterns of kinase signaling.

3.1.2. Various designs of kinase reporters

To fulfill their complex role in signal transduction, protein kinases are subjected to multiple levels of regulation, such as ligand binding, protein–protein interaction, and phosphorylation by other kinases or by themselves. To date, there are three families of biosensors, kinase activation biosensors, kinase activity reporters (KARs), and substrate-based reporters. A kinase activation biosensor usually consists of a kinase sandwiched between a FP pair, capturing the kinase "activation"; whereas a kinase activity reporter (KAR) monitors the activity of protein kinases by using the biosensor as a surrogate substrate for the kinase. Therefore, for the kinase activation biosensors, a linear relationship exists between the number of activated signaling molecules and the observed signals. However, one active kinase molecule can phosphorylate many KARs, and these biosensors thus benefit from enzymatic amplification and display a nonlinear, and typically logarithmic, relationship between the signal and activated kinase species. The third group, substrate-based biosensors employ similar design as KARs, however, report phosphorylation of a specific substrate by any of the multiple upstream kinases.

A kinase activation biosensor is generally constructed by flanking the kinase-of-interest with a fluorescent protein pair in such a way that the conformational change of the kinase induced by activation can be translated into a change in FRET signal. This strategy has been utilized to generate reporters for activation of kinases such as protein kinase C (PKC) (Schleifenbaum et al., 2004), Akt (Calleja et al., 2007), and extracellular regulated protein kinase (ERK) (Fujioka et al., 2006). As another example, the FRET-based reporter of Akt action (ReAktion) was constructed by flanking full-length Akt-1 with FP donor and acceptor, and the biosensor reports Akt-1 activation when it is phosphorylated on a threonine in the activation loop. This has led to a study of the regulatory role of this phosphorylation event in dissociation of Akt from the plasma membrane upon its activation (Ananthanarayanan et al., 2007). As the design of kinase activation biosensors usually requires incorporation of the full-length or a large portion of an active kinase, potential complication of this approach includes the perturbation of endogenous signal transduction due to presence of an active kinase. In this case, restricting the expression to a relatively low level becomes important. Alternatively, using a biologically inert mutant of the protein which retains its ability to undergo a conformational change is appropriate.

The second type of kinase reporters, KARs utilize an engineered molecular switch consisting of a phosphorylation-sensing domain (a substrate for the kinase-of-interest) with a suitable PAABD. There are three critical components to this design. First, it is critical to have a substrate domain that is specific for the kinase of interest. This is often designed based on the sequence of an endogenous substrate or the consensus substrate sequence for a kinase of interest, which is often derived from peptide library screening data. Second, a PAABD, such as phosphotyrosine binding Src homology 2 (SH2) domain, phosphoserine/threonine binding 14-3-3 and WW domain, or phosphothreonine binding forkhead-associated (FHA) domain, can be selected to be compatible with candidate substrate sequences. For instance, FHA1 domain specifically recognizes phosphothreonine with aspartate at the $+3$ position as respect to the phosphothreonine site, whereas WW domain recognizes phosphoserine/phosphothreonine with proline at the $+1$ position. Therefore, the substrate sequence may need to be further modified to fit the PAABD binding requirement by incorporating appropriate point mutations, but it is important to be sure that the kinase can still specifically recognize and phosphorylate the mutated substrate. For example, in Akt activity reporter (AktAR), the substrate sequence has been modified by mutating the residue at the $+3$ position with respect to the phosphothreonine site to aspartate for efficient binding to the FHA1 domain that is utilized in this reporter (Gao and Zhang, 2008). Because Akt recognizes a consensus sequence as RXRXXS/T, this change following the phosphoamino acid site unlikely affects the binding and phosphorylation of AktAR by Akt. A third component of KAR design is the linker regions between individual segments (fluorescent proteins, substrate, PAABD). Such linkers can be optimized to appropriate length to improve the dynamic range and intracellular stability of KARs. If structural information regarding the binding domain and phosphor-substrate is available, optimization of linker lengths may be guided by examining the relative positioning of these components in the bound form. Alternatively, if structural information is unavailable, linker lengths can be optimized by empirical testing in a trial-and-error manner or through screening of libraries of biosensors containing randomized linker lengths; as a general starting point, flexible linkers such as GGSGG are used. In one example, a long 72-glycine linker is utilized in a reporter for ERK activity (Harvey et al., 2008).

Other design characteristics of KARs are optional but help to improve specificity and sensitivity in detecting certain signaling events. First, to study localized signaling events, a targeting sequence can be added to the sequence of the biosensor. For example, by targeting AktAR to different microdomains within the plasma membrane, a faster and larger AktAR response was observed within membrane rafts compared to nonraft regions of plasma membrane, indicating that Akt activity is differentially regulated in discrete membrane microdomains (Gao and Zhang, 2008). As untargeted KARs diffuse faster than membrane targeted probes (Lu et al., 2008), addition of a targeting sequence improves the signal-to-noise ratio in detecting the localized kinase activities (Lim et al., 2008). Second, in order to increase the specificity of kinase activity detection, a kinase docking domain can be added to the biosensor. This is especially important for kinases such as MAPKs which are similar in terms of substrate consensus sequences and often use docking domains to enhance the specificity of phosphorylating their endogenous substrates. For instance, such docking domains are essential in establishing the specificity of JNKAR and EKAR for the c-Jun N-terminal Kinase (JNK) and ERK, respectively (Fosbrink et al., 2010; Harvey et al., 2008).

As alluded to above, an effective approach to increase the dynamic range of KARs is to design bimolecular kinase activity reporters (BimKAR). This class of biosensors utilizes a kinase-inducible bimolecular switch (KIBS) in which a kinase-specific substrate and a PAABD are each fused to a FP and expressed as independent peptide chains. When phosphorylated by endogenous kinase, the phosphopeptide is bound by the PAABD, resulting in intermolecular FRET. With this design, FRET change largely depends on the ability of the KIBS to bring the two parts into close proximity, as the basal FRET is low when they are fully separated. Based on this concept, A-Kinase and C-Kinase inducible bimolecular switches have been generated (Herbst et al., 2011). While this design strategy is suitable for any KAR which contains a two-component molecular switch, in the case of biosensors such as EKAR and JNKAR which also contain kinase docking domains (Fosbrink et al., 2010; Harvey et al., 2008), it will be important to keep the docking domain on the portion of the KIBS that contains the substrate sequence to facilitate substrate recognition. Further, as aforementioned, targeted KARs can be used to probe compartmentalized kinase activity. Notably, in BimKAR, the two components of the KIBS need not be expressed in the same subcellular compartment; to monitor local signaling dynamics, it is only necessary to target the substrate portion of the switch. It is with this design that the most sensitive nonraft region targeted PKA activity reporter (BimAKAR-CAAX) was developed (Herbst et al., 2011). In the case that a unimolecular biosensor does not already exist, it is still possible to construct a KIBS for a kinase of interest. In fact, as the KIBS design relieves some of the structural constraints of a unimolecular biosensor, it may be a more suitable starting point for KAR design. As the KIBS design is modular, a BimKAR for a kinase of interest can be designed by replacing the PKA-specific substrate with a substrate sequence specific to the kinase of interest, and this was shown with the generation of BimCKAR, a bimolecular reporter of PKC activity. However, as is the case when generating any new biosensor, it is critical to test the specificity of the BimKAR for the kinase of interest by using suitable inhibitors and through generation of mutants in which the phosphoacceptor residues are mutated to Ala.

The third group of biosensors, substrate-based reporters, is designed to report the phosphorylation of a specific substrate, which can be mediated by multiple upstream kinases. As a proof-of-concept, the newly developed indicator of CREB activation due to phosphorylation (ICAP) was constructed to report the critical phosphorylation at serine 133 of Ca^{2+}- and cAMP-responsive element-binding protein (CREB) that leads to its activation (Friedrich et al., 2010). Flanked by CFP/Citrine, the engineered molecular switch consists of a moiety from kinase-inducible domain (KID) of CREB, which contains the key residue serine 133 and a moiety from the KIX domain of CREB binding protein, which specifically recognizes phosphoserine 133. Upon given stimulations, the serine 133 in the sensor was phosphorylated by a number of upstream kinases, such as PKA and CaMK IV, leading to a conformational change in the molecular switch and a subsequent FRET change. Of note, this design has been used to generate functional sensors of ATF-1 or CREM activation by substitution of the KID domain of CREB in ICAP with the homologous KID domains of ATF-1 or CREM, respectively. Therefore, ICAP and analogous substrate-based reporters represent a new class of biosensors which allows the study of the combined control of phosphorylation of a specific protein, often a key node in signaling networks, by a group of protein kinases in living cells.

3.1.3. Application—Oscillations in kinase activity
Recently, KARs have revealed new modes of exquisite temporal controls of kinase activity in cells. It has been long acknowledged that the concentration of second messengers such as Ca^{2+} oscillates in a periodic fashion in response to certain stimuli (Berridge et al., 2003; Clapham, 2007), and fluorescent

kinase biosensors have begun to reveal additional oscillatory signals that are involved in regulating various important cellular processes (Dunn et al., 2006; Ni et al., 2011; Violin et al., 2003).

In one such example, Dunn et al. investigated how waves of spontaneous electrical activity that spread across the retinal ganglion cell (RGC) layer are coupled to the cAMP/PKA pathway by using a cAMP indicator, ICUE2, and a PKA activity reporter, AKAR2.2. Previously, it had been demonstrated that such retinal waves produced by RGCs play an essential role in the early stages of retinal development. Although the cAMP/PKA pathway had previously been implicated in establishment and refinement of RGC axonal projections, the link between the retinal waves and cAMP/PKA signaling had not been characterized. Using the biosensor imaging approach, cAMP and PKA activities were shown to oscillate in response to depolarization-induced Ca^{2+} influx caused by spontaneous electrical activity in RGCs (Dunn et al., 2006).

Also using fluorescence imaging, a highly integrated oscillatory circuit consisting of Ca^{2+}–cAMP–PKA has been discovered in MIN6 β cells. PKA is known to play important roles in regulating Ca^{2+}-dependent exocytosis, which is critical for the pulsatile insulin secretion in pancreatic cells. In this study, fura-2, a fluorescent Ca^{2+} indicator, was simultaneously imaged with the genetically encoded FRET-based biosensor for PKA (AKAR-GR, green and red FP version) to illustrate the precise temporal correlation between Ca^{2+} dynamics and the oscillations in PKA activity, or with the cAMP probe (ICUE-YR, yellow and red FP version) to uncover the coordination between Ca^{2+} and cAMP dynamics (Ni et al., 2011). To probe the temporal relationship between PKA and cAMP activities, a single-chain, dual-specificity biosensor, ICUEPID, was also employed to simultaneously monitor cAMP levels and PKA activity (Ni et al., 2011). The observed synchronized oscillations of Ca^{2+}, cAMP, and PKA, together with quantitative mathematic modeling data, demonstrate that these three signaling molecules form a highly integrated oscillatory circuit (Ni et al., 2011). In addition, PKA activity was found to be required for the oscillatory Ca^{2+}–cAMP–PKA circuit, and it was not only capable of initiating the signaling oscillations but also able to modulate their frequency such that input signals can be integrated and output signals can be diversified to exquisitely control substrate phosphorylation in a context dependent manner.

3.2. Small GTPases

3.2.1. Introduction

Small GTPases are enzymes that catalyze the hydrolysis of guanosine triphosphate (GTP) to guanosine diphosphate (GDP). As the most well-known members, Ras GTPases play essential roles in regulating cell growth, cell differentiation, cell migration, and lipid vesicle trafficking. Ras GTPases cycle between the active GTP-bound state and the inactive GDP-bound state; they are inactivated by GTPase activating proteins (GAP) and activated by guanine nucleotide exchange factors (GEF). The former activate the intrinsic GTPase activity of Ras GTPases, which hydrolyze GTP to GDP, while the latter cause dissociation of GDP from Ras GTPases and association of GTP. Characterizing the spatial and temporal regulation of small GTPases using traditional biochemical methods have proven difficult (Walker and Lockyer, 2004), but these complications have been overcome by the development of genetically encodable biosensors.

3.2.2. Different GTPase activation reporters

GTPase biosensors, like KARs, are based off of a modular design composed of a molecular switch and a FRET pair. When GTP binds to Ras, it induces a conformational change in the effector region of Ras and consequently recruits and activates downstream molecules, including the serine/threonine kinase Raf. The first genetically encodable FRET-based biosensor for GTPase activation, Raichu-Ras (Ras and interacting protein chimeric unit for Ras), used this interaction between activated Ras and Raf as the basis for its molecular switch. Specifically, Raichu-Ras used H-Ras as the sensing segment and Ras Binding Domain (RBD) of Raf as the receiving segment to form an engineered molecular switch, which is sandwiched by CFP and YFP. With this design, activation of Ras can be monitored in living cells by changes in FRET. By simply replacing the H-Ras with Rap1, a biosensor for Rap1 activity, Raichu-Rap1, was generated. Using these biosensors in COS-7 cells, epidermal growth factor-induced activation of Ras was shown to occur at the plasma membrane, and activation of Rap1 at the perinuclear region, indicating spatially distinct regulation. Further, by using the fluorescence recovery after photobleaching (FRAP) technique, high Ras activity at the extending neurite in PC12 cells following nerve growth factor treatment was found to be due to high GTP/GDP exchange rate and/or low GTPase activity, rather than the retention of the active Ras (Mochizuki et al., 2001).

A different approach has been employed to design a biosensor for RhoA, another small GTPase involved in cytoskeletal regulation. In addition to GAP and GEF, Rho-family G proteins are regulated by guanine nucleotide dissociation inhibitor (GDI). The interaction of GDIs with Rho GTPases not only prevents the release of GDP from GTPases but also retains Rho G proteins in the cytosol. To maintain a free C-terminus of RhoA for interacting with Rho GDIs, a biosensor is constructed by fusing, from N-terminus to C-terminus, Rho-binding domain of the effector rhotekin fragment, which specifically binds to GTP-RhoA, to CFP, YFP, and RhoA

(Pertz et al., 2006) (Fig. 16.1C). Thus this biosensor is sensitive to GDI and can be reversibly targeted from cytosol to membrane.

The bimolecular design has also been adopted to develop GTPase biosensors. For instance, a bimolecular molecular switch has been employed to generate activation reporter for Rac1, a GTPase that is involved in diverse cellular events including actin polymerization in cell adhesion. In this case, a fragment of p21-activated kinase (PAK) which binds to activated Rac1 was fused to YPet, and Rac1 was linked to CyPet, to generate a bimolecular FRET biosensor, Rac1 FLAIR.

3.2.3. Application—Coordination of small GTPases in migrating cells

Three GTPases, Rac, Rho, and cdc42, play essential roles in coordinating cytoskeleton dynamics during cell migration. Previously, all three GTPases were shown to activate at the front of migrating cells (Kraynov et al., 2000; Nalbant et al., 2004; Pertz et al., 2006; Ridley et al., 2003), but the spatiotemporal coordination between them is unknown. To unravel the activation dynamics of RhoA, Cdc42, and Rac1 during cell protrusion, first the activation of each GTPase was visualized in separate experiments using two single-chain biosensors for RhoA and cdc42, and a dual chain biosensor, Rac1 FLAIR for Rac. Then the activation profiles of the three GTPases were subsequently aligned using the timing of cell protrusion/retraction as a common reference and analyzed using a computational multiplexing methodology. This technique compares the activity maps of the three GTPases at different time points and different locations along the cell boundary and builds up correlations between distinct signaling activities and the morphological dynamics of the cell edge. Using this approach, it was demonstrated that RhoA activation at the cell edge is coincident with edge advancement, whereas cdc42 and Rac1 are activated 2 μm behind the edge with a 40 s time lag. The time shifts between GTPase activation suggested by separate imaging experiments with help of computational multiplexing was further confirmed by coimaging of a dye-labeled bimolecular Cdc42 biosensor and the unimolecular RhoA biosensor to directly reveal the temporal correlation between activation of these two signaling molecules (Machacek et al., 2009). These data have led to a model of distinct roles of the three GTPases in initiating, reinforcing, and stabilizing membrane protrusions with spatiotemporally coordinated antagonistic actions of Rac1 and RhoA involved. More recently, a similar approach has been utilized to elucidate the regulatory role of PKA as a pacemaker of protrusion–retraction cycle via phosphorylating RhoA and thus interfering with RhoA–RhoGDI interaction (Tkachenko et al., 2011), further demonstrating the great potential of the computationally aided biosensor imaging approach in dissecting complex signaling controls of cellular processes.

4. Example: A-kinase Activity Reporter (AKAR)

Here, we provide a specific example of A-kinase activity reporter (AKAR), with discussion of the chronicle of the design and development of AKARs, followed by a practical perspective of application of FRET-based enzyme biosensors.

4.1. Development of AKAR

The initial AKAR1 reporter consists of ECFP, a phosphoserine/threonine binding domain (14-3-3) (Aitken et al., 1995; Fu et al., 2000), a PKA-specific peptide sequence (modified Kemptide, LRRA★SLP), and Citrine (Zhang et al., 2001) (Fig. 16.2A). One flexible linker, GGTGGS, replaced the C-terminal tail of 14-3-3 without disruption of the 14-3-3/substrate binding, and another linker, GTGGSEL was inserted between the substrate peptide and the YFP. With this design, AKAR1 has been successfully used to study compartmentalized PKA activity. Because of tight 14-3-3/substrate binding, however, the substrate of AKAR1 is inaccessible to phosphatases and thus irreversibly phosphorylated. Consequently, it does not report decreases in PKA activity.

Efforts to generate a reversible AKAR began with the hypothesis that reducing the affinity of the PAABD for the phosphosubstrate would render the biosensor more accessible to phosphatases and thus reversibility. To this end, the lower affinity PAABD, FHA1 was used. However, to facilitate the usage of FHA1 in AKAR1, the PKA-specific substrate also had to be changed (Zhang et al., 2005). As was the case with AktAR, to accommodate the preference of FHA1 binding for an Asp in the $+3$ position (Gao and Zhang, 2008), the PKA substrate was modified to LRRA★TLVD to give rise to AKAR2. AKAR2 was further optimized by introducing the mutation A206K into the FPs which reduced dimerization tendency of the FPs. The resulting reporter, AKAR2.2, displayed enhanced reversibility compared to it AKAR2 predecessor, however, it still suffered from a low dynamic range (Zacharias et al., 2002; Zhang et al., 2005).

To improve the dynamic range of AKAR2, different variants of CFP (ECFP and CyPet) and YFP (Citrine, YPet, Venus, circular permutated cpVN144, cpVK156, cpVE172, and cpVL194) were tested systematically. The highest dynamic change was observed in the construct harboring ECFP and cpVE172 as FRET pair, which was named AKAR3 (Allen and Zhang, 2006).

Engineering of new FP variants continues to provide opportunities for further improving biosensors. For instance, the most sensitive unimolecular reporter for PKA, AKAR4, was generated by replacing the ECFP of

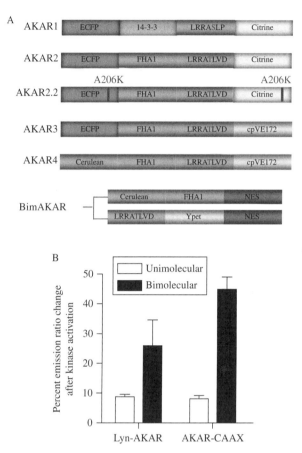

Figure 16.2 (A) Scheme representation of AKAR development; (B) Comparison of emission ratio change of plasma membrane targeted AKAR and BimAKAR following kinase activation. (For color version of this figure, the reader is referred to the Web version of this chapter.)

AKAR3 with Cerulean, a brighter variant of CFP. Compared to AKAR3, AKAR4 shows an improved dynamic range, with a response of 67% to the β-AR receptor agonist isoproterenol in HEK293T cells (Depry et al. 2011).

As has been emphasized throughout this review, switching from the unimolecular design to bimolecular design provides another strategy to improve the sensitivity of a KAR. In the case of PKA activity detection, generation of a bimolecular AKAR (BimAKAR) was achieved by fusing Cerulean to the N-terminus of FHA1 and the PKA-specific substrate to the N-terminus of YPet (Herbst et al., 2011) (Fig. 16.2A). In particular, two variants of BimAKAR, BimAKAR-CAAX and Lyn-BimAKAR, which report PKA activity in nonraft and lipid raft-like portions of the plasma membrane, respectively, show significantly enhanced dynamic range compared to their unimolecular counterparts. Specifically, in response to Fsk treatment, BimAKAR-CAAX shows a fivefold increase in dynamic range relative to AKAR4-CAAX while Lyn-BimAKAR shows a threefold increase in dynamic range compared to Lyn-AKAR4 (Herbst et al., 2011) (Fig. 16.2B).

4.2. Microscope setup

For wide-field fluorescence imaging, the instrumentation consists of an epifluorescence microscope with a shutter-controlled excitation light source, filter sets for the donor and acceptor, and a charge coupled device (CCD) camera. For a typical imaging setup, an Axiovert 200M microscopt (Carl Zeiss) with a cooled charged couple device (CCD) camera was controlled by Metafluor software (Universal imaging). For CFP and YFP, dual-emission ratio imaging uses two excitation filters (420DF20 and 495DF10), two dichroic mirrors (450DRLP and 515DRLP), and two emission filters (475DF40 for CFP and 535DF25 for YFP) switched by a Lambda 10-2 filter changer (Sutter Instruments).

4.3. Cellular expression of AKAR

Typically, HEK293T cells are maintained in Dulbecco's modified Eagle's medium (DMEM) supplemented with 10% fetal bovine serum at 37 °C with 5% CO_2. For imaging AKAR, cells are plated onto glass coverslip (No.1 thickness) in 35 mm dishes and transfected with a mammalian expression vector (pcDNA3) containing AKAR using Ca^{2+} phosphate-mediated transfection at 50–60% confluency, and then grown for approximately 24 h before imaging. In comparison, BimAKAR requires a double transfection. To increase the probability that both portions of the KIBS are expressed in cells, it is important to mix the DNA of both portions of the sensor prior to transfection. Typically, this is done at a 1:1 ratio.

To ensure consistency between imaging experiments, expression level of reporter should be examined before imaging. First, it is important to check if reporters are expressed in cells at sufficient level and also to ensure that cells have a healthy morphology. Second, proper subcellular localization of reporter should be verified. For example, uniform fluorescence throughout the cell should be expected for untargeted reporters, whereas targeted reporters should be expressed in their designed locale. Confirmation of localization can be verified by using colocalization markers. Finally, it is important to make sure that unimolecular reporters remain intact and that no severe proteolysis occurred. This also can be done by checking the intensity of both donor and acceptor fluorescence and by confirming that both are expressed in the same location.

4.4. Cell imaging

Before imaging, culture media is removed and cells are washed twice with Hank's balanced salt solution buffer and maintained in the dark. Generally, healthy cells exhibiting normal morphology and expressing intermediate to high levels of fluorescence are selected. Cells with exceptionally high fluorescence intensity have too much reporter expressed, which may interfere with endogenous signaling pathways. Conversely, cells that are too dim do not have sufficient reporter expression and will be less sensitive.

In the case of imaging BimAKAR, special considerations are required for cell selection. As it is important that both halves of the KIBS be expressed at similar levels, both the donor and acceptor channels must be checked prior to the start of the experiment to ensure sufficient expression and proper localization of both components. In addition, one practical strategy to help select cells that have similar expression levels is to establish a narrow range of starting Y/C ratio. It is important to note that the starting ratio for each BimKAR will vary as this parameter is dependent on the imaging setup, the FPs used, and the basal activity of the kinase in the locale under study. Nonetheless, once an appropriate starting ratio range is determined, it will facilitate the cell selection process, reducing the variability in responses between cells.

For a typical time course imaging for CFP–YFP FRET pair, three image channels are acquired using an acquisition time of 10–1000 ms at time interval of 10–60 s: (1) CFP direct (excitation of donor CFP and acquisition of CFP emission), (2) YFP FRET (excitation of donor CFP and measurement of acceptor YFP emission), and (3) an optional YFP direct (excitation of acceptor YFP and acquisition of YFP emission) as a control to check for photobleaching or cellular morphological changes. It is critical to optimize the excitation exposure time since long exposure time increases the potential of photobleaching and phototoxicity whereas short exposure results in poor signal-to-noise ratio. In addition, time interval should be adjusted to be fast enough to capture the kinase activity dynamics while avoiding too frequent excitation exposure. For AKAR and BimAKAR, a 500-ms exposure for both the FRET and CFP channels and a 50-ms exposure for YFP channel, as well as 30 s time-lapse interval is used (Herbst *et al.*, 2011).

4.5. Controls

An important control to verify that the emission ratio change originates from PKA phosphorylation of the designated threonine in the substrate region in AKAR is to generate and test an AKAR T/A mutant, in which the phosphorylation site is mutated to alanine. For example, all AKAR and BimAKAR T/A mutants do not show Fsk-induced FRET changes (Herbst *et al.*, 2011; Zhang *et al.*, 2005). Such a mutant should be used as a negative control to confirm that the FRET change is phosphorylation dependent when designing or testing any KAR.

It is also critical to test the specificity of KARs. In the case of AKAR and BimAKAR, this includes testing the cAMP-induced response of the sensors in the presence of PKA inhibitor. Specificity testing for a KAR should also be confirmed by testing the response of the biosensor to a compound which will activate cellular kinases other than the target kinase. For instance, BimAKAR does not show a FRET increase upon treatments which activate endogenous PKC, demonstrating the specificity of BimAKAR to detect PKA activity (Herbst et al., 2011). Such controls are critical to confirm the specificity of the biosensor for the event under study.

4.6. Data analysis and FRET quantification

After the imaging experiment, several cell regions, as well as control region without cells are selected within the images from each time point. For reporters which have fixed stoichiometry between the donor and acceptor fluorescent proteins, the change in emission ratio (YFP FRET/CFP) can be linked to FRET change, therefore representing the most experimental convenient readout as shown in Eq. (16.1), where F_A, F_D are the fluorescence intensity of acceptor emission upon donor excitation, and intensity of donor emission upon direct excitation, respectively. In the case of bimolecular reporters, the stoichiometry between the donor and acceptor fluorescent proteins is not fixed and will vary among different cells, thereby complicating the quantification and comparison between different cells. However, the emission ratio change (Eq. (16.1)) may still be used, for example, to get some kinetic information about acutely stimulated signaling activity. In addition, acceptor photobleaching can be used to quantify the FRET efficiency. By irreversibly photobleaching acceptor fluorophores, usually at the end of time-lapse experiments, the energy transfer from donor to acceptor is abolished and causes recovery of donor fluorescence intensity. Therefore, FRET efficiency can be represented by the relative fluorescence intensity of the donor before (F_D) and after (F_{DA}) acceptor photobleaching (Eq. (16.1)).

$$\text{FRET emission ratio} = F_A/F_D \qquad (16.1)$$

$$E = 1 - (F_{DA}/F_D) \qquad (16.2)$$

5. Summary and Perspectives

Genetically encoded FRET-based biosensors allow for the tracking of dynamic cellular signaling events in real time, thus representing versatile and powerful tools to study the regulation of signal transduction with unprecedented spatial and temporal resolution. The generalizable modular design has facilitated the development of an expanding repertoire of biosensors for visualizing the dynamics of several major classes of cellular signaling molecules, including protein kinases, GTPases, and second messengers, and have successfully uncovered specific molecular mechanisms regulating various signal transduction pathways in live cells.

Based on the highly modular and adaptable design strategies, it is anticipated that an increasing number of biosensors for signaling enzymes will be developed. By combining these versatile and powerful fluorescent biosensors with novel molecular tools that manipulate biological systems in a spatially and temporally restricted manner (Airan et al., 2009; Levskaya et al., 2009; Wu et al., 2009), biochemical events underlying signal transduction in the native context of cell or live organism can be studied. Moreover, quantitative fluorescence imaging data can be incorporated into mathematical models to obtain a systems understanding of the signaling network. Taken together, these technologies offer great potential to unravel the complex regulation of dynamic signaling processes in living biosystems.

ACKNOWLEDGMENTS

Work in the lab was supported by NIH grant R01 DK073368 (JZ).

REFERENCES

Ai, H. W., et al. (2008). Fluorescent protein FRET pairs for ratiometric imaging of dual biosensors. *Nat. Methods* **5**, 401–403.

Airan, R. D., et al. (2009). Temporally precise in vivo control of intracellular signalling. *Nature* **458**, 1025–1029.

Aitken, A., et al. (1995). 14-3-3 proteins: Biological function and domain structure. *Biochem. Soc. Trans.* **23**, 605–611.

Albertazzi, L., et al. (2009). Quantitative FRET analysis with the EGFP-mCherry fluorescent protein pair. *Photochem. Photobiol.* **85**, 287–297.

Allen, M. D., et al. (2006). Reading dynamic kinase activity in living cells for high-throughput screening. *ACS Chem Biol.* **1**, 371–376.

Allen, M. D., and Zhang, J. (2006). Subcellular dynamics of protein kinase A activity visualized by FRET-based reporters. *Biochem. Biophys. Res. Commun.* **348**, 716–721.

Ananthanarayanan, B., et al. (2005). Signal propagation from membrane messengers to nuclear effectors revealed by reporters of phosphoinositide dynamics and Akt activity. *Proc. Natl. Acad. Sci. USA* **102**, 15081–15086.

Ananthanarayanan, B., et al. (2007). Live-cell molecular analysis of Akt activation reveals roles for activation loop phosphorylation. *J. Biol. Chem.* **282**, 36634–36641.

Aoki, K., and Matsuda, M. (2009). Visualization of small GTPase activity with fluorescence resonance energy transfer-based biosensors. *Nat. Protoc.* **4**, 1623–1631.

Aye-Han, N. N., et al. (2009). Fluorescent biosensors for real-time tracking of post-translational modification dynamics. *Curr. Opin. Chem. Biol.* **13**, 392–397.

Berridge, M. J., et al. (2003). Calcium signalling: Dynamics, homeostasis and remodelling. *Nat. Rev. Mol. Cell Biol.* **4,** 517–529.

Calleja, V., et al. (2007). Intramolecular and intermolecular interactions of protein kinase B define its activation in vivo. *PLoS Biol.* **5,** e95.

Carrillo, L. D., et al. (2006). A cellular FRET-based sensor for beta-O-GlcNAc, a dynamic carbohydrate modification involved in signaling. *J. Am. Chem. Soc.* **128,** 14768–14769.

Chalfie, M., et al. (1994). Green fluorescent protein as a marker for gene expression. *Science* **263,** 802–805.

Clapham, D. E. (2007). Calcium signaling. *Cell* **131,** 1047–1058.

Cohen, L. B., et al. (1974). Changes in axon fluorescence during activity: Molecular probes of membrane potential. *J. Membr. Biol.* **19,** 1–36.

Cubitt, A. B., et al. (1995). Understanding, improving and using green fluorescent proteins. *Trends Biochem. Sci.* **20,** 448–455.

Day, R. N., and Schaufele, F. (2008). Fluorescent protein tools for studying protein dynamics in living cells: A review. *J. Biomed. Opt.* **13,** 031202.

Depry, C., et al. (2011). Visualization of PKA activity in plasma membrane microdomains. *Mol Biosyst.* **7,** 52–58.

DiPilato, L. M., et al. (2004). Fluorescent indicators of cAMP and Epac activation reveal differential dynamics of cAMP signaling within discrete subcellular compartments. *Proc. Natl. Acad. Sci. USA* **101,** 16513–16518.

Dunn, T. A., et al. (2006). Imaging of cAMP levels and protein kinase A activity reveals that retinal waves drive oscillations in second-messenger cascades. *J. Neurosci.* **26,** 12807–12815.

Forster, T. (1948). Intermolecular energy migration and fluorescence. *Ann. Phys.* **2,** 55–75.

Fosbrink, M., et al. (2010). Visualization of JNK activity dynamics with a genetically encoded fluorescent biosensor. *Proc. Natl. Acad. Sci. USA* **107,** 5459–5464.

Friedrich, M. W., et al. (2010). Imaging CREB activation in living cells. *J. Biol. Chem.* **285,** 23285–23295.

Fu, H., et al. (2000). 14-3-3 proteins: Structure, function, and regulation. *Annu. Rev. Pharmacol. Toxicol.* **40,** 617–647.

Fujioka, A., et al. (2006). Dynamics of the Ras/ERK MAPK cascade as monitored by fluorescent probes. *J. Biol. Chem.* **281,** 8917–8926.

Gao, X., and Zhang, J. (2008). Spatiotemporal analysis of differential Akt regulation in plasma membrane microdomains. *Mol. Biol. Cell* **19,** 4366–4373.

Grynkiewicz, G., et al. (1985). A new generation of Ca2+ indicators with greatly improved fluorescence properties. *J. Biol. Chem.* **260,** 3440–3450.

Harvey, C. D., et al. (2008). A genetically encoded fluorescent sensor of ERK activity. *Proc. Natl. Acad. Sci. USA* **105,** 19264–19269.

Henderson, J. N., et al. (2007). Structural basis for reversible photobleaching of a green fluorescent protein homologue. *Proc. Natl. Acad. Sci. USA* **104,** 6672–6677.

Herbst, K. J., et al. (2011). Luminescent kinase activity biosensors based on a versatile bimolecular switch. *J. Am. Chem. Soc.* **133,** 5676–5679.

Hodgson, L., et al. (2008). Design and optimization of genetically encoded fluorescent biosensors: GTPase biosensors. *Methods Cell Biol.* **85,** 63–81.

Karasawa, S., et al. (2004). Cyan-emitting and orange-emitting fluorescent proteins as a donor/acceptor pair for fluorescence resonance energy transfer. *Biochem. J.* **381,** 307–312.

Kraynov, V. S., et al. (2000). Localized Rac activation dynamics visualized in living cells. *Science* **290,** 333–337.

Levskaya, A., et al. (2009). Spatiotemporal control of cell signalling using a light-switchable protein interaction. *Nature* **461,** 997–1001.

Lim, C. J., et al. (2008). Integrin-mediated protein kinase A activation at the leading edge of migrating cells. *Mol. Biol. Cell* **19,** 4930–4941.

Lin, C. W., et al. (2004). Genetically encoded fluorescent reporters of histone methylation in living cells. *J. Am. Chem. Soc.* **126,** 5982–5983.

Lu, S., et al. (2008). The spatiotemporal pattern of Src activation at lipid rafts revealed by diffusion-corrected FRET imaging. *PLoS Comput. Biol.* **4,** e1000127.

Machacek, M., et al. (2009). Coordination of Rho GTPase activities during cell protrusion. *Nature* **461,** 99–103.

Markwardt, M. L., et al. (2011). An improved cerulean fluorescent protein with enhanced brightness and reduced reversible photoswitching. *PLoS One* **6,** e17896.

Miyawaki, A. (2003). Visualization of the spatial and temporal dynamics of intracellular signaling. *Dev. Cell* **4,** 295–305.

Miyawaki, A., et al. (1997). Fluorescent indicators for Ca2+ based on green fluorescent proteins and calmodulin. *Nature* **388,** 882–887.

Mochizuki, N., et al. (2001). Spatio-temporal images of growth-factor-induced activation of Ras and Rap1. *Nature* **411,** 1065–1068.

Nagai, T., et al. (2004). Expanded dynamic range of fluorescent indicators for Ca(2+) by circularly permuted yellow fluorescent proteins. *Proc. Natl. Acad. Sci. USA* **101,** 10554–10559.

Nalbant, P., et al. (2004). Activation of endogenous Cdc42 visualized in living cells. *Science* **305,** 1615–1619.

Newman, R. H., and Zhang, J. (2008). Visualization of phosphatase activity in living cells with a FRET-based calcineurin activity sensor. *Mol. Biosyst.* **4,** 496–501.

Newman, R. H., et al. (2011). Genetically encodable fluorescent biosensors for tracking signaling dynamics in living cells. *Chem. Rev.* **111,** 3614–3666.

Nguyen, A. W., and Daugherty, P. S. (2005). Evolutionary optimization of fluorescent proteins for intracellular FRET. *Nat. Biotechnol.* **23,** 355–360.

Ni, Q., et al. (2006). Analyzing protein kinase dynamics in living cells with FRET reporters. *Methods* **40,** 279–286.

Ni, Q., et al. (2011). Signaling diversity of PKA achieved via a Ca^{2+}-cAMP-PKA oscillatory circuit. *Nat. Chem. Biol.* **7,** 34–40.

Perroy, J., et al. (2004). Real-time monitoring of ubiquitination in living cells by BRET. *Nat. Methods* **1,** 203–208.

Pertz, O., et al. (2006). Spatiotemporal dynamics of RhoA activity in migrating cells. *Nature* **440,** 1069–1072.

Ridley, A. J., et al. (2003). Cell migration: Integrating signals from front to back. *Science* **302,** 1704–1709.

Rizzo, M. A., et al. (2004). An improved cyan fluorescent protein variant useful for FRET. *Nat. Biotechnol.* **22,** 445–449.

Ross, W. N., et al. (1974). A large change in dye absorption during the action potential. *Biophys. J.* **14,** 983–986.

Sasaki, K., et al. (2009). Real-time imaging of histone H4 hyperacetylation in living cells. *Proc. Natl. Acad. Sci. USA* **106,** 16257–16262.

Schleifenbaum, A., et al. (2004). Genetically encoded FRET probe for PKC activity based on pleckstrin. *J. Am. Chem. Soc.* **126,** 11786–11787.

Shaner, N. C., et al. (2005). A guide to choosing fluorescent proteins. *Nat. Methods* **2,** 905–909.

Tkachenko, E., et al. (2011). Protein kinase A governs a RhoA-RhoGDI protrusion-retraction pacemaker in migrating cells. *Nat. Cell Biol.* **13,** 661–668.

Tsien, R. Y. (1998). The green fluorescent protein. *Annu. Rev. Biochem.* **67,** 509–544.

Violin, J. D., et al. (2003). A genetically encoded fluorescent reporter reveals oscillatory phosphorylation by protein kinase C. *J. Cell Biol.* **161,** 899–909.

Walker, S. A., and Lockyer, P. J. (2004). Visualizing Ras signalling in real-time. *J. Cell Sci.* **117,** 2879–2886.

Wu, Y. I., *et al.* (2009). A genetically encoded photoactivatable Rac controls the motility of living cells. *Nature* **461,** 104–108.

Zaccolo, M., *et al.* (2000). A genetically encoded, fluorescent indicator for cyclic AMP in living cells. *Nat. Cell Biol.* **2,** 25–29.

Zacharias, D. A., *et al.* (2002). Partitioning of lipid-modified monomeric GFPs into membrane microdomains of live cells. *Science* **296,** 913–916.

Zhang, J., and Allen, M. D. (2007). FRET-based biosensors for protein kinases: Illuminating the kinome. *Mol. Biosyst.* **3,** 759–765.

Zhang, J., *et al.* (2001). Genetically encoded reporters of protein kinase A activity reveal impact of substrate tethering. *Proc. Natl. Acad. Sci. USA* **98,** 14997–15002.

Zhang, J., *et al.* (2002). Creating new fluorescent probes for cell biology. *Nat. Rev. Mol. Cell Biol.* **3,** 906–918.

Zhang, J., *et al.* (2005). Insulin disrupts beta-adrenergic signalling to protein kinase A in adipocytes. *Nature* **437,** 569–573.

CHAPTER SEVENTEEN

Live-Cell Imaging of Aquaporin-4 Supramolecular Assembly and Diffusion

A. S. Verkman,* Andrea Rossi,* and Jonathan M. Crane*

Contents

1. Aquaporin-4 (AQP4) and Orthogonal Arrays of Particles — 342
2. Approaches to Image AQP4 and OAPs — 343
3. AQP4 Diffusion and OAPs Studied by Quantum Dot Single Particle Tracking — 345
4. OAP Dynamics and Structure Studied with GFP-AQP4 Chimeras — 347
5. Single-Molecule Analysis Shows AQP4 Heterotetramers — 349
6. Photobleaching Reveals Post-Golgi Assembly of OAPs — 351
7. Super-Resolution Imaging of AQP4 OAPs — 351
References — 352

Abstract

Aquaporin-4 (AQP4) is a water channel expressed in astrocytes throughout the central nervous system, as well as in epithelial cells in various peripheral organs. AQP4 is involved in brain water balance, neuroexcitation, astrocyte migration, and neuroinflammation and is the target of pathogenic autoantibodies in neuromyelitis optica. Two AQP4 isoforms produced by alternative splicing, M1 and M23 AQP4, form heterotetramers that assemble in cell plasma membranes in supramolecular aggregates called orthogonal arrays of particles (OAPs). OAPs have been studied morphologically, by freeze-fracture electron microscopy, and biochemically, by native gel electrophoresis. We have applied single-molecule and high-resolution fluorescence microscopy methods to visualize AQP4 and OAPs in live cells. Quantum dot single particle tracking of fluorescently labeled AQP4 has quantified AQP4 diffusion in membranes, and has elucidated the molecular determinants and regulation of OAP formation. The composition, structure, and kinetics of OAPs containing fluorescent protein-AQP4 chimeras have been studied utilizing total internal reflection fluorescence microscopy, single-molecule photobleaching, and super-resolution imaging

* Departments of Medicine and Physiology, University of California, San Francisco, San Francisco, California, USA

methods. The biophysical data afforded by live-cell imaging of AQP4 and OAPs has provided new insights in the roles of AQP4 in organ physiology and neurological disease.

1. AQUAPORIN-4 (AQP4) AND ORTHOGONAL ARRAYS OF PARTICLES

Aquaporin-4 (AQP4) is a water-transporting, integral membrane protein cloned by our lab (Hasegawa *et al.*, 1994). AQP4 is expressed at the plasma membrane in astrocytes throughout the central nervous system (CNS), in epithelial cells in kidney, stomach, lung and exocrine glands, and in skeletal muscle (Frigeri *et al.*, 1995). Phenotype analysis of knockout mice lacking AQP4 (Ma *et al.*, 1997) has defined the roles of AQP4 in brain water balance (Manley *et al.*, 2000; Papadopoulos *et al.*, 2004), astrocyte migration and glial scar formation (Auguste *et al.*, 2007; Saadoun *et al.*, 2005), neuroexcitation (Binder *et al.*, 2006; Padmawar *et al.*, 2005) neurosensory signaling (Li and Verkman, 2001; Lu *et al.*, 2008), and neuroinflammation (Li *et al.*, 2011). AQP4 is also involved in the multiple sclerosis-like disease neuromyelitis optica (NMO), where pathogenic antibodies against AQP4 cause astrocyte damage, neuroinflammation, and demyelination, which leads to paralysis and blindness (Jarius *et al.*, 2008; Lennon *et al.*, 2005).

AQP4 monomers, each of ~30 kDa molecular size, contain six, membrane-spanning helical domains surrounding a narrow aqueous pore that confers water selectivity (Ho *et al.*, 2009). Like other aquaporins, AQP4 is present as tetramers in membranes. AQP4 is expressed in two isoforms produced by alternative splicing: a long (M1) isoform with translation initiation at Met-1, and a short (M23) isoform with translation initiation at Met-23 (Jung *et al.*, 1994; Lu *et al.*, 1996; Yang *et al.*, 1995). As discussed further below, the M1 and M23 isoforms associate in membranes as heterotetramers. AQP4 tetramers can further assemble into supramolecular aggregates called orthogonal arrays of particles (OAPs), which are square arrays of intramembrane particles seen in cell membranes by freeze-fracture electron microscopy (FFEM; Landis and Reese, 1974; Rash *et al.*, 1974; Wolburg *et al.*, 2011). Our lab discovered that AQP4 is the major constituent of OAPs based on the appearance of OAPs in AQP4-transfected cells (Yang *et al.*, 1996) and their absence in AQP4 knockout mice (Verbavatz *et al.*, 1997). As discussed further below, the M23 isoform of AQP4 forms OAPs, which can incorporate with AQP4-M1 by heterotetramer formation (Furman *et al.*, 2003; Silberstein *et al.*, 2004). The biological function of OAPs remains unclear, with proposed involvement of AQP4 OAPs in water transport (Fenton *et al.*, 2009; Silberstein *et al.*, 2004), cell–cell adhesion (Hiroaki *et al.*, 2006; Zhang and Verkman, 2008), and membrane polarization (Noell et al., 2009). Pathogenic autoantibodies in NMO bind preferentially to AQP4 OAPs (Crane et al., 2011a).

2. Approaches to Image AQP4 and OAPs

Figure 17.1A shows FFEM in cells transfected with AQP4-M1 (left) versus AQP4-M23 (right). While AQP4-M1 is largely dispersed as individual tetramers, AQP4-M23 assembles into large OAPs of >100 particles. AQP4 aggregates, which represent OAPs, have also been studied by blue native gel electrophoresis (BN-PAGE), in which AQP4 tetramers are separated from higher order AQP4 aggregates using a nondenaturing detergent. By BN-PAGE with AQP4 immunoblot, AQP4-M1 is seen as a dense tetramer band and a faint band at ~600 kDa, whereas M23-AQP4 is seen

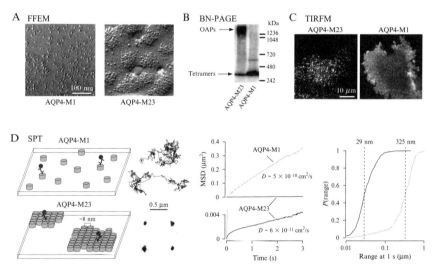

Figure 17.1 Approaches to visualize AQP4 and OAPs. (A) Freeze-fracture electron micrographs of the plasma membrane P-face of COS-7 cells expressing the M1 and M23 isoforms of AQP4. (B) AQP4 immunoblot following Blue-native gel electrophoresis of cell lysates from AQP4-expressing COS-7 cells. (C) Total internal reflection fluorescence micrographs of GFP-AQP4 chimeras. (D) (left) Schematic showing the organization of AQP4 tetramers (left) and examples of single particle trajectories of Qdot-labeled AQP4 molecules in the plasma membrane of AQP4-expressing COS-7 cells. Each cylinder represents one AQP4 tetramer in which a subset of AQP4 molecules are labeled with quantum dots (red) for single particle tracking. (Center) Combined mean squared displacement (MSD) versus time plots and averaged diffusion coefficients for AQP4-M1 (gray) and AQP4-M23 (black) in COS-7 cells. (Right) Cumulative probability distribution of range at 1 s (P(range)) deduced from SPT measurements, with dashed lines indicating median range. Adapted from Crane et al. (2008) and Tajima et al. (2010). (See Color Insert.)

as diffuse band migrating at > 1200 kDa, corresponding to large AQP4 aggregates, along with a smaller band at ~300 kDa, corresponding to AQP4 tetramers (Fig. 17.1B). FFEM and BN-PAGE, while informative, require cell fixation or detergent solubilization, precluding measurements in live cells.

Total internal reflection fluorescence microscopy (TIRFM) allows visualization of fluorescently labeled AQP4 tetramers and OAPs. Figure 17.1C shows TIRFM of cells expressing green fluorescent protein (GFP)-AQP4 chimeras. AQP4-M1 is fairly uniformly distributed over the cell surface, whereas AQP4-M23 is seen as distinct, diffraction-limited puncta, corresponding to dense OAPs. As discussed below, TIRFM of fluorescently labeled AQP4 is useful to investigate AQP4/OAP composition, structure, and kinetics.

Single particle tracking (SPT) provides an incisive single-molecule approach to study AQP4/OAP dynamics and assembly. In one implementation of SPT, a subset of AQP4 molecules is labeled at their extracellular surface by quantum dots, utilizing an antibody against an engineered extracellular epitope on AQP4, or an NMO autoantibody against native extracellular epitopes (Fig. 17.1D, left). We found that inclusion of a Myc or HA epitope in its second extracellular loop did not affect AQP4 expression, trafficking, or function (Crane *et al.*, 2008). The movement of individual quantum dots, consequent to AQP4 diffusion, is recorded by wide-field fluorescence microscopy, allowing reconstruction and analysis of single particle trajectories. We initially applied SPT to study the diffusion of a different aquaporin, AQP1, finding long-range free diffusion over a wide variety of conditions, indicating that AQP1 exists in the plasma membrane largely free of specific interactions (Crane and Verkman, 2008). In applying SPT to AQP4, we reasoned that individual AQP4 tetramers should be mobile, whereas AQP4 in large OAPs should be relatively immobile. Figure 17.1D (center) shows remarkably greater mobility of AQP4-M1 than AQP4-M23. SPT data can be quantified by a variety of approaches to resolve diffusion mechanisms and potential interactions (Jin and Verkman, 2007; Jin *et al.*, 2007). Figure 17.1D (right) quantifies the large difference in AQP4-M1 versus AQP4-M23 diffusion from mean squared displacement (MSD) and cumulative probability analyses. As reported, analysis of SPT data showed that AQP4-M1 diffuses freely, with diffusion coefficient $\sim 5 \times 10^{-10}$ cm^2/s, covering ~ 5 μm in 5 min. AQP4-M23 diffuses only ~ 0.4 μm in 5 min. Whereas actin modulation by latrunculin or jasplakinolide did not affect AQP4-M23 diffusion, deletion of the AQP4 C-terminal Post synaptic density protein (PSD95), Drosophila disc large tumor suppressor (Dlg1), and Zonula occludens-1 (ZO-1) (PDZ)-binding domain increased its range by approximately twofold over minutes. Biophysical analysis of short-range AQP4-M23 diffusion in OAPs indicated a spring-like potential with a restoring force of ~ 6.5 pN/μm. As described below, SPT has been applied to identify the molecular determinants of AQP4 assembly.

3. AQP4 Diffusion and OAPs Studied by Quantum Dot Single Particle Tracking

Figure 17.2A shows the amino acid sequence of AQP4, with indicated sites of translation initiation, epitope insertion, and PDZ-domain interactions, as well as the residues involved in OAP assembly. From measurements on AQP4 mutants and chimeras, we concluded that OAP formation by AQP4-M23 involves hydrophobic intermolecular interactions of N-terminal AQP4 residues just downstream of Met-23, and that the inability of

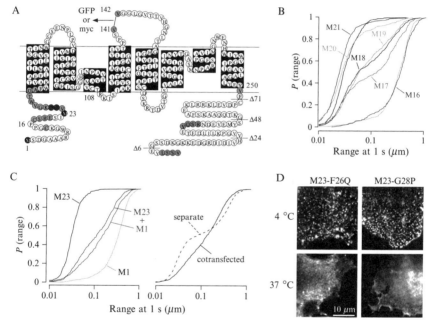

Figure 17.2 Quantum dot single particle tracking reveals determinants of AQP4 OAP assembly. (A) AQP4 sequence and topology showing site of GFP or epitope (myc) insertion in the second extracellular loop. Black: Met-1 and Met-23 translation initiation sites; blue: residues where mutations did not affect OAP assembly; red: mutations disrupt OAPs; pink: mutations mildly disrupt OAPs; yellow: mutations reduce plasma membrane expression; green: C-terminal PDZ-binding domains. (B) P(range) for indicated AQP4 truncation mutants. (C.) OAP modulation by coexpression of M1-AQP4 and M23-AQP4. (Left) P(range) for cells transfected with M23 only (black) or M1 only (gray), or cotransfected with M23 (red) and M1 (green) at M23-to-M1 ratios of 1:1. (Right) P(range) comparing M1/M23 cotransfection (solid) versus "separate" (dashed), computed by summing P(range) curves for separate transfections. (D) TIRFM of Alexa-labeled AQP4 in cells expressing M23-F26Q or M23-G28P and fixed at 4 or 37 °C. Adapted from Crane and Verkman (2009a, 2009b) and Crane et al. (2009). (See Color Insert.)

AQP4-M1 to form OAPs results from nonspecific blocking of N-terminal interactions by residues just upstream of Met-23 (Crane and Verkman, 2009a). Our study involved the generation of serial deletion mutants at the AQP4-M1 N-terminus. Deletions of up to 16 residues had little effect on diffusion. Further truncation, however, resulted in a large fraction of AQP4 with restricted diffusion (examples shown in Fig. 17.2B). Continued deletions up to Met-23 further increased the percentage of AQP4 in OAPs, whereas truncations or certain mutations downstream of Met-23 produced progressive loss of OAPs. Also, AQP1, which does not itself form OAPs, was induced to form OAPs upon replacement of its N-terminal domain with that of AQP4-M23. These findings defined the molecular determinants of AQP4 OAP assembly.

In a follow-on study, we investigated M1/M23 interactions and regulated OAP assembly by quantum dot SPT in live cells expressing differentially tagged AQP4 isoforms, and in primary glial cell cultures in which native AQP4 was labeled with a monoclonal recombinant NMO autoantibody (Crane et al., 2009). Diffusion of AQP4-M1 and AQP4-M23 were measured individually at different M23:M1 ratios. Upon coexpression with AQP4-M1, the diffusion of AQP4-M23 increased significantly, whereas expression of AQP4-M23 produced a marked reduction in AQP4-M1 diffusion (Fig. 17.2C, left). The similar diffusion of the two isoforms indicated a high level of interaction. Quite different from the experimental data are the dashed curves in Fig. 17.2C (right), which are the predicted curves if AQP4-M1 and AQP4-M23 do not comingle. Further, two-color SPT and BN-PAGE showed that mutants of AQP4-M23 that do not themselves form OAPs (M23-F26Q and M23-G28P) fully coassociated with native AQP4-M23 to form large immobile OAPs. In primary astrocyte cell cultures, which coexpress the M1 and M23 AQP4 isoforms, differential regulation of OAP assembly by palmitoylation, calcium, and protein kinase C activation was found. These results indicated the coassembly of AQP4-M1 and AQP4-M23 in OAPs, and the regulation of OAP assembly in astrocytes by specific signaling events.

Having established regulated assembly of AQP4 in OAPs, we carried out biophysical studies showing rapid and reversible temperature-dependent assembly into OAPs of certain weakly associating AQP4 mutants (Crane and Verkman, 2009b). By TIRFM, M23-F26Q, and M23-G28P formed few, if any OAPs at 37 °C, but assembled efficiently in OAPs at 10 °C (Fig. 17.4D). The computed free energy from temperature-dependence data was -13 ± 2 kcal/mol for OAP formation of M23-F26Q below 30 °C. OAP assembly by M23-F26Q (but not of native AQP4-M23) could also be modulated by reducing its membrane density. Interestingly, OAP assembly and disassembly occurred within seconds or less following temperature changes. These data further support the paradigm that AQP4 OAPs are dynamic, regulated supramolecular structures.

In related studies, quantum dot SPT was applied in investigations of NMO autoantibody binding to AQP4-M1 versus AQP4-M23 (Crane et al., 2011a) and to Arg-19 mutants of AQP4-M1 that have been associated with NMO (Crane et al., 2011b). In other studies, the biology of a third, longer isoform of AQP4, called AQP4-Mz, was studied by quantum dot SPT (Rossi et al., 2011a). AQP4-Mz is expressed in rat, with translation initiation 126-bp upstream from that of AQP4-M1. However, AQP4-Mz is not expressed in human or mouse because of in-frame stop codons. It was found that Mz, like M1, diffused rapidly in the cell plasma membrane and did not form OAPs. However, when coexpressed with M23, Mz associated in OAPs by forming heterotetramers with M23. Therefore, AQP4-Mz is unable to form OAPs on its own but able to associate with M23 AQP4 in heterotetramers.

4. OAP Dynamics and Structure Studied with GFP-AQP4 Chimeras

The ability of GFP-labeled AQP4-M23 to form OAPs allows visualization of OAP dynamics and structure in live cells (Tajima et al., 2010). Figure 17.3A (left) shows TIRFM of cells expressing GFP-labeled AQP4-M23. The distinct, well-demarcated spots correspond to individual OAPs. Because the actual sizes of OAPs are generally less than the x, y-spatial resolution of TIRFM, most OAPs appear as diffraction-limited fluorescent spots whose intensity is proportional to the number of GFP-AQP4 molecules in the OAP. Figure 17.3A (right) shows individual OAP trajectories from TIFRM time-lapse imaging done over 3 h. Most trajectories show slow Brownian motion with various reorganization events, such as OAP fusion and fission, occurring over tens of minutes. An example of OAP fission is shown in Fig. 17.3A (bottom), where a single fluorescent spot separates into at least four distinct fluorescent spots. Analysis of OAP trajectories gave a median diffusion coefficient of $\sim 10^{-12}$ cm^2/s.

TIRFM allows determination of the size distribution of AQP4 OAPs from computation of area-integrated, background-subtracted spot intensities. To study the dependence of OAP size on M23:M1 ratio, cells were transfected with GFP-labeled AQP4-M1 and AQP4-M23 at different ratios under conditions of low expression to enable visualization of individual fluorescent spots. OAP size was deduced from spot-integrated fluorescence intensities referenced to GFP standards (Jin et al., 2011). Figure 17.3B shows TIRFM (top) and number histograms (bottom) of single-spot fluorescence. The number histogram showed ∼fourfold greater intensity of individual M1 tetramers than monomeric GFP, as expected. Higher spot fluorescence intensities, corresponding to larger AQP4 aggregates (OAPs),

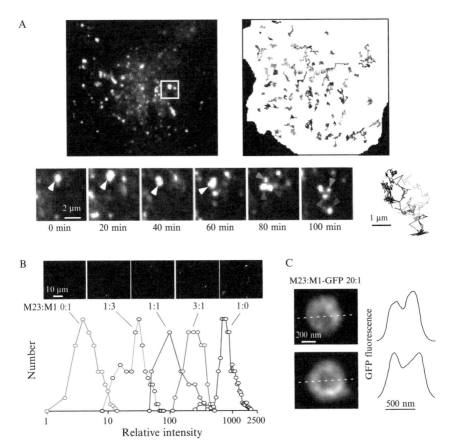

Figure 17.3 OAP dynamics and structure revealed by TIRFM of GFP-AQP4 chimeras. (A) TIRFM image (left) showing distinct fluorescent spots in cells expressing M23-AQP4, corresponding to OAPs, with deduced single OAP trajectories over 3 h shown at the right. (Bottom) High magnification of boxed region showing spontaneous OAP disruption events; trajectories of original OAP (black) and daughter OAPs (red, green, yellow, blue) shown at the right. (B) U87MG cells were transfected with GFP-M23 and GFP-M1 AQP4 at indicated ratios. Representative TIRF micrographs show fluorescent spots (top). Deduced number histograms of single-spot fluorescence (background-subtracted, area-integrated intensities), proportional to OAP size, shown at the bottom. Unity represents the intensity of monomeric GFP. (C) U87MG cells were transfected with GFP-M23 and (untagged) M1 AQP4 at a ratio of 20:1. TIRFM of two large AQP4 aggregates (left), showing relative concentration of fluorescence at the periphery. Line profiles (dashed white lines at the left) shown at the right. Adapted from Tajima *et al.* (2010) and Jin *et al.* (2011). (See Color Insert.)

were seen at greater M23:M1 ratios, with considerable heterogeneity seen at all ratios. These data agreed well with predictions of a mathematical model of AQP4 OAP assembly based on random heterotetrameric association of AQP4-M1 and AQP4-M23, inter-tetramer associations between AQP4-M23 and AQP4-M23 (but not between AQP4-M1 and AQP4-M23 or AQP4-M1), and a free-energy constraint limiting OAP size.

Another prediction of the AQP4 OAP model is that OAPs consist of an AQP4-M23 enriched core decorated by an AQP4-M1 enriched periphery. To test this prediction, cells were transfected with GFP-labeled AQP4-M1 and (untagged) AQP4-M23. Experiments were done at high M23:M1 ratio in order to generate large OAPs of size greater than the TIRFM x, y-resolution (diffraction limit). Figure 17.3C shows high magnification TIRFM and corresponding line scans of two large AQP4 aggregates in which GFP-AQP4-M1 fluorescence was greater at the periphery than at the core.

5. SINGLE-MOLECULE ANALYSIS SHOWS AQP4 HETEROTETRAMERS

Single-molecule analysis of GFP intensity and photobleaching demonstrated that AQP4-M1 and AQP4-M23 associate in heterotetramers (Tajima et al., 2010). The idea is that a homotetramer containing four GFP-labeled AQP4 molecules would undergo multistep photobleaching as each of the four GFPs are sequentially bleached, and have four times the intensity of a GFP monomer. If a tetramer consists of one GFP-labeled AQP4 molecule and three unlabeled AQP4 molecules, then photobleaching would occur in a single step and the spot intensity would be the same as that of a GFP monomer. Figure 17.4A shows the implementation of this approach, with multistep photobleaching and high spot intensity of GFP-labeled AQP4-M1, but single-step photobleaching and low spot intensity in the presence of excess unlabeled AQP4-M1. To investigate heterotetramer formation, photobleaching and intensity analysis were done on cells expressing GFP-labeled AQP4-M1 coexpressed with an excess of unlabeled AQP4-M1 or the non-OAP-forming AQP4-M23 mutant, M23-G28P. For these studies, it was necessary to use an M23 mutant that cannot form OAPs because single-spot photobleaching and intensity analysis can be applied only to physically distinct AQP4 tetramers (and not to AQP4 in OAPs). We found single-step photobleaching and low spot intensity for GFP-labeled AQP4-M1 together with excess of the AQP4-M23 mutant, providing evidence that AQP4-M1 and AQP4-M23 are able to form heterotetramers.

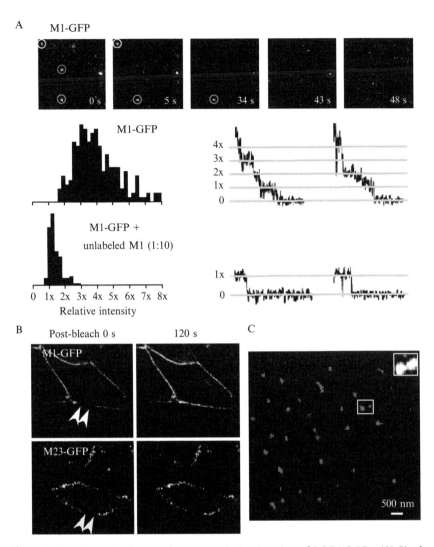

Figure 17.4 Photobleaching and super-resolution imaging of AQP4 OAPs. (A) Single molecule step-photobleaching and intensity analysis shows AQP4 heterotetrameric association. (Top) Serial TIRFM images of recombinant monomeric GFP-labeled AQP4-M1 showing multistep loss of fluorescence. (Bottom, left) Single-spot (background-subtracted) integrated fluorescence intensity histograms for GFP-labeled AQP4-M1 without or with excess unlabeled AQP4-M1. (Right) Single-spot intensities as a function of time during continuous illumination, showing single versus multistep photobleaching. (B) Diffusion of GFP-labeled M1 and M23 AQP4 at the plasma membrane in live cells. Arrowheads indicate bleached area. Adapted from Tajima *et al.* (2010). (C) Super-resolution image of AQP4 OAPs. PALM image of AQP4-M23 chimera containing PA-GFP at its C-terminus. Inset: TIRFM (non-super-resolution) of area in white box.

6. Photobleaching Reveals Post-Golgi Assembly of OAPs

Fluorescence recovery after photobleaching of GFP-AQP4 chimeras confirmed that AQP4 OAPs are present in cell plasma membranes but not in endoplasmic reticulum or Golgi (Rossi et al., 2011b; Tajima et al., 2010). Figure 17.4B shows confocal fluorescence micrographs of cells expressing GFP-labeled AQP4-M1 and AQP4-M23 at the cell surface. Following photobleaching of a circular spot, recovery was seen for AQP4-M1 but not for AQP4-M23, in agreement with the conclusion that AQP4-M1 is present mainly as tetramers and AQP4-M23 as OAPs. Similar photobleaching measurements in endoplasmic reticulum or Golgi-targeted AQP4-M1 and AQP4-M23 showed unrestricted and rapid diffusion of both isoforms, indicating absence of OAPs. The conclusions from photobleaching measurements were confirmed by BN-PAGE and freeze-fracture electron microscopy. It was found that AQP4 OAP formation in plasma membranes but not in Golgi was not related to AQP4 density, pH, membrane lipid composition, C-terminal PDZ-domain interactions, or α-syntrophin expression. However, fusion of AQP4-containing Golgi vesicles with (AQP4-free) plasma membrane vesicles produced OAPs, suggesting the involvement of plasma membrane factor(s) in AQP4 OAP formation. In investigating additional possible determinants of OAP assembly, we discovered membrane curvature-dependent OAP assembly, in which OAPs were disrupted by extrusion of plasma membrane vesicles to $\sim$110 nm diameter but not to $\sim$220 nm diameter. AQP4 supramolecular assembly in OAPs is thus a post-Golgi phenomenon involving plasma membrane-specific factor(s). Post-Golgi and membrane curvature-dependent OAP assembly may be important for vesicle transport of AQP4 in the secretory pathway and AQP4-facilitated astrocyte migration.

7. Super-Resolution Imaging of AQP4 OAPs

Though nearly all OAPs in cells coexpressing AQP4-M1 and AQP4-M23 are smaller than the diffraction limit of conventional light microscopy, many OAPs are suitable for super-resolution imaging by photoactivated localization microscopy (PALM), stochastic optical reconstruction microscopy (STORM), or other super-resolution methods. Figure 17.4C shows a PALM image of a chimera containing AQP4-M23 and GFP that can undergo photoactivation (PA-GFP). Various other fluorescent protein-AQP4 chimeras can be generated for multicolor PALM for simultaneous detection of AQP4-M1 and AQP4-M23. STORM imaging is possible as well, using fluorescently labeled antibodies against engineered extracellular epitopes on AQP4, as described above, or recombinant monoclonal NMO autoantibodies directed against extracellular epitopes on native AQP4 (Crane et al., 2011a).

REFERENCES

Auguste, K. I., Jin, S., Uchida, K., Yan, D., Manley, G. T., Papadopoulos, M. C., and Verkman, A. S. (2007). Greatly impaired migration of implanted aquaporin-4-deficient astroglial cells in mouse brain toward a site of injury. *Faseb J.* **21,** 108–116.

Binder, D. K., Yao, X., Zador, Z., Sick, T. J., Verkman, A. S., and Manley, G. T. (2006). Increased seizure duration and slowed potassium kinetics in mice lacking aquaporin-4 water channels. *Glia* **53,** 631–636.

Crane, J. M., and Verkman, A. S. (2008). Long-range nonanomalous diffusion of quantum dot-labeled aquaporin-1 water channels in the cell plasma membrane. *Biophys. J.* **94,** 702–713.

Crane, J. M., and Verkman, A. S. (2009a). Determinants of aquaporin-4 assembly in orthogonal arrays revealed by live-cell single-molecule fluorescence imaging. *J. Cell Sci.* **122,** 813–821.

Crane, J. M., and Verkman, A. S. (2009b). Reversible, temperature-dependent supramolecular assembly of aquaporin-4 orthogonal arrays in live cell membranes. *Biophys. J.* **97,** 3010–3018.

Crane, J. M., Van Hoek, A. N., Skach, W. R., and Verkman, A. S. (2008). Aquaporin-4 dynamics in orthogonal arrays in live cells visualized by quantum dot single particle tracking. *Mol. Biol. Cell* **19,** 3369–3378.

Crane, J. M., Bennett, J. L., and Verkman, A. S. (2009). Live cell analysis of aquaporin-4 M1/M23 interactions and regulated orthogonal array assembly in glial cells. *J. Biol. Chem.* **284,** 35850–35860.

Crane, J. M., Lam, C., Rossi, A., Gupta, T., Bennett, J. L., and Verkman, A. S. (2011a). Binding affinity and specificity of neuromyelitis optica autoantibodies to aquaporin-4 M1/M23 isoforms and orthogonal arrays. *J. Biol. Chem.* **286,** 16516–16524.

Crane, J. M., Rossi, A., Gupta, T., Bennett, J. L., and Verkman, A. S. (2011b). Orthogonal array formation by human aquaporin-4: Examination of neuromyelitis optica-associated aquaporin-4 polymorphisms. *J. Neuroimmunol.* **236**(1–2), 93–98.

Fenton, R. A., Moeller, H. B., Zelenina, M., Snaebjornsson, M. T., Holen, T., and MacAulay, N. (2009). Differential water permeability and regulation of three aquaporin 4 isoforms. *Cell. Mol. Life Sci.* **67,** 829–840.

Frigeri, A., Gropper, M. A., Turck, C. W., and Verkman, A. S. (1995). Immunolocalization of the mercurial-insensitive water channel and glycerol intrinsic protein in epithelial cell plasma membranes. *Proc. Natl. Acad. Sci. USA* **92,** 4328–4331.

Furman, C. S., Gorelick-Feldman, D. A., Davidson, K. G., Yasumura, T., Neely, J. D., Agre, P., and Rash, J. E. (2003). Aquaporin-4 square array assembly: Opposing actions of M1 and M23 isoforms. *Proc. Natl. Acad. Sci. USA* **100,** 13609–13614.

Hasegawa, H., Ma, T., Skach, W., Matthay, M. A., and Verkman, A. S. (1994). Molecular cloning of a mercurial-insensitive water channel expressed in selected water-transporting tissues. *J. Biol. Chem.* **269,** 5497–5500.

Hiroaki, Y., Tani, K., Kamegawa, A., Gyobu, N., Nishikawa, K., Suzuki, H., Walz, T., Sasaki, S., Mitsuoka, K., et al. (2006). Implications of the aquaporin-4 structure on array formation and cell adhesion. *J. Mol. Biol.* **355,** 628–639.

Ho, J. D., Yeh, R., Sandstrom, A., Chorny, I., Harries, W. E., Robbins, R. A., Miercke, L. J., and Stroud, R. M. (2009). Crystal structure of human aquaporin 4 at 1.8 A and its mechanism of conductance. *Proc. Natl. Acad. Sci. USA* **106**, 7437–7442.

Jarius, S., Aboul-Enein, F., Waters, P., Kuenz, B., Hauser, A., Berger, T., Lang, W., Reindl, M., Vincent, A., and Kristoferitsch, W. (2008). Antibody to aquaporin-4 in the long-term course of neuromyelitis optica. *Brain* **131**, 3072–3080.

Jin, S., and Verkman, A. S. (2007). Single particle tracking simulations of complex diffusion in membranes: Simulation and detection of barrier, raft, and interaction phenomena. *J. Phys. Chem. B* **111**, 3625–3632.

Jin, S., Haggie, P. M., and Verkman, A. S. (2007). Single particle tracking analysis of membrane protein diffusion in a potential: Simulation and application to confined diffusion of CFTR Cl$^-$ channels. *Biophys. J.* **93**, 1079–1088.

Jin, B. J., Rossi, A., and Verkman, A. S. (2011). Model of aquaporin-4 supramolecular assembly in orthogonal arrays based on heterotetrameric association of M1/M23 isoforms. *Biophys. J.* **100**, 2936–2945.

Jung, J. S., Bhat, R. V., Preston, G. M., Guggino, W. B., Baraban, J. M., and Agre, P. (1994). Molecular characterization of an aquaporin cDNA from brain: Candidate osmoreceptor and regulator of water balance. *Proc. Natl. Acad. Sci. USA* **91**, 13052–13056.

Landis, D. M., and Reese, T. S. (1974). Arrays of particles in freeze-fractured astrocytic membranes. *J. Cell Biol.* **60**, 316–320.

Lennon, V. A., Kryzer, T. J., Pittock, S. J., Verkman, A. S., and Hinson, S. R. (2005). IgG marker of optic-spinal multiple sclerosis binds to the aquaporin-4 water channel. *J. Exp. Med.* **202**, 473–477.

Li, J., and Verkman, A. S. (2001). Impaired hearing in mice lacking aquaporin-4 water channels. *J. Biol. Chem.* **276**, 31233–31237.

Li, L., Zhang, H., Varrin-Doyer, M., Zamvil, S. S., and Verkman, A. S. (2011). Proinflammatory role of aquaporin-4 in autoimmune neuroinflammation. *Faseb J.* **25**, 1556–1566.

Lu, M., Lee, M. D., Smith, B. L., Jung, J. S., Agre, P., Verdijk, M. A., Merkx, G., Rijss, J. P., and Deen, P. M. (1996). The human AQP4 gene: Definition of the locus encoding two water channel polypeptides in brain. *Proc. Natl. Acad. Sci. USA* **93**, 10908–10912.

Lu, D. C., Zhang, H., Zador, Z., and Verkman, A. S. (2008). Impaired olfaction in mice lacking aquaporin-4 water channels. *Faseb J.* **22**, 3216–3223.

Ma, T., Yang, B., Gillespie, A., Carlson, E. J., Epstein, C. J., and Verkman, A. S. (1997). Generation and phenotype of a transgenic knockout mouse lacking the mercurial-insensitive water channel aquaporin-4. *J. Clin. Invest.* **100**, 957–962.

Manley, G. T., Fujimura, M., Ma, T., Noshita, N., Filiz, F., Bollen, A. W., Chan, P., and Verkman, A. S. (2000). Aquaporin-4 deletion in mice reduces brain edema after acute water intoxication and ischemic stroke. *Nat. Med.* **6**, 159–163.

Noell, S., Fallier-Becker, P., Deutsch, U., Mack, A. F., and Wolburg, H. (2009). Agrin defines polarized distribution of orthogonal arrays of particles in astrocytes. *Cell Tissue Res.* **337**, 185–195.

Padmawar, P., Yao, X., Bloch, O., Manley, G. T., and Verkman, A. S. (2005). K$^+$ waves in brain cortex visualized using a long-wavelength K$^+$-sensing fluorescent indicator. *Nat. Methods* **2**, 825–827.

Papadopoulos, M. C., Manley, G. T., Krishna, S., and Verkman, A. S. (2004). Aquaporin-4 facilitates reabsorption of excess fluid in vasogenic brain edema. *Faseb J.* **18**, 1291–1293.

Rash, J. E., Staehelin, L. A., and Ellisman, M. H. (1974). Rectangular arrays of particles on freeze-cleaved plasma membranes are not gap junctions. *Exp. Cell Res.* **86**, 187–190.

Rossi, A., Crane, J. M., and Verkman, A. S. (2011a). Aquaporin-4 Mz isoform: Brain expression, supramolecular assembly and neuromyelitis optica antibody binding. *Glia* **59**, 1056–1063.

Rossi, A., Baumgart, F., Van Hoek, A. N., and Verkman, A. S. (2011b). Post-golgi supramolecular assembly of aquaporin-4 in orthogonal arrays. Traffic (in press). Epub 2011 Nov 8.

Saadoun, S., Papadopoulos, M. C., Watanabe, H., Yan, D., Manley, G. T., and Verkman, A. S. (2005). Involvement of aquaporin-4 in astroglial cell migration and glial scar formation. *J. Cell Sci.* **118,** 5691–5698.

Silberstein, C., Bouley, R., Huang, Y., Fang, P., Pastor-Soler, N., Brown, D., and Van Hoek, A. N. (2004). Membrane organization and function of M1 and M23 isoforms of aquaporin-4 in epithelial cells. *Am. J. Physiol. Renal Physiol.* **287,** F501–F511.

Tajima, M., Crane, J. M., and Verkman, A. S. (2010). Aquaporin-4 (AQP4) associations and array dynamics probed by photobleaching and single-molecule analysis of green fluorescent protein-AQP4 chimeras. *J. Biol. Chem.* **285,** 8163–8170.

Verbavatz, J. M., Ma, T., Gobin, R., and Verkman, A. S. (1997). Absence of orthogonal arrays in kidney, brain and muscle from transgenic knockout mice lacking water channel aquaporin-4. *J. Cell Sci.* **110,** 2855–2860.

Wolburg, H., Wolburg-Buchholz, K., Fallier-Becker, P., Noell, S., and Mack, A. F. (2011). Structure and functions of aquaporin-4-based orthogonal arrays of particles. *Int. Rev. Cell Mol. Biol.* **287,** 1–41.

Yang, B., Ma, T., and Verkman, A. S. (1995). cDNA cloning, gene organization, and chromosomal localization of a human mercurial insensitive water channel. Evidence for distinct transcriptional units. *J. Biol. Chem.* **270,** 22907–22913.

Yang, B., Brown, D., and Verkman, A. S. (1996). The mercurial insensitive water channel (AQP-4) forms orthogonal arrays in stably transfected Chinese hamster ovary cells. *J. Biol. Chem.* **271,** 4577–4580.

Zhang, H., and Verkman, A. S. (2008). Evidence against involvement of aquaporin-4 in cell–cell adhesion. *J. Mol. Biol.* **382,** 1136–1143.

CHAPTER EIGHTEEN

Coiled-Coil Tag–Probe Labeling Methods for Live-Cell Imaging of Membrane Receptors

Yoshiaki Yano, Kenichi Kawano, Kaoru Omae, *and* Katsumi Matsuzaki

Contents

1. Introduction 356
2. Various Principles Used for Tag–Probe Labeling 357
    2.1. Protein–ligand interactions 357
    2.2. Peptide–metal interactions 359
    2.3. Enzymatic reactions 359
3. Coiled-Coil Tag–Probe Labeling 360
    3.1. Design of tag-fused membrane proteins 361
    3.2. Expression of tagged membrane proteins in living cells 362
    3.3. Preparation of fluorophore-labeled probe peptides 363
    3.4. Labeling of E3-tagged proteins with K probes 364
4. Applications 366
    4.1. Receptor internalization 366
    4.2. Receptor oligomerization 366
Acknowledgments 368
References 368

Abstract

Tag–probe labeling methods have advantages over conventional fusion with fluorescent proteins in terms of smaller labels, surface specificity, availability of pulse labeling, and ease of multicolor labeling. With this method, the gene of the target protein is fused with a short tag sequence, expressed in cells, and the protein is labeled with exogenous fluorescent probes that specifically bind to the tag. Various labeling principles, such as protein–ligand interaction, peptide–peptide interaction, peptide–metal interaction, and enzymatic reactions, have been applied to the tag–probe labeling of membrane receptors. We describe our coiled-coil tag–probe method in detail, including the design and synthesis of the tag and probe, labeling procedures, and observations by confocal microscopy. Applications to the analysis of receptor internalization and oligomerization are also introduced.

Graduate School of Pharmaceutical Sciences, Kyoto University, Sakyo-ku, Kyoto, Japan

1. INTRODUCTION

Fluorescence imaging of proteins in living cells has become a conventional technique to study the intracellular localization and dynamic behavior of the target proteins. The discovery and development of the green fluorescent proteins (GFPs) have drastically changed the procedures to obtain fluorophore-labeled proteins in living cells (Chudakov et al., 2010; Giepmans et al., 2006). The expression of genetic fusion constructs with fluorescent proteins into cultured cells (Fig. 18.1A) enables facile protein-specific labeling in cells

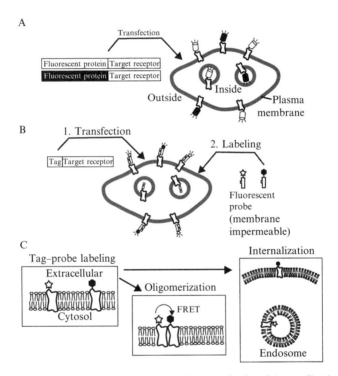

Figure 18.1 Principle of the tag–probe labeling method and its applications to the labeling of membrane proteins in living cells. (A) Labeling by genetic fusion with fluorescent proteins. The gene of the target protein fused with that of fluorescent protein is expressed in cells to obtain fluorophore-tagged target proteins. Multicolor labeling requires coexpression of multiple target genes fused with different tag proteins. Control of the labeling ratio is usually difficult. (B) Posttranslational labeling in the tag–probe method. The gene of the target protein is fused with a short tag sequence and expressed in cells. The expressed tag sequence is specifically labeled with an exogenous probe attached to a fluorophore. Surface labeling is achieved using a membrane-impermeable probe. Multicolor labeling is possible using probes labeled with different fluorophores. Precise control of the labeling ratio is possible. (C) Observation of internalization and oligomerization of membrane receptors by the tag–probe method. The surface specificity and easiness of multicolor labeling enable sensitive and accurate detection.

without laborious processes such as microinjections of target proteins into cells after expression, purification, and chemical labeling with exogenous fluorophores (Adams *et al.*, 1991). Cloning and mutational studies have established numerous new fluorescent and luminescent proteins with unique characteristics, including photoswitchable probes, expanding their spectral ranges and applications (Chudakov *et al.*, 2010; Fernandez-Suarez and Ting, 2008).

In spite of their excellent specificity, fluorescent/luminescent protein tags have several disadvantages. Their large size (e.g., $\sim$27 kDa for GFP) can disrupt the normal trafficking and function of target proteins (Hoffmann *et al.*, 2005; Lisenbee *et al.*, 2003). It is not easy to control the labeling ratio in multicolor labeling, which is crucial for the quantitative analysis of intermolecular fluorescence resonance energy transfer (FRET). Further, genetic fluorophores are not suitable for cell-surface-specific labeling of membrane proteins such as G protein-coupled receptors (GPCRs) because of uniform labeling of both cell surface and intracellular proteins.

To overcome these difficulties, a hybrid approach using a genetic tag and synthetic fluorescent (or other functional) probes that specifically bind to the tag has been actively studied (Lin and Wang, 2008; Yano and Matsuzaki, 2009; Fig. 18.1B). As shown in Table 18.1, tag–probe methods based on diverse interactions such as protein–ligand, peptide–peptide, and peptide–metal interactions have been reported. An alternative labeling principle is the use of enzymatic reactions. A more comprehensive list has been described elsewhere (Yano and Matsuzaki, 2009). By using tag–probe labeling, the size can be reduced to 2 kDa, although in many cases there is a trade-off between size and labeling specificity. Posttranslational labeling enables surface-specific labeling of membrane receptors using membrane-impermeable probes, pulse labeling with arbitrary timing, and easy control of the labeling ratio in multicolor labeling (Fig. 18.1B). Using these advantages, receptor internalization and oligomerization (Fig. 18.1C) have been quantitatively analyzed. In this chapter, we show examples of tag–probe methods useful for labeling membrane receptors, particularly focusing on procedures using the coiled-coil tag–probe method we developed.

2. Various Principles Used for Tag–Probe Labeling

2.1. Protein–ligand interactions

Labeling using interactions between a protein tag and a ligand probe can achieve high specificity without complicated labeling procedures, although the degree of miniaturization of the label is modest. Both noncovalent ligands and covalent ligands have been applied to the tag–probe labeling of proteins including membrane receptors. For example, labeling with

Table 18.1 Various tag–probe labeling methods

Labeling principle	Labeling system	Tag	Probe	Tag–probe size (approx.)	Cofactor/enzyme	Affinity	Typical labeling conditions	Material availability/note	Reference
Protein–ligand	SNAP-tag™	hAGT	Benzilguanine	20 kDa	No	Covalent	5 µM, 1 h	Commercial kit (NEB), intracellular labeling is also possible	Maurel et al. (2008)
Peptide–peptide	Coiled-coil	(EIAALKE)₃	(KIAALEK)₃ or (KIAALEK)₄	6 kDa	No	6 nM (K4), 64 nM (K3)	20 nM, 1 min	Commercial probe (Peptide Institute)	Yano et al. (2008)
Peptide–metal	Biarcenical-tetracycteine	FLNCCPG-CCMEP	Biarsenical fluorophores	2 kDa	Ethanedithiol	Covalent	0.5 µM, 1 h	Commercial probe (Invitrogen), intracellular labeling, washout is necessary	Hoffmann et al. (2010)
Enzymatic	ACP-tag™ MCP-tag™	Acyl carrier protein	Coenzyme A	9 kDa	PPTase	Covalent	5 µM, 40 min	Commercial kit (NEB)	Meyer et al. (2006)

SNAP-tag, which is based on the irreversible transfer of an alkyl group from O^6-alkylguanine-DNA to human O^6-alkylguanine-DNA alkyltransferase (Keppler et al., 2003), has been used to detect the internalization and oligomerization of GPCRs. Both membrane-permeable and impermeable probes are available from New England Biolabs (Ipswich, MA). The internalization of orexin and cannabinoid receptors following stimulation with their agonists was detected by labeling the receptors with a luminescent terbium chelate, resulting in an increase in detection sensitivity (Ward et al., 2011). Taking advantage of the surface-specific labeling, time-resolved FRET from europium chelate to organic fluorophores has also been measured to investigate the oligomerization of GPCRs in HEK 293 cell membranes (Maurel et al., 2008). The authors found that class C metabotropic glutamate receptors form strict dimers, whereas $GABA_B$ receptors can form dimer of dimers.

2.2. Peptide–metal interactions

Strong interactions between peptides and metals (or metalloids) are a fascinating tool for labeling. Organic fluorophores that contain metalloid atoms such as arsenic (Griffin et al., 1998) and boron (Halo et al., 2009) have been used to label tetracysteine and tetraserine motifs, respectively. The first tag–probe labeling system in living cells, reported in 1998, was based on a reversible covalent bond between an arsenic derivative of fluorescein (designated as FlAsh) and pairs of thiols (Griffin et al., 1998). Thereafter, development of the tag sequence (Martin et al., 2005), expansion of available fluorophores (Pomorski and Krezel, 2011), and refinement of labeling procedures (Hoffmann et al., 2010) were reported. FlAsh labeling in combination with a genetic fluorescent protein is useful to detect conformational changes of GPCRs by intramolecular FRET (Hoffmann et al., 2005).

The specific assembly of a chelator tag peptide and a chelator probe is possible through the coordination of metal cations, such as Ni^{2+} (Guignet et al., 2004) and Zn^{2+} (Hauser and Tsien, 2007; Ojida et al., 2006). An example is the membrane-impermeable HisZiFit probe that binds to a hexahistidine tag via Zn^{2+} coordination (Hauser and Tsien, 2007). Surface exposure of a membrane protein, stromal interaction molecule 1, from the endoplasmic reticulum in HEK293 cells was successfully detected using this method.

2.3. Enzymatic reactions

Enzymes that covalently attach a substrate to a specific site of a polypeptide have been applied to tag–probe labeling in living cells. A smaller size and tight labeling can be achieved with this approach, although a longer labeling time in the presence of excess probes (=substrates) is usually required for efficient labeling. An example is the use of phosphopantetheinyl transferase which transfers part of a phosphopantetheinyl probe to an acyl carrier protein (ACP) tag consisting of ~00 amino acids (George et al., 2004; Yin et al., 2004; available from New England Biolabs). The lateral organization of neurokinin-1 (NK_1) receptors in HEK293 cell membranes was investigated by FRET from Cy3 to Cy5 attached to the receptors by ACP labeling (Meyer et al., 2006). The authors analyzed FRET efficiency at various donor/acceptor ratios and expression levels, and concluded that NK_1 receptors are monomeric but locally concentrated in membrane microdomains to give FRET signals in HEK293 cell membranes. Consistent with the result, perturbation of the membrane microdomains by extraction of cholesterol with methyl-β-cyclodextrin decreased the FRET signal.

3. Coiled-Coil Tag–Probe Labeling

We have examined the utility of tag–probe labeling based on peptide–peptide interactions to achieve a reasonable balance between small size and labeling specificity. Relatively quick labeling will also be possible because of the simple principle involved (physicochemical binding). Coiled-coil tag–probe labeling is based on α-helical coiled-coil formation between negatively charged E_n peptides $(EIAALEK)_n$ and positively charged K_n peptides $(KIAAEKE)_n$ ($n = 3$ or 4), originally studied by Litowski and Hodges (2002). In addition to electrostatic attractions, leucine zipper-type hydrophobic interactions at the interface drive tight heterodimer formation (Fig. 18.2A). We found the E3 peptide to be suitable as the N-terminal extracellular tag of membrane proteins, which was specifically labeled with K3 and K4 peptide probes while retaining protein functions (Yano et al., 2008). For example, E3-tagged β2 adrenergic receptors ($β_2ARs$) transiently expressed in Chinese hamster ovary (CHO) cells were successfully labeled with K4 probes, as confirmed by costaining with fluorescent ligands (Fig. 18.2B). The labeling can be completed within 1 min, which is much faster than other tag–probe methods. Labeling of the E3 tag with the K3 probe ($K_d \sim 60$ nM) is reversible and the probe can be washed out whereas labeling with K4 is stronger ($K_d \sim 6$ nM) and therefore suitable for long-term observation. Because the charged peptides are membrane-impermeable, the label is surface-specific. Various fluorophores are available for the labeling (see Section 3.4), advantageous in multicolor labeling. Another peptide–peptide tag–probe pair based on heterotrimeric coiled-coil formation has also been reported (Tsutsumi et al., 2009).

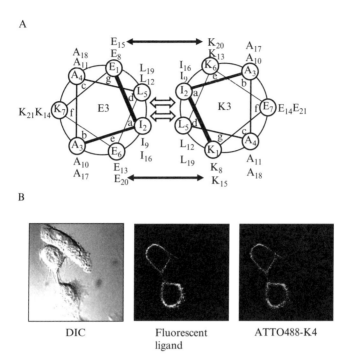

Figure 18.2 (A) Helical wheel representation of E3/K3 coiled-coil heterodimer. Hydrophobic and electrostatic interactions are indicated by white and black arrows, respectively. (B) Confocal imaging of E3-tagged β_2 adrenergic receptors (β_2ARs) transiently expressed in CHO cells. The receptors were colabeled with fluorescent ligands (CA 200689, 10 nM) and K4 probes (ATTO488-K4, 10 nM) for 10 min. Comparison with the differential interference contrast (DIC) image demonstrates specific labeling of cells that express the target receptors.

3.1. Design of tag-fused membrane proteins

In coiled-coil labeling, the gene of the target protein, which is cloned into mammalian expression vectors, is fused with the E3 sequence as an extracellular tag. Figure 18.3 shows an example of a plasmid map encoding the E3-tagged target protein (β_2AR) and the DNA sequence of the tag. The oligonucleotide encoding the E3 tag ($\sim$100 base pair) can be custom-made (e.g., Invitrogen, synthesis scale of 200 nmol with cartridge purification) and inserted at the N-terminal site immediately after the start codon by standard recombinant techniques. In the design of the tag sequence, a repetitive DNA sequence should be avoided to prevent undesired annealing. This is easily achieved by using different codons for the repeated amino acid sequence (EIAALEK)$_3$.

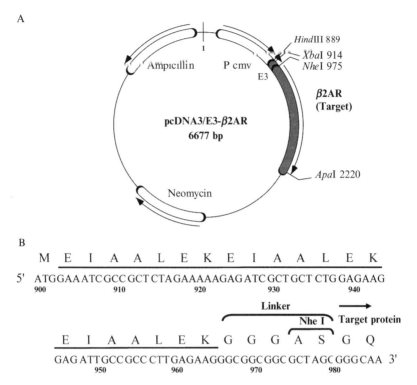

Figure 18.3 An example of design of the tag-fused receptor. (A) Plasmid map for E3-tagged human β2 adrenergic receptor (β$_2$AR). Representative restriction sites are shown. (B) DNA sequence for the E3 tag and a linker between the tag and target receptor.

A spacer (e.g., Gly-Gly-Gly-Ala-Ser) was inserted between the tag and the target protein. When the E3 tag is fused at the N-terminus of a protein that has a signal sequence (e.g., EGF receptors), the tag sequence should be inserted after the signal sequence instead of the start codon.

3.2. Expression of tagged membrane proteins in living cells

A variety of strategies are available to introduce genes into eukaryotic cells. We have used biochemical transfection reagents (Lipofectamine) according to the manufacturer's instructions. Both transient and stable expressions are available by using standard expression vectors. In transient expression, the cells are typically observed 24–48 h after transfection. The transfection reagents should be washed out after transfection (e.g., 5 h for Lipofectamine) and the cells incubated for 1 day to reduce the damage to cell membranes and nonspecific binding of probes.

- *Host cells*: It is important to select appropriate host cells for successful labeling using the coiled-coil method, because the positively charged K probes may nonspecifically bind to the surface of highly negatively charged cells due to electrostatic adsorption. In our experience, CHO cells have negligible nonspecific labeling and are suitable for coiled-coil labeling compared with other cell lines such as HEK-293, COS-7, and PC-12. The K4 probes labeled with negatively charged fluorophores (Alexa 568 and Alexa 647) can decrease nonspecific labeling for HEK 293 cells (K. Kawano *et al.*, unpublished observation).
- *Glass bottom dish*: For fluorescence imaging, the cells are seeded on a ϕ35-mm glass bottom dish 1 day before transient transfection or 1 day before confocal observation of cells stably expressing the target gene. A glass bottom dish coated with Advanced TC^{TM} polymer (#627965 advanced glass bottom; Greiner Bio-One, Frickenhausen, Germany) significantly suppresses nonspecific absorption of the probes on the glass. Other coating reagents (poly-L-lysine and collagen) did not improve the nonspecific binding compared with normal dishes, although normal glass bottom dishes (e.g., IWAKI #3911-035 and Matsunami #D110400) are also available for confocal imaging with coiled-coil labeling.

3.3. Preparation of fluorophore-labeled probe peptides

The K probe peptides are synthesized by a standard 9-fluorenylmethyloxycarbonyl (Fmoc)-based solid phase method (Amblard *et al.*, 2006). After elongation of the peptide sequence from the C-terminus on the resin, the N-terminal amino group can be used as a specific labeling site with amino-reactive fluorophores, followed by detachment of the peptide from the resin and deprotection of side chains by treatment with trifluoroacetic acid (TFA). Procedures for N-terminal labeling with fluorophores are given below.

Dried K4-attached resin ($\sim$1 μmol of peptide, which typically corresponds to 5–10 mg of resin) is placed in a regular 1.5-ml polypropylene tube, and swelled with 500 μl of *N*, *N*-dimethylformamide (DMF) overnight on a shaker. After precipitation of the resin by standing, the DMF is replaced, the resin washed by shaking, and as much of the supernatant removed as possible. The *N*-hydroxysuccinimide ester of fluorophores (1–1.5 μmol or 1 mg) is dissolved in a minimum amount of DMF ($<$100 μl) and added to the resin, followed by gentle stirring with a small magnetic bar for 48 h in the dark. Addition of a base (e.g., 5% *N*, *N*-diisopropylethylamine) may be required for efficient coupling, particularly when the fluorophores are provided as a salt form with acids such as HCl. After the coupling, wash the resin by repeated replacement and shaking for 5 min with organic solvents until the supernatant becomes colorless (typically, DMF 3 $\times$ followed by methanol 5 $\times$). Vacuum-dry the methanol-washed resin for deprotection.

The labeling of fluorophores that degrade in TFA may be difficult. In that case, a Cys residue for specific labeling should be introduced after deprotection of the peptide. Purify the crude peptide by HPLC using common reversed-phase octadecylsilyl ODS columns. Phenyl-based columns (e.g., PLRP-S series; Agilent Technologies) are also useful to separate fluorophore-labeled peptides from unlabeled peptides. After lyophilization of the purified peptide, dissolve it in *distilled water* to obtain a stock solution (CAUTION: do not use buffers for the stock solution). Depending on the attached fluorophore, the probe may not dissolve in water (e.g., the fluorescein-K4 probe aggregates in water). We found that 0.01 M NaOH could stably solubilize the fluorescein-K4 probe. Avoid exposure of the probes to light as much as possible. Determine the concentration of the probe by absorbance based on the extinction coefficients of the fluorophores. To avoid significant loss of the probes by adsorption on the tube, the concentration of stock solution should be above 10 μM. To prevent the degradation of the peptide, store small aliquots of the stock solution below $-20\,°C$ in the dark. In the case of multicolor labeling, preparation of a premixed probe solution is recommended. In time-lapse imaging, addition of quenchers for reactive oxygen species, such as ascorbic acid and crocetin (Tsien *et al.*, 2006), might be required to suppress the phototoxicity of the probes.

3.4. Labeling of E3-tagged proteins with K probes

Thaw the probe stock solutions and keep them on ice or in a refrigerator. Avoid exposure of the probes to light as much as possible. The thawed solutions should be used within a few days. Take out the glass bottom dish from a CO_2 incubator and replace the culture medium (αMEM or F-12) with 1 ml of PBS (+). When the labeling is performed in medium, the solution should be buffered (e.g., by adding 10 mM HEPES). Dilute the stock solution with PBS (+) to obtain a fresh labeling solution (final concentration: typically 20 nM K4), which should be prepared *immediately before* application to the cells. Replace PBS (+) with 1 ml of the labeling solution and incubate, for example, for 1–2 min with 20 nM probe or for 10–15 min with 2 nM probe before starting confocal imaging. If necessary, wash out the probes by replacing the solution with PBS (+) before observation, although background fluorescence from free probes can be negligible compared with that on cell membranes in many cases.

- *Confocal microscopy*: We usually use a 60× Plan Apochromat water-immersion objective lens to obtain confocal images of the target receptors on cell membranes, focusing 2–3 μm above the glass surface. The excitation laser power and sensitivity of the detector should be optimized to

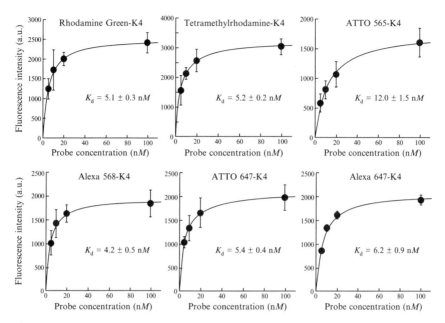

Figure 18.4 Determination of affinity between the tag and probe. The K4 probes labeled with various fluorophores were added to CHO cells stably expressing E3-β_2ARs at probe concentrations [P] of 5, 10, 20, and 100 nM. Averaged fluorescence intensity (F) at each probe concentration was determined from the confocal images by defining regions of interest at cell membranes ($n = 10$). The dissociation constant (K_d) was obtained from the fitting, $F = F_{max}*[P]/([P] + K_d)$, where F_{max} indicates maximal fluorescence intensity.

avoid rapid photobleaching and saturation of signals. The linearity between the concentration of the probe and detector response can be checked by titrating free probes in water.

Figure 18.4 shows the concentration-dependence of fluorescence intensity on cell membranes for the K4 probes labeled with various fluorophores, quantified from confocal images for E3-β_2ARs stably expressed on CHO cells. The dissociation constants (K_d) for the tag–probe pairs were estimated from the fitting. Most fluorophores (tetramethylrhodamine (TMR), Rhodamine green (RG), ATTO 647, Alexa 568, and Alexa 467) attached to K4 gave dissociation constants of around 5 nM, whereas the attachment of ATTO 565 slightly decreased the affinity ($K_d \sim$ 12 nM). Unexpectedly, the negatively charged fluorophore Alexa 647 (-3), which may partially neutralize positive charges of the K4 probe, did not reduce the formation of the coiled-coil. These results indicate that diverse fluorophores are available for the labeling.

4. Applications

4.1. Receptor internalization

Upon the stimulation of cell-surface receptors with ligands, sequestration to endosomes is often observed, which is a mechanism of receptor desensitization (Wolfe and Trejo, 2007). Because receptor internalization is a ubiquitous process irrespective of downstream signaling pathways, it is applicable to the monitoring of activities of a wide variety of membrane receptors including orphan receptors. Fluorescence imaging of receptors fused with fluorescent proteins has been used to visualize the internalization of target receptors (McLean and Milligan, 2000). However, fluorescent proteins inevitably label receptors even in intracellular compartments, which can partially obscure observations of internalization. On the other hand, the coiled-coil method has an advantage for the observation of internalization because of cell-surface labeling. Further, the quickness and reversibility of coiled-coil labeling enable pulse-chase labeling (Fig. 18.5). A shorter K3 probe ($K_d \sim 60$ nM) is useful for reversible labeling. After stimulation of the TMR-K3-labeled E3-β_2AR with an agonist, the receptors on the cell (receptors that had not been internalized and newly externalized receptors) could be labeled by the second probe FL-K4 after the washout of TMR-K3 (PBS, 10 times). The absence of TMR fluorescence on the cell surface indicates the reversibility of the labeling (Yano et al., 2008).

4.2. Receptor oligomerization

Protein–protein interactions in membrane environments drive the formation of noncovalent oligomeric structures for membrane proteins that are necessary for function. For example, potassium channels function as tightly assembled homotetramers (MacKinnon, 2003). Although rhodopsin (Bayburt et al., 2007) and β_2AR (Whorton et al., 2007) can function as monomers, many GPCRs are believed to form homo-oligomers and hetero-oligomers with other receptors. Experimental evidence of receptor oligomerization includes the cointernalization and coexpression of receptors (Sartania et al., 2007; Uberti et al., 2005). Also, recent single molecule studies using fluorescent ligands for the M1 muscarinic receptor (Hern et al., 2010) and the N-formyl peptide receptor (Kasai et al., 2011) revealed that the receptors diffuse on cell membranes accompanied by transient colocalization with each other, consistent with reversible receptor oligomerization. However, it is not clear from these results whether the receptors come directly into contact with each other or are locally concentrated into membrane domains (Jacquier et al., 2006; Meyer et al., 2006). Further, the relationship between

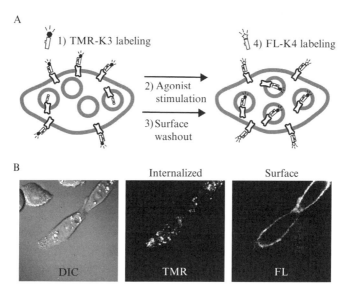

Figure 18.5 Selective labeling of internalized receptors and surface-remaining receptors. (A) The labeling scheme. CHO cells expressing E3-β_2AR were labeled with TMR-K3 (60 nM) for 2 min, and then stimulated with the agonist isoproterenol (10 μM) for 5 min. After the cells were washed with PBS, fluorescein-K4 (20 nM) was added and the cells were observed by confocal microscopy. (B) Images for DIC, TMR, and FL channels.

oligomerization of receptors and biological function is unknown, although pharmacological studies indicate the presence of allosteric regulatory mechanisms for GPCRs (Han et al., 2009). Detection of oligomerization by resonance energy transfer is direct evidence of close contact between receptors within ca. 5 nm. FRET and bioluminescent resonance energy transfer studies using GPCR fused with fluorescent/luminescent proteins have concluded that most receptors form oligomers. However, quantitative analysis of the oligomerization is not easy, as exemplified in controversial results for self-association of β_2AR reported from different research groups (Bouvier et al., 2007; James et al., 2006). Development of an improved method of analysis for receptor oligomerization is important.

Taking advantage of the ease of multicolor labeling of the coiled-coil method, we measured the self-association of metabotropic glutamate receptors (mGluRs), a class C GPCR that forms dimers (Maurel et al., 2008), by FRET from RG to TMR (critical transfer distance: $\sim$55 Å). To clarify fluorescence from the donor and acceptor, spectral imaging was performed using a Nikon C1Si confocal microscope. Figure 18.6A indicates fluorescence spectra for mGluR-expressing CHO cell membranes doubly labeled with RG-K4 and TMR-K4 excited at 488 nm, at which the donor RG is selectively excited. In addition to the donor RG fluorescence ($\sim$530 nm),

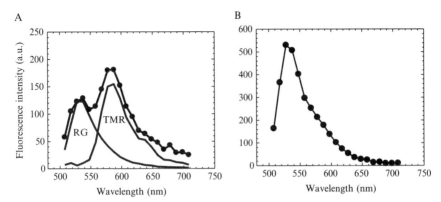

Figure 18.6 Detection of receptor oligomerization by FRET using fluorescence emission spectra in CHO cells expressing (A) E3-tagged metabotropic glutamate receptors (E3-mGluRs) and (B) E3-tagged glycophorin A G83I mutant. The target proteins were doubly labeled with a 1/1 mixture of RG-K4 and TMR-K4 (25 nM each), the donor RG was excited at 488 nm, and spectral images in the membranes were obtained. The fluorescence spectrum originating from direct excitation of TMR has been subtracted for clarity. In the spectrum for mGluR (A), contributions from RG and TMR fluorescence are also shown, which were determined by a least square fitting using the reference spectra.

sensitized emission from the acceptor TMR (~580 nm) was clearly observed, indicating strong FRET by oligomerization of the receptors. On the other hand, a monomeric control protein (Glycophorin A G83I mutant) did not show any sensitized emission (Fig. 18.6B). These results demonstrate that correct evaluation of receptor oligomerization is possible using coiled-coil labeling.

ACKNOWLEDGMENTS

We thank Dr. Shinya Oishi, Prof. Nobutaka Fujii, Prof. Yukihiko Sugimoto, Ms. Akiko Yano, and Prof. Gozoh Tsujimoto (Kyoto University) for technical advice. This work was financially supported by MEXT (Targeted Proteins Research Programs and Grant-in-Aid for Scientific Research (B) 21390007).

REFERENCES

Adams, S. R., Harootunian, A. T., Buechler, Y. J., Taylor, S. S., and Tsien, R. Y. (1991). Fluorescence ratio imaging of cyclic AMP in single cells. *Nature* **349,** 694–697.
Amblard, M., Fehrentz, J. A., Martinez, J., and Subra, G. (2006). Methods and protocols of modern solid phase peptide synthesis. *Mol. Biotechnol.* **33,** 239–254.
Bayburt, T. H., Leitz, A. J., Xie, G., Oprian, D. D., and Sligar, S. G. (2007). Transducin activation by nanoscale lipid bilayers containing one and two rhodopsins. *J. Biol. Chem.* **282,** 14875–14881.

Bouvier, M., Heveker, N., Jockers, R., Marullo, S., and Milligan, G. (2007). BRET analysis of GPCR oligomerization: Newer does not mean better. *Nat. Methods* **4**, 3–4.

Chudakov, D. M., Matz, M. V., Lukyanov, S., and Lukyanov, K. A. (2010). Fluorescent proteins and their applications in imaging living cells and tissues. *Physiol. Rev.* **90**, 1103–1163.

Fernandez-Suarez, M., and Ting, A. Y. (2008). Fluorescent probes for super-resolution imaging in living cells. *Nat. Rev. Mol. Cell Biol.* **9**, 929–943.

George, N., Pick, H., Vogel, H., Johnsson, N., and Johnsson, K. (2004). Specific labeling of cell surface proteins with chemically diverse compounds. *J. Am. Chem. Soc.* **126**, 8896–8897.

Giepmans, B. N., Adams, S. R., Ellisman, M. H., and Tsien, R. Y. (2006). The fluorescent toolbox for assessing protein location and function. *Science* **312**, 217–224.

Griffin, B. A., Adams, S. R., and Tsien, R. Y. (1998). Specific covalent labeling of recombinant protein molecules inside live cells. *Science* **281**, 269–272.

Guignet, E. G., Hovius, R., and Vogel, H. (2004). Reversible site-selective labeling of membrane proteins in live cells. *Nat. Biotechnol.* **22**, 440–444.

Halo, T. L., Appelbaum, J., Hobert, E. M., Balkin, D. M., and Schepartz, A. (2009). Selective recognition of protein tetraserine motifs with a cell-permeable, pro-fluorescent bis-boronic acid. *J. Am. Chem. Soc.* **131**, 438–439.

Han, Y., Moreira, I. S., Urizar, E., Weinstein, H., and Javitch, J. A. (2009). Allosteric communication between protomers of dopamine class A GPCR dimers modulates activation. *Nat. Chem. Biol.* **5**, 688–695.

Hauser, C. T., and Tsien, R. Y. (2007). A hexahistidine-Zn^{2+}-dye label reveals STIM1 surface exposure. *Proc. Natl. Acad. Sci. USA* **104**, 3693–3697.

Hern, J. A., Baig, A. H., Mashanov, G. I., Birdsall, B., Corrie, J. E., Lazareno, S., Molloy, J. E., and Birdsall, N. J. (2010). Formation and dissociation of M1 muscarinic receptor dimers seen by total internal reflection fluorescence imaging of single molecules. *Proc. Natl. Acad. Sci. USA* **107**, 2693–2698.

Hoffmann, C., Gaietta, G., Bunemann, M., Adams, S. R., Oberdorff-Maass, S., Behr, B., Vilardaga, J. P., Tsien, R. Y., Ellisman, M. H., and Lohse, M. J. (2005). A FlAsH-based FRET approach to determine G protein-coupled receptor activation in living cells. *Nat. Methods* **2**, 171–176.

Hoffmann, C., Gaietta, G., Zurn, A., Adams, S. R., Terrillon, S., Ellisman, M. H., Tsien, R. Y., and Lohse, M. J. (2010). Fluorescent labeling of tetracysteine-tagged proteins in intact cells. *Nat. Protoc.* **5**, 1666–1677.

Jacquier, V., Prummer, M., Segura, J. M., Pick, H., and Vogel, H. (2006). Visualizing odorant receptor trafficking in living cells down to the single-molecule level. *Proc. Natl. Acad. Sci. USA* **103**, 14325–14330.

James, J. R., Oliveira, M. I., Carmo, A. M., Iaboni, A., and Davis, S. J. (2006). A rigorous experimental framework for detecting protein oligomerization using bioluminescence resonance energy transfer. *Nat. Methods* **3**, 1001–1006.

Kasai, R. S., Suzuki, K. G., Prossnitz, E. R., Koyama-Honda, I., Nakada, C., Fujiwara, T. K., and Kusumi, A. (2011). Full characterization of GPCR monomer–dimer dynamic equilibrium by single molecule imaging. *J. Cell Biol.* **192**, 463–480.

Keppler, A., Gendreizig, S., Gronemeyer, T., Pick, H., Vogel, H., and Johnsson, K. (2003). A general method for the covalent labeling of fusion proteins with small molecules in vivo. *Nat. Biotechnol.* **21**, 86–89.

Lin, M. Z., and Wang, L. (2008). Selective labeling of proteins with chemical probes in living cells. *Physiology (Bethesda)* **23**, 131–141.

Lisenbee, C. S., Karnik, S. K., and Trelease, R. N. (2003). Overexpression and mislocalization of a tail-anchored GFP redefines the identity of peroxisomal ER. *Traffic* **4**, 491–501.

Litowski, J. R., and Hodges, R. S. (2002). Designing heterodimeric two-stranded α-helical coiled-coils. Effects of hydrophobicity and α-helical propensity on protein folding, stability, and specificity. *J. Biol. Chem.* **277,** 37272–37279.

MacKinnon, R. (2003). Potassium channels. *FEBS Lett.* **555,** 62–65.

Martin, B. R., Giepmans, B. N., Adams, S. R., and Tsien, R. Y. (2005). Mammalian cell-based optimization of the biarsenical-binding tetracysteine motif for improved fluorescence and affinity. *Nat. Biotechnol.* **23,** 1308–1314.

Maurel, D., Comps-Agrar, L., Brock, C., Rives, M. L., Bourrier, E., Ayoub, M. A., Bazin, H., Tinel, N., Durroux, T., Prezeau, L., Trinquet, E., and Pin, J. P. (2008). Cell-surface protein–protein interaction analysis with time-resolved FRET and snap-tag technologies: Application to GPCR oligomerization. *Nat. Methods* **5,** 561–567.

McLean, A. J., and Milligan, G. (2000). Ligand regulation of green fluorescent protein-tagged forms of the human β_1- and β_2-adrenoceptors: Comparisons with the unmodified receptors. *Br. J. Pharmacol.* **130,** 1825–1832.

Meyer, B. H., Segura, J. M., Martinez, K. L., Hovius, R., George, N., Johnsson, K., and Vogel, H. (2006). FRET imaging reveals that functional neurokinin-1 receptors are monomeric and reside in membrane microdomains of live cells. *Proc. Natl. Acad. Sci. USA* **103,** 2138–2143.

Ojida, A., Honda, K., Shinmi, D., Kiyonaka, S., Mori, Y., and Hamachi, I. (2006). Oligo-Asp tag/Zn(II) complex probe as a new pair for labeling and fluorescence imaging of proteins. *J. Am. Chem. Soc.* **128,** 10452–10459.

Pomorski, A., and Krężel, A. (2011). Exploration of biarsenical chemistry—Challenges in protein research. *Chembiochem* **12,** 1152–1167.

Sartania, N., Appelbe, S., Pediani, J. D., and Milligan, G. (2007). Agonist occupancy of a single monomeric element is sufficient to cause internalization of the dimeric β_2-adrenoceptor. *Cell. Signal.* **19,** 1928–1938.

Tsien, R. Y., Ernst, L., and Waggoner, A. (2006). Fluorophores for confocal microscopy: Photophysics and photochemistry. *In* "Biological Confocal Microscopy," (J. B. Pawley, ed.), pp. 338–352. Springer, New York.

Tsutsumi, H., Nomura, W., Abe, S., Mino, T., Masuda, A., Ohashi, N., Tanaka, T., Ohba, K., Yamamoto, N., Akiyoshi, K., and Tamamura, H. (2009). Fluorogenically active leucine zipper peptides as tag–probe pairs for protein imaging in living cells. *Angew. Chem. Int. Ed Engl.* **48,** 9164–9166.

Uberti, M. A., Hague, C., Oller, H., Minneman, K. P., and Hall, R. A. (2005). Heterodimerization with β_2-adrenergic receptors promotes surface expression and functional activity of α1D-adrenergic receptors. *J. Pharmacol. Exp. Ther.* **313,** 16–23.

Ward, R. J., Pediani, J. D., and Milligan, G. (2011). Ligand-induced internalization of the orexin OX_1 and cannabinoid CB_1 receptors assessed via N-terminal SNAP and CLIP-tagging. *Br. J. Pharmacol.* **162,** 1439–1452.

Whorton, M. R., Bokoch, M. P., Rasmussen, S. G., Huang, B., Zare, R. N., Kobilka, B., and Sunahara, R. K. (2007). A monomeric G protein-coupled receptor isolated in a high-density lipoprotein particle efficiently activates its G protein. *Proc. Natl. Acad. Sci. USA* **104,** 7682–7687.

Wolfe, B. L., and Trejo, J. (2007). Clathrin-dependent mechanisms of G protein-coupled receptor endocytosis. *Traffic* **8,** 462–470.

Yano, Y., and Matsuzaki, K. (2009). Tag–probe labeling methods for live-cell imaging of membrane proteins. *Biochim. Biophys. Acta* **1788,** 2124–2131.

Yano, Y., Yano, A., Oishi, S., Sugimoto, Y., Tsujimoto, G., Fujii, N., and Matsuzaki, K. (2008). Coiled-coil tag–probe system for quick labeling of membrane receptors in living cell. *ACS Chem. Biol.* **3,** 341–345.

Yin, J., Liu, F., Li, X., and Walsh, C. T. (2004). Labeling proteins with small molecules by site-specific posttranslational modification. *J. Am. Chem. Soc.* **126,** 7754–7755.

CHAPTER NINETEEN

Monitoring Protein Interactions in Living Cells with Fluorescence Lifetime Imaging Microscopy

Yuansheng Sun,[*,†,‡] Nicole M. Hays,[§] Ammasi Periasamy,[*,†,‡] Michael W. Davidson,[¶] *and* Richard N. Day[§]

Contents

1. Introduction	372
1.1. Overview of FRET microscopy	373
1.2. The FPs and FRET microscopy	375
1.3. Overview of FLIM	375
2. FD FLIM Measurements	376
2.1. The analysis of FD FLIM measurements	378
2.2. The calibration of a FLIM system	378
2.3. Testing the FLIM system using "FRET-standard" proteins	381
2.4. FLIM measurements of the FRET-standard proteins	383
3. Measuring Protein–Protein Interactions in Living Cells Using FLIM–FRET	385
3.1. Construction of the C/EBPα B Zip domain plasmids and transfection for FLIM–FRET	385
3.2. FLIM measurements to detect dimerized B Zip proteins	386
4. The Strengths and Limitations of FLIM	388
Acknowledgment	389
References	389

Abstract

Fluorescence lifetime imaging microscopy (FLIM) is now routinely used for dynamic measurements of signaling events inside single living cells, such as monitoring changes in intracellular ions and detecting protein–protein

---

[*] W.M. Keck Center for Cellular Imaging, University of Virginia, Charlottesville, Virginia, USA
[†] Department of Biology, University of Virginia, Charlottesville, Virginia, USA
[‡] Department of Biomedical Engineering, University of Virginia, Charlottesville, Virginia, USA
[§] Department of Cellular and Integrative Physiology, Indiana University School of Medicine, Indianapolis, Indiana, USA
[¶] National High Magnetic Field Laboratory and Department of Biological Science, The Florida State University, Tallahassee, Florida, USA

Methods in Enzymology, Volume 504 © 2012 Elsevier Inc.
ISSN 0076-6879, DOI: 10.1016/B978-0-12-391857-4.00019-7 All rights reserved.

interactions. Here, we describe the digital frequency domain FLIM data acquisition and analysis. We describe the methods necessary to calibrate the FLIM system and demonstrate how they are used to measure the quenched donor fluorescence lifetime that results from Förster Resonance Energy Transfer (FRET). We show how the "FRET-standard" fusion proteins are used to validate the FLIM system for FRET measurements. We then show how FLIM–FRET can be used to detect the dimerization of the basic leucine zipper (B Zip) domain of the transcription factor CCAAT/enhancer binding protein α in the nuclei of living mouse pituitary cells. Importantly, the factors required for the accurate determination and reproducibility of lifetime measurements are described in detail.

1. INTRODUCTION

The development of the many different genetically encoded fluorescent proteins (FPs), which now span the full visible spectrum, has sparked a revolution in optical imaging in biomedical research (Day and Davidson, 2009). These new FPs have expanded the repertoire of imaging applications from multi-color imaging of protein co-localization and behavior inside living cells to the detection of changes in intracellular activities, such as pH or ion concentration. However, it is the use of the genetically encoded FPs for Förster Resonance Energy Transfer (FRET) microscopy in living cells that has generated the most interest in these probes (Aye-Han *et al.*, 2009; Giepmans *et al.*, 2006; Piston and Kremers, 2007; Sun *et al.*, 2011; Vogel *et al.*, 2006).

FRET is the process by which energy absorbed by one fluorophore (the "donor") is transferred directly to another nearby molecule (the "acceptor") through a nonradiative pathway. This process depletes the excited-state energy of the donor molecule, quenching its fluorescence emission while causing increased (sensitized) emission from the acceptor. The quenching of the donor as well as the sensitized acceptor emission can be used to quantify energy transfer. Since the distance over which efficient energy transfer can occur is limited to less than 100 Å, FRET can be used to monitor protein–protein interactions inside living cells, tissues, and organisms (Miyawaki, 2011; Tsien, 2005; Yasuda, 2006; Zhang *et al.*, 2002). There are many different methods used to measure FRET, each with distinct advantages and disadvantages, and these have been reviewed elsewhere (Jares-Erijman and Jovin, 2003; Periasamy and Day, 2005).

The most accurate methods for detecting FRET are the approaches that measure the change in the donor fluorescence lifetime that results from the quenching by the acceptor (Periasamy and Clegg, 2010). The fluorescence lifetime is the average time that a molecule spends in the excited state before returning to the ground state, which is usually accompanied by the emission of a photon. The fluorescence lifetime is an intrinsic property of each fluorophore, and most probes used in biological studies have lifetimes ranging between 1 to about 10 ns. The fluorescence lifetime of the fluorophore also carries information about events in the local microenvironment that can affect its photophysical behavior. Because energy transfer is a quenching process affecting the excited state of the donor fluorophore, the donor fluorescence lifetime is shortened by FRET.

Fluorescence lifetime imaging microscopy (FLIM) maps the spatial distribution of probe lifetimes inside living cells and can accurately measure the shorter donor lifetimes that result from FRET (Periasamy and Clegg, 2010; Sun et al., 2010, 2011; Wouters and Bastiaens, 1999; Yasuda, 2006). Here, we describe in detail the procedures that are necessary to use the frequency domain (FD) FLIM method to measure FRET, allowing us to monitor protein–protein interactions inside living cells. The basic calibration procedures necessary for accurate lifetime determinations are described. We demonstrate the measurement of FRET using protein standards designed to validate microscope systems for the detection of FRET (Koushik et al., 2006). We then use digital FD FLIM to characterize the dimerization of the transcription factor CAATT/enhancer binding protein α (C/EBPα) in regions of centromeric heterochromatin in the living cell nucleus. The methodology described here can be adapted for other commercial FLIM systems.

1.1. Overview of FRET microscopy

FRET involves the direct transfer of excited-state energy from a donor fluorophore to nearby acceptor molecules through near-field electromagnetic dipole interactions. There are three basic requirements for the efficient transfer of energy to the acceptor (Forster, 1965; Lakowicz, 2006; Stryer, 1978). First, because energy transfer involves electromagnetic dipolar interactions between the excited-state donor and the acceptor fluorophores, the efficiency of energy transfer (E_{FRET}) varies as the inverse of the sixth power of the distance (r) that separates the fluorophores. This sixth power dependence is described by the Förster equation (19.1):

$$E_{FRET} = \frac{R_0^6}{R_0^6 + r^6} \qquad (19.1)$$

where R_0 is the Förster distance at which the efficiency of energy transfer is 50%. The relationship of E_{FRET} to the distance separating the fluorophores is illustrated in Fig. 19.1A. Because E_{FRET} varies as the inverse of the sixth power of the separation distance between the fluorophores, the efficiency of energy transfer falls off sharply over the range of 0.5 R_0 to 1.5 R_0 (shaded

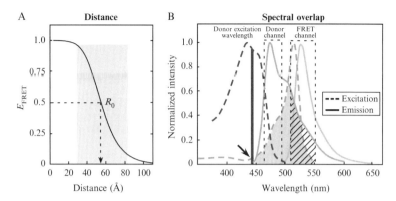

Figure 19.1 The distance dependence (A) and the spectral overlap (B) requirements for efficient FRET. (A) The efficiency of energy transfer, E_{FRET}, was determined using Eq. (19.1), and is plotted as a function of the separation distance in Å. The shaded region illustrates the range of $0.5\ R_0$ to $1.5\ R_0$ over which FRET can be accurately measured. (B) The excitation and emission spectra for the CFP (donor) and YFP (acceptor) FRET pair are shown, with the shaded region indicating the spectral overlap between the donor emission and acceptor excitation. The dashed boxes indicate the donor (480/40 nm) and FRET (530/43 nm) detection channels. The arrow indicates the direct acceptor excitation at the donor excitation wavelength, and the hatching shows donor SBT into the FRET channel.

area, Fig. 19.1A). This is why energy transfer between the FP-labeled proteins is limited to distances of less than approximately 100 Å.

The second requirement for efficient transfer of energy is the favorable alignment of the electromagnetic dipoles of the donor emission and acceptor absorption. The angular dependence of the donor and acceptor dipolar interaction is described by the orientation factor, κ^2. Depending on the relative orientation of the donor and acceptor, the value for κ^2 can range from 0 to 4 (Gryczynski et al., 2005). It is difficult, however, to determine κ^2 in most experimental systems. Fortunately, for many biological applications, where proteins labeled with the donor and acceptor fluorophores freely diffuse within cellular compartments and adopt a variety of conformations, the orientations of the FP tags randomize over the time scales of the measurements. Under these conditions, κ^2 is often assumed to be 2/3, which reflects the random orientations of the probes. However, if the donor and acceptor adopt unique orientations and are immobile on the time scale of FRET, then 2/3 will not accurately describe the dipolar orientation. Therefore, caution is necessary when assigning precise separation distances based on FRET measurements. The final requirement for FRET is that the fluorophores share a strong overlap between the donor emission spectrum and the absorption spectrum of the acceptor (see

Fig. 19.1B). In this regard, it is desirable to select a donor fluorophore with a high quantum yield that shares significant (>30%) spectral overlap with the acceptor. The quantum yield of the acceptor is not important when the donor fluorescence lifetime is used to measure FRET, but it is critical that the acceptor has a high extinction coefficient. When these basic requirements for energy transfer are met, the quantification of FRET by microscopy can provide Angstrom-scale measurements of the spatial relationship between the fluorophore-labeling proteins inside living cells.

1.2. The FPs and FRET microscopy

The cloning of *Aequorea* green FP and the subsequent engineering to finetune its spectral characteristics yielded many different FPs with fluorescence emissions ranging from the blue to the yellow regions of the visible spectrum (Shaner et al. 2005; Tsien 1998). Currently, the cyan (CFP) and yellow (YFP) FPs are the most widely used for FRET-based imaging studies because of their significant spectral overlap (shaded area, Fig. 19.1B). The monomeric (m) mCerulean CFPs (Rizzo et al., 2004) used in combination with either the mVenus (Nagai et al., 2002) or mCitrine (Griesbeck et al., 2001) YFP are among the most popular FRET pairings. When energy is transferred from mCerulean to mVenus, the mCerulean emission signal detected in the donor channel is quenched, and there is increased (sensitized) emission from the acceptor, which is detected in the FRET channel (Fig. 19.1B). However, a consequence of the strong spectral overlap, which is required for FRET, is significant background fluorescence (called spectral bleed through, SBT) that is also detected in the FRET channel. The SBT results from the direct excitation of the acceptor by the donor excitation wavelengths (arrow, Fig. 19.1B) and the donor emission signal that bleeds into the FRET detection channel (hatching, Fig. 19.1B). Therefore, the accurate measurement of FRET by detecting sensitized emission signals requires correction methods that define and remove these different SBT components (Periasamy and Day, 2005). Significantly, the accuracy of these SBT correction methods is degraded as the spectral overlap between the FPs is increased to the point where the SBT components overwhelm the FRET signal (Berney and Danuser, 2003).

1.3. Overview of FLIM

The alternative to the SBT correction methods is to measure the effect of FRET on the donor fluorophore. Importantly, because the SBT background is detected in the acceptor emission (FRET) channel (see Fig. 19.1B), these artifacts can be avoided when making measurements in the donor channel. Since the transfer of energy quenches the emission from donors participating in FRET, the fluorescence lifetime of that donor population becomes shorter. Therefore, measuring the change in the donor fluorescence lifetime in the presence of the acceptor is the most direct method for monitoring FRET, and this is what FLIM does. FLIM is particularly useful for biological applications, since the donor fluorescence lifetime is a time measurement that is not affected by variations in the probe concentration, excitation intensity, and other factors that can limit intensity-based measurements.

The first measurements of the nanosecond decay of fluorescence using optical microscopy were made in 1959 (Venetta, 1959). The FLIM methodologies have evolved significantly in recent years and now encompass biological, biomedical, and clinical research applications (Periasamy and Clegg, 2010). The FLIM techniques are broadly subdivided into the time domain (TD) and the FD methods. The physics that underlies these two different methods is identical, only the analysis of the measurements differs (Clegg, 2010). The TD method uses a pulse-d-laser source to excite the specimen. For probes with nanosecond lifetimes, femtosecond to picosecond pulse durations are used. The laser is synchronized to high-speed detectors that can record the arrival time of a single photon relative to the excitation pulse. The photons are accumulated at different time bins relative to the excitation pulse to build a histogram, which is analyzed to determine the fluorescence decay profile, providing an estimate of the fluorescence lifetime. In contrast, the FD method uses a light source modulated at high radio frequencies to excite a fluorophore and then measures the modulation and phase of the emission signals. The fundamental modulation frequency is chosen depending on the lifetime of the fluorophore and is usually between 10 and 100 MHz for the measurement of nanosecond decays. The emission signal is then analyzed for changes in phase and amplitude relative to the excitation source to extract the fluorescence lifetime of the fluorophore.

2. FD FLIM Measurements

The FD FLIM system described in this paper employs the technique called FastFLIM (Colyer et al., 2008), which measures the phase delays and modulation ratios at multiple frequencies simultaneously. Because of the finite lifetime of the excited state, there is a phase delay (Φ) and a change in the modulation (M) of the emission relative to the excitation waveform (Fig. 19.2). The lifetime of the fluorophores can be directly determined by

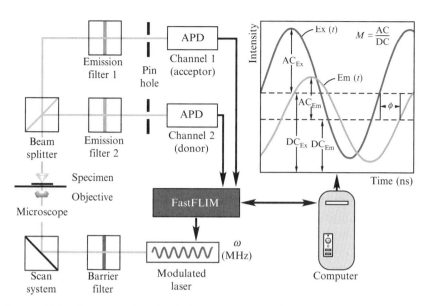

Figure 19.2 Frequency-domain (FD) FLIM setup. The excitation source for the FD FLIM system in this study is a 448-nm diode laser that is directly modulated by the ISS FastFLIM module at the fundamental frequency of 20 MHz. The modulated laser is coupled to the ISS scanning system that is attached to an Olympus IX71 microscope. The emission signals from the specimen travel through the scanning system (de-scanned detection) and are then routed by a beam splitter through the donor and acceptor emission channel filters to two identical APD detectors. The phase delays (Φ) and modulation ratios of the emission relative to the excitation are measured at up to seven modulation frequencies ($\omega = 20$–140 MHz) for each XY raster scanning location. The basic principle of the FD FLIM method is illustrated, showing the phase delay (Φ) and modulation ratio ($M = AC/DC$) of the emission (Em) relative to the excitation (Ex) that are used to estimate the fluorescence lifetime.

measuring either the phase delays (the phase lifetime τ_Φ, Eq. 19.2) or the modulation ratio (the modulation lifetime τ_M, Eq. 19.3) at different modulation frequencies (ω).

$$\tau_\Phi = \frac{\tan\Phi}{\omega} \qquad (19.2)$$

$$\tau_M = \sqrt{\frac{1-M^2}{(M\omega)}} \qquad (19.3)$$

If the fluorescence decay from the fluorophores is best fit to a single exponential, then τ_Φ and τ_M will be the same at all ω. Conversely, if the decay is best fit by a multi-exponential, then $\tau_\Phi < \tau_M$ and their values will depend on the ω. Therefore, the FD FLIM method provides immediate information regarding lifetime heterogeneity within a system.

2.1. The analysis of FD FLIM measurements

The measurements from the FD FLIM method are commonly analyzed using the polar plot (also called a phasor plot) method. This method was originally developed as a way to analyze transient responses to repetitive perturbations, such as dielectric relaxation experiments, and can be applied to any system with frequency characteristics (Cole and Cole, 1941; Redford and Clegg, 2005). The polar plot directly displays the modulation fraction and the phase of the emission signal in every pixel in a FLIM image, allowing determination of the lifetime. The frequency characteristics at each image pixel are displayed with the coordinates $x = M(\omega) \times \cos \Phi(\omega)$ and $y = M(\omega) \times \sin \Phi(\omega)$. The relationship $x^2 + y^2 = x$ defines a "universal" semicircle with a radius of 0.5 that is centered at $\{0.5, 0\}$. This semicircle describes the lifetime trajectory for any single lifetime component, with longer lifetimes to the left (0, 0 is infinite lifetime) and shorter lifetimes to right (Redford and Clegg, 2005). Most important, the polar plot does not require a fitting model to determine fluorescence lifetime distributions, but rather expresses the overall decay in each pixel in terms of the polar coordinates on the universal semicircle. A population of fluorophores that has only one lifetime component will result in a distribution of points that fall directly on the semicircle. In contrast, a population of fluorophores with multiple lifetime components will have a distribution of points that fall inside the semicircle.

2.2. The calibration of a FLIM system

Before imaging biological samples, it is necessary to calibrate the FLIM system using a fluorescence lifetime standard. The fluorescence lifetime for many fluorophores has been established under standard conditions (an online source is available at: http://www.iss.com/resources/reference/data_tables/LifetimeDataFluorophores.html), and any of these probes can be used for calibration of the FLIM system. Since the fluorescence lifetime of a fluorophore is sensitive to its environment, it is critical to prepare the standards according to the conditions specified in the literature, including the solvent and the pH. It is also important to choose a standard fluorophore with excitation, emission, and fluorescence lifetime properties that are similar to those of the fluorophore used in the biological samples. For example, the dye Coumarin 6 dissolved in ethanol (peak excitation and emission of 460 and 505 nm, respectively), with a reference lifetime of ~ 2.5 ns, is often used as the calibration standard for the CFPs. It is important to note that if the excitation wavelength is changed, it is necessary to recalibrate with an appropriate lifetime standard.

2.2.1. Required materials
2.2.1.1. Device and materials for FLIM measurements Figure 19.2 shows the basic diagram of the digital FD FLIM system used here. The ISS ALBA FastFLIM system (ISS Inc., Champaign, IL) is coupled to an Olympus IX71 microscope equipped with a $60 \times /1.2$ NA water-immersion objective lens. A Pathology Devices (Pathology Devices, Inc., Exton, PA) stage top environmental control system is used to maintain temperature at $36°$ C and CO_2 at 5%. A 5 mW, 448-nm diode laser is modulated by the FastFLIM module of the ALBA system at the fundamental frequency of 20 MHz with up to seven sinusoidal harmonics. The modulated laser is coupled to the ALBA scanning system, which is controlled by ISS VistaVision software (http://www.iss.com/microscopy/software/vistavision.html). The fluorescence signals emitted from the specimen are routed by a 495 nm long-pass beam splitter through the 530/43 nm (channel 1, acceptor emission) and the 480/40 nm (channel 2, donor emission) band-pass emission filters. The pinholes for each channel are set at 50 micrometer, and the signals are then detected using two identical avalanche photodiodes (APDs). The phase delays and modulation ratios of the emission relative to the excitation are measured at six modulation frequencies (20, 40, 60, 80, 100, 120 MHz) for each pixel of an image.

2.2.1.2. Additional materials
- Coumarin 6 (Sigma-Aldrich Inc., cat. # 546283)
- HPTS (8-Hydroxypyrene-1,3,6-trisulfonic acid, trisodium salt from AnaSpec Inc., cat. # 84610)
- Chambered cover glass (4 or 8 well; Thermo Scientific, cat. # 155382, 155409).

2.2.2. Calibration and validation
The calibration of the system with the Coumarin 6 dye provides the software with the reference standard that will be used to estimate the lifetime values from the experimental data. It is essential that the calibration is done before each experiment, and it is prudent to repeat the calibration several times during the experiment. Additionally, we use a second reference standard to check that the system is accurately reporting the fluorescence lifetime of a known sample. Here, we use HPTS dissolved in phosphate buffer (PB) pH 7.8 (peak excitation and emission at 454 and 511 nm, respectively), which has an expected lifetime of 5.3 ns.

The measurement of the fluorescence lifetimes for these two different probes is illustrated in Fig. 19.3. Following calibration of the system, we

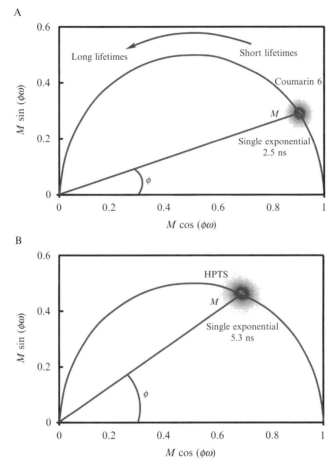

Figure 19.3 The polar plot analysis of the fluorescence lifetime of (A) Coumarin 6 in ethanol and (B) HPTS in PB, pH 7.8. The polar plot analysis was made using the first harmonic (20 MHz). (A) For Coumarin 6, the frequency characteristics for each pixel in the 256 × 256 image (65,536 points) are displayed on the polar plot with the coordinates $x = M(\omega) \times \cos \Phi(\omega)$, and $y = M(\omega) \times \sin \Phi(\omega)$. The vector length is determined from the modulation (M), and the phase delay (Φ) determines the angle. The centroid of the distribution falls on the universal semicircle, representing a single exponential lifetime of 2.5 ns. (B) For HPTS, each pixel in the 256 × 256 image is displayed on the polar plot, illustrating how the distribution for the longer lifetime probe is shifted to the left along the semicircle. The centroid of the distribution falls on the semicircle, indicating a single exponential lifetime of 5.3 ns.

used the FD FLIM method to determine the fluorescence lifetime of Coumarin 6 in ethanol. A 256 × 256 pixel FLIM image was obtained from a chambered cover glass containing 50 µM Coumarin 6 in ethanol. Frame averaging is used to accumulate approximately 100 counts per pixel, and the distribution of the lifetimes for all the pixels in the image (a total of 65,536 pixels) is represented on the polar plot, allowing determination of an average lifetime of 2.5 ns (Fig. 19.3A). The distribution falls directly on the semicircle, indicating a single lifetime component for Coumarin 6. To verify that the system is accurately calibrated, measurements are then made of a second dye solution containing 10 mM HPTS dissolved in PB. The polar plot for the HPTS sample, shown in Fig. 19.3B, clearly demonstrates that the position of its lifetime distribution is shifted to the left along semicircle relative to Coumarin 6 (compare Fig. 19.3A and B). Again, the distribution falls directly on the semicircle and indicates a single lifetime component of 5.3 ns for HPTS.

2.3. Testing the FLIM system using "FRET-standard" proteins

Before making FLIM measurements from biological samples expressing FP-labeled proteins of interest that have not been characterized in FRET-based assays, it is advisable to test the system (both the cellular model and the microscope system) using "FRET-standard" proteins. The Vogel laboratory (Koushik et al., 2006; Thaler et al., 2005) created genetic constructs that encoded mCerulean directly coupled to mVenus through protein linkers of different lengths. For example, they generated a genetic construct that encoded mCerulean separated from mVenus by a five amino acid linker (Cerulean-5AA-Venus), producing a fusion protein with consistently high FRET efficiency. Since the fluorescence lifetime is very sensitive to the probe environment, an identical genetic construct that encodes Cerulean linked to a chromophore mutant of Venus, called Amber, was also generated (Koushik et al., 2006). Amber is a non-fluorescent form of Venus where the chromophore tyrosine was changed to cysteine, producing a protein that folds correctly, but does not act as a FRET acceptor. The Cerulean-5AA-Amber protein recapitulates the local environment of the Cerulean donor fluorophore without the quenching that results from energy transfer.

The Vogel laboratory also made genetic constructs with a larger linkers. For example, a linker that encoded the 229-amino acid tumor necrosis factor receptor-associated factor (TRAF) domain was inserted between the mCerulean and mVenus proteins (Cerulean-TRAF-Venus). This produced a fusion protein with low FRET efficiency. They then used three different techniques to measure FRET in cells that expressed the "FRET-standard" fusion proteins. For each of the fusion proteins tested, there was consensus in the results obtained by the different FRET methods, demonstrating that these genetic constructs could serve as FRET standards (Thaler et al., 2005). Most important, these genetic constructs have been freely distributed to many other laboratories, where they are routinely used to verify and evaluate FRET measurements obtained in different experimental systems.

2.3.1. Transfection of cells by electroporation

The mouse pituitary GHFT1 cells (Lew *et al.*, 1993) used in our laboratory are efficiently transfected by electroporation. We established the following conditions by transfecting the cells with a plasmid containing the cytomegalovirus promoter driving the luciferase reporter gene (Promega, Madison, WI) using a range of different electroporation voltages. The highest level of luciferase protein expression in the GHFT1 cells was achieved at 200 V.

2.3.2. Required materials

2.3.2.1. Device and materials for electroporation There are several commercially available systems for gene transfer by electroporation. We use the BTX ECM 830 device (Harvard Apparatus, Holliston, MA)

- Mouse pituitary GHFT1 cells (Lew *et al.*, 1993) are maintained as monolayer cultures in Dulbecco's modified Eagle's medium (DMEM) containing 10% newborn calf serum (NBCS) at 37 °C in a 5% CO_2 incubator
- 0.05% Trypsin/0.53 mM ethylenediaminetetraacetic acid (EDTA) (Fisher Scientific Inc., cat. # MT25-051-Cl)
- 2-mm Gap electroporation cuvettes (BTX, cat. # 45-0125)
- Chambered cover glass (2 well; Thermo Scientific, cat. # 155379)
- Dulbecco's phosphate-buffered saline (PBS)—calcium and magnesium free (Fisher, Cellgro, cat. # 21-031-CV)
- BioBrene (Applied Biosystems Inc., cat. #400385)
- Imaging media: Phenol red-free F12:DMEM (1:1) containing 10% NBCS (Sigma, cat. # D2906-10x1L). Since phenol red may cause background fluorescence signals, it is important to use a phenol red-free medium during imaging
- Cerulean-5AA-Amber plasmid construct (provided by Steven S. Vogel at NIH/NIAAA, and serves as the donor-alone control for the FRET)
- Cerulean-5AA-Venus plasmid construct (provided by Steven S. Vogel at NIH/NIAAA, and is used for verifying FLIM–FRET)
- *Optional reagent:* Cerulean-TRAF-Venus plasmid construct (provided by Steven S. Vogel at NIH/NIAAA, and is used as a low FRET efficiency standard for FLIM–FRET)

2.3.3. Electroporation of GHFT1 cells with the FRET-standard constructs

2.3.3.1. Performing the electroporation The GHFT1 cells are maintained in 150-cm^2 culture flasks and are harvested at about 80% confluence (approximately 2×10^7 cells). The cells are washed with PBS and briefly treated with trypsin (0.05% in 0.53 mM EDTA). The trypsin–EDTA solution is removed, and the flasks are returned to the incubator. When cells begin to release from the surface of the flask (usually in about 5 min), they are recovered in culture medium containing serum. The cells are washed two times by centrifugation in Dulbecco's calcium–magnesium-free PBS. The cells are then resuspended in Dulbecco's calcium–magnesium-free PBS with 0.1% glucose and 0.1 ng/ml BioBrene at a final concentration of approximately 1×10^7 cells/ml. Exactly 400 μl of the cell suspension is transferred to each 0.2-cm gap electroporation cuvette containing the plasmid DNA. We typically use between 5 and 10 μg of purified plasmid DNA per cuvette. The contents of the cuvettes are gently mixed and then pulsed at 200 V in the BTX electroporator at a capacitance of 1200 μF, yielding pulse durations of 9–10 ms. The cells are immediately recovered from the cuvette and diluted in phenol red-free tissue culture medium containing serum. The suspension is used to inoculate a sterile 2-well-chambered cover glass (approximately 2 ml per chamber), which are then placed in an incubator (37 °C and 5% CO_2) overnight prior to imaging the following day.

2.4. FLIM measurements of the FRET-standard proteins

The FLIM imaging method is applied to the cells expressing the Cerulean-5AA-Amber (unquenched donor), Cerulean-TRAF-Venus (low FRET), or Cerulean-5AA-Venus (high FRET) "FRET-standard" proteins. The chambered coverglass with the transfected cells is placed into the stage top incubator to maintain temperature at 36° C and CO_2 at 5%. The cells expressing the

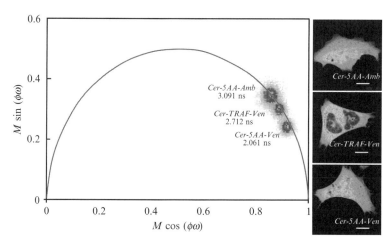

Figure 19.4 The polar plot analysis of the donor (Cerulean) lifetime for cells expressing Cerulean-5AA-Amber (unquenched donor), Cerulean-TRAF-Venus, or Cerulean-5AA-Venus (see text for details). The intensity image for each cell is shown in the right panels; the calibration bar indicates 10 μm. The lifetime distribution for all pixels in each image is displayed on the polar plot. The distribution of lifetime for Cerulean-5AA-Amber falls directly on the semicircle, indicating a single exponential decay with an average lifetime of 3.09 ns. In contrast, the distributions for the cells expressing Cerulean-TRAF-Venus and Cerulean-5AA-Venus fall inside the polar plot, indicating lifetime heterogeneity within the cells, with average quenched lifetimes of 2.71 and 2.06 ns, respectively.

fusion proteins are first identified under epi-fluorescence illumination. The power to the 448 nm-modulated laser is then adjusted to acquire a confocal image at the same count rate that was used to calibrate the system (section 2.2.2.). Representative results from individual cells expressing the indicated proteins are shown in Fig. 19.4. An intensity image (256 × 256 pixels) was acquired of each different cell expressing the indicated fusion protein (right panels, Fig. 19.4), and FD FLIM was used to determine the lifetime at every pixel of those images. The distribution of lifetimes for the three different cells is displayed on a common polar plot, allowing direct comparison of the donor lifetimes in each cell. The lifetime distribution for the cell expressing the Cerulean-5AA-Amber protein falls directly on the semicircle indicating it fits well to a single component lifetime of 3.09 ns (Fig. 19.4). In contrast, the average donor lifetime for the cell expression Cerulean-TRAF-Venus was 2.71 ns, while the average donor lifetime for Cerulean-5AA-Venus was 2.06 ns, indicating the quenching of Cerulean by Venus. Note that the lifetime distributions for the cells expressing the Cerulean-TRAF-Venus and Cerulean-5AA-Venus fusion proteins fall inside the semicircle (Fig. 19.4). This result indicates heterogeneity in the lifetimes for the Cerulean-Venus fusion proteins, which likely reflect the distribution of different donor–acceptor dipole orientations adopted by the expressed fusion proteins.

2.4.1. The interpretation of the FRET-standard FLIM data

As described above (Section 2.1), measuring the change in the donor fluorescence lifetime that occurs with energy transfer to an acceptor is the most direct method for quantifying FRET. The FRET efficiency (E_{FRET}) is determined from the ratio of the donor lifetime in the presence (τ_{DA}) and absence (τ_D) of acceptor (Equation 19.4).

$$E_{FRET} = 1 - \left(\frac{\tau_{DA}}{\tau_D}\right) \quad (19.4)$$

Table 19.1 FLIM–FRET measurements of the Cerulean "FRET-standard" fusion proteins

Fusion protein	τ_m (ns)[a]	E_{FRET} (%)[b]
Cer-5AA-Amber (n = 12)	3.05 ± 0.061	NA
Cer-TRAF-Venus (n = 10)	2.83 ± 0.061	7.2
Cer-5AA-Venus (n = 20)	2.18 ± 0.078	28.5
Cer-C/EBP B Zip (n = 10)	3.07 ± 0.073	NA
Cer-C/EBP B Zip + Ven-CEBP B Zip (n = 20)	2.78 ± 0.141 (range 2.5–3.0)	9.4 (range 2.2–18.6)

[a] ±SD.
[b] Determined by $E = 1 - (\tau_{DA}/\tau_D)$; see text.

The τ_D (unquenched donor lifetime) is determined from the cells that express only the donor-labeled protein, in this case the Cerulean-5AA-Amber. The τ_{DA} (donor lifetime quenched by the acceptor) is determined from the cells that express both the donor and acceptor; here, the fusion proteins with Cerulean directly linked to Venus. The donor fluorescence lifetime determined from populations of mouse GHFT1 cells expressing each of the FRET-standard fusion proteins (described above) is presented in Table 19.1. The average fluorescence lifetime of the unquenched donor (Cerulean-5AA-Amber) was 3.05 ns ($n = 12$ cells). On average, the donor lifetime for the Cerulean-TRAF-Venus fusion protein was shortened to 2.83 ns ($n = 10$ cells), corresponding to mean FRET efficiency of 7.2%. The average fluorescence lifetime for the Cerulean-5AA-Venus fusion protein was 2.18 ns ($n = 20$ cells), resulting in a mean FRET efficiency of 28.5%.

3. Measuring Protein–Protein Interactions in Living Cells Using FLIM–FRET

The biological model used here to demonstrate FLIM–FRET measurements is the basic region-leucine zipper (B Zip) domain of the C/EBPα transcription factor. C/EBPα acts to direct programs of cell differentiation and plays key roles in the regulation of genes involved in energy metabolism (Johnson, 2005). The B Zip family proteins form obligate dimers through their leucine-zipper domains, which positions the basic region residues for binding to specific DNA elements. Immunocytochemical staining of mouse cells showed that the endogenous C/EBPα protein was preferentially bound to satellite DNA-repeat sequences located in regions of centromeric heterochromatin (Schaufele *et al.*, 2001; Tang and Lane, 1999). The B Zip domain alone is necessary and sufficient for targeting the DNA-binding dimer to centromeric heterochromatin (Day *et al.*, 2001). The dimerized protein bound to well-defined structures in the mouse cell nucleus is an ideal tool to test detection of FLIM–FRET in a relevant biological model.

3.1. Construction of the C/EBPα B Zip domain plasmids and transfection for FLIM–FRET

The FP-labeled C/EBPα B Zip domain plasmid constructs were prepared using the cDNA sequence for the rat C/EBPα (Landschulz *et al.*, 1988). This cDNA was used as template for polymerase chain reaction (PCR) amplification of the carboxyl-terminal sequence encoding the B Zip domain using primers that incorporated restriction enzyme sites. The PCR product was purified and digested with *Bsp*EI and *Bam*HI, and this fragment was inserted into the mCerulean C1 vector or the mVenus C1 vector in the correct reading-frame starting with the methionine at position

237 and ending with the stop codon of C/EBPα. All vectors were confirmed by direct sequencing.

3.1.1. Required materials (in addition to those described in Section 2.3.2)

- Cerulean-B Zip plasmid construct (available from the Day laboratory, serves as the donor-alone control for the FRET)
- Venus-B Zip plasmid construct (available from the Day laboratory, when expressed with the Cerulean-B Zip, serves as a intermolecular FRET probe).

3.1.2. Transfection of GHFT1 cells with plasmids encoding the FP-labeled B Zip domain

3.1.2.1. Performing the electroporation The GHFT1 cells were transfected as described above (Section 2.3.1). Since the Cerulean-B Zip and Venus-B Zip plasmids are co-transfected into the cells, the proteins will be produced in the cells independently of one another. Therefore, the acceptor-to-donor ratio in the transfected cells will vary from cell to cell. The acceptor-to-donor ratio influences the FRET efficiency, with cells expressing high levels of donor relative to acceptor (i.e., most dimers are donor–donor pairs) having low FRET efficiency. To improve the detection of the proteins involved in FRET (i.e., dimers composed of donor–acceptor pairs), the Cerulean-B Zip and Venus-B Zip plasmid DNAs are mixed in the 0.2-cm gap electroporation cuvette at a 1:4 ratio using a total of 10 μg of purified plasmid DNA per cuvette. The handling of the cells after transfection is the same as described above (Section 2.3.1).

3.2. FLIM measurements to detect dimerized B Zip proteins

The FLIM imaging approach was used to determine the fluorescence lifetime of the Cerulean-B Zip (unquenched donor) fusion protein localized in the nuclei of the mouse GHFT1 cells. The results for an individual cell are shown in Fig. 19.5A, demonstrating the localization of the B Zip protein to regions of centromeric heterochromatin. The FD FLIM was used to determine the lifetime distribution from every pixel of the image, indicating an average lifetime of 3.1 ns for this cell, with an average lifetime of 3.07 ns determined for 10 different cells (Table 19.1). When co-expressed with the Venus-B Zip fusion protein, the Cerulean-B Zip lifetime was shortened (single cell measurement in Fig. 19.5A). The average lifetime for Cerulean-B Zip in the presence of Venus-B Zip for the population of cells was 2.78 ns, yielding a mean FRET efficiency of 9.4% (Table 19.1). As noted above, the acceptor-to-donor ratio varies for each transfected cell, leading to a broad range of measured FRET

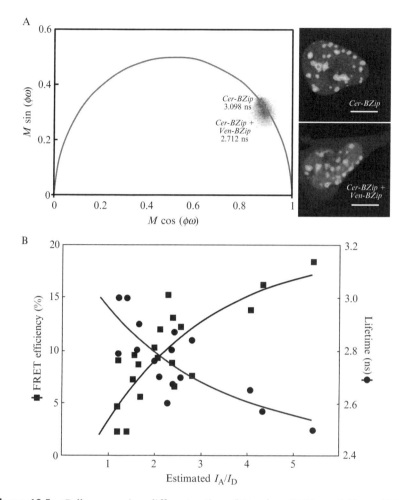

Figure 19.5 Cells expressing different ratios of Cerulean-B Zip and Venus-B Zip were subjected to FLIM–FRET, and the data were analyzed as described in the - Section 3.2. (A) Single cell measurements for a cell expressing the Cerulean-B Zip protein alone (unquenched donor, top right panel; the calibration bar is 10 μm) or a cell co-expressing the Cerulean- and Venus-B Zip proteins (lower right panel). The lifetime distribution for all pixels in each image is displayed on the common polar plot. (B) The average FRET efficiency (left axis) and donor fluorescence lifetime (right axis) were determined for the cell nucleus for 20 cells separate cells, and the results are plotted as a function of the estimated acceptor-to-donor ratio (I_A/I_D) for each cell.

efficiencies when measured in many different cells (2.2–18.6%, $n = 20$; see Table 19.1).

Since FLIM measurements are made in the donor channel (see Fig. 19.2), only the donor–donor and donor–acceptor pairs are detected (the acceptor–acceptor pairs are detected in the separate acceptor channel).

Therefore, as the acceptor-to-donor ratio is increased, more of the B Zip dimers detected in the donor channel will be Cerulean-B Zip and Venus-B Zip pairs. The highest FRET efficiencies (lowest donor lifetimes) are achieved in cells with a highest level of acceptor protein relative to the donor protein. This relationship is illustrated in Fig. 19.5B, where the FRET efficiencies (■) and donor fluorescence lifetimes (●) are plotted as a function of the estimated acceptor-to-donor ratio (I_A/I_D) for 20 different cells. This result demonstrates the importance of determining the relative donor and acceptor levels in FRET experiments involving independently expressed fusion proteins.

4. THE STRENGTHS AND LIMITATIONS OF FLIM

Fluorescence lifetime measurements are insensitive to factors that commonly limit measurements by intensity-based imaging. For example, FLIM measurements are not affected by changes in fluorophore concentration, excitation intensity, or light scattering. Further, since the fluorescence lifetime of a fluorophore is sensitive to its environment, FLIM can be a good choice for visualizing signal changes from biosensor probes that report ion binding, pH, or protein phosphorylation. In this regard, the genetically encoded FRET-standard proteins (Section 2.3) are useful tools for optimizing cell culture conditions for FRET measurements and for evaluating imaging systems for the detection of FRET. The low FRET efficiency standards are especially useful for assessing the background noise in the system.

However, for FRET-based experiments, it is important to recognize that any amount of exogenous protein that is produced in a cell is by definition overexpressed relative to its endogenous counterpart. The transfection approach can yield very high levels of the fusion proteins in the target cells, especially when strong promoters are used. This can result in improper protein distribution and protein dysfunction that could lead to erroneous interpretations of protein activities. It is critical to verify that the fusion proteins retain the functions of the endogenous protein, and that they have the expected subcellular localization. However, even with careful assessment of FP–fusion protein function and biochemical demonstrations of protein interactions, false-negative results are common in FRET imaging studies, because it can be difficult to achieve the required spatial relationships for FRET. Further, positive FRET results from single cells are, by themselves, not sufficient to characterize the associations between proteins in living cells. Although the FRET measurements, when collected and quantified properly, are remarkably robust, there is still heterogeneity in the measurements. As was shown here (Section 3.2), the data must be collected and statistically analyzed from multiple cells to prevent the user from reaching false conclusions from a non-representative measurement.

Because the fluorescence lifetime is exquisitely sensitive to probe environment, the fluorescence lifetime will change unpredictably for probes in fixed specimens, so FLIM is limited to live specimens. It is critical to identify the sources of noise in FRET–FLIM measurements to determine the reliability of the data analysis. For instance, a donor fluorophore whose intrinsic lifetime has multiple components may not be suitable for FLIM–FRET, since it will complicate the data analysis. Here, the FRET-standard proteins are valuable tools, since they should report the same range of FRET signals each time they are used, and will effectively reveal problems in the imaging system. Although the analysis of FLIM data has become routine with the advanced software that is available, an understanding of the physics that underlies the changes in fluorescence lifetime is necessary for processing of the FLIM data and the interpretation of the results. Finally, the acquisition of FRET–FLIM data is typically slow, depending on the expression level of the labeled proteins. For example, acquiring sufficient photon counts to assign lifetimes using the FD method described above required about 30 s, which limits its application for monitoring dynamic events. As the technology improves, it is expected that the FLIM data acquisition time will decrease (Buranachai *et al.*, 2008).

ACKNOWLEDGMENT

The authors acknowledge funding from NIH 2RO1DK43701 and 3RO1 DK43701-15S1 (R.N.D.), and the University of Virginia and National Center for Research Resources NCRR-NIH RR027409 (A.P.). The authors thank Dr. Steven Vogel (NIH/NIAAA) for providing the FRET-standard constructs, and Drs. Beniamino Barbieri and Shih-Chu Liao (ISS) for valuable feedback.

REFERENCES

Aye-Han, N. N., Ni, Q., and Zhang, J. (2009). Fluorescent biosensors for real-time tracking of post-translational modification dynamics. *Curr. Opin. Chem. Biol.* **13,** 392–397.

Berney, C., and Danuser, G. (2003). FRET or no FRET: A quantitative comparison. *Biophys. J.* **84,** 3992–4010.

Buranachai, C., Kamiyama, D., Chiba, A., Williams, B. D., and Clegg, R. M. (2008). Rapid frequency-domain FLIM spinning disk confocal microscope: Lifetime resolution, image improvement and wavelet analysis. *J. Fluoresc.* **18,** 929–942.

Clegg, R. M. (2010). Fluorescence lifetime-resolved imaging what, why, how—A prologue. *In* "FLIM Microscopy in Biology and Medicine," (R. M. Clegg and A. Periasamy, eds.), pp. 3–34. CRC Press, London.

Cole, K. S., and Cole, R. H. (1941). Dispersion and absorption in dielectrics. I. Alternating current characteristics. *J. Chem. Phys.* **9,** 341–351.

Colyer, R. A., Lee, C., and Gratton, E. (2008). A novel fluorescence lifetime imaging system that optimizes photon efficiency. *Microsc. Res. Tech.* **71,** 201–213.

Day, R. N., and Davidson, M. W. (2009). The fluorescent protein palette: Tools for cellular imaging. *Chem. Soc. Rev.* **38,** 2887–2921.

Day, R. N., Periasamy, A., and Schaufele, F. (2001). Fluorescence resonance energy transfer microscopy of localized protein interactions in the living cell nucleus. *Methods* **25,** 4–18.

Forster, T. (1965). Delocalized excitation and excitation transfer. *In* "Modern Quantum Chemistry," (O. Sinanoglu, ed.) Vol. 3, pp. 93–137. Academic Press, New York.

Giepmans, B. N., Adams, S. R., Ellisman, M. H., and Tsien, R. Y. (2006). The fluorescent toolbox for assessing protein location and function. *Science* **312,** 217–224.

Griesbeck, O., Baird, G. S., Campbell, R. E., Zacharias, D. A., and Tsien, R. Y. (2001). Reducing the environmental sensitivity of yellow fluorescent protein. Mechanism and applications. *J. Biol. Chem.* **276,** 29188–29194.

Gryczynski, Z., Gryczynski, I., and Lakowicz, J. R. (2005). Basics of fluorescence and FRET. *In* "Molecular Imaging: FRET Microscopy and Spectroscopy," (A. Periasamy and R. N. Day, eds.), pp. 21–56. Oxford University Press, New York.

Jares-Erijman, E. A., and Jovin, T. M. (2003). FRET imaging. *Nat. Biotechnol.* **21,** 1387–1395.

Johnson, P. F. (2005). Molecular stop signs: Regulation of cell-cycle arrest by C/EBP transcription factors. *J. Cell Sci.* **118,** 2545–2555.

Koushik, S. V., Chen, H., Thaler, C., Puhl, H. L., 3rd, and Vogel, S. S. (2006). Cerulean, Venus, and VenusY67C FRET reference standards. *Biophys. J.* **91,** L99–L101.

Lakowicz, J. R. (2006). Principles of Fluorescence Spectroscopy. 3rd edn. Springer, New York.

Landschulz, W. H., Johnson, P. F., Adashi, E. Y., Graves, B. J., and McKnight, S. L. (1988). Isolation of a recombinant copy of the gene encoding C/EBP. *Genes Dev.* **2,** 786–800.

Lew, D., Brady, H., Klausing, K., Yaginuma, K., Theill, L. E., Stauber, C., Karin, M., and Mellon, P. L. (1993). GHF-1-promoter-targeted immortalization of a somatotropic progenitor cell results in dwarfism in transgenic mice. *Genes Dev.* **7,** 683–693.

Miyawaki, A. (2011). Development of probes for cellular functions using fluorescent proteins and fluorescence resonance energy transfer. *Annu. Rev. Biochem.* **80,** 357–373.

Nagai, T., Ibata, K., Park, E. S., Kubota, M., Mikoshiba, K., and Miyawaki, A. (2002). A variant of yellow fluorescent protein with fast and efficient maturation for cell-biological applications. *Nat. Biotechnol.* **20,** 87–90.

Periasamy, A., and Clegg, R. M. (2010). FLIM Microscopy in Biology and Medicine. Taylor & Francis, Boca Raton.

Periasamy, A., and Day, R. N.American Physiological Society (1887-) (2005). Molecular Imaging: FRET Microscopy and Spectroscopy. Published for the American Physiological Society byOxford University Press, Oxford; New York.

Piston, D. W., and Kremers, G. J. (2007). Fluorescent protein FRET: The good, the bad and the ugly. *Trends Biochem. Sci.* **32,** 407–414.

Redford, G. I., and Clegg, R. M. (2005). Polar plot representation for frequency-domain analysis of fluorescence lifetimes. *J. Fluoresc.* **15,** 805–815.

Rizzo, M. A., Springer, G. H., Granada, B., and Piston, D. W. (2004). An improved cyan fluorescent protein variant useful for FRET. *Nat. Biotechnol.* **22,** 445–449.

Schaufele, F., Enwright, J. F., 3rd, Wang, X., Teoh, C., Srihari, R., Erickson, R., MacDougald, O. A., and Day, R. N. (2001). CCAAT/enhancer binding protein alpha assembles essential cooperating factors in common subnuclear domains. *Mol. Endocrinol.* **15,** 1665–1676.

Shaner, N. C., Steinbach, P. A., and Tsien, R. Y. (2005). A guide to choosing fluorescent proteins. *Nat. Methods* **2,** 905–909.

Stryer, L. (1978). Fluorescence energy transfer as a spectroscopic ruler. *Annu. Rev. Biochem.* **47,** 819–846.

Sun, Y., Wallrabe, H., Booker, C. F., Day, R. N., and Periasamy, A. (2010). Three-color spectral FRET microscopy localizes three interacting proteins in living cells. *Biophys. J.* **99,** 1274–1283.

Sun, Y., Wallrabe, H., Seo, S. A., and Periasamy, A. (2011). FRET microscopy in 2010: The legacy of Theodor Forster on the 100th anniversary of his birth. *Chemphyschem* **12,** 462–474.

Tang, Q. Q., and Lane, M. D. (1999). Activation and centromeric localization of CCAAT/ enhancer-binding proteins during the mitotic clonal expansion of adipocyte differentiation. *Genes Dev.* **13,** 2231–2241.

Thaler, C., Koushik, S. V., Blank, P. S., and Vogel, S. S. (2005). Quantitative multiphoton spectral imaging and its use for measuring resonance energy transfer. *Biophys. J.* **89,** 2736–2749.

Tsien, R. Y. (1998). The green fluorescent protein. *Annu. Rev. Biochem.* **67,** 509–544.

Tsien, R. Y. (2005). Indicators based on fluorescence resonance energy transfer. *In* "Imaging in Neuroscience and Development," (R. Yuste and A. Konnerth, eds.), pp. 549–556. Cold Spring Harbor Laboratory Press, Cold Spring Harbor, NY.

Venetta, B. D. (1959). Microscope phase fluorometer for determining the fluorescence lifetimes of fluorochromes. *Rev. Sci. Instrum.* **30,** 450–457.

Vogel, S. S., Thaler, C., and Koushik, S. V. (2006). Fanciful FRET. *Sci. STKE* **2006,** re2.

Wouters, F. S., and Bastiaens, P. I. (1999). Fluorescence lifetime imaging of receptor tyrosine kinase activity in cells. *Curr. Biol.* **9,** 1127–1130.

Yasuda, R. (2006). Imaging spatiotemporal dynamics of neuronal signaling using fluorescence resonance energy transfer and fluorescence lifetime imaging microscopy. *Curr. Opin. Neurobiol.* **16,** 551–561.

Zhang, J., Campbell, R. E., Ting, A. Y., and Tsien, R. Y. (2002). Creating new fluorescent probes for cell biology. *Nat. Rev. Mol. Cell Biol.* **3,** 906–918.

CHAPTER TWENTY

Open Source Tools for Fluorescent Imaging

Nicholas A. Hamilton

Contents

1. Why Open Source Software?	394
1.1. Flexible, innovative, and verifiable	394
1.2. Modularity, reuse, and interoperability	396
1.3. Documentation, support, and quality	401
1.4. Common platforms for bio-image analysis to build upon	402
2. Open Source Software for Microscopy Imaging and Analysis	406
2.1. Image capture	407
2.2. Storage	408
2.3. Image processing and quantification	409
2.4. High-throughput pipelines	410
2.5. Automated image classification	411
2.6. Visualization, data analysis, and modeling	412
3. The Future	414
Acknowledgments	415
References	415

Abstract

As microscopy becomes increasingly automated and imaging expands in the spatial and time dimensions, quantitative analysis tools for fluorescent imaging are becoming critical to remove both bottlenecks in throughput as well as fully extract and exploit the information contained in the imaging. In recent years there has been a flurry of activity in the development of bio-image analysis tools and methods with the result that there are now many high-quality, well-documented, and well-supported open source bio-image analysis projects with large user bases that cover essentially every aspect from image capture to publication. These open source solutions are now providing a viable alternative to commercial solutions. More importantly, they are forming an interoperable and interconnected network of tools that allow data and analysis methods to be shared between many of the major projects. Just as researchers build on,

Division of Genomics Computational Biology, Institute for Molecular Bioscience, The University of Queensland, St. Lucia, Brisbane, Queensland, Australia transmit, and verify knowledge through publication, open source analysis methods and software are creating a foundation that can be built upon, transmitted, and verified. Here we describe many of the major projects, their capabilities, and features. We also give an overview of the current state of open source software for fluorescent microscopy analysis and the many reasons to use and develop open source methods.

1. Why Open Source Software?

Quantitative and automated analysis methods are becoming essential both to keep pace with the scale of the data now being generated by modern microscopes as well as to extract the maximum information and meaning from the imaging data. For instance, automated fluorescent microscope imaging technologies enable the experimental determination of a protein's localization and its dynamic trafficking within a range of contexts. These approaches generate vast numbers of images including multiple fluorophores under a variety of experimental conditions and are leading to a rapid growth in bio-image data sets in need of analysis on a scale comparable to that of the genomic revolution (Murphy, 2006; Wollman and Stuurman, 2007). In any rapidly developing field requiring computational support, software to implement new methodologies is inevitably a significant problem.

Many researchers may be initially drawn to open source software because of cost. It should be noted that open source software is not necessarily distributed for free, though the majority probably is. Commercial image analysis and storage solutions can cost thousands or tens of thousands of dollars with regular renewals for version upgrades and support. While cost can be a serious concern, particularly in developing countries, there are a wide range of reasons to want to use and contribute to open source bio-image analysis software. With high-quality open source software now available to perform every step from image acquisition through to publishable results, projects such as ImageJ, µManager, Cell Profiler, and the Open Microscopy Environment are now beginning to provide a powerful alternative with many other advantages.

1.1. Flexible, innovative, and verifiable

Microscopy imaging hardware as well as fluorophore technologies are advancing rapidly. To keep pace with these developments as well as to fully exploit and realize the potential of the imaging data there needs to be a parallel set of rapid developments in analysis and quantification. Indeed, new analysis methods are continually extending the range of information that can be extracted from fluorescent imaging (Hamilton, 2009).

One significant advantage of open source is that it is rapidly adaptable. In developing a new analysis protocol, existing software will often *nearly* do what is required but not quite provide an adequate solution. If a particular parameter in an existing analysis algorithm can be accessed and "tweaked," then it might enable the required information extract to be extracted. With open source code and some knowledge of the programming language it is usually straightforward to modify and incorporate changes to the code and hence the algorithms being applied to the imaging. While code modification may be beyond the expertise of many users there are a growing number of researchers with these skills who can then distribute modifications and improvements that they have made. And as is noted below, there are many skill levels at which useful contributions can be made.

Another point is that new analysis methods might initially or indeed always have only a small niche user base. Hence, it may not be financially viable for a commercial image analysis company to first identify a need for a small group of users, then develop, test, debug, and release supporting software. Even if it is financially viable it might take a year or more for a product to appear. In general the researcher is best placed to identify needs and new methods required and given the right tools, develop them. Many high-quality packages for the ImageJ image processing software have been produced and documented by groups with a special interest (see below).

Verifiability of methods is also a significant issue with scientific research. Closed source software is in many ways a black box. For a journal publication, it would not be acceptable to not fully describe the experimental protocol used. The difficulty with closed source software is twofold. First, to verify the analysis in a paper the (potentially very expensive) software used is required. Second, it is difficult to know exactly what the software is doing. Again, this is on two levels. First, the description in the software manual of a particular method will typically be at such a high level that it does not capture the numerous design decisions that can impact on the result that need to be made when implementing even simple algorithms. Second, all software contains bugs at some level; it is only a question of whether the bugs it contains impact on the analysis or not. Many methods of software development, including those for safety critical systems, have been developed to try and minimize software bugs with varying degrees of success. However, with open source software the code is typically freely available, the exact implementation of the algorithm can be seen in the code, and bugs may be reduced by having "many eyes" looking at the code. Hence, a greater degree of verifiability and reproducibility is achieved.

Just as researchers build on, transmit, and verify knowledge through publication, it is essential that analysis methods and software can be built upon, transmitted, and verified through common open source projects such as are described in the following.

1.2. Modularity, reuse, and interoperability

A common feature of many of the larger open source bio-image software is a foundation in the Java programming language. While not open source, Java is freely available and has excellent cross-platform support so that code created on one system will typically run on all of MacOSX, Linux, and Windows. It also has extensive advanced libraries such as OpenGL for 3D graphics and visualization. Another advantage of Java is that, it can be used to create web applets which can then offer a service or functionality via a web page. Hence, ImageJ, Fiji, VisBio, OMERO, Bio-Formats, μManager (see Table 20.1; Section 2), and many other projects are coded in Java.

As an object-orientated language where data and methods are encapsulated into objects, Java encourages modular programming. This means that image analysis methods can often either be directly utilized in one package from another or be reused with minor code modifications.

Another language that is becoming a standard in open source bio-imaging is Python (Python, 2011). While initially viewed as a simple scripting language and Perl replacement, Python has many features that make it attractive to use. It is relatively simple to learn in comparison with many other languages and quick to program or prototype in; it has a wide variety of libraries for computation and visualization; and language bridges can be used to implement computationally intensive algorithms in fast compiled languages such as C. In the Fiji image processing software Python is used as an intermediate level of programming between simple macros and more complex Java plug-ins. This is via Jython (Jython, 2011), a Java implementation of Python 2.1 that enables Python scripts to be run in Java programs. Version 2 of the CellProfiler high-throughput image analysis software (Kamentsky *et al.*, 2011) has been largely written in Python but exploits a Python/Java bridge so that the Bio-Formats libraries for reading and writing a large number of image formats can be seamlessly used.

These common languages, language bridges, as well as common data standards mean that a high degree of interoperability has been enabled between the major projects. Hence, CellProfiler has an interface to ImageJ so that the many image processing options available in ImageJ can be used to process high-throughput imaging; an ImageJ plug-in can download and upload images to and from the OMERO image database; there is a link between OMERO and CellProfiler so that images from OMERO can be input to CellProfiler, processed, and returned to OMERO; the μManager microscope control system is actually available as an ImageJ plug-in; the Bisque database has an ImageJ plug-in, BQBisqueJ, that allows direct access to the Bisque database system from ImageJ; nearly, every bio-image software project uses the Bio-Formats image library for image reading and writing; and so on. Being open source it is straightforward for the major projects to cooperate to ensure the seamless movement of data and methods between each environment.

Table 20.1 A selection of open source software tools for microscopy image analysis, storage, visualization, and modeling

Name	OS	Notes	Reference	URL
Micro-manager	Win/Mac/Linux	Microscope control and image capture.	Stuurman et al. (2007)	tiny.cc/aysd6
OMERO	Win/Mac/Linux	Client-server storage, visualization, and analysis of microscopy imaging. Includes a metadata model to describe imaging experiments. ImageJ client plug-in available.	Swedlow et al. (2009)	tiny.cc/2ih64
Bisque	Web-based Win/Mac/Linux	Image management, quantitative analysis, and visualization with a semantic data foundation.	Kvilekval et al. (2009)	tiny.cc/iz87f
Bio-Formats	Win/Mac/Linux	Java library for reading and writing bio-image files. Used in many open source projects such as Bisque, V3D, and ImageJ.	Swedlow et al. (2009)	tiny.cc/2ih64
ImageJ	Win/Mac/Linux	Image quantification and analysis with a very large number of plug-ins. Many online tutorials and a very active user community.	Abramoff et al. (2004)	tiny.cc/kh0oc

(Continued)

Fiji	Win/Mac/Linux	Based on ImageJ but organized differently and with a large number of selected high-quality plug-ins preinstalled.		fiji.sc
FARSIGHT TOOLKIT	Win/Mac/Linux	Sophisticated collection of software modules, which can be tied together with Python scripts, to preprocess segment based on morphology, analyze, and visualize multi-fluorophore 2D, 3D, and time-based imaging.	Bjornsson *et al.* (2008)	tiny.cc/q56oo
ICY	Win/Mac	A new image quantification and analysis package with advanced 3D visualization via the VTK libraries. Has a plug-in architecture and can use some ImageJ plug-ins.		tiny.cc/phb09
CellProfiler	Win/Mac/Linux	High-throughput image analysis pipelining. Can utilize ImageJ functions.	Carpenter *et al.* (2006)	tiny.cc/srnrh
CellProfiler Analyst	Win/Mac/Linux		Jones *et al.* (2008)	tiny.cc/srnrh

Table 20.1 (*Continued*)

Name	OS	Notes	Reference	URL
Micro-Pilot	Win	Data exploration and machine classification for cell phenotypes. Cell type/phenotype identification and classification. Relies on LabView.	Conrad et al. (2011)	tiny.cc/k3d6g
Murphy Lab	Various	A range of software for automated phenotype classification, clustering, and other applications.	Murphy (2006)	tiny.cc/ru74w
Cell ID	Win/Linux	Cell image quantification, tracking, and analysis.	Gordon et al. (2007)	tiny.cc/p3a35
BioView3D	Win/Mac/Linux	Visualization of multi-fluorophore 3D image stacks.	Kvilekval et al. (2009)	tiny.cc/hqvxs
V3D	Win/Mac/Linux	Very large bio-image and surface visualization and analysis of 3D/4D/5D data. Extendable via a plug-in interface.	Peng et al. (2009, 2010)	tiny.cc/mndq0
IMOD	Win/Mac/Linux	Interactive generation of 3D surface models from 3D imaging.	Kremer et al. (1996)	tiny.cc/42jpk

(*Continued*)

iCluster	Win/Mac/Linux	High-throughput image and data visualization.	Hamilton et al. (2009)	tiny.cc/ovif0
ImageSurfer	Win	Visualize and analyze 3D multichannel imaging.	Feng et al. (2007)	tiny.cc/6idn3
Virtual Cell	Web based	Mathematical modeling and simulation where the topology can be based on experimental imaging. Not strictly open source, but freely available as a Java web application.		vcell.org
Octave	Win/Mac/Linux	Interpreted language for numerical and mathematical computation. Similar to Matlab.	Eaton (2002)	tiny.cc/68b0p
R	Win/Mac/Linux	Statistical analysis and data visualization.	R Development Core Team (2008)	tiny.cc/nvl5a
Tulip		Advanced relational data visualization	Delest et al. (2003)	tiny.cc/59a0d

1.3. Documentation, support, and quality

In the past open source software has sometimes suffered from poor documentation or support. Indeed, after an initial flurry of interest in a new software method, authors may change research focus, lose interest in the project, or no longer have the time, incentive, or financial resources to maintain or distribute the software. However, this is changing. Online open source software repositories such as Sourceforge (http://sourceforge.net) and indexes such as Freshmeat (http://freshmeat.net) mean it is easy to create, version control, and distribute open source software. Such repositories ensure that software is archived and always available, and often report recent activity or ratings of a project which can be useful for evaluation before downloading and installing. Further, if the author can no longer develop the software it can easily be taken up and developed or modified by the next interested party. As Eric S. Raymond said in his seminal discussion of open source, The Cathedral and the Bazaar, "when you lose interest in a program, your last duty to it is to hand it onto a competent successor" (Raymond, 2000). Hence, the problem of distribution and archiving of open source software has largely been solved.

Maintenance, bug fixes, and continued development can still be a problem. However, funding agencies such as NIH and larger institutes are realizing the importance of continued support and funding for software initiatives. This has ensured that many of the major projects have a strong commitment to continuing in the future and the user can be reassured that their investment in learning a system will not be wasted. As described above, these projects are forming stable common platforms to which users can contribute new methods and extensions via plug-ins and macros, and their contribution can be preserved in the project archives.

Documentation and support has also greatly improved in recent times, though it can still be variable in quality. Larger projects such as ImageJ and CellProfiler have extensive online manuals, documentation, online tutorials, video tutorials, example data, training courses, conferences, and very active user communities. These often cater to a range of expertise levels from new user to active developer. The user communities are an invaluable resource. With large searchable archives of previous questions and frequently asked questions (FAQs), most problems can be solved by searching online user communities with selected keywords. And with user communities spread across the world, questions posted will sometimes be answered in a matter of a few hours. With active and engaged user communities, bug reports are generated quickly and since the code is open, suggested fixes or code patches may be submitted with the report. This leads to faster bug fix and release cycles and higher quality software.

Broadly, the success of GNU Linux and many other GNU projects has demonstrated that high-quality and stable open source software can be developed and maintained over long time periods. While there have not been many direct comparisons of features of open source to commercial bio-image analysis software, though in Mittag et al. (2011) it was shown that CellProfiler gives comparable results to commercial cytometry software; there can be little doubt that the larger open source projects now offer comparable documentation, features, and reliability to the commercial solutions.

1.4. Common platforms for bio-image analysis to build upon

As noted above, an important part of the scientific process is to be able to verify and build upon the research of others. One of the great benefits of many of the open source bio-image projects is that they enable researchers to contribute at various levels of expertise.

At the simplest level, many open source bio-image projects such as ImageJ, Fiji, Bisque, Micro-manager, and IMOD support macros, that is, a simple text description of a sequence of actions that can be run by the software. Often macros can be recorded as the user uses menus to apply a sequence of operations on some images and replayed to automate processing of a large number of images. ImageJ has a particularly well-documented and powerful macro language that is easy to learn and can be written directly by the user as well as recorded. Most macro languages are quite similar in syntax and structure meaning that skills learnt in one macro language are usually transferable others. Besides automation, macros are very useful for establishing a fixed analysis protocol to ensure identical processing for comparison.

Another approach to creating analysis protocols is through menus and simple user interfaces. While in the past open source software has sometimes suffered from bad or absent user interface design, there have been significant improvements. Hence, CellProfiler (see below) offers complete high-throughput cell analysis pipeline creation via simple menu selections and parameter entry. Such pipelines can then be saved to disk, distributed to other users, or added as supplementary material to a paper, thus increasing verifiability. Similarly, ImageJ now has a macro user interface plug-in (see Table 20.2) that allows image analysis pipelines to be created by dragging, dropping, and connecting boxes that process the images. This greater attention to interfaces can perhaps be viewed as part of the growing acknowledgment of the importance of data visualization within bio-imaging and the biosciences (Walter et al., 2010).

The more advanced level of extension is through plug-ins. Many projects such as ImageJ, CellProfiler, V3D, and Bisque have a modular architecture that supports user development of plug-ins to add new functionality and methods. Typically a plug-in is written in a high-level programming language such Java or C. The advantage here is that plug-ins can often

Table 20.2 A selection of plug-ins available for ImageJ and Fiji

Plug-in name	Description	Reference	URL
JACoP	All of the major colocalization methods in a single plug-in.	Bolte and Cordelieres (2006)	tiny.cc/sy9vo
OBCOL	3D multi-fluorophore automated object segmentation and colocalization analysis.	Woodcroft et al. (2009)	tiny.cc/b6qb0
Segmentation Editor	Manual segmentation of 3D imaging.		tiny.cc/sdkgy
Trainable Segmentation	The plug-in learns from user selected/classed regions of multiple classes and classifies regions accordingly.		tiny.cc/yjxic
Tubeness	Filters a 2D or 3D image to enhance tubular structures such as neurons or subcellular tubules.	Sato et al. (1998)	tiny.cc/fzty0
Watershed	Segmentation into regions/particles based on a "flooding" algorithm.		tiny.cc/pf047
Active Snakes	Semiautomated segmentation/region selection by active contours.		tiny.cc/n1668
Particle Tracker	Track multiple point-like structures in 2D video-microscopy.	Sbalzarini and Koumoutsakos (2005)	tiny.cc/4fild
Spot Tracker	Track a particle over time in noisy imaging.	Sage et al. (2005)	tiny.cc/m9ao5
Analyze Particles	Count discrete object in imaging and filter the results based on size and shape.		tiny.cc/ghypr
FRETCalc	Analysis of FRET by acceptor photobleaching.	Stepensky (2007)	tiny.cc/4mjj0

(Continued)

Name	Description	Reference	URL
StackReg	Register/align the images in a stack recursively starting from a reference image. Very useful to align and quantify regions of interest.	Thevenaz et al. (1998)	tiny.cc/6kspz
3D Viewer	Visualize multi-fluorophore 3D imaging in 3D.		tiny.cc/t5bqw
Auto Threshold & Local Threshold	Select foreground/background objects using some 16 different auto-threshold schemes for each plug-in. View results simultaneously to select the most appropriate to the imaging.		tiny.cc/0cita
Calculator Plus	Add, subtract, divide, or multiply images.		tiny.cc/t79dh
QuickPALM	Super-resolution particle detection and image reconstruction from PALM & STORM imaging in 2D/3D/4D.	Henriques et al. (2010)	tiny.cc/z0zni
FlowJ	Detect and quantify optical flows in imaging.	Abramoff et al. (2000)	tiny.cc/2fesu
MosaicJ	Create mosaic images from multiple overlapping images.	Thévenaz and Unser (2007)	tiny.cc/j44og
TransformJ	Geometric transformations on up to 4D imaging.		tiny.cc/7z16t
FeatureJ	Many methods of feature and edge detection such as derivatives, Laplacians, and Canny edge detection.		tiny.cc/vgf6n
Sync Windows	Synchronized mouse position in multiple windows. Surprisingly useful.		tiny.cc/q8v17

Table 20.2 (Continued)

Plug-in name	Description	Reference	URL
Straighten	Straighten images of curved lines.	Kocsis et al. (1991)	tiny.cc/uob96
TrakEM2	Sophisticated plug-in to segment, create mesh models, describe spatial relations, annotate, and register 3D imaging.	Schmid et al. (2010)	tiny.cc/ei22w
ImageFlow	Create macros using a drag/drop/connect interface.		tiny.cc/3t9kl
Documentation & Tutorials	Many tutorials and documents on using ImageJ, writing macros and plug-ins and examples.		tiny.cc/fsan0

utilize much more of the underlying functionality of the core package and will run faster than a macro. However, the cost is that the user needs to have relatively advanced programming skills.

A plug-in architecture enables a much less centralized model of development. Researchers with specialized interests can utilize the core functionality of packages such as ImageJ, thus avoiding reinventing the wheel, and create advanced additions such as Spot Tracker to automatically track a moving spot in noisy imaging (see Table 20.2). The plug-in approach reiterates the decentralized software development model that has proved so successful in the GNU Linux operating system and as well as many of the major projects that build on GNU Linux (Raymond, 2000).

While developing programming skills can take a substantial investment in time, the fact that Java and Python are somewhat standards in the field means that skills are very transferable. Many bioinformatics courses now provide extensive training in these languages with the result that there is an increasing pool of researchers with these skills as well as understanding of the biological problems to which they are being applied. Even without formal training, much can be learnt from reading and adapting other researcher's open source code.

Projects such as ImageJ, CellProfiler, and OMERO are becoming accumulation points for methods and knowledge in the bio-image analysis field. They are enabling contributions from researchers at all levels from new users to advanced bio-imaging programmers, and providing a verifiable, distributable foundation on which to build. Further, substantial cooperation between many of the projects is leading to increasing interoperability of methods, data, and metadata. Together these projects are forming an increasingly advanced operating system for bio-image informatics.

2. OPEN SOURCE SOFTWARE FOR MICROSCOPY IMAGING AND ANALYSIS

It is worth noting that there are a number of different open source license types, each with distinct features. Several are described in Swedlow and Eliceiri (2009). However, from the point of view of the researcher wishing to use software for noncommercial purposes, the differences are minimal. Here we do not distinguish between them, though they can be a subject of intense debate among proponents. It is interesting to note that many of the open licenses, such as GPL, are often said to grant rights to the user such as the freedom to change the software to their own needs or to redistribute them rather than to restrict them.

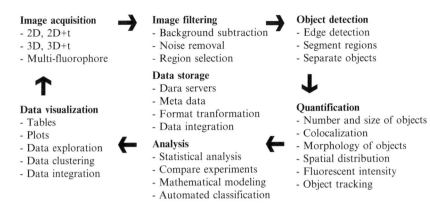

Figure 20.1 Stages in fluorescent image analysis. An image analysis cycle typically has many stages and may use several software methods or specialized packages. In creating an analysis cycle, it is to be recommended to test and adjust the capture and pipeline, possibly several times, on small set of images before commencing on large-scale image capture. It has been observed that "tweaking microscope settings for 5 min could save months of tweaking algorithms" (Auer *et al.*, 2007). Throughout the process, data storage is critical to store both source, processed, derived, and meta-data associated with the analysis pipeline. Adapted from Hamilton (2009).

There are often many steps between generating a hypothesis to test, acquiring imaging, and confirming or disproving the hypothesis, some of which are shown in Fig. 20.1. Typically several separate software packages will be used to perform elements of an analysis cycle. In the following we will outline many of the major open source softwares available to perform key steps in the cycle. Table 20.1 also gives a brief summary of some of the projects, their capabilities, and web links to further information.

2.1. Image capture

The first step in any pipeline is image capture. While microscope manufacturers usually provide control software with their products, they can sometimes be limited in the options for image capture and difficult to control to do nonstandard imaging procedures. **Micro-manager** or **μManager** (Stuurman *et al.*, 2007) is open source software for microscope control. Besides a user interface, it supports programming of microscope control from many high-level languages such as C++ and Java as well as scripting languages such as Python and Perl. It is also fully integrated with ImageJ thus enabling on-the-fly processing of imaging as it is acquired. Hence intelligent acquisition is possible where selected imaging is acquired and stored based on the content of the image (Jackson *et al.*, 2011). For instance, low probability cellular events might be found, selected, tracked, and imaged over time from large fields of cells without the need to store potentially terabytes of largely redundant data. Tests against major commercial solutions have shown μManager to give comparable performance when benchmarked on acquisition speed on nine types of common experimental imaging (Biehlmaier *et al.*, 2010). A comprehensive introduction to using μManager can be found in Edelstein *et al.* (2010).

2.2. Storage

Image storage is becoming an increasingly significant problem as the high-throughput capabilities of recent microscopes are exploited, and there is a strong case to be made for open and public repositories (Swedlow, 2011). The **OMERO** system as part of the long running Open Microscopy Environment (OME) project provides a server–client model for microscopy data storage and provision (Moore *et al.*, 2008; Schiffmann *et al.*, 2006). Image data may be accessed via web browser-based clients or alternatively, directly in ImageJ via an OMERO plug-in for ImageJ. It provides many tools for storage, visualization, and annotation of microscopy imaging. Two particularly important components of OME are Bio-Formats and OME-XML. Bio-Formats provides a Java-based library for reading and writing of a very wide range of microscopy image formats and has been incorporated into many open source bio-image softwares. OME-XML is a standardized format for describing the meta-data associated with imaging such as microscope settings (Linkert *et al.*, 2010). With proprietary meta-data formats information may be lost or corrupted when transforming to other formats; hence open standards are essential to ensure the provenance of data and enable its future reuse and reanalysis.

A more recent project in this space is **Bisque**, the Bio-Image Semantic Query User Environment (Kvilekval *et al.*, 2009). At its core is database storage for up to 5D imaging and associated annotations. All data have a Uniform Resource Locator (URL) which can be used to access it from a server via a web browser or to combine data in novel combinations ("mash-ups"). What makes Bisque unique is that it supports a flexible user-defined data model via tagging which can easily be extended. Typically image analysis will produce derived data from the imaging such as number of objects or average protein expression. These can readily be tagged and the data model for the imaging extended to include them. Like OMERO, Bisque provides an ImageJ plug-in, BQBisqueJ, that allows direct access to the Bisque database images from ImageJ. Bisque also provides a wide range of tools such as a digital notebook, a 3D multi-fluorophore image viewer as well as specific tools for image analysis such as algorithms to detect nuclei in 3D imaging.

2.3. Image processing and quantification

The key open source project for bio-image processing and analysis is the long running **ImageJ** (Abramoff et al., 2004). Built with Java, it runs on all major operating systems and provides a set of core set of algorithms for applications such as noise reduction, filtering, thresholding, measurement, masking, selection, geometric transformation, color channel, and image stack manipulation and image visualization, among many others. Automation of multiple image processing steps may be achieved by macros which can be either recorded from a sequence of menu selections or written directly by the user. More sophisticated algorithms may be implemented via a Java-based plug-in architecture to extend the functionality. With many tutorials and (open source) examples, it is relatively straightforward to learn how to use and extend ImageJ. With some 300 macros and 500 plug-ins contributed by the research user community, it has a range of applications from the generic to the highly specialized. For instance, JACoP (Bolte and Cordelieres, 2006) is the definitive tool for colocalization analysis because it combines and allows comparison of essentially every major 2D colocalization method. Also available is **Fiji** (Fiji is Just ImageJ) which takes the ImageJ core and adds a curated selection of plug-ins organized into submenus by category such as "Segmentation" (see Table 20.1). A selection of ImageJ user-contributed plug-ins is shown in Table 20.2. Typical tasks that can be performed in ImageJ include manual or automated segmentation of imaging, particle tracking, colocalization analysis, edge detection, noise removal, and Fluorescence resonance energy transfer (FRET) analysis. As well as extensibility, interoperability with many other major bio-image projects such as OMERO, CellProfiler, Bisque, VizBio, and μManager is a key feature. ImageJ has become a model for high-quality open source projects.

There are also now numerous stand-alone software packages that are designed to solve specific problems such as **3D nuclear detection** (Santella et al., 2010), **cell tracking** in 3D (Rabut and Ellenberg, 2004), **3D image deconvolution** (Sun et al., 2009), segmentation **and tracking of filamentous structures** in 2D and 3D imaging (Smith et al., 2010), and **tracking centrosomes** (Jaensch et al., 2010). While such methods will often perform at a very high level on the problem they are designed for, the disadvantage is that the user needs to learn a new interface and how to interact with each piece of software. A survey of applications of quantification of microscopy imaging with references to methods and software publications can be found in Hamilton (2009).

A recent trend has been to develop packages to quantify, analyze, and visualize large 3D imaging. The **V3D** software (Peng et al., 2010) supports image visualization and analysis for 3D, 4D, and 5D imaging. It has a plug-in architecture with some 100 plug-ins to perform operations such as cell segmentation and neuron tracing. A key feature is the ability to render and visualize very large (gigabyte) size multidimensional image sets. The **ICY** project (see Table 20.1) is a recent project with substantial image and data visualization capabilities such as multichannel volume rendering built upon a high-quality interactive user environment. It utilizes the Bio-Formats libraries for reading and writing images and can also use many ImageJ as well as user written plug-ins. **BioImageXD** (http://www.bioimagexd. net) supports advanced 3D rendering, 3D segmentation, colocalization analysis and has plans for features such as deconvolution to remove out-of-focus signal, spectral unmixing for more robust colocalization analysis, and particle tracking. The **FARSIGHT Toolkit** takes a very modular approach to bio-image analysis in that rather than being a single software environment, it is a collection of software modules that can be glued together to create pipelines using the Python language (Bjornsson et al., 2008). It utilizes the OMERO image database (see above) and has modules for image preprocessing, segmentation, manual objects editing, quantification, and visualization and has been applied to 5D analysis of cell migration, cell cycle analysis, and 3D segmentation of vasculature among other applications.

Several of the above have either automated or interactive methods of creating surface models from 3D imaging. When automated image segmentation methods fail, another useful package is **IMOD** (Kremer et al., 1996). Though initially designed to facilitate manual segmentation by hand-drawn contours of tomography imaging, **IMOD** can readily be used to generate surface mesh models from fluorescent imaging. These can then form a foundation for quantification or visualization of volumes and surfaces as well as structures within which to mathematically model processes.

2.4. High-throughput pipelines

Another class of image analysis software that is now coming to the fore is that for analysis of large volumes of 2D imaging. A typical application would be to detect phenotypic changes in (fluorescently tagged) protein's expression in cells under a large range of conditions or treatments. For instance, in Collinet et al. (2010), large-scale RNA interference screens were performed and multiple parameters measured for selected proteins expressed in cells to identify new components in endocytic trafficking. For such screens multiple fluorophores to delineate nuclear, cytoplasmic, and protein of interest (POI) are usually used. The nuclear and cytoplasmic markers are used to automatically segment individual cells within which to quantify the POI in a variety of ways such as spatial distribution.

While in theory it would be possible to create such a pipeline in ImageJ, the data management aspects of the problem, which nuclear image matches which POI image and so forth, and the relatively specialized nature of the segmentation and quantification problems involved require dedicated software solutions. The clear leader in this class is **CellProfiler** (Carpenter et al., 2006). Key features include: a simple user interface to create analysis pipelines using menu functions; many different algorithms for segmentation and quantification; many examples and tutorials that may be readily adapted to the user's needs; integration with the Bio-Formats library to read/write a very wide range of bio-image formats; integration with ImageJ to utilize many of ImageJ's macros and plug-ins; integration with the OMERO image database is underway; the ability to extend the methods via Python or Java; integration with the CellProfiler Analyst software for data exploration and analysis; and batch modes for processing large numbers of images (Jones et al., 2008). Another high-throughput analysis software package of note is **Blobfinder** (Allalou and Wahlby, 2009) which while having less functionality than CellProfiler is perhaps simpler to use for the new user.

2.5. Automated image classification

Besides high-throughput quantification, another key problem is automated phenotypic or class identification of fluorescent imaging, particularly at the cellular level. The approach is to generate some form of statistical quantification of the imaging under consideration that distinguishes classes. In a training phase, images, or rather their associated statistics, of known researcher-supplied classification are used to train a machine learning technique such as a support vector machine (Cortes and Vapnik, 1995) or neural network (Bishop, 1995). Large numbers of images may then be rapidly automatically classified using the trained system.

In recent years there has been enormous progress on this problem with near-perfect classification possible on many problems (Hamilton et al., 2007) and several open source automated classification systems now available. **CellProfiler Analyst**, which integrates with CellProfiler, supports the easy creation of automated image classifiers (Jones et al., 2009). First, using CellProfiler, the user selects a range of quantitative image statistics to generate a set of images. A subset of these images are then interactively presented and placed into classes by the user. These human classifications are used to train a classifier using the GentleBoosting algorithm, which can then automatically classify large number of images. **Enhanced CellClassifier** (Misselwitz et al., 2010) takes a similar approach by utilizing image statistics generated in CellProfiler, but then applies a support vector machine to learn the problem. The **Micro-Pilot** system (Conrad et al., 2011) combines automated image capture, object of interest selection, and tracking with machine learning techniques. Utilizing μManager (see above), fields of cells are imaged, initially at low resolution. An automated classifier then selects "cells of interest" from the field of view. These can then be tracked and imaged over time at high resolution. Morphology and texture measures are then applied to the (segmented) high-resolution imaging to give fine-grained automated classification of the time series, all of which happens in real time at the time of imaging. For the more expert, **Murphy lab** provides Matlab code for wide range of cutting edge image classification techniques (Boland *et al.*, 1998; Chen *et al.*, 2007; Newberg and Murphy, 2008; see also Table 20.1).

2.6. Visualization, data analysis, and modeling

Both the throughput and the dimensions of imaging data obtained from fluorescent microscopy are increasing rapidly as new automation and fluorophore technologies become available. As such, the ability to visualize and extract insights from the massive image and image-derived data sets is becoming a serious problem and potential bottleneck in any analysis pipeline. Hence visualization and analysis technologies are now receiving substantial interest and resources to keep pace with the data flows (Walter *et al.*, 2010).

Many of the more recent projects described above such as **V3D, ICY,** and **FARSIGHT** have sophisticated volume-rendering visualization methods for 3D imaging data and surface/mesh rendering for surfaces created from such data. Within ImageJ the **3D Viewer** plug-in (Table 20.2) renders multi-fluorophore 3D imaging to enable the user to explore the data. These tools are often based upon the **Visualization Toolkit (VTK;** http://vtk.org), a high-level open source library for 3D graphics, image processing, and visualization which allows the rapid creation of advanced visualizations. VTK acts as a usable layer on top of the powerful, but somewhat low-level **OpenGL** libraries commonly used for computer graphics and 3D games.

There are now many open source systems exploiting the capabilities of these libraries. One such is **VANO** (Peng *et al.*, 2009) for volume image object annotation. VANO takes 3D-segmented objects (produced by some other system) and allows the user to explore, annotate, delete, split, and merge objects in a 3D-interactive environment. The Bisque project has developed **BioView3D** (Table 20.1) which supports both volume and texture rendering and will generate 3D "fly-through" movies of data for presentation and publication. The **VisBio** project (Rueden *et al.*, 2004), from the group that also developed the Bio-formats library, supports 3D volume rendering, arbitrary slicing to visualize 2D sections of 3D data, movie making, multiple simultaneous data visualization and can interact with OMERO for storage and ImageJ for image processing. **Voxx** (Clendenon *et al.*, 2002) supports similar 3D rendering of imaging and movies and will also generate anaglyph stereo (using red/green or cyan red "3D" glasses). **3D Slicer** (Pieper *et al.*, 2006) is an advanced 3D visualization and image analysis system that also has segmentation and surface generation algorithms, measurement systems, image registration, and user extensions. Though largely for medical imaging such as MRI and CT, it also has a module for multichannel confocal imaging visualization. The **iCluster** system (Hamilton et al., 2009) supports simultaneously visualizing hundreds of 2D fluorescent images of proteins in cells by arranging the images in 3D space in such a way that those that have similar features are spatially close thus allowing the principle patterns of distribution in large image sets to be observed.

Besides image set visualization, analysis of data derived from imaging is also of increasing importance. Perhaps due to the diversity of imaging and flexibility required for the associated data analysis, data analysis software specific to data derived from microscopy are scarce. A notable exception is **CellProfiler Analyst** (Jones et al., 2008) which is designed to analyze, visualize, and explore data derived from high-throughput imaging experiments. The **ICY** project described above also has some data visualization facilities built in for data plots. Though not specific to image-derived data, **Bioconductor** (http://bioconductor.org) provides tools for bioinformatic analyses of high-throughput genomic data with some 460 packages for a range of analyses written in the statistical programming language R. Extensions (Chernomoretz et al., 2008) have also been written in R to analyze and visual data generated using the **Cell ID** software (see Table 20.1).

While some dedicated analyses and visualization packages are now becoming available, it is more typical to use generic systems for analyzing derived data. For simple analyzes, **OpenOffice** (http://OpenOffice.org) supports spreadsheets with a similar level of functionality to Excel. For more in-depth analyses, **R** (R Development Core Team, 2008) is a sophisticated open source package for statistical analysis and visualization of data with many libraries and user contributed code available. For advanced visualization and analysis of relational data such as interactions and networks, there is the **Tulip** system (Delest et al., 2003). And **Octave** (Eaton, 2002) is an advanced mathematical package for numerical computations that is similar to Matlab. Scripts written in Octave will often run in Matlab without modification and vice versa.

The next major stage in the imaging revolution will be to put data together into predictive mathematical models of the processes being imaged and observed over time. For instance, the geometry of subcellular compartments may readily be determined from 3D fluorescent imaging and used as a foundation for modeling molecular interactions on or in. However, systems currently available to do modeling based on experimental imaging, while usually free for academic use, are often not open source making them difficult to integrate with the data being generated from projects such as those described above. A survey of tools for kinetic modeling of biochemical networks, some of which are open source, can be found in Alves et al. (2006). In the longer term, as the aim becomes to model whole systems and integrate data for such modeling from multiple sources, interoperability and easy workflow generation will become increasingly important to allow the rapid and accurate generation of such models.

3. The Future

Here we have described a selection of the software currently available for capture, quantification, analysis, storage, visualization, and modeling from fluorescent microscopy imaging. Of course there are many other fields that rely on quantitative methods for imaging. Indeed, many have more developed methods for analysis of imaging perhaps due partly to being older fields of study or perhaps because they are intrinsically more computational methodologies such X-ray crystallography. For instance, the **OsiriX** Imaging Software is comprehensive and well-developed software for quantification, segmentation, annotation, surface rendering, and advanced visualization of medical imaging, many methods of which would be of use in microscopy imaging. However, it is largely focused on the DICOM image format or monochromatics imaging and so is not easily usable for fluorescent microscopy imaging. Similarly, **BioImage Suite** is another advanced software package geared toward biomedical imaging (http://bioimagesuite.org). Neuroscience Informatics Tools and Resources Clearinghouse (NITRC) (http://www.nitrc.org/) maintains lists of useful neuroimaging analysis software, much of it open source, but again much of it is focused on monochromatic imaging. While these tools can be useful for fluorescent microscopy as they are, with a relatively small amount of resources, many of these open source softwares could be adapted and extended to provide powerful new tools for fluorescent microscopy.

Two key areas for future development within open source bio-imaging applications will be increased interoperability and ease of use for the non-expert user. Interoperability is important to both allow the rapid transfer of data and methods between applications to create seamless data analysis pipelines, but also to ensure data and metadata provenance and integrity is preserved. In particular, there are many new open source software methods being made available and where possible authors should be encouraged to build upon or at least interoperate with ImageJ, OMERO, and other projects rather than developing stand alone systems. It is the interoperability that is enabling rapid generation of novel applications by putting together open source building blocks. Besides easing use by building upon elements familiar to the user, this should help ensure the long-term use and availability of projects. Hence, there are strong incentives for collaboration between developers of new methods. Given the rapidly changing types of data being generated by fluorescent microscopy, developing easy to use systems that are flexible and adaptable for the not necessarily expert user is essential. While some compartmentalization of skills is unavoidable, if the researcher capturing data is also creating the analyses pipeline, this will usually result in better outcomes since the analyst then has a better understanding of the data as well as the knowledge of how to "tweak" the experiment to create more robust analyses. Hence, more attention to user interface and ease of use issues will ensure better analyses and the widest possible uptake of methods.

ACKNOWLEDGMENTS

The author acknowledges the support of the Institute for Molecular Bioscience and The University of Queensland. Partially supported by National Health and Medical Research Council of Australia Project Grants 1002748 and 631584. The author also offers apologies to the many other high-quality open source bio-image analysis projects that there was no space to include in this brief overview.

REFERENCES

Abramoff, M. D., et al. (2000). Objective quantification of the motion of soft tissues in the orbit. *IEEE Transmed. Imaging* **19**(10), 986–995.

Abramoff, M. D., et al. (2004). Image processing with ImageJ. *Biophotonics Int.* **11**(7), 36–42.

Allalou, A., and Wahlby, C. (2009). Blobfinder, a tool for fluorescence microscopy image cytometry. *Comput. Methods Programs Biomed.* **94**(1), 58–65.

Alves, R., et al. (2006). Tools for kinetic modeling of biochemical networks. *Nat. Biotechnol.* **24**(6), 667–672.

Auer, M., et al. (2007). Development of multiscale biological image data analysis: Review of 2006 international workshop on multiscale biological imaging, data mining and informatics, Santa barbara, USA (bii06). *BMC Cell Biol.* **8**(Suppl. 1), S1.

Biehlmaier, O., et al. (2010). Acquisition speed comparison of microscope software programs. *Microsc. Res. Tech.* **74**(6), 539–545.

Bishop, C. M. (1995). Neural Networks for Pattern Recognition. Oxford University Press, Oxford.

Bjornsson, C. S., et al. (2008). Associative image analysis: A method for automated quantification of 3D multi-parameter images of brain tissue. *J. Neurosci. Methods* **170**(1), 165–178.

Boland, M. V., et al. (1998). Automated recognition of patterns characteristic of subcellular structures in fluorescence microscopy images. *Cytometry* **33**(3), 366–375.

Bolte, S., and Cordelieres, F. P. (2006). A guided tour into subcellular colocalization analysis in light microscopy. *J. Microsc.* **224**(3), 213–232.

Carpenter, A., et al. (2006). CellProfiler: Image analysis software for identifying and quantifying cell phenotypes. *Genome Biol.* **7**(10), R100.

Chen, S.-C., et al. (2007). Automated image analysis of protein localization in budding yeast. *Bioinformatics* **23**(13), i66–i71.

Chernomoretz, A., et al. (2008). Using cell-id 1.4 with r for microscope-based cytometry. *Curr. Protoc. Mol. Biol.Chapter 14(14): Unit 14 18*.

Clendenon, J. L., et al. (2002). Voxx: A pc-based, near real-time volume rendering system for biological microscopy. *Am. J. Physiol. Cell Physiol.* **282**(1), C213–C218.

Collinet, C., et al. (2010). Systems survey of endocytosis by multiparametric image analysis. *Nature* **464**(7286), 243–249.

Conrad, C., et al. (2011). Micropilot: Automation of fluorescence microscopy-based imaging for systems biology. *Nat. Methods* **8**(3), 246–249.

Cortes, C., and Vapnik, V. (1995). Support vector networks. *Mach. Learn.* **20**, 273–297.

Delest, M., et al. (2003). Tulip: A huge graph visualisation framework. In "Graph Drawing Softwares, Mathematics and Visualization," (P. Mutzel and M. Jünger, eds.), pp. 105–126. Springer-Verlag, Berlin.

Eaton, J. W. (2002). Gnu Octave Manual. Network Theory Limited, Bristol.

Edelstein, A., et al. (2010). Computer control of microscopes using μmanager. John Wiley & Sons, Inc., New York.

Feng, D., et al. (2007). Stepping into the third dimension. *J. Neurosci.* **27**(47), 12757–12760.

Gordon, A., et al. (2007). Single-cell quantification of molecules and rates using open-source microscope-based cytometry. *Nat. Methods* **4**(2), 175–181.

Hamilton, N. (2009). Quantification and its applications in fluorescent microscopy imaging. *Traffic* **10**(8), 951–961.

Hamilton, N. A., et al. (2007). Fast automated cell phenotype image classification. *BMC Bioinformatics* **8**, 110.

Hamilton, N. A., et al. (2009). Statistical and visual differentiation of subcellular imaging. *BMC Bioinformatics* **10**, 94.

Henriques, R., et al. (2010). Quickpalm: 3D real-time photoactivation nanoscopy image processing in Imagej. *Nat. Methods* **7**(5), 339–340.

Jackson, C., et al. (2011). Model building and intelligent acquisition with application to protein subcellular location classification. *Bioinformatics(Advanced access, publication, May 9 2011)*.

Jaensch, S., et al. (2010). Automated tracking and analysis of centrosomes in early caenorhabditis elegans embryos. *Bioinformatics* **26**(12), i13–i20.

Jones, T., et al. (2008). CellProfiler Analyst: Data exploration and analysis software for complex image-based screens. *BMC Bioinformatics* **9**(1), 482.

Jones, T. R., et al. (2009). Scoring diverse cellular morphologies in image-based screens with iterative feedback and machine learning. *Proc. Natl. Acad. Sci. USA* **106**(6), 1826–1831.

Jython (2011). Jython: Python for the java platform home page. Retrieved 11th May, 2011, from http://www.jython.org/.

Kamentsky, L., et al. (2011). Improved structure, function and compatibility for CellProfiler: Modular high-throughput image analysis software. *Bioinformatics* **27**(8), 1179–1180.

Kocsis, E., et al. (1991). Image averaging of flexible fibrous macromolecules: The clathrin triskelion has an elastic proximal segment. *J. Struct. Biol.* **107**(1), 6–14.

Kremer, J. R., et al. (1996). Computer visualization of three-dimensional image data using imod. *J. Struct. Biol.* **116**(1), 71–76.

Kvilekval, K., et al. (2009). Bisque: A platform for bioimage analysis and management. *Bioinformatics* **26**(4), 544–552.

Linkert, M., et al. (2010). Metadata matters: Access to image data in the real world. *J. Cell Biol.* **189**(5), 777–782.

Misselwitz, B., et al. (2010). Enhanced cellclassifier: A multi-class classification tool for microscopy images. *BMC Bioinformatics* **11**(30), 30.

Mittag, A., et al. (2011). Cellular analysis by open-source software for affordable cytometry. *Scanning* **33**(1), 33–40.

Moore, J., et al. (2008). Open tools for storage and management of quantitative image data. *Methods Cell Biol.* **85**, 555–570. Academic Press.

Murphy, R. F. (2006). Putting proteins on the map. *Nat. Biotechnol.* **24**, 1223–1224.

Newberg, J., and Murphy, R. (2008). A framework for the automated analysis of subcellular patterns in human protein atlas images. *J. Proteome Res.* **7**(6), 2300–2308.

Peng, H., et al. (2009). Vano: A volume-object image annotation system. *Bioinformatics* **25**(5), 695–697.

Peng, H., et al. (2010). V3D enables real-time 3D visualization and quantitative analysis of large-scale biological image data sets. *Nat. Biotechnol.* **28**(4), 348–353.

Pieper, S., et al. (2006). The NA-MIC kit: ITK, VTK, pipelines, grids and 3D slicer as an open platform for the medical image computing community. *Proc. IEEE Intl. Symp. Biomed. Imaging ISBI* 698–701.

Python (2011). Official website of the python programming language. Retrieved 11th May, 2011, from http://www.python.org/.

R Development Core Team (2008). R: A language and environment for statistical computing. R Foundation for Statistical Computing, Vienna, Austria.

Rabut, G., and Ellenberg, J. (2004). Automatic real-time three-dimensional cell tracking by fluorescence microscopy. *J. Microsc.* **216**(Pt. 2), 131–137.

Raymond, E. S. (2000). The cathedral and the bazaar (version 3.0). http://www.catb.org/~esr/writings/cathedral-bazaar/. Retrieved 10th May, 2011 from.

Rueden, C., et al. (2004). VisBio: A computational tool for visualization of multidimensional biological image data. *Traffic* **5**(6), 411–417.

Sage, D., et al. (2005). Automatic tracking of individual fluorescence particles: Application to the study of chromosome dynamics. *IEEE Trans. Image Process.* **14**(9), 1372–1383.

Santella, A., et al. (2010). A hybrid blob-slice model for accurate and efficient detection of fluorescence labeled nuclei in 3D. *BMC Bioinformatics* **11**(580), 580.

Sato, Y., et al. (1998). Three-dimensional multi-scale line filter for segmentation and visualization of curvilinear structures in medical images. *Med. Image Anal.* **2**(2), 143–168.

Sbalzarini, I. F., and Koumoutsakos, P. (2005). Feature point tracking and trajectory analysis for video imaging in cell biology. *J. Struct. Biol.* **151**(2), 182–195.

Schiffmann, D. A., et al. (2006). Open microscopy environment and findspots: Integrating image informatics with quantitative multidimensional image analysis. *Biotechniques* **41**(2), 199–208.

Schmid, B., et al. (2010). A high-level 3D visualization Api for Java and ImageJ. *BMC Bioinformatics* **11**, 274.

Smith, M. B., et al. (2010). Segmentation and tracking of cytoskeletal filaments using open active contours. *Cytoskeleton (Hoboken)* **67**(11), 693–705.

Stepensky, D. (2007). FRETcalc plugin for calculation of FRET in non-continuous intracellular compartments. *Biochem. Biophys. Res. Commun.* **359**(3), 752–758.

Stuurman, N., et al. (2007). Micro-manager: Open source software for light microscope imaging. *Microsc. Today* **15**(3), 42–43.

Sun, Y., et al. (2009). An open-source deconvolution software package for 3-D quantitative fluorescence microscopy imaging. *J. Microsc.* **236**(3), 180–193.

Swedlow, J. R. (2011). Finding an image in a haystack: The case for public image repositories. *Nat. Cell Biol.* **13**(3), 183.

Swedlow, J. R., and Eliceiri, K. W. (2009). Open source bioimage informatics for cell biology. *Trends Cell Biol.* **19**(11), 656–660.

Swedlow, J. R., et al. (2009). Bioimage informatics for experimental biology. *Annu. Rev. Biophys.* **38**, 327–346.

Thévenaz, P., and Unser, M. (2007). User-friendly semiautomated assembly of accurate image mosaics in microscopy. *Microsc. Res. Tech.* **70**(2), 135–146.

Thevenaz, P., et al. (1998). A pyramid approach to subpixel registration based on intensity. *IEEE Trans. Image Process.* **7**(1), 27–41.

Walter, T., et al. (2010). Visualization of image data from cells to organisms. *Nat. Methods* **7**(3), S26–S41.

Wollman, R., and Stuurman, N. (2007). High throughput microscopy: From raw images to discoveries. *J. Cell Sci.* **120**, 3715–3722.

Woodcroft, B. J., et al. (2009). Automated organelle-based colocalization in whole-cell imaging. *Cytometry A* **75**(11), 941–950.

CHAPTER TWENTY-ONE

Nanoparticle PEBBLE Sensors in Live Cells

Yong-Eun Koo Lee *and* Raoul Kopelman

Contents

1. Introduction	420
2. PEBBLE Sensor Designs	424
3. Preparation/Characterization	430
4. Examples	431
4.1. Dissolved oxygen sensing PEBBLEs	432
4.2. Reactive oxygen sensing PEBBLEs	439
4.3. Glucose sensing PEBBLEs	443
4.4. pH sensing PEBBLEs	444
4.5. Other cation sensing PEBBLEs	451
4.6. Anion sensing PEBBLEs	456
4.7. PEBBLEs for enzymatic intracellular processes	457
4.8. PEBBLEs for physical properties	459
5. Summary and Critical Issues	461
Acknowledgments	462
References	462

Abstract

Live cell studies are of fundamental importance to the life sciences and their medical applications. Nanoparticle (NP)-based sensor platforms have many advantages as sensors for intracellular measurements, due to their flexible engineerability, noninvasive nature (due to their nano-size and nontoxic matrix), and, for some of the NPs, intrinsic optical properties. NP-based fluorescent sensors for intracellular measurements, so called PEBBLE sensors, have been developed for many important intracellular analytes and functions, including ions, small molecules, reactive oxygen species, physical properties, and enzyme activities, which are involved in many chemical, biochemical, and physical processes taking place inside the cell. PEBBLE sensors can be used with a standard microscope for simultaneous optical imaging of cellular structures and sensing of composition and function, just like investigations performed with molecular probes. However, PEBBLE sensors of any design and

Department of Chemistry, University of Michigan, Ann Arbor, Michigan, USA matrix can be delivered into cells by several standard methods, unlike dye molecules that need to be cell permeable. Furthermore, new sensing possibilities are enabled by PEBBLE nanosensors, which are not possible with molecular probes. This review summarizes a variety of designs of the PEBBLE sensors, their characteristics, and their applications to cells.

1. INTRODUCTION

The cell is the structural and functional basic unit of living organisms and the building block that forms tissues and organs. Understanding behaviors of live cells (cellular reactions, metabolism, growth, division, and death) is fundamental for understanding the complex functions of the highly developed organisms such as plants, animals and humans; therefore live cell studies are of fundamental importance to the life sciences and their medical applications. Cells are composed of water, of inorganic ions and molecules, and of organic molecules and macromolecules (lipids, carbohydrates, nucleic acids, proteins). These ions and molecules are either a part of the cellular structure or involved in cellular functional activities as enzymes, substrates, catalysts, and metabolites. The cell is in a dynamic steady state, far from equilibrium with its surroundings, and this steady state is maintained by responding quickly to changes in the external environment. Consequently, tracking the dynamic processes involving the live cell's ionic and molecular components, as well as its physical properties, with high temporal and spatial resolution, is crucial for the study of live cells' behaviors. The majority of cells are microscopic—their typical size is 10–20 μm—and can be observed only through a microscope, primarily an optical microscope for live cells. Therefore, optical sensors—which can be easily monitored with an optical microscope—are preferred, because optical imaging of cellular and subcellular structures and sensing within them can be performed simultaneously. Moreover, the sensors should be significantly smaller than the size of cells, and chemically inert so as not to reduce the optical resolution or to impose physical perturbations and chemical interferences on the cells during measurements. Therefore, nanosensors made of biofriendly matrices are ideal intracellular sensors.

Fluorescent probe molecules ("dyes") have so far played a major role in intracellular sensing and imaging, and thus constitute the current standard method for the quantification of intracellular analytes (Haugland, 2005; Lakowicz, 2006). Their fast response, their intense signal against relatively low background noise, their physical noninvasiveness, and the relatively simple instrumental setup required, have made the dye-based fluorescence technique a nice match for real-time measurements in cells. However, these molecular probes have several drawbacks that affect the reliability of their intracellular measurements. First, not all of these dyes are chemically inert. The cytotoxicity or chemical perturbation effects of the available dyes are often a problem, as the mere presence of these dye molecules may chemically interfere with the cellular processes. Second, the dyes must be water dispersible and cell permeable, which restricts the applications of some of the most highly sensitive probes. Note that chemical modification of the probe molecules often interferes with their function. Third, the dye signals may change as they bind to proteins or other cell components. Fourth, they are often sequestered into specific organelles inside the cell, resulting in biased results. Fifth, fluorescent dyes often suffer from fast photobleaching as well as from interference by the autofluorescence of cellular components. Sixth, most of the probe molecules are not ratiometric and therefore are effective only as qualitative or semiquantitative sensors, in intensity-based measurements, because the measured fluorescence intensity depends upon probe concentration as well as analyte concentration. Furthermore, there are other possible artifacts, owing to variations in excitation intensity, emission collection efficiency, and optical loss due to the intracellular environment. Note that merely loading a separate reference dye into the cell, for ratiometric measurements, is not a solution because of the aforementioned sequestration and nonspecific binding effects. Therefore, implementation of technologically more demanding techniques, such as lifetime and phase-sensitive detection, has often been required (Lakowicz, 2006).

In an attempt to solve the above problems, nanoparticle (NP)-based optical sensors were first developed under the name of "PEBBLE" (photonic explorer for biomedical use with biologically localized embedding) (Clark et al., 1998; Sasaki et al., 1996). Since then research with NP-based optical sensors (nano "PEBBLEs") has grown, using the many advantages of the NP as a powerful building block for intracellular sensors. These NPs are in the dimension range of 1–1000 nm, that is, from a few atoms to mitochondria size, thus resulting in minimal physical interference to cells. Many advantageous sensing characteristics and designs can be realized due to the NP's engineerability, nontoxic matrix, size, and intrinsic optical properties, as summarized in Table 21.1.

Furthermore, unlike dye molecules, any PEBBLE sensors, irrespective of NP matrix type and design, can be delivered into cells by several standard methods, as shown in Fig. 21.1. Note that the molecular probes are delivered into cells only by diffusion and therefore the probe's cell permeability is a prerequisite for intracellular measurements. However, NPs usually do not diffuse through cell membranes, due to their size. Instead, the NPs enter the cells by nonspecific or receptor-mediated endocytosis. Nonspecific endocytosis means cellular uptake of the NPs by macropinocytosis (and/or phagocytosis for phagocytic cells, like macrophages and neutrophils), the efficiency of which depends on the surface charge, size, and matrix elasticity of the NPs (Banquy et al., 2009; Chithrani et al., 2006;

Table 21.1 Advantages of NP for constructing intracellular sensors

NP properties	Resultant advantages as intracellular sensors
Size	Causing minimal physical interference to cells during measurements[a] A high surface-to-volume ratio, resulting in high accessibility of analytes to the sensing components
Engineerability	(1) High loading of dyes or sensing components: enhancement of the signal-to-background ratio[b] (2) Loading of multiple sensing components (dyes, enzymes, smaller NPs) within NP matrix and on its surface. This enables ratiometric measurements, multiplex sensing, and diverse designs (3) Surface modification with targeting moieties for selective delivery to specific cells or to subcellular organelles[c]
Nontoxic matrix	Preventing chemical interferences (1) Protection of cellular contents from any potential toxic effects of the incorporated sensing dyes or other components (2) Protection of the incorporated sensing components from potential interferents in the cellular environment, for example, nonspecific binding to proteins and/or membrane/organelle sequestration. As a result, the calibration of the signal in a standard solution is still valid for intracellular measurements
Intrinsic optical properties	Some types of NPs, such as quantum dots (QDs), conjugated polymer dots (CPdots), and metal NPs, possess unique but controllable (fluorescent or nonfluorescent) optical properties[d] that can be utilized for constructing various sensor designs

[a] For example, a single spherical NP measuring 200 nm in diameter is about five orders of magnitude smaller in volume than a typical mammalian cell (500 fL) (Woods et al., 2005).
[b] In some cases, high loaded amounts of dyes in close proximity to each other, either within the NP matrix or on its surface, can allow each analyte to interact with multiple dyes, resulting in amplification of sensitivity (Frigoli et al., 2009; Montalti et al., 2005).
[c] Ligands that target selective cells include antibodies to the receptors on the cells' surface, peptides, and aptamers (Koo Lee et al., 2010a). Ligands for selective organelles include antibodies to markers of cellular compartments (Watson et al., 2005) or peptides targeting nuclei and mitochondria (Derfus et al., 2004; Rozenzhak et al., 2005).
[d] For instance, metal NPs have localized surface plasmon resonance (LSPR), induce surface-enhanced Raman scattering (SERS), and also quench fluorescence of fluorophores in close proximity (<5 nm) (Aslan et al., 2005).

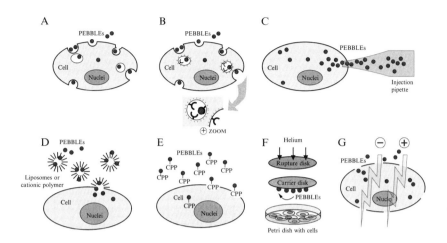

Figure 21.1 Intracellular delivery methods of NPs. (A) Endocytosis (pinocytosis and phagocytosis), (B) receptor-mediated endocytosis, (C) microinjection, (D) liposome and cationic polymer, (E) cell-penetrating peptide (CPP), (F) gene-gun, (G) electroporation. (For color version of this figure, the reader is referred to the Web version of this chapter.)

Harush-Frenkel et al., 2008; Jin et al., 2009; Sarkar et al., 2009). Receptor-mediated endocytosis means a selective cellular uptake of the NPs through interaction between receptors, overly expressed on the surface of specific cells, and ligands that are specific to the receptors. Receptor-mediated endocytosis requires surface modification of the NPs with specific ligands (Chan and Nie, 1998; Koo Lee et al., 2010b; Myc et al., 2007). In addition to these endocytosis pathways, NPs can also be delivered into cells by techniques that are commonly used for oligonucleotide delivery. These techniques include: (1) complexing with various chemical or biochemical agents—such as cell-penetrating peptides like TAT (Kim et al., 2010; Koo Lee et al., 2010b) and PEP-1 (Rozenzhak et al., 2005), liposome or lipid transfection reagents (Clark et al., 1999a,b; Graefe et al., 2008; Hammond et al., 2008; Josefsen et al., 2010), and cationic polymers like polyethyleneimine (PEI) (Duan and Nie, 2007; Guice et al., 2005; Kim et al., 2006, 2007a; Vetrone et al., 2010); (2) physical techniques such as microinjection (Clark et al., 1999a; Gota et al., 2009; Medintz et al., 2010; Schmälzlin et al., 2005), gene-gun (or a biolistic particle delivery system) (Brasuel et al., 2001, 2002, 2003; Clark et al., 1999a; Koo et al., 2004; Park et al., 2003; Wang et al., 2010a; Xu et al., 2001), and electroporation (Coogan et al., 2010; Nielsen et al., 2010; Slotkin et al., 2007). The perturbation caused by chemicals and physical forces used in these techniques may not be completely negligible. However, under optimized conditions, these techniques have been successful in delivering the NPs into various types of cells without causing any deleterious effect on the survival of cells. For instance,

PEI-coated silica NPs (5.5% of PEI, w/w) were delivered into COS-7 cells at NP concentrations of 1–1000 μg/mL, without affecting cell viability (Fuller et al., 2008). Also, the viability of neuroblastoma cells after gene-gun delivery of polyacrylamide (PAA) PEBBLEs was found to be about 98% compared to that of control cells (Clark et al., 1999a).

Advances in nanotechnology have made many types of NPs available as platforms, as well as parts of the sensing components, for the construction of PEBBLE sensors. Indeed, a wide variety of PEBBLE sensors have been developed, for an increased number of analytes that are involved in chemical, biochemical, and physical processes inside cell. For a majority of them, fluorescence is the optical signal for sensing. This review focuses on PEBBLE sensors with fluorescence-based detection and their intracellular applications. For NP sensors based on nonfluorescent optical detection—surface-enhanced Raman scattering and absorption/scattering of localized surface plasmon resonance—please refer to other reviews (Kneipp et al., 2010; Koo Lee et al., 2009; Wang et al., 2010b). It should be noted that the labeled NP—that consists of a reporter molecule attached to the outside of the fluorescent NP and its fluorescence does not change in response to changes in the concentration of a specific analyte—is not considered here as a sensor and therefore is not covered in this review. It should also be noted that in this review, "fluorescence" is defined in a broad sense, that is, the same as "luminescence", as the emission of electromagnetic radiation, especially of visible light and near-infrared (NIR) light, stimulated in a substance by the absorption of incident radiation, covering not only fluorescence in a narrow sense (very short delay time between absorption and emission, below 100 ns) but also phosphorescence (long delay time between absorption and emission, $>\mu s$) and chemiluminescence.

In the following sections, the design, properties, and intracellular delivery methods of PEBBLE sensors are described, and examples of intracellular measurements and future perspective are presented.

2. PEBBLE SENSOR DESIGNS

The basic structure of a sensor is made of two components: an analyte recognizer and a signal transducer. In case of fluorescent molecular probes, the two components are inseparable, that is, the dye serves as an analyte recognizer as well as a fluorescent signal transducer. NP PEBBLE sensors, however, can be designed to have the two as separate components. Moreover, as the NPs are easily engineerable platforms, in terms of composition and architecture, very efficient noninvasive ratiometric nanosensors can be designed that are tailor made to the desired application. For instance, the coimmobilization of a reference fluorophore providing an analyte-independent signal and the surface functionalization with biomolecules are just two examples. NPs made of a variety of matrixes—both inert polymer NPs and NPs with intrinsic optical properties such as quantum dots (QDs), conjugated polymer dots (CPdots), and metal NPs—have been utilized to produce NP PEBBLEs.

In general, PEBBLE designs can be classified into two types: TYPE 1 PEBBLE that uses a single sensing entity, serving as both analyte recognizer and signal transducer and TYPE 2 PEBBLE where the analyte recognizer and optical transducer are distinct (Koo Lee and Kopelman, 2009). The fluorescent PEBBLE sensors show quantifiable changes in fluorescence, for example, fluorescence intensity quenching/enhancement, fluorescence lifetime change, or fluorescence peak shifts, in response to the changes in analyte.

TYPE 1 PEBBLE sensors have been developed for measurements of ions (pH, Ca^{2+}, Cu^{1+}, Cu^{2+}, Fe^{3+}, Mg^{2+}, K^+, Na^+ Pb^{2+}, Zn^{2+}, Cl^-), small molecules (oxygen, singlet oxygen, H_2O_2), radical ($\cdot OH$), enzymatic intracellular processes (apoptosis), and physical properties (temperature, electric field), and they can be classified into three different designs. *The first* and the most common TYPE 1 PEBBLE design (TYPE 1 single PEBBLE) (Fig. 21.2A) is the NP, made of a matrix of homogeneous single composition, that is loaded with not only a single sensing entity—mostly a fluorescent dye (Borisov and Klimant, 2009; Borisov *et al.*, 2008a,b; Cao *et al.*, 2004, 2005; Cheng and Aspinwall, 2006; Clark *et al.*, 1999a,b; Coogan *et al.*, 2010; Cui *et al.*, 2011; Cywinski *et al.*, 2009; Graefe *et al.*, 2008; Hammond *et al.*, 2008; Hornig *et al.*, 2008; Josefsen *et al.*, 2010; Kim *et al.*, 2010; King and Kopelman, 2003; Koo Lee *et al.*, 2010b; Koo *et al.*, 2004; Liu and Liu, 2011; Nakayama-Ratchford *et al.*, 2007; Park *et al.*, 2003; Peng *et al.*, 2007, 2010a,b; Poulsen *et al.*, 2007; Ray *et al.*, 2011; Sarkar *et al.*, 2009; Schmälzlin *et al.*, 2005; Schulz *et al.*, 2010; Seo *et al.*, 2010; Stanca *et al.*, 2010; Sumner *et al.*, 2002, 2005; Sun *et al.*, 2006, 2009; Teolato *et al.*, 2007; Tyner

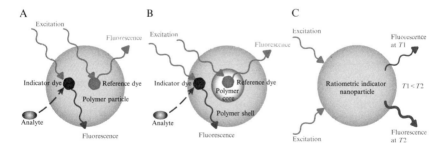

Figure 21.2 Schematics of TYPE 1 PEBBLE sensors. (A) TYPE 1 single PEBBLE (single core NP loaded with both sensing and reference components), (B) TYPE 1 core–shell PEBBLE (core–shell NPs with spatially separated sensing and reference components), (C) TYPE 1 T-sensitive NP PEBBLE (NP matrix as a sensing unit). (For color version of this figure, the reader is referred to the Web version of this chapter.)

et al., 2007; Wang *et al.*, 2011a; Xu *et al.*, 2001), but also a fluorescent protein (Sumner *et al.*, 2005; Sun *et al.*, 2008), or a Förster (fluorescence) resonance energy transfer (FRET) agent containing a peptide (Myc *et al.*, 2007). Often the NP is coloaded with a reference dye, if the sensing entity does not produce two peaks, for ratiometric intensity measurements. *The second* TYPE 1 PEBBLE design (TYPE 1 core–shell PEBBLE) (Fig. 21.2B) is the core–shell-type NP that has the core containing a reference dye, or the core made of fluorescent NPs, and the shell containing sensing dyes (Allard and Larpent, 2008; Arduini *et al.*, 2007; Brown and McShane, 2005; Burns *et al.*, 2006; Guice *et al.*, 2005; He *et al.*, 2010; Schulz *et al.*, 2010, 2011; Son *et al.*, 2010; Wang *et al.*, 2011b). The core–shell design is adopted so as to provide the greatest possible surface area for the sensor's interactions with analytes and so as to sequester the reference dyes away from any interfering ions and molecules, as well as to avoid FRET between the two dyes. *The third* TYPE 1 PEBBLE design (TYPE 1 Tsensitive NP PEBBLE) (Fig. 21.2C) is when the NP matrix itself serves as a sensing unit, as in the case of a temperature PEBBLE sensor made of lanthanide-doped NPs (Vetrone *et al.*, 2010). In the first and second designs, the NP matrix serves as an inert platform for loading sensing and reference components. The inert NP matrix types used for TYPE 1 PEBBLEs include PAA (Clark *et al.*, 1999a, b; Coogan *et al.*, 2010; Graefe *et al.*, 2008; Josefsen *et al.*, 2010; Koo Lee *et al.*, 2010b; Nielsen *et al.*, 2010; Poulsen *et al.*, 2007; Schulz *et al.*, 2010; Sumner and Kopelman, 2005; Sumner *et al.*, 2002, 2005; Sun *et al.*, 2006, 2008, 2009), silica/organically modified silica (ormosil) (Arduini *et al.*, 2007; Burns *et al.*, 2006; Cao *et al.*, 2005; Hammond *et al.*, 2008; He *et al.*, 2010; Kim *et al.*, 2010; Koo *et al.*, 2004; Peng *et al.*, 2007; Sarkar *et al.*, 2009; Seo *et al.*, 2010; Teolato *et al.*, 2007; Wang *et al.*, 2011b; Xu *et al.*, 2001), Silica shell over iron oxide (Son *et al.*, 2010), silica–dextran core–shell (Schulz *et al.*, 2011), polydecylmethacrylate (PDMA) (Cao *et al.*, 2004), polyelectrolyte layer, poly(styrene sulfonate)/poly(allylamine hydrochloride), assembled on the surface of a fluorescent NP (Brown and McShane, 2005; Guice *et al.*, 2005), polystyrene (Allard and Larpent, 2008; Schmälzlin *et al.*, 2005; Wang *et al.*, 2011a), poly(styrene-block-vinylpyrrolidone) (Borisov and Klimant, 2009; Borisov *et al.*, 2008a,b), dextran (Hornig *et al.*, 2008), gold stabilized with poly(vinyl alcohol) and polyacetal shell (Stanca *et al.*, 2010), carbon nanotube (Nakayama-Ratchford *et al.*, 2007), poly(ionic liquid) (Cui *et al.*, 2011), hydrophilic copolymer made of PEG and 2-(2-methoxyethoxy)ethyl-methacrylate (PEG-b-P(MEO$_2$MA)) (Liu and Liu, 2011), polymerized liposome (Cheng and Aspinwall, 2006), polymerized micelle (Tyner *et al.*, 2007), and dendrimer (Myc *et al.*, 2007).

TYPE 2 PEBBLE sensors have been developed for measurements of ions (pH, Cu^{2+}, Zn^{2+}, Hg^{2+}, K^+, Na^+, Cl^-), small molecules (oxygen, H_2O_2, glucose), physiological processes (apoptosis, phosphorylation, protease/DNase/DNA polymerase activity), and physical properties (temperature).

TYPE 2 PEBBLEs enable a synergistic signal and selectivity enhancement as well as sensitivity control that cannot be achieved with free molecular probes. They can be classified into four types. *The first* and the most common TYPE 2 design (TYPE 2 ET PEBBLE) (Fig. 21.3A) is based on energy transfer between analyte-recognition elements (fluorescent/nonfluorescent sensing

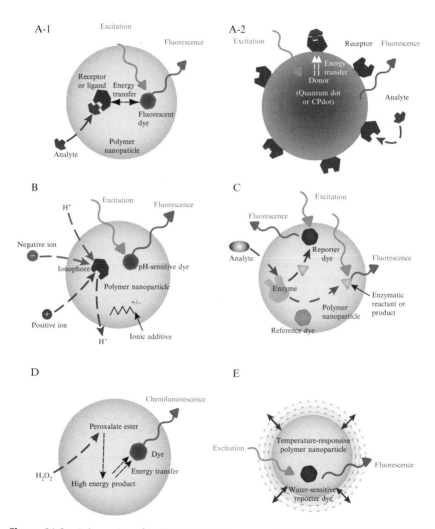

Figure 21.3 Schematics of TYPE 2 PEBBLE sensors. (A) TYPE 2 ET PEBBLE: (A-1) inert polymer NP containing energy donor and acceptor dyes, (A-2) fluorescent semiconductor NP conjugated with energy acceptor indicator dye or receptor molecule, (B) TYPE 2 IC PEBBLE, (C) TYPE 2 enzyme PEBBLE, (D) TYPE 2 CL PEBBLE, (E) TYPE 2 TW PEBBLE. (For color version of this figure, the reader is referred to the Web version of this chapter.)

dyes or receptors) and fluorescent reporters (Arduini et al., 2005; Brasola et al., 2003; Chan et al., 2011; Chen and Rosenzweig, 2002; Chen et al., 2008, 2009; Cordes et al., 2006; Frigoli et al., 2009; Gattas-Asfura and Leblanc, 2003; Gill et al., 2008; Gouanvé et al., 2007; He et al., 2005; Kim et al., 2006, 2007a,b; Krooswyk et al., 2010; Lee et al., 2008; Ma et al., 2011; Meallet-Renault et al., 2004, 2006; Medintz et al., 2010; Montalti et al., 2005; Oishi et al., 2009; Peng et al., 2010c; Rampazzo et al., 2005; Ruedas-Rama and Hall, 2008a,b, 2009; Ruedas-Rama et al., 2011; Shiang et al., 2009; Snee et al., 2006; Suzuki et al., 2008; Tang et al., 2008; Wang et al., 2010a,c; Wu et al., 2009a,b; Yin et al., 2009, 2010). In most cases, the fluorescent reporter itself does not have analyte recognizing ability and therefore has a steady fluorescence, irrespective of changes in analyte. However, with the analyte-recognition elements in close proximity (on the surface or within the NP), the energy transfer efficiency between the recognition elements and the reporters is affected by binding of a specific analyte to the analyte-recognition elements, producing a change in the fluorescence of the reporters. The fluorescent reporters include single dyes (Arduini et al., 2005; Brasola et al., 2003; Chen et al., 2009; Gouanvé et al., 2007; Kim et al., 2006, 2007a; Lee et al., 2008; Meallet-Renault et al., 2004, 2006; Montalti et al., 2005; Oishi et al., 2009; Rampazzo et al., 2005; Yin et al., 2009) or multiple fluorescent dyes (Frigoli et al., 2009; Kim et al., 2007b; Ma et al., 2011; Peng et al., 2010c; Wang et al., 2011c; Yin et al., 2010) as well as intrinsically fluorescent semiconducting NPs such as QDs (inorganic semiconductor) (Chen and Rosenzweig, 2002; Chen et al., 2008; Cordes et al., 2006; Gattas-Asfura and Leblanc, 2003; Gill et al., 2008; Krooswyk et al., 2010; Medintz et al., 2010; Ruedas-Rama and Hall, 2008a,b, 2009; Ruedas-Rama et al., 2011; Snee et al., 2006; Suzuki et al., 2008; Tang et al., 2008; Wang et al., 2010a; Wu et al., 2009b) and CPdots (organic semiconductor) (Chan et al., 2011; Wu et al., 2009a). These semiconducting NPs are brighter and more resistant to photobleaching than organic fluorescent dyes. Being NPs, the QDs and CPdots can serve not only as a fluorescent reporter or reference but also as a platform to load other components and therefore do not require another matrix to construct a PEBBLE sensor. These NPs, especially QDs, have been utilized for constructing PEBBLE sensors for various analytes. The fluorescence of QDs arises from the recombination of excitons so that any changes in charge or composition of the QD surface could affect the efficiency of the core electron–hole recombination, and consequently the luminescence efficiency (Wang et al., 2010a). Further, the broad excitation and narrow emission bands of QDs mark them as having excellent potential as donors for FRET, and in principle, differently colored QDs could be excited simultaneously for simultaneous detection of multiple analytes (Suzuki et al., 2008). A typical design of QD-based PEBBLE sensors is a QD that is surface modified with various dyes, receptors, and/or polymers. The design of CPdot-based PEBBLEs is similar to that of fluorescent dye-based PEBBLEs where fluorescent molecular probes are loaded into the CPdots instead of into inert NPs. We note that there are two kinds of nonfluorescent NPs that have been utilized as sensing components in TYPE 2 ET PEBBLEs, namely, gold and poly(N-isopropylacrylamide) (PNIPAM). The gold NPs (GNPs) served either as fluorescence quenchers (He et al., 2005; Lee et al., 2008; Oishi et al., 2009) or as luminescence particles, like QDs, after they were surface bound with molecules like 11-mercaptoundecanoic acid (Shiang et al., 2009). The PNIPAM NPs change their volume in response to temperature and/or chemical analytes (glucose, Cu^{2+}, and K^+), affecting the FRET efficiency between the FRET acceptor and the donor within the NP matrix (Wang et al., 2011c; Wu et al., 2009b; Yin et al., 2009, 2010). NP matrixes that have been used for loading sensing components, in TYPE 2 ET PEBBLEs, include silica (Arduini et al., 2005; Brasola et al., 2003; Montalti et al., 2005; Rampazzo et al., 2005; Wang et al., 2010a), polyurethane (Peng et al., 2010c), polystyrene (Frigoli et al., 2009), polymethylmethacrylate (PMMA) (Chen et al., 2009), PNIPAM (Wang et al., 2011c; Wu et al., 2009b; Yin et al., 2009, 2010), latex (Gouanvé et al., 2007; Meallet-Renault et al., 2004, 2006), PEI (Kim et al., 2006, 2007b), PEI and polyaspartic acid (Kim et al., 2007a), and polydiethylamino ethyl methacrylate (Oishi et al., 2009).

The second TYPE 2 design (TYPE 2 IC PEBBLE) (Fig. 21.3B) is based on ion-correlation mechanisms and has been used specifically for sensing ions such as K^+, Na^+, and Cl^- with high sensitivity and selectivity (Brasuel et al., 2001, 2002, 2003; Dubach et al., 2007a,b; Ruedas-Rama and Hall, 2007). Here, the hydrophobic NP is typically loaded with three lipophilic components: a highly selective but optically silent ionophore, a pH-sensitive fluorescent reporter dye, and an ionic additive. When a selective ionophore binds the ions of interest, ion exchange (for sensing cations) or ion coextraction (for sensing anions) occurs simultaneously, due to thermodynamic equilibrium. This changes the local pH in (logarithmic) proportion to the concentration of the ion of interest within the NP matrix, a change which is optically detected by the pH-sensitive dye. The ionic additive is used to maintain the ionic strength. The PEBBLEs' sensing characteristics, dynamic range, selectivity, and response time, can be controlled by the composition of the matrix, as well as by the ratios of the three sensor components (Brasuel et al., 2001; Ruedas-Rama and Hall, 2007). Moreover, in a modified design containing QD as a core (Dubach et al., 2007b), the sensitivity of the PEBBLE sensors was enhanced compared to those without QD core (Dubach et al., 2007a), as the ultrabright fluorescence of the QD, instead of that of the pH dye, is the reporting fluorescence, due to the spectral overlap between the QD and the pH dye. The NP matrixes for TYPE 2 IC PEBBLE sensors include PDMA (Brasuel et al., 2001, 2002, 2003), poly n-butyl acrylate (PnBA) (Ruedas-Rama and Hall, 2007), and poly(vinyl chloride) (PVC) (Dubach et al., 2007a,b). Due to the hydrophobic nature of the NP matrix, the aqueous suspendability of the TYPE 2 IC

PEBBLE sensors is often poor, unless there are additional hydrophilic polymer layers (Dubach et al., 2007a,b).

The third TYPE 2 design (TYPE 2 enzyme PEBBLE) (Fig. 21.3C) is based on a NP that is loaded with enzymes. The latter catalyze the reaction involving the analytes. The fluorescent reporter is a reactant (Poulsen et al., 2007), a product (Kim et al., 2005) of the enzyme reaction, or a probe that responds to concentration of a reactant (Xu et al., 2002) or a product (Gill et al., 2008) of the enzyme reaction. The NP matrixes with enzyme inside have been hydrogels, such as PAA (Poulsen et al., 2007; Xu et al., 2002) or PEG (Kim et al., 2005), that are prepared under a mild ("green") condition, so that the activity of the incorporated enzymes is preserved after PEBBLE preparation. QDs were also used as a platform to anchor enzymes on its surface (Gill et al., 2008).

The fourth TYPE 2 design (TYPE 2 CL PEBBLE) (Fig. 21.3D) is based on chemiluminescence. Here, the NP is either made of peroxalate or contains peroxalate and is loaded with fluorescent reporter dyes. The peroxalate ester groups react specifically with the analyte H_2O_2, generating a high-energy dioxetanedione, which then chemically excites the encapsulated fluorescent dyes, leading to chemiluminescence (Lee et al., 2007; Lim et al., 2010).

The fifth TYPE 2 design (TYPE 2 TW PEBBLE) (Fig. 21.3E) is the one for temperature-sensing PEBBLEs, for example, the one made of temperature-responsive PNIPAM and water-sensitive fluorescent dyes. Here, the NP's volume changes, in response to temperature changes, inducing changes in the water availability at the local environment surrounding the reporter dyes, thus resulting in changes in the fluorescence of reporter dyes (Gota et al., 2009; Liu and Liu, 2011).

3. Preparation/Characterization

Preparation of PEBBLEs involves preparation of NPs and loading of sensing and reference components into the NPs. The NP preparation methods are based on wet chemistry synthetic methods such as emulsion, microemulsion, and sol–gel, although they vary with types of NPs (Koo Lee and Kopelman, 2011; Monson et al., 2003). It should be noted that the sensing properties of the PEBBLE sensors, such as sensitivity, dynamic range, and selectivity, are mostly determined by the incorporated sensing elements. However, they are also affected by the NP matrix properties (porosity, charge, and hydrophobicity) and interactions between sensing probes and NP matrix (Koo Lee et al., 2010b; Yin et al., 2009). Therefore, selecting the right kind of NP matrix and proper loading of the sensing elements are important for producing a reliable sensor. In the case of the

TYPE 2 ET PEBBLEs that use dye-doped NPs, the sensing ability is also affected by the particle size. Note that the typical operating distance for efficient energy transfer is short (less than 100 Å) (Sapsford et al., 2006). Owing to their higher surface-to-volume ratio, smaller NPs might ensure remarkable efficiencies (Gouanvé et al., 2007). The loading of the single or multiple sensing components into the NPs is done at the beginning of, or during, the synthesis, or after the formation of the NPs. This is done by encapsulation, covalent linkage, bioaffinity interaction such as streptavidine-biotin, or physical adsorption through charge–charge or hydrophobic interaction. The dyes loaded into a NP by the encapsulation and physical adsorption methods tend to leach out of the NP matrix. The covalent linkage and bioaffinity interaction can prevent leaching-related problems and often increase the dye loading, although they require a relatively complicated preparation procedure. The covalent linkage is typically made by simple coupling reactions between the NP that is functionalized (with amine-, carboxyl- or thiol groups) and the molecules that contain complimentary functional groups.

Characterization of PEBBLEs involves not only the measurements of typical sensing characteristics—sensitivity, dynamic range, selectivity, response time, stability and reversibility—but also the characterization of the NPs' physicochemical properties, as these properties affect their functional efficiency. The physicochemical characterization of the NPs includes measurements of size, surface charge, and the amount of loaded dyes. The size and morphology of the dried NPs are determined usually not only by scanning electron microscopy or transmission electron microscopy but also by atomic force microscopy. The size and extent of aggregation of the nanoplatforms in aqueous solution, as well as the surface charge (zeta potential), are determined by light-scattering techniques such as dynamic light scattering (DLS) and quasi-elastic light scattering. The loaded amount of dyes is usually determined by measuring the absorbance or fluorescence of the prepared nanoplatform sample solution and comparing it with a calibration curve constructed from the mixture of free dye and blank NPs of known concentrations.

4. Examples

PEBBLE sensors have been developed for measurements of cations (Ca^{2+}, Cu^{2+}, Cu^{1+}, Fe^{3+}, H^+(pH), Hg^+, K^+, Mg^{2+}, Na^+, Pb^{2+}, Zn^{2+}), anions (Cl^-, phosphate), radicals ($OH\cdot$), small molecules (glucose, H_2O_2, oxygen, singlet oxygen), physiological processes (apoptosis, phosphorylation, protease/DNase/DNA polymerase activity) and physical properties (temperature, electric field).

4.1. Dissolved oxygen sensing PEBBLEs

Oxygen is the key metabolite in aerobic systems and plays an important role in the energy metabolism of living organisms. Maintaining oxygen homeostasis is critical for cellular function, proliferation, and survival, and an oxygen concentration below or above the proper level for each cellular organ could be a possible cause of disease as well as an indicator for disease. Despite its important biological roles, oxygen is an intracellular analyte that is difficult to quantify by available fluorescent molecular probes. First of all, many of the available oxygen-sensitive molecular probes are hydrophobic and therefore not useful for direct intracellular applications. Even if the probes are soluble in water, none of the probes contain two peaks that respond differently to oxygen concentrations. They can be utilized for quantitative determinations using lifetime measurements but are not good for ratiometric intensity measurements. Moreover, their response is often interfered by local factors such as pH and proteins (Koo Lee et al., 2010b; Xu et al., 2001). By introducing NPs as platform to construct oxygen sensors (see Table 21.1 for NPs' advantages), most of these problems may be avoided. As a result, oxygen sensing may be one of the intracellular sensing areas that may benefit the most by the NP PEBBLE technology.

Oxygen sensing PEBBLE designs that have been developed so far mostly belong to TYPE 1 single PEBBLEs (Borisov and Klimant, 2009; Borisov et al., 2008a,b; Cao et al., 2004; Cheng and Aspinwall, 2006; Coogan et al., 2010; Cywinski et al., 2009; Koo Lee et al., 2010b; Koo et al., 2004; Schmälzlin et al., 2005; Wang et al., 2011a; Xu et al., 2001), except a few cases, for instance, one that belongs to TYPE 1 core–shell PEBBLEs (Guice et al., 2005; Wang et al., 2011b) and one belonging to TYPE 2 ET PEBBLE (Wu et al., 2009a) (see Table 21.2). Such TYPE 1 PEBBLEs were prepared by loading the oxygen-sensitive probes into NPs or NP shells. The oxygen-sensitive probes include lipophilic ones—Iridium complex ($Ir(CS)_2(acac)$, $Ir(CN)_2(acac)$, $Ir(CS-Jul)_2(acac)$) (Borisov and Klimant, 2009), Pt(II) octaethylporphine (PtOEP) (Koo et al., 2004), Pt(II) octaethylporphine ketone (PtOEPK) (Cao et al., 2004; Koo et al., 2004), platinum(II) and palladium(II) benzoporphyrins (PtTPTBPF or PdTPTBPF) (Borisov et al., 2008b); moderate lipophilic/hydrophilic ones—Pt(II) meso-tetra-pentafluorophenyl-porphyrin (PtTFPP) (Borisov and Klimant, 2009; Borisov et al., 2008a; Cywinski et al., 2009; Schmälzlin et al., 2005; Wang et al., 2011a), Ru(II)-tris(4,7-diphenyl-1,10-phenanthroline) (Rudpp) dichloride (Borisov and Klimant, 2009; Cheng and Aspinwall, 2006; Guice et al., 2005; Wang et al., 2011b; Xu et al., 2001); and hydrophilic ones—Pd-tetra-(4-carboxyphenyl) tetrabenzoporphyrin dendrimer (G2) (Koo Lee et al., 2010b) and $Rudpp(SO_3Na)_2)_3]Cl_2$ (Coogan et al., 2010). The NP matrices or shells that have been used for loading the probes include both hydrophobic ones (organically modified silica (ormosil) (Koo et al., 2004), poly(decyl

Table 21.2 Fluorescent PEBBLE sensors for oxygen

Sensor type	Sensing signal	NP matrix loaded with sensing (S) and reference (R) fluorescent dyes	Q_{DO}	Size	Intracellular application (cell type, delivery method)	References
TYPE 1 single	Int_{ratio}	Silica NP with encapsulated Ru(dpp)Cl$_2$ (S) and Oregon Green 488-dextran (R)	80%	20–300 nm (SEM)	C6 glioma cell, gene-gun, measured intracellular oxygen concentration ($O_{2,intra}$) at ambient temperature was 7.9 ± 2.1 ppm for cells in air-saturated buffer, and 6.5 ± 1.7 and <1.5 for cells in N$_2$-saturated buffer after 25 and 120 s, respectively	Xu et al. (2001)
TYPE 1 single	Int_{ratio}	Ormosil NP with encapsulated PtOEP (S) and 3,3′-dioctadecyloxacarbocyanine perchlorate (R); Ormosil NP with PtOEPK (S) and OEP (R)	97%	120 nm (SEM)	C6 glioma cell, gene-gun, ($O_{2,intra}$) at ambient temperature was 7.8 ppm	Koo et al. (2004)
TYPE 1 single	Int_{ratio}	PDMA NP with encapsulated PtOEPK (S) and OEP (R)	97.5%	150–250 nm (SEM)	—	Cao et al. (2004)
TYPE 1 single	Lifetime	Polystyrene NP with encapsulated PtTFPP	N/A	300 nm–1 μm	Chara corallina cells, microinjection, $O_{2,intra}$ at 23 °C was 250 μM (8 ppm)	Schmälzlin et al. (2005)

(*Continued*)

Table 21.2 (Continued)

Sensor type	Sensing signal	NP matrix loaded with sensing (S) and reference (R) fluorescent dyes	Q_{DO}	Size	Intracellular application (cell type, delivery method)	References
TYPE 1 single	Int_{ratio}	DOPC phospholipid polymerized with methacrylate, encapsulated Ru(dpp)Cl$_2$ (S) and NBD-PE (R)	76%	150 nm (DLS)	—	Cheng and Aspinwall (2006)
TYPE 1 single[a]	Lifetime	Poly(styrene-block-vinylpyrrolidone) with encapsulated PtTPTBPF, PdTPTBPF, or PtTFPP	N/A	245 nm (DLS)	—	Borisov et al. (2008a,b)
TYPE 1 single[a]	Int_{ratio} and lifetime	Poly(styrene-block-vinylpyrrolidone) NP with encapsulated lipophilic fluorescent oxygen indicator dyes (S): Ru(dpp) Cl$_2$, Ir(C$_S$)$_2$(acac), Ir(C$_N$)$_2$(acac), Ir(C$_{S-Jul}$)$_2$(acac), PtPFPP, PdPFPP. For ratiometric intensity measurements, C$_{S-Jul}$ (R) was coencapsulated	N/A	245 nm (DLS)	—	Borisov and Klimant (2009)
TYPE 1 single	Int_{ratio}	Polystyrene NP with encapsulated PtTFPP (S) and S13 (R)	—	21 nm (AFM)	—	Cywinski et al. (2009)

TYPE 1 single	Lifetime	Polyacrylamide NP with encapsulated with Rudpp [(SO$_3$Na)$_2$)$_3$]Cl$_2$·6H$_2$O	N/A	45 nm (DLS)	Yeast by electroporation, Mammary adenosarcoma MCF-7 by endocytosis, the lifetime of the PEBBLEs in air and nitrogen were 1.81 and 3.88 μs, respectively, that of electroporated MCF-7 cells were 2.92 and 4.06 μs, respectively	Coogan et al. (2010)
TYPE 1 single	Int$_{ratio}$	Polyacrylamide NP with covalently linked G2 (S) and Hilyte 680 (R)	96%	30 nm (SEM) 51 nm (DLS)	C6 glioma and A549 human lung adenocarcinoma; endocytosis, TAT peptide, and gene-gun: 4.9 ± 1.2 ppm in the C6 glioma cells and 4.5 ± 1.0 ppm in the A549 cells	Koo Lee et al. (2010b)
TYPE 1 single	Int$_{ratio}$	Amino-modified polystyrene NP with postloaded PtTF$_{20}$PP (S) and NCPNBE (R)	74%	410–430 nm (TEM)	Epithelial normal rat Kidney (NRK) cells; Endocytosis	Wang et al. (2011a)
TYPE 1 core–shell	Int$_{ratio}$	Polyelectrolyte layers postloaded with Ru(dpp) Cl$_2$ (S) over commercial fluorescent NP (R)	60%	100 nm (SEM)	Human dermal fibroblasts; PEI (cationic polymer)	Guice et al. (2005)

(*Continued*)

Table 21.2 (Continued)

Sensor type	Sensing signal	NP matrix loaded with sensing (S) and reference (R) fluorescent dyes	Q_{DO}	Size	Intracellular application (cell type, delivery method)	References
TYPE 1 core–shell	Int_{single} and lifetime	Silica core–shell with shell containing covalently linked Rudpp	53.2%	230 nm (core 185 nm; shell 45 nm) (SEM)	—	Wang et al. (2011b)
TYPE 2 ET	Int_{ratio}	CPdot (R), made of PDHF or PFO, doped with PtOEP (S)	95%	25 nm for PDHF CPdot; 50 nm for PFO CPdot (AFM)	J774A1 cells (a macrophage-like murine cell line); endocytosis	Wu et al. (2009a)

Abbreviations: N/A, not available.

Chemical acronym: DiO, 3,3′-dioctadecyloxacarbocyanine perchlorate; DOPC, 1,2-dioleoyl-sn-glycero-3-phosphocholine; G2, Pd-tetra-(4-carboxyphenyl) tetrabenzoporphyrin dendrimer; Ir(C$_S$)$_2$(acac), Iridium(III) complex with 3-(benzothiazol-2-yl)-7-(diethylamino)-coumarin and acac, 2,4-pentanedione; Ir(C$_N$)$_2$(acac), Iridium(III) complex with 3-(N-methyl-benzoimidazol-2-yl)-7-(diethylamino)-coumarin; Ir(C$_{S-Ful}$)(acac), Iridium(III) with C$_{S-Jul}$, 10-(2-benzothiazolyl)-2,3,6,7-tetrahydro-1,1,7,7-tetramethyl-1H, 5H, 11H-(1)benzopyropzrano(6,7-8i,j)quinolizin-11-one; NCPNBE, (N-(5-carboxypentyl)-4-piperidino-1,8-naphthalimide butyl ester; NBD-PE, 1,2-dioleoyl-sn-glycero-3-phosphoethanol amine-N-(7-nitro-2-1,3-benzoxadiazol-4-yl; OEP, octaethylporphine; ormosil, organically modified silica; PDHF, poly(9,9-dihexylfluorene); PDMA, poly(decyl methacrylate); PFO, poly(9,9-dioctylfluorene); PdTPTBPF, Pd(II) meso-tetra(4-fluorophenyl)tetrabenzoporphyrin; PtTF$_{20}$PP, platinum(II) meso-tetrakis-(penta-fluorophenyl)porphyrinato; PtTPTBPF, Pt(II) meso-tetra-(4-fluorophenyl)tetrabenzoporphyrin; PtTFPP, Pt(II) 5,10,15,20-tetrakis-(2,3,4,5,6-pentafluorophenyl)-porphyrin; PdTFPP, Pd(II) 5,10,15,20-tetrakis-(2,3,4,5,6-pentafluorophenyl)-porphyrin; PtOEP, platinum(II) octaethylporphine; PtOEPK, platinum(II) octaethylporphine ketone; Rudpp, Ru(II)-tris(4,7-diphenyl-1,10-phenanthroline); S13, N,N′-bis(1-hexylheptyl)perylene-3,4:9,10-bis-(dicarboximide).

[a] Poly(styrene-block-vinylpyrrolidone) can be considered a core–shell type particle with hydrophobic polystyrene core and hydrophilic vinylpyrrolidone shell. However, for oxygen sensing, lipophilic oxygen probes (as well as lipophilic reference dyes in case of ratiometric intensity measurements) were loaded in the core. Therefore, these PEBBLEs are classified as TYPE 1 single PEBBLE.

methacrylate) (PDMA) (Cao et al., 2004), polystyrene (Cywinski et al., 2009; Schmälzlin et al., 2005; Wang et al., 2011a), polymerized liposome (Cheng and Aspinwall, 2006)) and hydrophilic ones (silica (Wang et al., 2011b; Xu et al., 2001), PAA (Coogan et al., 2010; Koo Lee et al., 2010b), polyelectrolyte shell that is coated over a commercial fluorescent NP (Guice et al., 2005)). The oxygen sensing TYPE 2 ET PEBBLE was prepared by loading an oxygen-sensitive dye, platinum(II) octaethylporphine (PtOEP), into CPdots made of polyfluorene derivatives poly(9,9-dihexylfluorene) (PDHF) or poly (9,9-dioctylfluorene) (PFO) (Wu et al., 2009a). Note that the CPdot matrix is hydrophobic. Upon light excitation, the polymer efficiently transfers energy to the PtOEP dye, which results in a bright phosphorescence that is highly sensitive to the concentration of dissolved oxygen.

It is noteworthy that some of the oxygen PEBBLEs have been developed using oxygen probes with NIR emission (Borisov et al., 2008b; Koo Lee et al., 2010b), aiming at oxygen measurements either in highly scattering media such as subcutaneous tissue or in media containing fluorescent substances such as chlorophyll. These PEBBLEs may also enable more accurate intracellular measurements because the fluorescent signals of these PEBBLEs are not overlapped by cellular autofluorescence.

The oxygen sensing principle of all of the oxygen PEBBLEs is based on quenching of the probes' phosphorescence by oxygen, which has been reported to be temperature dependent (Borisov and Wolfbeis, 2006). Consequently, the temperature effects need to be compensated for if measurements are performed at variable temperatures. Several oxygen PEBBLEs were studied for their temperature dependence, exhibiting higher sensitivity to oxygen at elevated temperatures (Borisov et al., 2008a; Koo Lee et al., 2010b).

It should be noted that most of the oxygen PEBBLEs have been prepared based on hydrophobic matrixes. This may be due to the following reasons: (1) there are more available hydrophobic oxygen probes, and a hydrophobic matrix helps enhance the loading of the probes; (2) A hydrophobic matrix, in general, has high oxygen solubility and permeability. Table 21.2 shows that the PEBBLEs based on hydrophilic matrixes, in - general, have lower oxygen sensitivity (represented by Q_{DO}) than those based on hydrophobic matrixes. Q_{DO}—where DO denotes dissolved oxygen—represents the overall quenching response between a nitrogen- and oxygen-saturated condition, which is defined as $Q_{DO-O2} = (R_{N2} - R_{O2})/R_{N2} \times 100$, where R_{N2} is the fluorescence intensity of the indicator dye or the indicator/reference intensity ratio, in fully deoxygenated water, and R_{O2} is that in fully oxygenated water. However, it should be pointed out that the PEBBLEs made of hydrophobic matrixes are less biocompatible than those made of hydrophilic matrixes due to their poor water suspendability. A solution for such problem may be the modification of the hydrophobic NP matrix with hydrophilic components, as was done in some of the PEBBLEs, either by introducing amine groups (Wang et al., 2011a) or by combining with a hydrophilic copolymer (Borisov and Klimant, 2009; Borisov et al., 2008a,b). Another solution may be the modification of the hydrophilic matrix for higher oxygen permeability and higher probe loading, as was shown in oxygen PEBBLEs made of hydrophilic PAA NP and hydrophilic G2 (Koo Lee et al., 2010b). Here, the NP matrix composition and the interaction between the matrix and oxygen probes were modulated so as to optimize both the brightness and the sensitivity of the PEBBLEs. The optimized PEBBLEs did achieve a very high oxygen sensitivity (96% Q_{DO}), close to the highest previously reported sensitivity of oxygen PEBBLEs made of hydrophobic matrixes (Cao et al., 2004; Koo et al., 2004), despite the use of a hydrophilic matrix.

Many of these PEBBLEs were introduced into cells by a variety of delivery methods, including endocytosis (Coogan et al., 2010; Koo Lee et al., 2010b; Wang et al., 2011a; Wu et al., 2009a), receptor-mediated endocytosis (Koo Lee et al., 2010b), TAT peptide (Koo Lee et al., 2010b), microinjection (Schmälzlin et al., 2005), gene-gun (Koo et al., 2004; Koo Lee et al., 2010b; Xu et al., 2001), cationic polymer PEI (Guice et al., 2005), and electroporation (Coogan et al., 2010). The investigated cell lines included A549 human lung adenocarcinoma, C6 glioma, Chara coralline cells, epithelial normal rat kidney cells, human dermal fibroblasts, J774A1 cells (a macrophage-like murine cell line), MDA-MB-435, MCF-7, and yeast. The efficiency of intracellular delivery of the PEBBLEs may depend upon the introduction methods and the cell type. For instance, a study with yeast and MCF-7 cells showed that endocytosis, performed by incubation of cells with the PEBBLEs, may be the simplest method for the mammalian cell but is not applicable to yeast, while the electroporation was useful for both cell lines (Coogan et al., 2010). The same study also shows that intracellular delivery efficiency of electroporation into yeast cells depends on the type of yeast—it is more effective with the fission yeast, *Schizosaccharomyces pombe*, than for the budding yeast, *Saccharomyces cerevisiae*.

These studies on cells demonstrated that the PEBBLEs were located within cells and that there were no cytotoxic effects due to the PEBBLEs. For instance, the C6 glioma cells with endocytosed PAA NPs with covalently linked G2 and Hilyte 680 dyes continued to divide (Fig. 21.4) after being incubated with the PEBBLEs up to 20 h at concentration of 4 mg/mL (Koo Lee et al., 2010b). MTT assay results also confirmed that the nanosensors did not affect cell viability (Koo Lee et al., 2010b).

In several applications, intracellular oxygen concentration was measured either at ambient temperature (Coogan et al., 2010; Koo et al., 2004; Schmälzlin et al., 2005; Wang et al., 2011a; Xu et al., 2001) or at 37 °C (Koo Lee et al., 2010b), which revealed that the average intracellular concentration is lower than the oxygen concentration of air-saturated water. Further, the changes in intracellular oxygen concentration were monitored in real time when the extracellular oxygen concentration was changed due to the use of a

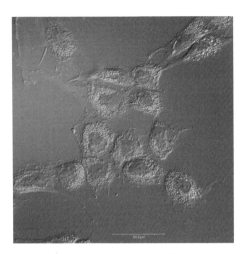

Figure 21.4 A confocal image of live C6 glioma cells loaded with oxygen PEBBLEs. The PEBBLE was made of a PAA NP with covalently linked G2 and Hilyte 680 dyes. (Reprinted with permission from Koo Lee *et al.*, 2010b. Copyright 2010 American Chemical Society.) (For color version of this figure, the reader is referred to the Web version of this chapter.)

medium saturated with nitrogen gas (Coogan *et al.*, 2010; Xu *et al.*, 2001); gradual oxygen depletion occurred either by respiration (Koo *et al.*, 2004) or by the reaction of glucose oxidase (GOx) with glucose (Koo Lee *et al.*, 2010b), or by the oxygen production by photosynthesis (Schmälzlin *et al.*, 2005). Interestingly, the changes in intracellular oxygen concentrations were found to be much lower than those in extracelluar oxygen concentrations, according to one study in which the same oxygen PEBBLEs were used to monitor the oxygen concentrations both inside and outside the cells as well as in phosphate buffer solution (PBS) without cells (Koo Lee *et al.*, 2010b) (Fig. 21.5). Moreover, the depletion rate of the extracellular oxygen concentration is much slower than that of the oxygen concentration in PBS, indicating that the presence of the cells subdues the changes of oxygen concentration not only inside but also outside the cells (Koo Lee *et al.*, 2010b) (Fig. 21.5). Oxygen PEBBLEs were also applied to measure the oxygen concentrations in biological samples such as serum (Cao *et al.*, 2004) and in a growing culture of bacteria (*Escherichia coli*) (Borisov and Klimant, 2009) or yeast (Cywinski *et al.*, 2009).

4.2. Reactive oxygen sensing PEBBLEs

The reactive oxygen species (ROS) are a group of short-lived, reactive oxygen containing small molecules, which include nitric oxide (NO·), peroxynitrite ($ONOO^-$), hypochlorous acid (HOCl), superoxide (O_2^-), hydroxyl radical (·OH), hydrogen peroxide (H_2O_2), and singlet oxygen

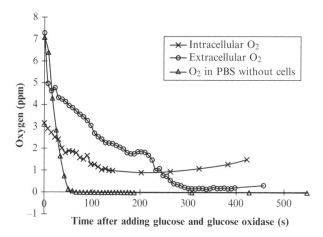

Figure 21.5 Response of oxygen PEBBLEs (PAA NPs with covalently linked G2 and Hilyte 680 dyes) inside and outside A549 cells as well as in PBS after D-glucose and glucose oxidase are added. (Reprinted with permission from Koo Lee et al., 2010b. Copyright 2010 American Chemical Society.)

($^1O_2$). PEBBLEs have been developed for detection of ROS such as hydrogen peroxide, singlet oxygen, and hydroxyl radical. These PEBBLEs are designed to show irreversible responses toward the ROS, due to the high reactivities and short lifetimes of the ROS.

4.2.1. Hydrogen peroxide sensing PEBBLEs

Hydrogen peroxide is a neutral molecule and the most abundant ROS in cells, and thereby the most studied ROS by PEBBLE techgnology. H_2O_2 sensing PEBBLE designs include TYPE 1 single (Hammond et al., 2008; Kim et al., 2010), TYPE 2 ET (Gill et al., 2008; Shiang et al., 2009), TYPE 2 enzyme (Kim et al., 2005; Poulsen et al., 2007), and TYPE 2 CL PEBBLEs (Lee et al., 2007; Lim et al., 2010).

The TYPE 1 single PEBBLEs include silica NP containing smaller silica NP with covalently linked 5(6)-carboxyfluorescein diacetate (Hammond et al., 2008) and ormosil NP with postloaded DCFDA (2′,7′-dichlorofluorescin diacetate, a precursor of the H_2O_2 probe, 2′,7′-dichlorofluorescin (DCFH) (Kim et al., 2010). The silica PEBBLE is responsive to most of the ROS including H_2O_2, hydroxyl radical, nitric oxide, peroxynitrile, and superoxide anion, with a very similar detectable range ($\sim$1–30 nM). However, the ormosil-DCFH PEBBLE shows exclusive selectivity toward H_2O_2 over other ROS because the NP matrix provides effective protection of the probe DCFH from all other ROS (Fig. 21.6). The probe puts up effective barriers, of different nature, against each interferent species.

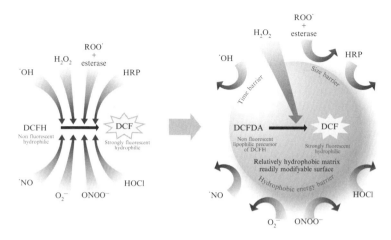

Figure 21.6 Schematic representation of induced H_2O_2 selectivity by encapsulating DCFDA into relatively hydrophobic ormosil NP. The left scheme shows that DCFH (activated form) can be oxidized by a variety of interfering agents. The right scheme shows that a relatively hydrophobic nanoparticle matrix encapsulating DCFDA (precursor form) can filter the interferences, thus only allowing penetration of H_2O_2, resulting in high selectivity toward this specific analyte. (Reprinted with permsision from Kim *et al.*, 2010. Copyright 2010 American Chemical Society.) (For color version of this figure, the reader is referred to the Web version of this chapter.)

TYPE 2 ET PEBBLEs were constructed using either glutathione (GSH)-capped QDs (Gill *et al.*, 2008) or 11-mercaptoundecanoic acid-bound gold nanodots (11-MUA–Au NDs) (Shiang *et al.*, 2009). In case of the QD-based TYPE 2 ET PEBBLE, fluorescein isothiocyanate (FITC)-modified avidin was conjugated to the QD's surface so as to serve as a reference dye for ratiometric intensity measurements (Gill *et al.*, 2008). The same QD PEBBLE also works for lifetime measurements. In case of the GNP-based TYPE 2 ET PEBBLE, the PEBBLE produces luminescence whose intensity depends upon the amount of surface-attached MUA that is removed from the surface due to oxidation of AuS bonds by H_2O_2.

Two kinds of H_2O_2 sensing TYPE 2 enzyme PEBBLEs have been constructed using horseradish peroxidase (HRP) so as to utilize the HRP's role in oxidation of many substrates at the expense of H_2O_2. In the first case, the PEG NP containing HRP was prepared and used together with externally introduced Amplex Red (10-acetyl-3,7-dihydroxyphenoxazine) that reacts with H_2O_2, in a 1:1 stoichiometry, to produce the red fluorescent oxidation product, resorufin (Kim *et al.*, 2005). In the second case, a PAA NP was prepared with encapsulated HRP and fluorescein (Poulsen *et al.*, 2007).

The H_2O_2 sensing of TYPE 2 CL PEBBLEs is based on the reaction between hydrogen peroxide and peroxalate ester groups, generating a high-energy dioxetanedione, which then chemically excites encapsulated fluorescent dyes, leading to chemiluminescence (Lee et al., 2007). Two kinds of TYPE 2 CL PEBBLEs have been developed. One is made of peroxalate ester NPs and encapsulated fluorescent dyes (Lee et al., 2007). The other one is made of Pluronic (F-127) NPs (FPOC NPs) encapsulating bis[3,4,6-trichloro-2-(pentyloxycarbonyl)phenyl] oxalate (CPPO) as a concentrated peroxyoxalate CL fuel with high reactivity, a small amount of 3,3'-diethylthiadicarbocyanine iodide (Cy5) as a NIR dye, and poly(lactic-co-glycolic acid) as a biocompatible polymeric binder for the nanostructure stability (Lim et al., 2010).

It should be noted that the response of some of these H_2O_2 PEBBLEs is interfered by pH either due to the probes, that is, the pH-dependent fluorescein derivatives (Gill et al., 2008; Kim et al., 2010; Poulsen et al., 2007) or due to the surface-capped acid molecules as in the case of the GNP-based TYPE 2 PEBBLE (Shiang et al., 2009). The H_2O_2 detectable range of the PEBBLEs is reported to be between nanomolar and millimolar and is useful for measuring H_2O_2 intracellular concentration: 10nM–100 µM for ormosil-DCFH PEBBLEs (Kim et al., 2010), 0.1–10 mM for GSH-QD PEBBLEs (Gill et al., 2008), 100 nM–1 mM for 11-MUA–Au NDs (Shiang et al., 2009), 0.5–10 µM for PAA-HRP-fluorescein PEBBLEs (Poulsen et al., 2007), 250 nM–10 µM for peroxalate ester NP PEBBLEs (Lee et al., 2007), and 1 nM–10 µM for FROC-CPPO-Cy5 PEBBLEs (Lim et al., 2010). Note that the concentration of H_2O_2 inside most cells is approximately 1 nM, but its concentration inside activated neutrophils was measured to be as high as 36–38 µM (Makino et al., 1986). Intracellular delivery of the H_2O_2 sensing PEBBLEs was conducted in RAW264.7 murine macrophages by phagocytosis (Kim et al., 2005, 2010) and TAT (Kim et al., 2010), and bovine oviduct cells by liposome (Hammond et al., 2008). The PEBBLEs in macrophages were found to respond to exogenous H_2O_2 (Kim et al., 2005) as well as endogenous peroxide induced by lipopolysaccharide (LPS) (Kim et al., 2005) or by N-formyl-methionyl-L-leucyl-L-phenylalanine (Kim et al., 2010). The two TYPE 2 CL PEBBLEs were applied for in vivo imaging of hydrogen peroxide, externally injected (Lee et al., 2007) as well as endogenously produced during LPS-induced inflammatory response (Lee et al., 2007; Lim et al., 2010).

4.2.2. PEBBLEs for other ROS

Construction of PEBBLEs for highly reactive ROS is a challenging task as the NP matrix may block the entry of short-lived reactive ROS. Ratiometric TYPE 1 single PEBBLEs have been developed for sensing hydroxyl radical (King and Kopelman, 2003) and singlet oxygen (Cao et al., 2004), respectively. The hydroxyl radical sensing PEBBLE was prepared by attaching the hydroxyl indicator dye, coumarin-3-carboxylic acid, onto the PAA NP surface while encapsulating the reference dye deep inside it (King and Kopelman, 2003). Such design enables highly reactive OH· to reach the probes without being destroyed while protecting the reference dye inside, which is a similar concept as that of the TYPE 1 core–shell type PEBBLE developed later. Note that the hydroxyl radical is the most reactive ROS with a lifetime in aqueous phase on the order of 1 ns (Roots and Okada, 1975). The singlet oxygen sensing PEBBLE was prepared using ormosil and encapsulated singlet oxygen probe, 9,10-dimethylanthracene, and a reference dye, octaethylporphine (Cao et al., 2004). The much longer lifetime of singlet oxygen in the ormosil matrix, compared to aqueous solutions—in addition to the relatively high singlet oxygen solubility because of the highly permeable structure and the hydrophobic nature of the ormosil NPs—results in an excellent overall response to singlet oxygen.

4.3. Glucose sensing PEBBLEs

Glucose sensing PEBBLE designs include TYPE 2 enzyme (Gill et al., 2008; Lim et al., 2010; Shiang et al., 2009; Xu et al., 2002) and TYPE 2 ET PEBBLEs (Cordes et al., 2006; Tang et al., 2008; Wang et al., 2011b; Wu et al., 2009b). Glucose sensing TYPE 2 enzyme PEBBLE designs utilize GOx that catalyzes the oxidation of β-D-glucose through a two-step reaction (Scheme 21.1), measuring glucose concentration indirectly either by oxygen depletion (Xu et al., 2002) or by hydrogen peroxide production (Gill et al., 2008; Lim et al., 2010; Shiang et al., 2009).

In the case of oxygen-based glucose PEBBLEs, GOx, an oxygen-sensitive ruthenium-based dye, and a reference dye, were incorporated within a PAA NP (Xu et al., 2002). The dynamic range of thus prepared glucose sensing PEBBLE was found to be ~ 0.3–8 mM. In case of a H_2O_2-based glucose PEBBLE, GOx was conjugated to the H_2O_2 sensing QD-based TYPE 2 ET PEBBLE (Gill et al., 2008). The dynamic range was 1–10 mM. In anther two cases, an H_2O_2 sensing PEBBLE—FROC NP-based TYPE 2 CL PEBBLE (Lim et al., 2010) or GNP-based TYPE 2 ET PEBBLE (Shiang et al., 2009)—was mixed with GOx for glucose measurements. The FROC NP PEBBLE showed the dynamic range of 0–5 mM and was applied to glucose measurements in serum and *in vivo* (Lim et al., 2010).

$$\beta\text{-D-glucose} + \text{GOx-FAD} \rightleftharpoons \text{GOx-FADH}_2 + \delta\text{-D-luconolactone}$$

$$\text{GOx-FADH}_2 + O_2 \longrightarrow \text{GOx-FAD} + H_2O_2$$

Scheme 21.1 Glucose oxidation by glucose oxidase (GOx). Here, FAD is the the glucose oxidase cofactor, flavin adenine dinucleotide, and FADH$_2$ is the reduced form of FAD.

TYPE 2 ET PEBBLEs for glucose have been constructed using QDs (Cordes et al., 2006; Tang et al., 2008; Wu et al., 2009b) and PNIPAM NP (Wang et al., 2011c). In one QD-based design, QD forms a complex with boronic acid-substituted viologen quenchers and its fluorescence is quenched (Cordes et al., 2006). When glucose binds with the boronic acid substituted quencher, the quencher–QD interaction is broken, inducing a robust fluorescence recovery. In another QD-based design, a CdTe QD conjugated with concanavalin A (ConA)—a glucose-recognizing protein—forms an assembly with thiolated β-cyclodextrins (beta-SH-CDs)-modified GNPs (beta-CDs-AuNPs) (Tang et al., 2008). Here the CdTe QD is an energy donor and GNP is an energy acceptor. In the presence of glucose, the AuNPs-beta-CDs segment of the assembly is displaced by glucose, which competes with beta-CDs on the binding sites of ConA, resulting in the fluorescence recovery of the quenched QDs. The QDs-ConA-beta-CDs-AuNPs showed a dynamic range of 0.1–50 μM with a detection limit of 50 nM. In the other QD-based design, CdS QDs are incorporated in the copolymer microgel NPs of poly(N-isopropylacrylamide-acrylamidephenylboronic acid) [p(NIPAM-AAm-PBA)] (Wu et al., 2009b). The QDs' fluorescence is reversibly quenched and dequenched when the microgel undergoes swelling and deswelling in response to the glucose concentration change. The hydrodynamic radius of the PEBBLE in PBS of pH 8.8 at 22 °C changes from 135 to 165 nm when the glucose concentration varies from 1 to 25 mM. The PNIPAM-based TYPE 2 ET PEBBLE for glucose is made of temperature-responsive PNIPAM microgels containing covalently incorporated glucose-recognizing moieties, N-acryloyl-3-aminophenylboronic acid; FRET donor dyes, 4-(2-acryloyloxyethylamino)-7-nitro-2,1,3-benzoxadiazole (NBDAE); and rhodamine B-based FRET acceptors (RhBEA) (Wang et al., 2011c). The distance between FRET donors and acceptors within microgels—hence, the FRET efficiency—can be tuned via the thermo-induced microgels' collapse or glucose-induced NP swelling at appropriate pH and temperatures, enabling dual ratiometric fluorescent sensing for glucose and temperature. The PNIPAM-based glucose PEBBLE showed a higher glucose sensitivity at 37 °C than 25 °C, with a dynamic range of 0–100 mM.

Although the detectable ranges of the glucose PEBBLEs are within physiological glucose concentrations (2.5–20 mM) (Cordes et al., 2006), the PEBBLEs have not yet been applied for intracellular measurements. They were applied for measurements in serum (Lim et al., 2010; Tang et al., 2008) and *in vivo* (Lim et al., 2010).

4.4. pH sensing PEBBLEs

The positively charged hydrogen ion (H^+) is one of the most important intracellular analytes and has been studied the most by PEBBLE technology, as shown in Table 21.3. The designs of pH PEBBLE sensors include

Table 21.3 Fluorescent PEBBLE sensors for pH

Sensor type	Sensing signal	NP matrix with loaded components	pH range	Size	Intracellular application (cell type; delivery method)	References
TYPE 1 single	Int$_{ratio}$	PAA NP with encapsulated sensing (S) and reference (R) fluorescent dyes CNF (S) CDMF (S) + SR (R) BCPCF (S) + SR (R) FSA (S) + SR (R) SNAFL (S)	 7.0–7.7 6.2–7.4 6.2–7.2 5.8–7.0 7.2–8.0	20–100 nm (SEM)	Mouse oocyte; microinjection	Clark et al. (1999a)
TYPE 1 single	Int$_{ratio}$	PAA NP with covalently linked fluorescein (S) and RhB (R)[a]	5.8–7.2	50 nm (DLS)	—	Sun et al. (2006)
TYPE 1 single	Int$_{ratio}$	Silica NPs with covalently linked FITC (S) and encapsulated RuBPY (R)	4–7	42 nm (TEM); 45 nm (DLS)	Murine macrophage and Hela cell; endocytosis; pH$_{ic}$ was 4.8 for macrophage and 7.2 for Hela cell	Peng et al. (2007)
TYPE 1 single	Int$_{single}$	SWNTs bound with fluorescein-PEG-NHS	5.6–8.4	158 nm (AFM)	BT474 breast cancer cells; endocytosis	Nakayama-Ratchford et al. (2007)
TYPE 1 single	Int$_{ratio}$	Dextran NPs with covalently linked FITC (S) and SRB (R)	4.9–8.2	500 nm (SEM)	Human foreskin fibroblasts; endocytosis	Hornig et al. (2008)

(Continued)

Table 21.3 (Continued)

Sensor type	Sensing signal	NP matrix with loaded components	pH range	Size	Intracellular application (cell type; delivery method)	References
TYPE 1 single	Int$_{ratio}$	Positively charged PAA NP[b] with covalently linked dyes Fluorescein (S) + RhB (R) Oregon green (S) + RhB (R)	4.1–5.7 5.8–7.5	60–140 nm (DLS)	HepG2 cell; endocytosis	Sun et al. (2009)
TYPE 1 single	Int$_{ratio}$	PAA NP with covalently linked SR (R) and naphthalimide-based pH indicator (S)		28 nm (DLS)	Primary human foreskin fibroblasts; endocytosis	Schulz et al. (2010)
TYPE 1 single	Int$_{ratio}$	PVA-polyacetal shell with a gold core that is covalently linked with FITC (S) and RBITC (R)	5–8	20–30 nm (TEM)	CHO cells; endocytosis	Stanca et al. (2010)
TYPE 1 single	Int$_{ratio}$	PAA NP with encapsulated HPTS, a two-photon excitable ratiometric pH indicator dye	6–8	68 nm (DLS)	9 L glioma cell; endocytosis and receptor-mediated endocytosis; pH$_{ic}$ was 6.3 by endocytosis, and 7.1 by receptor-mediated endocytosis	Ray et al. (2011)
TYPE 1 core–shell	Int$_{ratio}$	Silica core–shell NPs: core containing covalently linked TRITC (R) and shell containing covalently linked FITC (S)	5–7.4	70 nm (50 nm core) (SEM)	Rat basophilic leukemia mast cells (RBL-2H3); endocytosis after co incubation with PDB; pH$_{ic}$ was 5.1–6.6	Burns et al. (2006)

TYPE 1 core–shell	Int$_{ratio}$	Polystyrene core–shell type NPs: core containing post-embedded DPA (R) and shell containing covalently linked FITC (S)[a]	4–6.5	18–20 nm (QELS)	—	Allard and Larpent (2008)
TYPE 1 core–shell	Int$_{ratio}$	PAA core–shell NPs: core containing covalently linked SR (R) and shell containing covalently linked naphthalimide-based pH indicator (S)[a]	5.0–8.2	33–44 nm (DLS)	Primary human foreskin fibroblasts; endocytosis	Schulz et al. (2010)
TYPE 2 ET	Int$_{ratio}$	CdSe/ZnS QD (620 nm em) coated with a hydrophobically modified poly(acrylic acid) and conjugated with a pH-sensitive squaraine dye	6–10	—[c]	—	Snee et al. (2006)
TYPE 2 ET	Int$_{single}$	QD (490 nm em) conjugated with fluorescein maleimide	5.5–7.0 and 7.5–8.5	—[c]	—	Suzuki et al. (2008)
TYPE 2 ET	Int$_{ratio}$	CdSe/ZnS QD (455 nm em) coated with an	5–8	—[c]	—	Chen et al. (2008),

(Continued)

Table 21.3 (Continued)

Sensor type	Sensing signal	NP matrix with loaded components	pH range	Size	Intracellular application (cell type; delivery method)	References
		amphiphilically modified poly(acrylic acid) and then conjugated with fluorescein maleimide				Krooswyk et al. (2010)
TYPE 2 ET	Lifetime and Int$_{ratio}$	CdSe/ZnS QD (550 nm em), surface modified with self-assembled dopamine–peptide containing (His)$_6$ sequence	4.8–10.1 (Int$_{ratio}$); 6.5–11.5 (Lifetime)	—^{c}	COS-1 cells; microinjection	Medintz et al. (2010)
TYPE 2 ET	Int$_{ratio}$	Polyurethane NP with encapsulated pH indicator bromothymol blue and pH insensitive fluorophores (coumarin 6 and Nile red)	6–8	20–30 nm (SEM); 127–137 nm (DLS)	Epithelial normal rat Kidney cells; endocytosis	Peng et al. (2010c)
TYPE 2 ET	Int$_{ratio}$	Silica with two CdSe/ZnS QDs: encapsulated QD (540 nm em) (R); and surface-conjugated QD (610 nm) that is coated with MPA and Nile blue	4.1–8.9	300 nm (TEM)	T-REx293 cells; gene-gun delivery	Wang et al. (2010a)

| | TYPE 2 ET | Lifetime | CdSe/ZnS QD coated with MPA | 5.2–6.9 | — | —[c] | | Ruedas-Rama et al. (2011) |
| | TYPE 2 ET | Int$_{ratio}$ | PPE CPdots (440 nm em) conjugated with fluorescein | 5–8 | | 20–30 nm (TEM) | HeLa cell; endocytosis; pH$_{ic}$ was 4.8–5.0 | Chan et al. (2011) |

Chemical acronyms: BCPCF, 2′,7′-bis-(2-carboxypropyl)-5-(and 6)carboxyfluorescein; CDMF, 5-(and 6-)carboxy-4′,5′-dimethylfluorescein; CNF, 5-(and 6-) carboxynaphthofluorescein; DPA, 1,9-diphenylanthracene; FITC, fluorescein isothiocyanate; FSA, fluorescein-5-(and 6)sulfonic acid; HPTS, 8-hydroxypyrene-1,3,6-trisulfonic acid; PDB, phorbol-12,13-dibutyrate; PEG, polyethylene glycol; PVA, poly(vinyl alcohol); PPE, poly(2,5-di(3′,7′-dimethyloctyl)phemylene-1,4-ethynylene); RhB, rhodamine B; RuBPY, rhodamine B isothiocyanate; RuBPY, tris(2,2′-bipyrdyl)dichlororuthenium(II) hexahydrate; SNAFL, 5-(and 6)-carboxy SNAFL-1; SR, sulforhodamine; SRB, sulforhodamine B; TRITC, tetramethylrhodamine isothiocyanate.

[a] Use two different excitation lights for sensing dye and reference dye.

[b] Positively charged polyacrylamide was prepared from acrylamide monomer mixture containing 4.5–24% of positively charged monomer (3-acrylamidopropyl) trimethylammonium chloride.

[c] The typical size of QDs is less than 10 nm.

TYPE 1 single PEBBLEs made of PAA (Clark et al., 1999a,b; Ray et al., 2011; Schulz et al., 2010; Sun et al., 2006, 2009), silica (Peng et al., 2007; Wang et al., 2010a), dextran (Hornig et al., 2008), gold core with poly(vinyl alcohol)-polyacetal shell (Stanca et al., 2010), and single-wall carbon nanotubes (Nakayama-Ratchford et al., 2007); TYPE 1 core–shell PEBBLEs made of PAA (Schulz et al., 2010), silica (Burns et al., 2006), and polystyrene (Allard and Larpent, 2008); and TYPE 2 ET PEBBLEs made of QDs (Chen et al., 2008; Krooswyk et al., 2010; Medintz et al., 2010; Ruedas-Rama et al., 2011; Snee et al., 2006; Suzuki et al., 2008; Wang et al., 2010a), CPdots (Chan et al., 2011), and polyurethane (Peng et al., 2010c). The pH sensing has been mostly based on fluorescence intensity ratio (Int_{ratio}) but also on single peak fluorescence intensity (Int_{single}) and lifetime. It is noteworthy that two-photon excited fluorescence, using NIR excitation, was used in one reported case (Ray et al., 2011), so as to avoid interference from cellular autofluorescence in the intensity-based measurements.

For TYPE 1 PEBBLEs, fluorescent pH indicators and reference dyes are loaded into inert NPs. For TYPE 2 PEBBLEs, there is one design that uses inert NPs and dyes as in the case of TYPE 1 PEBBLEs. Instead of fluorescent pH indicator sensing and reference dyes, three dyes—a nonfluorescent pH indicator dye (bromothymol blue) and two pH insensitive fluorescent dyes that are FRET pairs to the bromothymol blue—are loaded into inert polyurethane NPs for ratiometric measurements (Peng et al., 2010c). All the rest of TYPE 2 ET PEBBLE designs use fluorescent semiconducting NPs, QDs or CPdots. It should be noted that the fluorescence of QDs arises from the recombination of excitons so that any changes in charge or composition of the QD surface could affect the efficiency of the core electron–hole recombination and consequently the fluorescence efficiency (Wang et al., 2010a). For instance, water solubilized QDs capped with small molecules, including cysteine and carboxylic acids such as mercaptopropionic acid (MPA), are generally very pH sensitive (Ruedas-Rama et al., 2011). However, such QDs capped with non-fluorescent ligands produce only a single fluorescence peak, and these QDs alone are not suitable for ratiometric intensity measurements. In one case, separate fluorescent reference nanospheres were used together with these QDs (Medintz et al., 2010), and in another case, these pH-sensitive QDs were surface conjugated to bigger, inert polymer NPs that contain pH insensitive QDs inside (Wang et al., 2010a). However, the most common form of TYPE 2 ET PEBBLEs based on QDs are QDs coated/conjugated with fluorescent pH indicator dyes that are a FRET pair to the QDs, producing two pH-sensitive fluorescent peaks. A similar design was applied to organic semiconductor fluorescent NPs—CPdots made of poly(2,5-di(3′,7′-dimethyloctyl)phenylene-1,4-ethynylene) (PPE) were conjugated with fluorescent pH indicator dye for ratiometric pH measurements (Chan et al., 2011).

Fluorescein derivatives are the most commonly used pH indicator dyes, both for TYPE 1 and for TYPE 2 ET PEBBLEs. As the measurable pH range is determined mostly by the incorporated pH indicator dyes, the pH range of 5–8 is the most common for the pH sensing PEBBLEs developed so far. Many of these PEBBLEs were introduced to cells, mostly by endocytosis (Burns et al., 2006; Chan et al., 2011; Clark et al., 1999a; Hornig et al., 2008; Nakayama-Ratchford et al., 2007; Peng et al., 2007, 2010c; Ray et al., 2011; Schulz et al., 2010; Stanca et al., 2010; Sun et al., 2009) but also by microinjection (Medintz et al., 2010), and by gene-gun delivery (Wang et al., 2010a), demonstrating intracellular delivery of the PEBBLEs and the absence of cytotoxic effects due to PEBBLEs, even up to a period of 22 days (Hornig et al., 2008). The endocytosis of the PEBBLEs was typically performed by incubation of cells with PEBBLEs, but in one case, a compound known to increase endocytotic activity, phorbol-12,13-dibutyrate, was coincubated to facilitate the endocytosis of silica NPs (Burns et al., 2006). In several applications, the intracellular pH (pH_{ic}) was measured (Burns et al., 2006; Chan et al., 2011; Peng et al., 2007; Ray et al., 2011) and also changes in pH_{ic}, accompanying external stimulations by drugs (Medintz et al., 2010; Peng et al., 2007), were monitored in real time. The PEBBLEs delivered through endocytosis and receptor-mediated endocytosis seem to remain in the endosome—with the measured pH_{ic} values being less than 7.4, a typically known cytoplasmic pH—although the endocytosed PEBBLEs may not end up in the lysosomes (Ray et al., 2011; Stanca et al., 2010).

4.5. Other cation sensing PEBBLEs

4.5.1. Copper ion sensing PEBBLEs

The copper ion, mostly Cu^{2+} except for one reported case (Sumner et al., 2005) including both Cu^{2+} and Cu^+, is the second most studied ion by PEBBLEs. This is probably because the copper ion is a powerful fluorescence quencher and is selectively recognized by many ligands. Copper ion sensing PEBBLE designs include TYPE 1 single and TYPE 2 ET PEBBLEs. TYPE 1 single PEBBLEs contain Cu^+/Cu^{2+} sensitive fluorescent dyes or proteins—encapsulated DsRed (Sumner et al., 2005), (2-(4-amino-2-hydroxyphenyl) benzothiazole derivative (Cui et al., 2011) and lucifer yellow-CH (Borisov et al., 2008a), and conjugated diethyl iminodiacetate fluorescein (Seo et al., 2010)—inside or on the surface of NPs made of PAA (Sumner et al., 2005), poly(ionic liquid) (Cui et al., 2011), poly(styrene-block-vinylpyrrolidone) (Borisov et al., 2008a), and silica (Seo et al., 2010). TYPE 2 ET PEBBLEs were prepared typically by loading nonfluorescent copper ion chelating ligands—such as cyclam (Frigoli et al., 2009; Gouanvé et al., 2007; Meallet-Renault et al., 2004, 2006), picolinamide (Arduini et al., 2005; Brasola et al., 2003; Rampazzo et al., 2005), picolinamine (Yin et al., 2009), polyamine (Montalti et al., 2005), PEI (Chen et al., 2009), peptide

(Gly-His-Leu-Leu-Cys) (Gattas-Asfura and Leblane, 2003), and thioglycerol (Chen and Rosenzweig, 2002)—and fluorescent reporter dyes inside or on the surface of NPs made of latex (Gouanvé et al., 2007; Meallet-Renault et al., 2004, 2006), polystyrene (Frigoli et al., 2009), PMMA (Chen et al., 2009), PNIPAM (Yin et al., 2009), and silica (Arduini et al., 2005; Brasola et al., 2003; Montalti et al., 2005; Rampazzo et al., 2005)—or using QDs as fluorescent reporters (Chen and Rosenzweig, 2002; Gattas-Asfura and Leblane, 2003). A Cu^{2+} sensing TYPE 2 ET PEBBLE was also prepared using GNPs coordinated with the fluorescent chromophore-containing pyridyl moieties in which the fluorescence quenching changes in the presence of Cu^{2+} because of the stronger coordination ability of the Cu^{2+} ion with the pyridyl moiety, in comparison with that of the GNPs (He et al., 2005).

The copper ion detection of these PEBBLEs is based on a single intensity peak except one design each for TYPE 1 and TYPE 2 ET PEBBLEs. In the case of the TYPE 1 ratiometric PEBBLE, a reference dye was co-loaded with the copper sensing DsRed protein into the PAA NPs (Sumner et al., 2005). In the case of the TYPE 2 ET ratiometric PEBBLE, a pair of fluorescent dyes that are energy donor and acceptor, 9,10-diphenylanthracene (DPA) as the donor (D) (374 nm ex, 411 nm em) and pyrromethene 567 (PM567) (536 nm em) as the acceptor (A), were postloaded into polystyrene NPs grafted with cyclams (Frigoli et al., 2009). In the absence of cupric ions, when the donor DPA is excited, the energy is transferred to the proximal acceptor PM567 within the cyclam-grafted NPs, leading to emission of PM567. In the presence of Cu^{2+}, a copper–cyclam complex forms at the surface and acts as a quencher (or a secondary acceptor). As a result, the sensitized emission of the primary acceptor (PM567) is efficiently attenuated through FRET to the quencher. The remaining emission of the primary donor (DPA) is less affected, enabling ratiometric measurements.

There is one reported case of an intracellular application in which the TYPE 1 PEBBLEs made of 10 nm silica (detection limit of 0.5 μM) were endocytosed into HeLa cells and a high copper concentration was simulated, by incubating the cells with 5.0 μM $Cu(ClO_4)$ for 30 min, so as to observe the response of the PEBBLEs (Seo et al., 2010). It should be noted that while copper is a biochemically essential metal for life and relatively abundant in cells (~ 10 μM total) (Finney and O'Halloran, 2003), the normal unbound ("free") copper ion level is only ~ 1 nM (Changela et al., 2003), which is below the detectable ranges of copper ions by the PEBBLEs reported so far, which are in the ranges of millimolar (Borisov et al., 2008a), micromolar (Arduini et al., 2005; Brasola et al., 2003; Chen and Rosenzweig, 2002; Chen et al., 2009; Cui et al., 2011; Gattas-Asfura and Leblane, 2003; Gouanvé et al., 2007; Montalti et al., 2005; Rampazzo et al., 2005; Seo et al., 2010), or nanomolar (Frigoli et al., 2009; Meallet-Renault et al., 2004, 2006; Sumner et al., 2005; Yin et al., 2009). In order to study the copper ion homeostasis under normal conditions, PEBBLEs with higher sensitivity need to be developed.

4.5.2. Zinc ion sensing PEBBLEs

Zinc ion (Zn^{2+}) sensing PEBBLE designs include TYPE 1 single, TYPE 1 core–shell, and TYPE 2 ET PEBBLEs. Examples of TYPE 1 single PEBBLEs for Zn^{2+} sensing include PAA NPs with encapsulated fluorescent dyes, Newport green (for sensing), and Texas Red-dextran (for reference) (Sumner et al., 2002); silica with covalently linked dyes, 6-methoxy-8-(p-toluensulfonamido)-quinoline (sensing) and a coumarin dye (reference) (Teolato et al., 2007); silica covalently linked with a dialdehyde fluorescent Zn^{2+} sensing chromophore, 4-methyl-2,6-diformyl phenol (Sarkar et al., 2009); and double hydrophilic block thermoresponsive copolymers (PEG-b-P(MEO$_2$MA-co-OEGMA) bearing quinoline-based Zn^{2+}-recognizing fluorescent moieties (ZQMA) (Liu and Liu, 2011). The PEBBLEs based on the ZQMA bearing thermoresponsive NP was also a TYPE 2 TW PEBBLE for temperature, showing enhanced fluorescence intensity ($\sim$6 times) when the temperature changed from 20 to 37 °C (Liu and Liu, 2011).

In case of TYPE 1 core–shell PEBBLE for zinc sensing, a core–shell type silica matrix was used, with a core covalently linked with a reference dye and a shell covalently linked with a zinc sensing dye, AQZ (carboxamido-quinoline with an alkoxyethylamino chain) (He et al., 2010). This PEBBLE was also utilized for ratiometric detection of $H_2PO_4^-$—after forming a complex with Zn^{2+}—based on the fact that $H_2PO_4^-$ is a well-known Zn^{2+} binder. The detection range was 6500 μM of $H_2PO_4^-$. The Zn^{2+} sensing TYPE 2 ET PEBBLEs were prepared using QDs. In one example, CdSe/ZnS QDs were covalently linked with three different azamacrocycles, nonfluorescent Zn^{2+} ligands: TACN (1,4,7-triazacyclononane), cyclen (1,4,7,10-tetraazacyclododecane), and cyclam (1,4,8,11-tetraazacyclotetradecane) (Ruedas-Rama and Hall, 2008a). As the surface-conjugated azamacrocycles disrupt the radiative recombination process of the QDs, the QDs' fluorescence is quenched. The binding of Zn^{2+} with the azamacrocycles switches on the QD emission and a dramatic increase of the fluorescence intensity results. It should be noted that cyclam has been used as Cu^{2+} ligand in several Cu^{2+} sensing PEBBLEs (Frigoli et al., 2009; Gouanvé et al., 2007; Meallet-Renault et al., 2004, 2006). Interference studies showed that Cu^{2+} at 0.0001 mM, Fe^{3+}/Fe^{2+} at 0.25 mM, and Co^{2+} at 0.001 mM produced quenching of the QD-azamacrocycle conjugates' fluorescence (Ruedas-Rama and Hall, 2008a). However, these free metal cations do not have a big footprint in physiological measurements, as in the majority of the biological systems they are present at a much smaller concentration than the interfering concentration. In another example, CdSe/dS QDs were coated with zincon, a nonfluorescent zinc ion sensing dye, by a layer-by-layer technique (Ruedas-Rama and Hall, 2009). Fluorescence of the PEBBLE (QD–zincon complex) is reduced, compared to that of the QD, due to charge transfer, electron tunneling, and/or the overlap of molecular orbitals. Upon binding with Zn^{2+}, energy transfer occurs from the QD as a donor to the zincon–Zn^{2+} complex, as acceptor, resulting in further fluorescence quenching of the QD–zincon complex. The same PEBBLE also shows sensitivity toward Mn^{2+}. In this case, the QD–zincon interaction is disrupted as a result of formation of a zincon–Mn^{2+} complex, causing recovery of the QD's fluorescence. The only ion that strongly interfered with the detection of Zn^{2+} and Mn^{2+} was Cu^{2+}, showing quenching of the fluorescence.

The detection range of most of the Zn^{2+} sensing PEBBLEs is in the micromolar range (He et al., 2010; Ruedas-Rama and Hall, 2008a, 2009; Sarkar et al., 2009; Sumner et al., 2002), but a nanomolar detection range was also achieved for several designs (Liu and Liu, 2011; Teolato et al., 2007). These Zn^{2+} sensing PEBBLEs may not be sensitive enough for measuring intracellular Zn^{2+} concentration under normal conditions. Just like for copper ion, the intracellular concentration of free Zn^{2+} is strictly controlled in cells and was recently reported to be only ~ 0.4 nM (Vinkenborg et al., 2009), despite the high total concentration of zinc inside such cells (~ 100 μM).

Several of the Zn^{2+} sensing TYPE 1 PEBBLEs have been introduced into HeLa (He et al., 2010; Liu and Liu, 2011; Sarkar et al., 2009) and A375 human melanoma cells (Sarkar et al., 2009) by endocytosis. In one case, the silica-based PEBBLEs were prepared in two different sizes, 120 and 960 nm, and compared for endocytosis efficiency (Sarkar et al., 2009). Only the 120 nm PEBBLEs were efficiently endocytosed into the cells. The changes in intracellular Zn^{2+} concentration were monitored after incubation of the cells with a medium of varying Zn^{2+} concentrations (He et al., 2010; Liu and Liu, 2011; Sarkar et al., 2009). In one case, a Zn^{2+} ionophore, pyrithione (2-mercaptopyridine N-oxide), was also used in order to bring extracellular Zn^{2+} into the cytoplasm (Sarkar et al., 2009).

4.5.3. Ca^{2+} sensing PEBBLEs

The Ca^{2+} sensing PEBBLEs developed so far include TYPE 1 single PEBBLEs (Clark et al., 1999a,b; Josefsen et al., 2010) and TYPE 1 core–shell PEBBLEs (Schulz et al., 2011). The Ca^{2+} sensing TYPE 1 single PEBBLEs were prepared using PAA NPs with encapsulated calcium-sensitive dyes—calcium green (Clark et al., 1999a,b; Josefsen et al., 2010) and calcium orange (Clark et al., 1999b)—and reference dyes. The Ca^{2+} sensing TYPE 1 core–shell PEBBLEs were prepared using silica–dextran core–shell particles with a core covalently linked with rhodamine-based reference dye (rhodamine B isothiocyanate) and a shell containing the covalently linked, Ca^{2+} sensitive dye Fluo-4 (Schulz et al., 2011). The detectable Ca^{2+} concentration range of these PEBBLEs is in the submicromolar (Clark et al., 1999a,b; Josefsen et al., 2010) or micromolar range (Clark et al., 1999b; Schulz et al., 2011). The typical free calcium concentration in the cytosol of resting cells is 50–200 nM (Fedrizzi et al., 2008), while the total calcium content in cells is typically more than three orders of magnitude higher than the free Ca^{2+} (Chandra et al., 1989). However, the developed PEBBLEs have been applied to monitor the increase in the calcium ion concentration induced by a toxin (*m*-dinitrobenzene) (Clark *et al.*, 1999b), mitogen (ConA) (Clark *et al.*, 1999a) or photodynamic therapy (Josefsen *et al.*, 2010). The investigated cell lines included C6 glioma (Clark *et al.*, 1999b), human embryonic kidney cells (HEK 293) (Josefsen *et al.*, 2010), rat alveolar macrophage (Clark *et al.*, 1999a), and SY5Y human neuroblastoma (Clark *et al.*, 1999b; Josefsen *et al.*, 2010). The PEBBLEs were delivered into cells by liposomes (Clark *et al.*, 1999b; Josefsen *et al.*, 2010) and phagocytosis (Clark *et al.*, 1999a).

4.5.4. K^+, Na^+ sensing PEBBLEs

The designs of K^+ sensing PEBBLEs include TYPE 1 core–shell (Brown and McShane, 2005), TYPE 2 IC (Brasuel *et al.*, 2001; Ruedas-Rama and Hall, 2007), and TYPE 2 ET PEBBLEs (Yin *et al.*, 2010); those of Na^+ sensing PEBBLEs include TYPE 1 single (Stanca *et al.*, 2010) and TYPE 2 IC PEBBLEs (Brasuel *et al.*, 2002; Dubach *et al.*, 2007a,b). The K^+ sensing TYPE 1 core–shell PEBBLE was made of fluorescent europium core NP with a polyelectrolyte shell containing potassium-binding benzofuran isophthalate. The Na^+ sensing TYPE 1 PEBBLE was made of a GNP stabilized with poly(vinyl alcohol) and a polyacetal shell containing Sodium Green as a Na^+ sensing probe, and rhodamine B isothiocyanate as a reference dye (Stanca *et al.*, 2010). The TYPE 2 IC PEBBLEs have been made of PDMA (Brasuel *et al.*, 2001, 2002), PnBA (Ruedas-Rama and Hall, 2007), and PVC (Dubach *et al.*, 2007a,b), embedded with three sensing components: a non-fluorescent ionophore; a fluorescent hydrogen ion selective dye, alone or together with QDs as a reporter; and a lipophilic additive that maintains ionic strength. In these TYPE 2 IC PEBBLEs, the characteristics of the sensors—such as its dynamic range, selectivity, sensitivity, and response time—were modulated by the composition of the three components (ionophore, pH sensitive dye, and lipophilic additive) (Brasuel *et al.*, 2001; Ruedas-Rama and Hall, 2007) and incorporation of QDs (Dubach *et al.*, 2007b). The K^+ sensing TYPE 2 ET PEBBLE was made of thermoresponsive PNIPAM microgel NPs covalently incorporated with K^+-recognizing 4-acrylamidobenzo-18-crown-6 residues (B18C6Am), fluorescence resonance energy transfer (FRET) donor dyes, NBDAE, and RhBEA by utilizing K^+-induced changes in microgel volume phase transition (VPT) temperature (Yin *et al.*, 2010). B18C6Am moieties within the NPs can preferentially capture K^+ via the formation of 1:1 molecular recognition complexes, resulting in the enhancement of microgel hydrophilicity and elevated VPT temperature, making the NPs responsive to temperature as well as K^+. The swelling/deswelling of the PEBBLEs is monitored by changes in fluorescence intensity ratios, that is, FRET efficiencies.

The K^+ and Na^+ sensing TYPE 2 IC PEBBLEs made of PDMA were introduced into C6 glioma cells by a gene-gun (Brasuel *et al.*, 2001, 2002). The response of the PEBBLE sensors inside the cells during ion-channel stimulation by kainic acid, a K^+ channel opening agonist, showed an increase in K^+ and Na^+ concentrations.

4.5.5. PEBBLEs for Fe^{3+}, Mg^{2+}, Hg^{2+}, and Pb^{2+}

TYPE 1 single PEBBLEs for sensing Mg^{2+} (Park et al., 2003) and Fe^{3+} (Sumner and Kopelman, 2005) have been developed using PAA NPs with encapsulated sensing dyes—coumarin 343 for Mg^{2+} and Alexa Fluor 488 for Fe^{3+}—and reference dyes (Texas Red). It is interesting to note that coumarin 343 is a cell-impermeable hydrophilic dye and therefore the dye itself cannot be used for intracellular measurements despite its superior selectivity for Mg^{2+} over Ca^{2+}—the most common interfering intracellular ion for most of available Mg^{2+} probes. The Mg^{2+} PEBBLEs were delivered into C6 glioma cells, by a gene-gun, and into human macrophage cells, by phagocytosis, for monitoring intracellular changes in Mg^{2+} concentration (Park et al., 2003).

Pb^{2+} sensing PEBBLEs have been developed based on TYPE 1 core–shell PEBBLE design. One design is made of a four-layered silica–core containing a covalently linked reference dye (methoxynaphthalene), the first shell made of pure silica for separation, the second shell containing covalently linked signal transducer dye (dansyl dye linked triethoxy silane) and the outermost third shell containing (mercaptopropyl)triethoxysilane (MPS) (Arduini et al., 2007). Here, the surface thiols, from MPS, significantly enhanced the sensitivity. The PEBBLEs showed a very similar sensitivity toward Cu^{2+}, making Cu^{2+} a serious interference. In another design, a Fe_3O_4 core with a silica shell is prepared and then coated with an additional silica layer containing conjugated Pb^{2+} sensitive BODIPY dyes (Son et al., 2010). The Fe_3O_4 core/silica shell PEBBLEs were introduced into Hela cells by endocytosis, demonstrating the PEBBLEs' response and reversibility in live cells by switching the media containing Pb^{2+} and ethylenediaminetetraacetic acid.

A Hg^{2+} sensing TYPE 2 ET PEBBLE was prepared using core/corona micelles formed by a poly(ethylene oxide)-b-polystyrene diblock copolymer (Ma et al., 2011). Here, a hydrophobic fluorescein derivative (FLS-C12), which serves as the energy transfer donor, is incorporated into the micelle core during the micelle formation, and a spirolactam-rhodamine derivative (RhB-CS) as a probe for mercury ions is located at the micelle core/corona interface. An efficient ring-opening reaction of RhB-CS induced by mercury ions generates the long-wavelength rhodamine B fluorophore which can act as the energy acceptor. The detection limit was 0.1 μM in water.

4.6. Anion sensing PEBBLEs

For anion sensing, PEBBLEs have been developed for the detection of only two anions: phosphate and chloride. A phosphate sensing TYPE 1 single PEBBLE was developed by embedding phosphate-sensitive fluorescent reporter proteins (FLIPPi) in PAA NPs (Sun et al., 2008). The sensor activity and protein loading efficiency varied with NP composition, that is, the total monomer content and the cross-linker content. Chloride sensing PEBBLEs have been constructed based on TYPE 1 single (Graefe et al., 2008), TYPE 2 ET (Ruedas-Rama and Hall, 2008b), and TYPE 2 IC (Brasuel et al., 2003) PEBBLE designs. The Cl^- TYPE 1 single PEBBLE was prepared using a PAA NP with the encapsulated Cl^--sensitive dye, lucigenin, and a reference dye, sulforhodamine derivative, and its detection range was 0–18.2 mM (Graefe et al., 2008). The Cl^- TYPE 2 ET PEBBLE was prepared by coating MPA-capped CdSe–ZnS QD, with lucigenin, and is based on an electrostatic interaction through negative-charged QDs and the positive charge of lucigenin (Ruedas-Rama and Hall, 2008b). Mutual quenching of the lucigenin and QD was observed due to formation of the QD–MPA–lucigenin conjugate probably due to spin–orbit coupling or electron transfer between the QD and the lucigenin dication (Luc^{2+}). The conjugate luminescence is restored by adding chloride ions. The PEBBLE showed a very good linearity in the range 1–250 mM, with a detection limit of 0.29 mM. The Cl^- TYPE 2 IC PEBBLE was made of PDMA NPs containing the optically silent chloride ionophore III (ETH 9033) and the chromoionophore III (ETH 5350). It showed a linear dynamic range of 0.4–190 mM Cl^-, with a detection limit of 0.2 mM Cl^- at a pH of 7.2. All the three types of Cl^- sensing PEBBLEs have a dynamic range that is suitable for intracellular measurements. Note that chloride concentrations range from 2 mM in skeletal muscle, 20–40 mM in epithelial cells, to 90 mM in erythrocytes (Graefe et al., 2008). The TYPE 1 single and TYPE 2 IC PEBBLEs were applied for intracellular measurements. The TYPE 1 single Cl^- PEBBLEs were delivered into Chinese hamster ovary (CHO) cells and mouse fibroblasts by liposomal delivery (Graefe et al., 2008). The PEBBLE responded to changes in chloride concentrations in the cell. In this work, the intracellular [Cl^-] was set equal to the extracellular [Cl^-] by using tributyltin and the ionophore nigericin. Tributyltin acts as a Cl^-/OH^- antiporter, which exchanges Cl^- for OH^- ions. To prevent a shift of the intracellular pH, which would have been invariably caused by the Cl^-/OH^- exchange, the K^+/H^+ antiporter nigericin was added, which clamped the intracellular pH to the extracellular pH. The TYPE 2 IC PEBBLEs were delivered into C6 glioma cells, utilizing a gene-gun, and intracellular chloride levels were monitored during ion-channel stimulation by kainic acid.

4.7. PEBBLEs for enzymatic intracellular processes

Intracellular processes are often triggered by enzymes. PEBBLEs have been developed for detecting activities of the enzymes, including caspases for triggering apoptosis (Kim et al., 2006; Myc et al., 2007; Oishi et al., 2009), protein kinases (PKAs) for phosphorylation (Kim et al., 2007a,b), and other enzymes like proteases (Lee et al., 2008; Suzuki et al., 2008), DNase (Suzuki et al., 2008), and DNA polymerase (Suzuki et al., 2008).

4.7.1. Apoptosis sensing PEBBLEs

Apoptosis, a physiological process that consists of a programmed or suicidal cell death, is triggered by a series of enzymes named caspases (cysteine proteases). PEBBLEs for detecting apoptosis have been developed utilizing the Asp-Glu-Val-Asp (DEVD) peptide sequence that is selectively cleaved by caspases. The apoptosis PEBBLE designs include TYPE 1 single and TYPE 2 ET PEBBLEs. A TYPE 1 single PEBBLE for apoptosis is made of G5 poly(amidoamine) dendrimer conjugated with a FRET-based apoptosis agent, PhiPhiLux G1D2 that contains the protease recognition sequence, DEVDGI (Myc et al., 2007). TYPE 2 ET PEBBLEs for apoptosis were prepared in two ways. The first one is the NP made of PEI modified with deoxycholic acid (DOCA) hydroxysuccinimide ester and conjugated with Cy5.5-labeled DEVD peptides (Cy5.5-DEVD-PEI-DOCA NP) (Kim et al., 2006). The second one is PEGylated poly diethylamino ethyl methacrylate nanogel (80 nm by DLS) that contains GNPs (8.5 nm) (fluorescence quenchers) in the cross-linked gel core and FITC-labeled DEVD peptides at the tethered PEG chain ends (FITC–DEVD–nanogel–GNP) (Oishi et al., 2009). Both PEBBLEs initially showed very little fluorescence through FRET process but strong fluorescence in the presence of caspases, caspase-3 (Kim et al., 2006; Oishi et al., 2009) and/or caspase-7 (Kim et al., 2006). In case of the FITC–DEVD–nanogel–GNP, the restored fluorescence was estimated to be only $\sim$25% of the intensity recovered after cyanide etching treatment, indicating restricted accessibility of caspase-3 toward DEVD peptide, most likely due to the steric hindrance of the tethered PEG chains of the nanogel (Oishi et al., 2009). The three apoptosis PEBBLEs were all applied for intracellular studies. The dendrimer PEBBLEs, conjugated with folic acids, were delivered into KB cells (overexpressing folate receptors) through receptor-mediated endocytosis (Myc et al., 2007). The Cy5.5-DEVD-PEI-DOCA NPs were delivered into Hela cells by PEI (Kim et al., 2006), while the FITC–DEVD–nanogel–GNPs were delivered into human hepatocyte (HuH-7) cells or HuH-7 multicellular tumor spheroids by endocytosis (Oishi et al., 2009). The cells loaded with the PEBBLEs, after being treated with apoptosis-inducing agents, staurosporine (Myc et al., 2007; Oishi et al., 2009) or TRAIL (Kim et al., 2006), showed strong fluorescence signals, visualizing caspase-dependent apoptosis.

4.7.2. Phosphorylation sensing PEBBLEs

TYPE 2 ET PEBBLEs were also developed for detecting phosphorylation by PKAs (Kim et al., 2007a,b). The PEBBLEs were prepared using a self-assembly of PEI and poly(aspartic acid) (PASA), where FITC (Kim et al., 2007a,b) or Cy5.5 (Kim et al., 2007a) labeled PKA-specific substrates (Leu–Arg–Arg–Ala–Ser–Leu–Gly, kemptide) were conjugated to PEI. The self-assembled NPs are dissociated by phosphorylation—because negatively charged phosphate groups are incorporated into the serine residue of kemptide, resulting in polyelectrolyte solubilization—restoring the originally quenched fluorescence. In one design, a reference dye, tetramethyl rhodamine isothiocyanate, was conjugated to PASA, for ratiometric measurements (Kim et al., 2007b). The PEBBLEs were delivered to PKA-overexpressing CHO-K1 cells by PEI-assisted endocytosis, visualizing PKAs' activities (Kim et al., 2007a).

4.7.3. PEBBLEs for sensing activities of other enzymes

A TYPE 2 PEBBLE for detecting the activity of matrix metalloproteases (MMPs) was prepared using GNPs (20 nm) stabilized with a Cy5.5 substrate, namely, Cy5.5-Gly-Pro-Leu-Gly-Val-Arg-Gly-Cys-(amide), where the peptide sequence of Pro-Leu-Gly-Val-Arg shows selectivity for MMP (Lee et al., 2008). MMPs are a family of zinc-dependent endopeptidases that play key roles in several biological processes, including promotion of cancer progression. When the target proteases meet the PEBBLEs, cleavage of the Cy5.5 substrate occurs as a consequence of the specific substrate recognition by the protease, restoring the NIR fluorescence of the Cy5.5. TYPE 2 PEBBLEs for detecting the activity of a protease (e.g., trypsin), DNase, and DNA polymerase were developed using QDs (Suzuki et al., 2008). These enzymes interact with moieties bound to QDs—cleavage of a GFP variant with an inserted sequence recognized by a protease to release GFP from the QD surface, digestion by DNase of dsDNA (labeled with fluorescent dUTP) bound to a QD, incorporation of fluorescently labeled dUTPs into ssDNA on a QD by extension with DNA polymerase—resulting in fluorescence intensity changes. The MMP-responsive PEBBLEs were applied for *in vivo* measurements in mice bearing SCC7 tumors (SCC7: squamous cell carcinoma, a MMP-2 expressed tumor cell line). The PEBBLEs produced a high NIR fluorescence signal—which significantly reduced in case of administration of the MMP-2 inhibitor, visualizing the activity of MMP-2 in tumor.

4.8. PEBBLEs for physical properties

4.8.1. Temperature sensing PEBBLEs

Temperature is the physical property that has been studied the most by PEBBLEs. Measuring cellular temperature may explain intricate biological processes and contribute to achieving optimal therapeutic results in photothermal cancer therapy. The temperature was measured by PEBBLEs, based on fluorescence lifetime (Borisov et al., 2008a; Peng et al., 2010a), or fluorescence intensity (Gota et al., 2009), or the ratio of two peaks (Peng et al., 2010b; Vetrone et al., 2010). The PEBBLE designs include TYPE 1 single (Borisov et al., 2008a; Peng et al., 2010a,b), TYPE 1 T-sensitive NP

(Vetrone et al., 2010), and TYPE 2 TW PEBBLEs (Gota et al., 2009). The TYPE 1 single PEBBLEs are prepared in two ways: (1) a core–shell NP, core made of mixture of 2-bis(trimethoxysilyl)decane and PMMA and shell made of silica, containing a temperature probe, europium(III)-tris(dinaphthoylmethane)-bis-(trioctylphosphine oxide) in the core (Peng et al., 2010a,b), with (Peng et al., 2010a) or without (Peng et al., 2010b) a reference dye; (2) poly(styrene-block-vinylpyrrolidone) doped with europium(III) tris(thenoyltrifluoroacetonate) dipyrazoltriazine complex (Borisov et al., 2008a). The temperature detection range of these TYPE 1 single PEBBLEs was 0–50 °C. A TYPE 1 single T-sensitive NP PEBBLE was made of lanthanide-doped fluorescent NPs—$NaYF_4:Er^{3+},Yb^{3+}$ NPs, where the intensity ratio of the green fluorescence bands of the Er^{3+} dopant ions ($^2H_{11/2} \rightarrow {}^4I_{15/2}$ and $^4S_{3/2} \rightarrow {}^4I_{15/2}$) changes with temperature—coated with PEI for water dispersibility (Vetrone et al., 2010). It should be noted that the NPs are capable of upconverting long-wavelength NIR light (920–980 nm) to shorter-wavelength light, via energy transfer from Yb^{3+} ions to the fluorescent Er^{3+} ions, avoiding autofluorescence of cellular components and allowing deep tissue penetration in vivo. Note that the $NaYF_4:Er^{3+},Yb^{3+}$ NP has multiple peaks, enabling intensity-based ratiometric measurements. The temperature detection range was 25–60 °C. A TYPE 2 TW PEBBLE for temperature was prepared using a poly(NIPAM-co-DBD-AA) NP made from copolymerization of N-isopropylacrylamide (NIPAM) and N-{2-[(7-N,N-dimethylaminosulfonyl)-2,1,3-benzoxadiazol-4-yl](methyl)amino}ethyl-N-methylacrylamide(DBD-AA) (Gota et al., 2009). At a lower temperature, the poly(NIPAM-co-DBD-AA) NP swells by absorbing water into its interior, where the fluorescence of the water-sensitive DBD-AA units is quenched by the neighboring water molecules. When heated, the NP shrinks with the release of water molecules, resulting in restoration of fluorescence from the DBD-AA units.

A couple of PEBBLEs were applied for temperature measurements inside cells. First example is the poly(NIPAM-co-DBD-AA) NPs that were introduced into COS7 cells by microinjection (Gota et al., 2009). The temperature resolution of the intracellular poly(NIPAM-co-DBD-AA) NP was 0.29–0.50 °C over the range 27–33 °C. The PEBBLEs were successfully used to monitor intracellular temperature variations induced by external chemical stimuli such as carbonyl cyanide 4-(trifluoromethoxy)phenylhydrazone (FCCP), an uncoupler of oxidative phosphorylation in mitochondria, causing heat production by respiration, and camptothecin, a DNA topoisomerase I inhibitor. A second example are the $NaYF_4:Er^{3+}$, Yb^{3+} NPs that were delivered into Hela cells by PEI (Vetrone et al., 2010). The PEBBLEs measured the internal temperature of the living cell from 25 °C to its thermally induced death at 45 °C.

4.8.2. Electric field sensing PEBBLEs

A TYPE 1 PEBBLE to determine the electric field inside any live cell or cellular compartment, called E-PEBBLE, was developed using polymerized micelles (Tyner et al., 2007). The E-PEBBLE is prepared by encasing the fast response, voltage sensitive dye di-4-ANEPPS inside the hydrophobic core of a silane-capped (polymerized) mixed micelle, which provides a uniform environment for the molecules and therefore allows for universal calibration. The PEBBLEs were introduced into DITNC astrocytes by endocytosis and enabled, for the first time, complete 3-dimensional electric field profiling throughout the entire volume of living cells (not just inside membranes).

5. Summary and Critical Issues

NPs made of biofriendly matrixes are an ideal platform for the construction of sensors for intracellular measurements, due to the NP's engineerability, nontoxic matrix, nano-size, and intrinsic optical properties. A wide variety of NP-based fluorescent sensors for intracellular measurements, so called PEBBLE or nanoPEBBLE sensors, have been developed for an increasing number of analytes that are involved in chemical, biochemical, and physical processes inside cells. PEBBLE designs range from a simple one, made of an NP and a molecular probe that recognizes the analyte and produces a responsive fluorescent signal, to sophisticated synergistic ones involving several sensing components that interact with each other so as to produce the desired signals. NPs made of a variety of matrixes—both inert polymer NPs as well as NPs with intrinsic optical properties such as QDs, CPdots, and GNPs—have been utilized to produce PEBBLEs.

PEBBLEs have been applied for real-time intracellular measurements. While these studies have demonstrated the utility of the PEBBLEs as a potent intracellular sensor, they have also indicated that there are critical issues for improving on the PEBBLE designs and their intracellular/subcellular delivery. First, most of the studies confirmed the delivery of the PEBBLEs into cells and demonstrated no changes in cells' morphology and viability due to the presence of the PEBBLEs. However, not all the intracellular studies with PEBBLEs have measured the analytes' absolute concentrations inside cells. Mostly, except for several analytes such as pH and oxygen, the PEBBLEs were used to measure the highly elevated concentrations of analytes under stimulated conditions, because the PEBBLEs were not sensitive enough to measure the normal intracellular concentrations. One way to improve the intracellular sensitivity of the PEBBLEs is to avoid interference from cellular autofluorescence. There are several approaches toward this goal, including the use of NIR fluorescent probes/reporters, two-photon-based detection, and the development of "MOON (modulated optical nanoprobe)" type PEBBLEs. MOONs are metallically half-capped fluorescent NPs (Anker and Kopelman, 2003; Anker et al., 2003), enabling the separation of the magnetically modulated sensor signal from the static background. They have been shown to improve signal/background by up to 4000 times.

Second, getting chemical or physical information from a single cell or a specific location within a single cell would be one of the important future applications of PEBBLE sensors. PEBBLEs have been delivered into cells by several standard methods. In order to deliver to a specific cellular organelle, one should first deliver into the cell, using one of these methods, and then to the subcellular organelle, using organelle-specific ligands, conjugated to the PEBBLEs' surface, or through remote steering means such as magnetic or laser tweezers. Endocytosis has been the most frequently used delivery method. For most of the cases, endocytosis of PEBBLEs results in sequestration of the majority of the NPs in the endosomal or endolysosomal compartment so that they become unavailable for subsequent intracellular trafficking (Derfus et al., 2004). However, studies show that cellular uptake and accumulation of the NPs into lysosomal compartments of the cells are affected by the PEBBLE's size, surface charge, and softness of the NP matrix (Banquy et al., 2009). Delivery with liposome/cationic polymers, gene-gun, and electroporation are efficient schemes for delivering NPs to the cytoplasm of a large population of cells, yet they often form large aggregates that can restrict subsequent trafficking (Coogan et al., 2010; Derfus et al., 2004). In contrast, microinjection delivers NPs to the cell interior as monodisperse NPs but requires each cell to be individually manipulated. More studies are required for enabling intracellular relocalization of the endocytosed PEBBLEs or for more facile and improved delivery methods.

ACKNOWLEDGMENTS

This work was supported by NIH/NCI Grant R33CA125297 (RK). We thank Minna Koo for several schematic diagrams in Figs. 21.1–21.3.

REFERENCES

Allard, E., and Larpent, C. (2008). Core-shell type dually fluorescent polymer nanoparticles for ratiometric pH-sensing. *J. Polym. Sci. A: Polym. Chem.* **46,** 6206–6213.

Anker, J. N., and Kopelman, R. (2003). Magnetically modulated optical nanoprobes. *Appl. Phys. Lett.* **82,** 1102–1104.

Anker, J. N., Behrend, C., and Kopelman, R. (2003). Aspherical magnetically modulated optical nanoprobes (MagMOONs). *J. Appl. Phys.* **93,** 6698–6700.

Arduini, M., Marcuz, S., Montolli, M., Rampazzo, E., Mancin, F., Gross, S., Armelao, L., Tecilla, P., and Tonellato, U. (2005). Turning fluorescent dyes into Cu(II) nanosensors. *Langmuir* **21,** 9314–9321.
Arduini, M., Mancin, F., Tecilla, P., and Tonellato, U. (2007). Self-organized fluorescent nanosensors for ratiometric Pb^{2+} detection. *Langmuir* **23,** 8632–8636.
Aslan, K., Gryczynski, I., Malicka, J., Matveeva, E., Lakowicz, J. R., and Geddes, C. D. (2005). Metal-enhanced fluorescence: An emerging tool in biotechnology. *Curr. Opin. Biotechnol.* **16,** 55–62.
Banquy, X., Suarez, F., Argaw, A., Rabanel, J.-M., Grutter, P., Bouchard, J.-F., Hildgen, P., and Giasson, S. (2009). Effect of mechanical properties of hydrogel nanoparticles on macrophage cell uptake. *Soft Matter* **5,** 3984–3991.
Borisov, S. M., and Klimant, I. (2009). Luminescent nanobeads for optical sensing and imaging of dissolved oxygen. *Microchim. Acta* **164,** 7–15.
Borisov, S. M., and Wolfbeis, O. S. (2006). Temperature-sensitive europium(III) probes and their use for simultaneous luminescent sensing of temperature and oxygen. *Anal. Chem.* **78,** 5094–5101.
Borisov, S. M., Mayr, T., and Klimant, I. (2008a). Poly(styrene-block-vinylpyrrolidone) beads as a versatile material for simple fabrication of optical nanosensors. *Anal. Chem.* **80,** 573–582.
Borisov, S. M., Nuss, G., and Klimant, I. (2008b). Red light-excitable oxygen sensing materials based on platinum(II) and palladium(II) benzoporphyrins. *Anal. Chem.* **80,** 9435–9442.
Brasola, E., Mancin, F., Rampazzo, E., Tecilla, P., and Tonellato, U. (2003). A fluorescence nanosensor for Cu^{2+} on silica particles. *Chem. Commun.* **24,** 3026–3027.
Brasuel, M., Kopelman, R., Miller, T. J., Tjalkens, R., and Philbert, M. A. (2001). Fluorescent nanosensors for intracellular chemical analysis: Decyl methacrylate liquid polymer matrix and ion-exchange-based potassium PEBBLE sensors with real-time application to viable rat C6 glioma cells. *Anal. Chem.* **73,** 2221–2228.
Brasuel, M., Kopelman, R., Kasman, I., Miller, T. J., and Philbert, M. A. (2002). Ion concentrations in live cells from highly selective ion correlations fluorescent nano-sensors for sodium. *Proc. IEEE Sensors* **1,** 288–292.
Brasuel, M. G., Miller, T. J., Kopelman, R., and Philbert, M. A. (2003). Liquid polymer nano-PEBBLES for Cl^- analysis and biological applications. *Analyst* **128,** 1262–1267.
Brown, J. Q., and McShane, M. J. (2005). Core-referenced ratiometric fluorescent potassium ion sensors using self-assembled ultrathin films on europium nanoparticles. *IEEE Sensors J.* **5,** 1197–1205.
Burns, A., Sengupta, P., Zedayko, T., Baird, B., and Wiesner, U. (2006). Core/shell fluorescent silica nanopartictes for chemical sensing: Towards single-particle laboratories. *Small* **2,** 723–726.
Cao, Y., Koo, Y.-E. L., and Kopelman, R. (2004). Poly (decyl methacrylate)-based fluorescent PEBBLE swarm nanosensors for measuring dissolved oxygen in biosamples. *Analyst* **129,** 745–750.
Cao, Y., Koo, Y.-E. L., Koo, S., and Kopelman, R. (2005). Ratiometric singlet oxygen nano-optodes and their use for monitoring photodynamic therapy nanoplatforms. *Photochem. Photobiol.* **81,** 1489–1498.
Chan, W. C. W., and Nie, S. M. (1998). Quantum dot bioconjugates for ultrasensitive nonisotopic detection. *Science* **281,** 2016–2018.
Chan, Y. H., Wu, C. F., Ye, F. M., Jin, Y. H., Smith, P. B., and Chiu, D. T. (2011). Development of ultrabright semiconducting polymer dots for ratiometric pH sensing. *Anal. Chem.* **83,** 1448–1455.
Chandra, S., Gross, D., Ling, Y. C., and Morrison, G. H. (1989). Quantitative imaging of free and total intracellular calcium in cultured cells. *Proc. Natl. Acad. Sci. USA* **86,** 1870–1874.

Changela, A., Chen, K., Xue, Y., Holschen, J., Outten, C. E., O'Halloran, T. V., and Mondragon, A. (2003). Molecular basis of metal-ion selectivity and zeptomolar sensitivity by CueR. *Science* **301,** 1383–1387.

Chen, Y. F., and Rosenzweig, Z. (2002). Luminescent CdS quantum dots as selective ion probes. *Anal. Chem.* **74,** 5132–5138.

Chen, Y., Thakar, R., and Snee, P. T. (2008). Imparting nanoparticle function with size-controlled amphiphilic polymers. *J. Am. Chem. Soc.* **130,** 3744–3745.

Chen, J., Zeng, F., Wu, S., Su, J., Zhao, J., and Tong, Z. (2009). A facile approach for cupric ion detection in aqueous media using polyethyleneimine/PMMA core-shell fluorescent nanoparticles. *Nanotechnology* **20,** 365502.

Cheng, Z. L., and Aspinwall, C. A. (2006). Nanometre-sized molecular oxygen sensors prepared from polymer stabilized phospholipid vesicles. *Analyst* **131,** 236–243.

Chithrani, B. D., Ghazani, A. A., and Chan, W. C. W. (2006). Determining the size and shape dependence of gold nanoparticle uptake into mammalian cells. *Nano Lett.* **6,** 662–668.

Clark, H. A., Barker, S. L. R., Brasuel, M., Miller, M. T., Monson, E., Parus, S., Shi, Z.-Y., Song, A., Thorsrud, B., Kopelman, R., Ade, A., Meixner, W., et al. (1998). Subcellular optochemical nanobiosensors: Probes encapsulated by biologically localised embedding (PEBBLEs). *Sens. Act. B: Chem.* **51,** 12–16.

Clark, H. A., Hoyer, M., Parus, S., Philbert, M. A., and Kopelman, R. (1999a). Opto-chemical nanosensors and subcellular applications in living cells. *Mikrochim. Acta* **131,** 121–128.

Clark, H. A., Hoyer, M., Philbert, M. A., and Kopelman, R. (1999b). Optical nanosensors for chemical analysis inside single living cells. 1. Fabrication, characterization, and methods for intracellular delivery of PEBBLE sensors. *Anal. Chem.* **71,** 4831–4836.

Coogan, M. P., Court, J. B., Gray, V. L., Hayes, A. J., Lloyd, S. H., Millet, C. O., Pope, S. J. A., and Lloyd, D. (2010). Probing intracellular oxygen by quenched phosphorescence lifetimes of nanoparticles containing polyacrylamide-embedded [Ru(dpp (SO$_3$Na)$_2$)$_3$]Cl^{-2}. *Photochem. Photobiol. Sci.* **9,** 103–109.

Cordes, D. B., Gamsey, S., and Singaram, B. (2006). Fluorescent quantum dots with boronic acid substituted viologens to sense glucose in aqueous solution. *Angew. Chem. Int. Ed.* **45,** 3829–3832.

Cui, K., Lu, X., Cui, W., Wu, J., Chen, X., and Lu, Q. (2011). Fluorescent nanoparticles assembled from a poly(ionic liquid) for selective sensing of copper ions. *Chem. Commun.* **47,** 920–922.

Cywinski, P. J., Moro, A. J., Stanca, S. E., Biskup, C., and Mohr, G. J. (2009). Ratiometric porphyrin-based layers and nanoparticles for measuring oxygen in biosamples. *Sens. Act. B: Chem.* **135,** 472–477.

Derfus, A. M., Chan, W. C. W., and Bhatia, S. N. (2004). Intracellular delivery of quantum dots for live cell labeling and organelle tracking. *Adv. Mater.* **16,** 961–966.

Duan, H., and Nie, S. (2007). Cell-penetrating quantum dots based on multivalent and endosome-disrupting surface coatings. *J. Am. Chem. Soc.* **129,** 3333–3338.

Dubach, J. M., Harjes, D. I., and Clark, H. A. (2007a). Fluorescent ion-selective nanosensors for intracellular analysis with improved lifetime and size. *Nano Lett.* **7,** 1827–1831.

Dubach, J. M., Harjes, D. I., and Clark, H. A. (2007b). Ion-selective nano-optodes incorporating quantum dots. *J. Am. Chem. Soc.* **129,** 8418–8419.

Fedrizzi, L., Lim, D., and Carafoli, E. (2008). Calcium and signal transduction. *Biochem. Mol. Biol. Educ.* **36,** 175–180.

Finney, L. A., and O'Halloran, T. V. (2003). Transition metal speciation in the cell: Insights from the chemistry of metal ion receptors. *Science* **300,** 931–936.

Frigoli, M., Ouadahi, K., and Larpent, C. (2009). A cascade FRET-mediated ratiometric sensor for Cu^{2+} ions based on dual fluorescent ligand-coated polymer nanoparticles. *Chem. Eur. J.* **15,** 8319–8330.

Fuller, J. E., Zugates, G. T., Ferreira, L. S., Ow, H. O., Nguyen, N. N., Wiesner, U. B., and Langer, R. S. (2008). Intracellular delivery of core-shell fluorescent silica nanoparticles. *Biomaterials* **29**, 1526–1532.

Gattas-Asfura, K. M., and Leblane, R. M. (2003). Peptide-coated CdS quantum dots for the optical detection of copper(II) and silver(I). *Chem. Commun.* **21**, 2684–2685.

Gill, R., Bahshi, L., Freeman, R., and Willner, I. (2008). Optical detection of glucose and acetylcholine esterase inhibitors by H_2O_2-sensitive CdSe/ZnS quantum dots. *Angew. Chem. Int. Ed.* **47**, 1676–1679.

Gota, C., Okabe, K., Funatsu, T., Harada, Y., and Uchiyama, S. (2009). Hydrophilic fluorescent nanogel thermometer for intracellular thermometry. *J. Am. Chem. Soc.* **131**, 2766–2767.

Gouanvé, F., Schuster, T., Allard, E., Meallet-Renault, R., and Larpent, C. (2007). Fluorescence quenching upon binding of copper ions in dye-doped and ligand-capped polymer nanoparticles: A simple way to probe the dye accessibility in nano-sized templates. *Adv. Funct. Mater.* **17**, 2746–2756.

Graefe, A., Stanca, S. E., Nietzsche, S., Kubicova, L., Beckert, R., Biskup, C., and Mohr, G. J. (2008). Development and critical evaluation of fluorescent chloride nanosensors. *Anal. Chem.* **80**, 6526–6531.

Guice, K. B., Caldorera, M. E., and McShane, M. J. (2005). Nanoscale internally referenced oxygen sensors produced from self-assembled nanofilms on fluorescent nanoparticles. *J. Biomed. Opt.* **10**, 064031.

Hammond, V. J., Aylott, J. W., Greenway, G. M., Watts, P., Webster, A., and Wiles, C. (2008). An optical sensor for reactive oxygen species: Encapsulation of functionalised silica nanoparticles into silicate nanoprobes to reduce fluorophore leaching. *Analyst* **133**, 71–75.

Harush-Frenkel, O., Rozentur, E., Benita, S., and Altschuler, Y. (2008). Surface charge of nanoparticles determines their endocytic and transcytotic pathway in polarized MDCK cells. *Biomacromolecules* **9**, 435–443.

Haugland, R. P. (2005). The Handbook: A Guide to Fluorescent Probes and Labeling Technologies. 10th edn. Molecular Probes Inc., Eugene.

He, X., Liu, H., Li, Y., Wang, S., Li, Y., Wang, N., Xiao, J., Xu, X., and Zhu, D. (2005). Gold nanoparticle-based fluorometric and colorimetric sensing of copper(II) ions. *Adv. Mater.* **17**, 2811–2815.

He, C., Zhu, W., Xu, Y., Zhong, Y., Zhou, J., and Qian, X. (2010). Ratiometric and reusable fluorescent nanoparticles for Zn^{2+} and $H_2PO_4^-$ detection in aqueous solution and living cells. *J. Mater. Chem.* **20**, 10755–10764.

Hornig, S., Biskup, C., Grafe, A., Wotschadlo, J., Liebert, T., Mohr, G. J., and Heinze, T. (2008). Biocompatible fluorescent nanoparticles for pH-sensoring. *Soft Matter* **4**, 1169–1172.

Jin, H., Heller, D. A., Sharma, R., and Strano, M. S. (2009). Size-dependent cellular uptake and expulsion of single-walled carbon nanotubes: Single particle tracking and a generic uptake model for nanoparticles. *ACS Nano* **3**, 149–158.

Josefsen, L. B., Aylott, J. W., Beeby, A., Warburton, P., Boyle, J. P., Peers, C., and Boyle, R. W. (2010). Porphyrin-nanosensor conjugates. New tools for the measurement of intracellular response to reactive oxygen species. *Photochem. Photobiol. Sci.* **9**, 801–811.

Kim, S.-H., Kim, B., Yadavalli, V. K., and Pishko, M. V. (2005). Encapsulation of enzyme within polymer spheres to create optical nanosensors for oxidative stress. *Anal. Chem.* **77**, 6828–6833.

Kim, K., Lee, M., Park, H., Kim, J. H., Kim, S., Chung, H., Choi, K., Kim, I. S., Seong, B. L., and Kwon, I. C. (2006). Cell-permeable and biocompatible polymeric nanoparticles for apoptosis imaging. *J. Am. Chem. Soc.* **128**, 3490–3491.

Kim, J. H., Lee, S., Park, K., Nam, H. Y., Jang, S. Y., Youn, I., Kim, K., Jeon, H., Park, R., Kim, I., Choi, K., and Kwon, I. C. (2007a). Protein-phosphorylation-responsive polymeric nanoparticles for imaging protein kinase activities in single living cells. *Angew. Chem. Int. Ed.* **46,** 5779–5782.

Kim, J. H., Lee, S., Kim, K., Jeon, H., Park, R. W., Kim, I. S., Choi, K., and Kwon, I. C. (2007b). Polymeric nanoparticles for protein kinase activity. *Chem. Commun.* **13,** 1346–1348.

Kim, G., Koo Lee, Y. E., Xu, H., Philbert, M. A., and Kopelman, R. (2010). Nanoencapsulation method for high selectivity sensing of hydrogen peroxide inside live cells. *Anal. Chem.* **82,** 2165–2169.

King, M., and Kopelman, R. (2003). Development of a hydroxyl radical ratiometric nanoprobe. *Sens. Act. B: Chem.* **90,** 76–81.

Kneipp, J., Kneipp, H., Wittig, B., and Kneipp, K. (2010). Novel optical nanosensors for probing and imaging live cells. *Nanomedicine* **6,** 214–226.

Koo Lee, Y.-E., and Kopelman, R. (2009). Optical nanoparticle Sensors for quantitative intracellular imaging. *WIREs Nanomed. Nanobiotechnol.* **1,** 98–110.

Koo Lee, Y.-E., and Kopelman, R. T. (2011). Targeted hydrogel nanoparticles for imaging, surgery and therapy of cancer. In "Multifunctional Nanoparticles for Drug Delivery: Imaging, Targeting and Delivery," (R. K. Prud'homme and S. Svenson, eds),. Springer, New York(in press).

Koo Lee, Y.-E., Smith, R., and Kopelman, R. (2009). Nanoparticle PEBBLE sensors in live cells and in vivo. *Annu. Rev. Anal. Chem.* **2,** 57–76.

Koo Lee, Y.-E., Orringer, D. A., and Kopelman, R. (2010a). Polymer-based nanosensors for medical applications. In "Polymer-Based Nanostructures: Medical Applications," (P. Broz, ed.), pp. 333–353. RSC Publishing, Cambridge, UK.

Koo Lee, Y. E., Ulbrich, E., Kim, G., Hah, H., Strollo, C., Fan, W., Gurjar, R., Koo, S. M., and Kopelman, R. (2010b). Near infrared luminescent oxygen nanosensors with nanoparticle matrix tailored sensitivity. *Anal. Chem.* **82,** 8446–8455.

Koo, Y.-E. L., Cao, Y., Kopelman, R., Koo, S. M., Brasuel, M., and Philbert, M. A. (2004). Real-time measurements of dissolved oxygen inside live cells by ormosil (organically modified silicate) fluorescent PEBBLE nanosensors. *Anal. Chem.* **76,** 2498–2505.

Krooswyk, J. D., Tyrakowski, C. M., and Snee, P. T. (2010). Multivariable response of semiconductor nanocrystal-dye sensors: The case of pH. *J. Phys. Chem. C* **114,** 21348–21352.

Lakowicz, J. R. (2006). Principles of Fluorescence Spectroscopy. 3rd edn. Springer, New York.

Lee, D., Khaja, S., Velasquez-Castano, J. C., Dasari, M., Sun, C., Petros, J., Taylor, W. R., and Murthy, N. (2007). In vivo imaging of hydrogen peroxide with chemiluminescent nanoparticles. *Nat. Mater.* **6,** 765–769.

Lee, S., Cha, E.-J., Park, K., Lee, S.-Y., Hong, J.-K., Sun, I.-C., Kim, S. Y., Choi, K., Kwon, I. C., Kim, K., and Ahn, C.-H. (2008). A near-infrared-fluorescence-quenched gold-nanoparticle imaging probe for in vivo drug screening and protease activity determination. *Angew. Chem. Int. Ed.* **47,** 2804–2807.

Lim, C.-K., Lee, Y.-D., Na, J., Oh, J. M., Her, S., Kim, K., Choi, K., Kim, S., and Kwon, I. C. (2010). Chemiluminescence-generating nanoreactor formulation for near-infrared imaging of hydrogen peroxide and glucose level in vivo. *Adv. Funct. Mater.* **20,** 2644–2648.

Liu, T., and Liu, S. Y. (2011). Responsive polymers-based dual fluorescent chemosensors for Zn^{2+} ions and temperatures working in purely aqueous media. *Anal. Chem.* **83,** 2775–2785.

Ma, B. L., Xu, M. Y., Zeng, F., Huang, L., and Wu, S. (2011). Micelle nanoparticles for FRET-based ratiometric sensing of mercury ions in water, biological fluids and living cells. *Nanotechnology* **22**, 065501.

Makino, R., Tanaka, T., Iizuka, T., Ishimura, Y., and Kanegasaki, S. (1986). Stoichiometric conversion of oxygen to superoxide anion during the respiratory burst in neutrophils. *J. Biol. Chem.* **261**, 11444–11447.

Meallet-Renault, R., Pansu, R., Amigoni-Gerbier, S., and Larpent, C. (2004). Metal-chelating nanoparticles as selective fluorescent sensor for Cu^{2+}. *Chem. Commun.* **20**, 2344–2345.

Meallet-Renault, R., Herault, A., Vachon, J. J., Pansu, R. B., Amigoni-Gerbier, S., and Larpent, C. (2006). Fluorescent nanoparticles as selective Cu(II) sensors. *Photochem. Photobiol. Sci.* **5**, 300–310.

Medintz, I. L., Stewart, M. H., Trammell, S. A., Susumu, K., Delehanty, J. B., Mei, B. C., Melinger, J. S., Blanco-Canosa, J. B., Dawson, P. E., and Mattoussi, H. (2010). Quantum-dot/dopamine bioconjugates function as redox coupled assemblies for in vitro and intracellular pH sensing. *Nat. Mater.* **9**, 676–684.

Monson, E., Brasuel, M., Philbert, M., and Kopelman, R. (2003). PEBBLE nanosensors for in vitro bioanalysis. *In* "Biomedical Photonics Handbook," (T. Vo-Dinh, ed.), pp. 1–14. CRC Press, Boca Raton.

Montalti, M., Prodi, L., and Zaccheroni, N. (2005). Fluorescence quenching amplication in silica nanosensors for metal ions. *J. Mater. Chem.* **15**, 2810–2814.

Myc, A., Majoros, I. J., Thomas, T. P., and Baker, J. R. (2007). Dendrimer-based targeted delivery of an apoptotic sensor in cancer cells. *Biomacromolecules* **8**, 13–18.

Nakayama-Ratchford, N., Bangsaruntip, S., Sun, X. M., Welsher, K., and Dai, H. J. (2007). Noncovalent functionalization of carbon nanotubes by fluorescein-polyethylene glycol: Supramolecular conjugates with pH-dependent absorbance and fluorescence. *J. Am. Chem. Soc.* **129**, 2448–2449.

Nielsen, L. J., Olsen, L. F., and Ozalp, V. C. (2010). Aptamers embedded in polyacrylamide nanoparticles: A tool for in vivo metabolite sensing. *ACS Nano* **4**, 4361–4370.

Oishi, M., Tamura, A., Nakamura, T., and Nagasaki, Y. (2009). A smart nanoprobe based on fluorescence-quenching PEGylated nanogels containing gold nanoparticles for monitoring the response to cancer therapy. *Adv. Func. Mater.* **19**, 827–834.

Park, E. J., Brasuel, M., Behrend, C., Philbert, M. A., and Kopelman, R. (2003). Ratiometric optical PEBBLE nanosensors for real-time magnesium ion concentrations inside viable cells. *Anal. Chem.* **75**, 3784–3791.

Peng, J. F., He, X. X., Wang, K. M., Tan, W., Wang, Y., and Liu, Y. (2007). Noninvasive monitoring of intracellular pH change induced by drug stimulation using silica nanoparticle sensors. *Anal. Bioanal. Chem.* **388**, 645–654.

Peng, H., Stich, M. I. J., Yu, J., Sun, L.-N., Fischer, L. H., and Wolfbeis, O. S. (2010a). Luminescent europium(III) nanoparticles for sensing and imaging of temperature in the physiological range. *Adv. Mater.* **22**, 716–719.

Peng, H.-S., Huang, S. H., and Wolfbeis, O. S. (2010b). Ratiometric fluorescent nanoparticles for sensing temperature. *J. Nanoparticle Res.* **12**, 2729–2733.

Peng, H.-S., Stolwijk, J. A., Sun, L.-N., Wegener, J., and Wolfbeis, O. S. (2010c). A nanogel for ratiometric fluorescent sensing of intracellular pH values. *Angew. Chem. Int. Ed.* **49**, 4246–4249.

Poulsen, A. K., Scharff-Poulsen, A. M., and Olsen, L. F. (2007). Horseradish peroxidase embedded in polyacrylamide nanoparticles enables optical detection of reactive oxygen species. *Anal. Biochem.* **366**, 29–36.

Rampazzo, E., Brasola, E., Marcuz, S., Mancin, F., Tecilla, P., and Tonellato, U. (2005). Surface modification of silica nanoparticles: A new strategy for the realization of self-organized fluorescent chemosensors. *J. Mater. Chem.* **15**, 2687–2696.

Ray, A., Koo Lee, Y.-E., Epstein, T., Kim, G., and Kopelman, R. (2011). Two-photon nano-PEBBLE sensors: Subcellular pH measurements. *Analyst* **136,** 3616–3622.

Roots, R., and Okada, S. (1975). Estimation of life times and diffusion distances involved in the X-ray induced DNA strand breaks or killing of mammalian cells. *Radiat. Res.* **64,** 306–320.

Rozenzhak, S. M., Kadakia, M. P., Caserta, T. M., Westbrook, T. R., Stone, M. O., and Naik, R. R. (2005). Cellular internalization and targeting of semiconductor quantum dots. *Chem. Commun.* **17,** 2217–2219.

Ruedas-Rama, M. J., and Hall, E. A. H. (2007). K^+-selective nanospheres: Maximizing response range and minimizing response time. *Analyst* **131,** 1282–1291.

Ruedas-Rama, M. J., and Hall, E. A. H. (2008a). Azamacrocycle activated quantum dot for zinc ion detection. *Anal. Chem.* **80,** 8260–8268.

Ruedas-Rama, M. J., and Hall, E. A. H. (2008b). A quantum dot-lucigenin probe for Cl^-. *Analyst* **133,** 1556–1566.

Ruedas-Rama, M. J., and Hall, E. A. H. (2009). Multiplexed energy transfer mechanisms in a dual-function quantum dot for zinc and manganese. *Analyst* **134,** 159–169.

Ruedas-Rama, M. J., Orte, A., Hall, E. A. H., Alvarez-Pez, J. M., and Talavera, E. M. (2011). Quantum dot photoluminescence lifetime-based pH nanosensor. *Chem. Commun.* **47,** 2898–2900.

Sapsford, K. E., Berti, L., and Medintz, I. L. (2006). Materials for fluorescence resonance energy transfer analysis: Beyond traditional donor-acceptor combinations. *Angew. Chem. Int. Ed.* **45,** 4562–4588.

Sarkar, K., Dhara, K., Nandi, M., Roy, P., Bhaumik, A., and Banerjeeet, P. (2009). Selective zinc(II)-ion fluorescence sensing by a functionalized mesoporous material covalently grafted with a fluorescent chromophore and consequent biological applications. *Adv. Funct. Mater.* **19,** 223–234.

Sasaki, K., Shi, Z. Y., Kopelman, R., and Masuhara, H. (1996). Three-dimensional pH microprobing with an optically-manipulated fluorescent particle. *Chem. Lett.* **25,** 141–142.

Schmälzlin, E., van Dongen, J. T., Klimant, I., Marmodée, B., Steup, M., Fisahn, J., Geigenberger, P., and Löhmannsröben, H. G. (2005). An optical multifrequency phase-modulation method using microbeads for measuring intracellular oxygen concentrations in plants. *Biophys. J.* **89,** 1339–1345.

Schulz, A., Wotschadlo, J., Heinze, T., and Mohr, G. J. (2010). Fluorescent nanoparticles for ratiometric pH-monitoring in the neutral range. *J. Mater. Chem.* **20,** 1475–1482.

Schulz, A., Woolley, R., Tabarin, T., and McDonagh, C. (2011). Dextran-coated silica nanoparticles for calcium-sensing. *Analyst* **136,** 1722–1727.

Seo, S., Lee, H. Y., Park, M., Lim, J. M., Kang, D., Yoon, J., and Jung, J. H. (2010). Fluorescein-functionalized silica nanoparticles as a selective fluorogenic chemosensor for Cu^{2+} in living cells. *Eur. J. Inorg. Chem.* **6,** 843–847.

Shiang, Y. C., Huang, C. C., and Chang, H. T. (2009). Gold nanodot-based luminescent sensor for the detection of hydrogen peroxide and glucose. *Chem. Commun.* **23,** 3437–3439.

Slotkin, J. R., Chakrabarti, L., Dai, H. N., Carney, R. S., Hirata, T., Bregman, B. S., Gallicano, G. I., Corbin, J. G., and Haydar, T. F. (2007). In vivo quantum dot labeling of mammalian stem and progenitor cells. *Dev. Dyn.* **236,** 3393–3401.

Snee, P. T., Somers, R. C., Nair, G., Zimmer, J. P., Bawendi, M. G., and Nocera, D. G. (2006). A ratiometric CdSe/ZnS nanocrystal pH sensor. *J. Am. Chem. Soc.* **128,** 13320–13321.

Son, H., Lee, H. Y., Lim, J. M., Kang, D., Han, W. S., Lee, S. S., and Jung, J. H. (2010). A highly sensitive and selective turn-on fluorogenic and chromogenic sensor based on BODIPY-functionalized magnetic nanoparticles for detecting lead in living cells. *Chem. Eur. J.* **16,** 11549–11553.

Stanca, S. E., Nietzsche, S., Fritzsche, W., Cranfield, C. G., and Biskup, C. (2010). Intracellular ion monitoring using a gold-core polymer-shell nanosensor architecture. *Nanotechnology* **21**, 055501.

Sumner, J. P., and Kopelman, R. (2005). Alexa Fluor 488 as an iron sensing molecule and its application in PEBBLE nanosensors. *Analyst* **130**, 528–533.

Sumner, J. P., Aylott, J. W., Monson, E., and Kopelman, R. (2002). A fluorescent PEBBLE nanosensor for intracellular free zinc. *Analyst* **127**, 11–16.

Sumner, J. P., Westerberg, N., Stoddard, A. K., Fierke, C. A., and Kopelman, R. (2005). Cu^+ and Cu^{2+} sensitive PEBBLE fluorescent nanosensors using Ds Red as the recognition element. *Sens. Act. B: Chem.* **113**, 760–767.

Sun, H., Scharff-Poulsen, A. M., Gu, H., and Almdal, K. (2006). Synthesis and characterization of ratiometric, pH sensing nanoparticles with covalently attached fluorescent dyes. *Chem. Mater.* **18**, 3381–3384.

Sun, H., Scharff-Poulsen, A. M., Gu, H., Jakobsen, I., Kossmann, J. M., Frommer, W. B., and Almdal, K. (2008). Phosphate sensing by fluorescent reporter proteins embedded in polyacrylamide nanoparticles. *ACS Nano* **2**, 19–24.

Sun, H., Andresen, T. L., Benjaminsen, R. V., and Almdal, K. (2009). Polymeric nanosensors for measuring the full dynamic pH range of endosomes and lysosomes in mammalian cells. *J. Biomed. Nanotechnol.* **5**, 676–682.

Suzuki, M., Husimi, Y., Komatsu, H., Suzuki, K., and Douglas, K. T. (2008). Quantum dot FRET biosensors that respond to pH, to proteolytic or nucleolytic cleavage, to DNA synthesis, or to a multiplexing combination. *J. Am. Chem. Soc.* **130**, 5720–5725.

Tang, B., Cao, L., Xu, K., Zhuo, L., Ge, J., Li, Q., and Yu, L. (2008). A new nanobiosensor for glucose with high sensitivity and selectivity in serum based on fluorescence resonance energy transfer (FRET) between CdTe quantum dots and Au nanoparticles. *Chem. Eur. J.* **14**, 3637–3644.

Teolato, P., Rampazzo, E., Arduini, M., Mancin, F., Tecilla, P., and Tonellato, U. (2007). Silica nanoparticles for fluorescence sensing of Zn^{II}: Exploring the covalent strategy. *Chem. Eur. J.* **13**, 2238–2245.

Tyner, K. M., Kopelman, R., and Philbert, M. A. (2007). "Nanosized voltmeter" enables cellular-wide electric field mapping. *Biophys. J.* **93**, 1163–1174.

Vetrone, F., Naccache, R., Zamarron, A., de la Fuente, A. J., Sanz-Rodriguez, F., Maestro, L. M., Rodriguez, E. M., Jaque, D., Sole, J. G., and Capobianco, J. A. (2010). Temperature sensing using fluorescent nanothermometers. *ACS Nano* **4**, 3254–3258.

Vinkenborg, J. L., Nicolson, T. J., Bellomo, E. A., Koay, M. S., Rutter, G. A., and Merkx, M. (2009). Genetically encoded FRET sensors to monitor intracellular Zn^{2+} homeostasis. *Nat. Methods* **6**, 737–740.

Wang, X. J., Boschetti, C., Ruedas-Rama, M. J., Tunnacliffe, A., and Hall, E. A. H. (2010a). Ratiometric pH-dot ANSors. *Analyst* **135**, 1585–1591.

Wang, G., Stender, A. S., Sun, W., and Fang, N. (2010b). Optical imaging of non-fluorescent nanoparticle probes in live cells. *Analyst* **135**, 215–221.

Wang, X.-D., Gorris, H. H., Stolwijk, J. A., Meier, R. J., Groegel, D. B. M., Wegener, J., and Wolfbeis, O. S. (2011a). Self-referenced RGB colour imaging of intracellular oxygen. *Chem. Sci.* **2**, 901–906.

Wang, S., Li, B., Zhang, L., Liu, L., and Wang, Y. (2011b). Photoluminescent and oxygen sensing properties of core-shell nanospheres based on a covalently grafted ruthenium(II) complex. *Appl. Organometal. Chem.* **25**, 21–26.

Wang, D., Liu, T., Yin, J., and Liu, S. (2011c). Stimuli-responsive fluorescent poly(N-isopropylacrylamide) microgels labeled with phenylboronic acid moieties as multifunctional ratiometric probes for glucose and temperatures. *Macromolecules* **44**, 2282–2290.

Watson, P., Jonesb, A. T., and Stephens, D. J. (2005). Intracellular trafficking pathways and drug delivery: Fluorescence imaging of living and fixed cells. *Adv. Drug Deliv. Rev.* **57,** 43–61.

Woods, L. A., Powell, P. R., Paxon, T. L., and Ewing, A. G. (2005). Analysis of mammalian cell cytoplasm with electrophoresis in nanometer inner diameter capillaries. *Electroanalysis* **17,** 1192–1197.

Wu, C., Bull, B., Christensen, K., and McNeill, J. (2009a). Ratiometric single-nanoparticle oxygen sensors for biological imaging. *Angew. Chem. Int. Ed.* **48,** 2741–2745.

Wu, W., Zhou, T., Shen, J., and Zhou, S. (2009b). Optical detection of glucose by CdS quantum dots immobilized in smart microgels. *Chem. Commun.* **14,** 4390–4392.

Xu, H., Aylott, J. W., Kopelman, R., Miller, T. J., and Philbert, M. A. (2001). A real-time ratiometric method for the determination of molecular oxygen inside living cells using sol–gel-based spherical optical nanosensors with applications to rat C6 glioma. *Anal. Chem.* **73,** 4124–4133.

Xu, H., Aylott, J. W., and Kopelman, R. (2002). Fluorescent nano-PEBBLE sensors designed for intracellular glucose imaging. *Analyst* **127,** 1471–1477.

Yin, J., Guan, X., Wang, D., and Liu, S. (2009). Metal-chelating and dansyl-labeled poly(N-isopropylacrylamide) microgels as fluorescent Cu^{2+} sensors with thermo-enhanced detection sensitivity. *Langmuir* **25,** 11367–11374.

Yin, J., Li, C., Wang, D., and Liu, S. (2010). FRET-derived ratiometric fluorescent K^+ sensors fabricated from thermoresponsive poly(N-isopropylacrylamide) microgels labeled with crown ether moieties. *J. Phys. Chem. B* **114,** 12213–12220.

Author Index

Note: Page numbers followed by "f" indicate figures, and "t" indicate tables.

A

Aach, J., 194–195
Aaron, J., 99
Abe, F., 69
Abe, S., 358
Aboul-Enein, F., 340
Abramoff, M. D., 395t, 401t, 407
Achilefu, S., 61
Acufia, A. U., 66t
Adachi, K., 92–93
Adams, M. C., 151, 298
Adams, M. D., 153
Adams, S. R., 354–355, 357, 370
Adam, V., 164–166, 165t
Adashi, E. Y., 383
Ade, A., 419
Adler, J., 68
Agapow, P. M., 188
Agard, D. A., 40, 46
Agre, P., 340–341
Aguet, F., 184
Aguilar, R. C., 256–258, 259–260, 259t
Aharon, M., 46
Ahn, C.-H., 424–427, 455, 457
Aida, G. P., 173
Ai, H. W., 164–166, 318
Airan, R. D., 335
Aitken, A., 330
Aker, J., 64t
Akhmanova, A., 186–187, 188f
Akiyoshi, K., 358
Albertazzi, L., 318
Alberti, S., 202, 203f
Ali, M. Y., 92
Alivisatos, A. P., 93–94, 98–99
Alkilany, A. M., 84, 100–101
Al-Kofahi, Y., 189t
Allalou, A., 408–409
Allard, E., 423–427, 428–429, 442–448, 443t, 449–450, 451–452
Allen, M. D., 322–323, 330
Allen, P. B., 64t
Allison, P. A., 238
Almdal, K., 423–424, 442–448, 443t, 449, 454–455
Almers, W., 303
Altschuler, Y., 419–422
Alvarez-Pez, J. M., 424–427, 442–448, 443t
Alves, R., 411–412

Amanatides, P. G., 153
Amblard, M., 361
Ameer-Beg, S. M., 110–111
Amemiya, S., 238
Amigoni-Gerbier, S., 424–427, 449–450, 451–452
Ananthanarayanan, B., 319–321, 323–324
Anderson, K. I., 74
Ando, R., 165t
Andresen, T. L., 123, 423–424, 442–448, 443t, 449
Andresen, V., 109–128
Andreu, N., 273–274
Andrey, P., 48
Aniento, F., 164–166, 173, 174f
Anker, J. N., 459–460
Antunes, F., 58–59, 60f, 63–67, 66t, 70–71, 72
Aoki, K., 322–323
Appelbaum, J., 357
Appelbe, S., 364–365
Applegate, K. T., 189t, 303
Appleyard, D. C., 92
Aptel, F., 110
Araya, C. L., 188
Arduini, M., 423–427, 449–450, 452, 454
Aresta-Branco, F., 60f, 63–67, 66t, 70–71, 72
Argaw, A., 419–422, 460
Arighi, C. N., 259t
Arimura, S., 173
Ariola, F. S., 63–67, 64t
Armelao, L., 424–427, 449–450
Armendariz, K. P., 238
Arora, R., 283–285
Arridge, S. R., 111–112
Asahi, T., 86
Ashby, M. C., 129–130, 129t, 132–133, 137, 139, 142
Aslan, K., 420t
Aspinwall, C. A., 423–424, 430–435, 431t
Athale, C., 189t
Athanassiou, C., 194
Auer, M., 405f
Auguste, K. I., 340
Avdulov, N. A., 64t
Axelrod, D., 39–40, 102, 139, 141–142, 173–175, 238
Aye-Han, N. N., 319–321, 370
Aylott, J. W., 419–422, 423–424, 428, 430–435, 431t, 436–437, 438, 440, 441, 451, 452–453

471

Ayoub, M. A., 355–357, 365–366
Azevedo, R. B. R., 188

B

Baak, J. P., 113–114
Babbey, C. M., 294
Bacher, C. P., 189t
Bachir, A. I., 90–91
Baek, K., 166–167
Baer, T. M., 272–273
Bahnson, A., 194
Bahshi, L., 424–427, 428, 438, 439, 440, 441
Baig, A. H., 364–365
Baird, B., 131f, 139–140, 423–424, 442–448, 443t, 449
Baird, G. S., 373
Bajenoff, M., 110
Bakal, C., 194–195
Baker, J. R., 419–422, 423–424, 455, 456
Bakker, G.-J., 109–128
Balci, H., 84
Baldeyron, C., 189t
Baldwin, T. C., 166t
Balkin, D. M., 357
Ballesteros Katemann, B., 239–240
Ball, G., 29–58
Baluska, F., 166t
Banerjeeet, P., 419–422, 423–424, 451, 452
Bangsaruntip, S., 423–424, 442–448, 443t, 449
Banquy, X., 419–422, 460
Bao, Z., 188, 189t
Baraban, J. M., 340–341
Barbero, P., 259t
Barber, P. R., 110–111
Bard, A. J., 238
Barisic, M., 147–163
Barker, S. L. R., 419
Barragan, V., 223–224
Barroso, M., 110–111
Barton, K., 164
Bastiaens, P. I., 110–111, 371
Bastos, A. E. P., 57–85
Bates, G. W., 166–167
Bates, I. R., 90–91
Bates, M., 40, 42, 222
Baudisch, B., 164
Bauer, B., 90
Baumgart, F., 349
Baumgart, T., 68
Bawendi, M. G., 424–427, 442–448, 443t
Baxter, S. C., 84
Bayburt, T. H., 364–365
Bazin, H., 355–357, 365–366
Beaurepaire, E., 110
Beckert, R., 419–422, 423–424, 454–455
Bedioui, F., 242, 246

Beeby, A., 419–422, 423–424, 452–453
Behr, B., 272–273, 355, 357
Behrend, C., 419–422, 423–424, 454, 459–460
Behrens, M. A., 258
Belcher, A. M., 92
Bell, J. D., 69
Bellomo, E. A., 452
Beltman, J. B., 192–193, 194, 196, 197
Belyakov, O. V., 5
Benda, A., 113–114, 116–119
Benita, S., 419–422
Benjaminsen, R. V., 423–424, 442–448, 443t, 449
Bennett, J. L., 340–341, 343f, 344, 345, 349–350
Bent, A. F., 169
Bentolila, L. A., 84
Ben-Yakar, A., 90
Berciaud, S., 86f, 87–88
Berezin, M. Y., 61
Berger, J., 58–59
Berger, T., 340
Bergey, E. J., 90
Berlemont, S., 184, 196
Berney, C., 373
Berns, M. W., 8–9
Berns, W., 7–8
Berridge, M. J., 326–327
Berti, L., 428–429
Best, K. B., 69
Betzig, E., 42, 222, 223
Bewersdorf, J., 40
Bewley, J. D., 164–166
Beznoussenko, G. V., 202
Bezstarosti, K., 8–9
Bhatia, S. N., 420t, 460
Bhat, R. V., 340–341
Bhaumik, A., 419–422, 423–424, 451, 452
Biehlmaier, O., 405–406
Binder, D. K., 340
Binder, W. H., 223–224
Bird, S., 166–167
Birdsall, B., 364–365
Birdsall, N. J., 364–365
Bisby, R. H., 10–11
Bishop, C. M., 409
Biskup, C., 419–422, 423–424, 430–435, 431t, 436–437, 442–448, 443t, 449, 453, 454–455
Bista, M., 166t
Bittman, R., 68
Bittova, L., 259t
Bivand, R. S., 48
Bjelkmar, P., 90–91
Bjornsson, C. S., 189t, 395f, 407–408
Blab, G. A., 86f, 87–88
Blake, J. A., 277
Blanco-Canosa, J. B., 419–422, 424–427, 442–448, 443t, 449
Blank, P. S., 379

Bloch, O., 340
Block, S. M., 84, 92, 96
Bloembergen, N., 275–276
Bloom, K. S., 150–151
Blumenthal, R., 142
Boehme, S., 164–166, 165t
Boehm, M., 258
Böhme, S., 165t
Bokoch, G. M., 292
Bokoch, M. P., 364–365
Boland, M. V., 409–410
Bollen, A. W., 340
Bolte, S., 48, 401t, 407
Bonifacino, J. S., 222, 256–258, 259–260, 259t, 262, 263–264, 265–267
Bonneau, S., 186–187
Bonn, M., 90, 283–285
Booker, C. F., 371
Book, L. D., 277
Borges, C., 58–59
Borisov, S. M., 423–424, 430–437, 431t, 449–450, 457–458
Borisy, G. G., 149–150, 292
Born, M., 85
Borri, P., 89–90, 280–281, 282–285
Borst, J. W., 62, 64t
Boschetti, C., 419–422, 424–427, 442–448, 443t, 449
Bosgraaf, L., 189t
Bossert, N. L., 272–273
Botchway, S. W., 4–5, 6–7, 8–9, 10–11, 21–22
Bouchard, J.-F., 419–422, 460
Boulanger, J., 44–45
Bouley, R., 340–341
Bourgeois, D., 164–166, 165t
Bourrier, E., 355–357, 365–366
Bouschet, T., 129–130, 132–133
Bouvier, M., 364–365
Bowey, A. G., 5
Bowser, S. S., 149–150
Boyle, J. P., 419–422, 423–424, 452–453
Boyle, R. W., 419–422, 423–424, 452–453
Boyle, T. J., 188, 189t
Brach, T., 166t
Bradke, F., 166t
Bradley, J., 111
Brady, H., 379–380
Braga, J., 173–175
Brasola, E., 424–427, 449–450
Brasselet, S., 90–91
Brasuel, M. G., 419–422, 423–424, 427–429, 430–437, 431t, 453–455
Braun, V., 188
Bregman, B. S., 419–422
Bretschneider, S., 41
Bretschneider, T., 189t
Brickley, K., 90–91

Brincat, J., 166t
Brock, A., 110–111
Brock, C., 355–357, 365–366
Bröcker, E. B., 113
Brown, D. A., 223–224, 340–341
Browning, L. M., 100–101
Brown, J. Q., 423–424
Bruggenwirth, H. T., 8–9
Brunet, A., 242
Brunner, C., 88
Brust-Mascher, I., 131f, 139–140
Brust–Mascher, I., 150–151
Buades, A., 44–45
Buccione, R., 202, 203f
Buechler, Y. J., 354–355
Bulinski, J. C., 149–150, 151, 152–153
Bull, B., 424–427, 430–435, 431t, 436
Bunemann, M., 355, 357
Buranachai, C., 84, 386–387
Burbank, K. S., 151, 157–158
Burns, A., 423–424, 442–448, 443t, 449

C

Caiolfa, V. R., 112, 119–120, 123
Cairo, C. W., 165t
Cai, X., 164
Caldorera, M. E., 419–422, 423–424, 430–435, 431t, 436
Calleja, V., 319, 323–324
Cameron, L. A., 149–151, 152–153, 154–155, 155f, 157–158
Campagnola, P. J., 273–274
Campbell, R. E., 164–166, 165t, 370, 373
Canman, J. C., 150–151, 157–158
Cao, L., 424–427, 441, 442
Cao, Y., 419–422, 423–424, 430–437, 431t, 440–441
Capobianco, J. A., 419–422, 423–424, 457–458
Carafoli, E., 452–453
Carlson, E. J., 340
Carlton, P. M., 42, 44–45
Carmo, A. M., 364–365
Carmo-Fonseca, M., 173–175
Carney, R. R., 91
Carney, R. S., 419–422
Carpenter, A. E., 189t, 395t, 408–409
Carragher, N. O., 110–111
Carrière, M., 5
Carrillo, L. D., 322–323
Carter, A. P., 92
Carter, B. C., 186, 195
Carter, R. E., 259–260, 259t
Caserta, T. M., 419–422, 420t
Castellino, F., 110
Castro, B. M., 70–71, 72
Cebecauer, M., 224, 231

Celniker, S. E., 153
Cha, E.-J., 424–427, 455, 457
Chakrabarti, L., 419–422
Chalfie, M., 30–31, 292, 317–318
Chan, A., 164
Chandra, S., 452–453
Changela, A., 450
Chang, H. T., 424–427, 438, 439, 440, 441
Chang, J. H., 189t
Chang, W. S., 93–94, 273–274
Chan, P., 340
Chan, W. C. W., 419–422, 420t, 460
Chan, Y. H., 424–427, 442–448, 443t, 449
Charpak, S., 111
Chattopadhyay, A., 64t, 71
Chaumont, F., 166t
Cheerambathur, D. K., 150–151
Cheezum, M. K., 48, 186, 195
Chemla, D. S., 92–93
Chen, B. P. C., 8–9
Chen, D. J., 8–9
Cheng, J. X., 90, 275–276, 277, 285–287
Cheng, T., 194
Cheng, Z. L., 423–424, 430–435, 431t
Chen, H., 371, 379
Chen, J., 424–427, 449–450
Chen, K., 450
Chen, P. C., 92, 99
Chen, S.-C., 409–410
Chen, S. S., 189t
Chen, X., 252–253, 423–424, 449–450
Chen, Y. F., 47, 110–111, 424–427, 442–448, 443t, 449–450
Chepurnykh, T. V., 164–166, 165t
Chernomoretz, A., 411
Chhun, B. B., 41, 152
Chiba, A., 386–387
Chieppa, M., 110
Chin, E., 150–151
Chithrani, B. D., 419–422
Chiu, D. T., 64t, 424–427, 442–448, 443t, 449
Chklovskaia, E., 58–59
Choi, K., 419–422, 424–427, 428, 438, 440, 441, 442, 455, 456–457
Choma, M., 258, 259t, 266–267, 268
Choquet, D., 86f, 87–88, 90–91, 128
Chorny, I., 340–341
Chou, Y. H., 87, 92
Christensen, K., 424–427, 430–435, 431t, 436
Christie, R., 110
Chua, N. H., 166t
Chudakov, D. M., 165t, 304–305, 354–355
Chung, H., 419–422, 424–427, 455, 456
Church, G., 194–195
Churchman, L. S., 47
Cimini, D., 150–151, 157–158
Clapham, D. E., 326–327
Clarke, C., 189t

Clark, H. A., 419–422, 423–424, 427–428, 442–448, 443t, 449, 452–453
Clegg, R. M., 73–74, 370–371, 374, 376, 386–387
Clendenon, J. L., 410–411
Clough, S. J., 169
Coghlan, L., 99
Cognet, L., 86f, 87–88, 90–91
Cohan, C. S., 149–151, 298
Cohen, L. B., 316
Cohen, L. D., 186–187
Cole, K. S., 376
Cole, R. H., 376
Coll, B., 44–45
Colles, S., 64t
Collier, T., 99
Collinet, C., 408
Collingridge, G. L., 129–130, 129t, 132–133
Collins, A. R., 20
Colyer, R. A., 374–375
Comps-Agrar, L., 355–357, 365–366
Compton, D. A., 149–150
Conchello, J. A., 292–293, 294
Conein, E., 8, 21
Conlan, R. S., 238
Conrad, C., 395t, 409–410
Coogan, M. P., 419–422, 423–424, 430–435, 431t, 436–437, 460
Coppey-Moisan, M., 186–187
Corbin, J. G., 419–422
Cordeiro, A. M., 60f, 63–67, 66t, 70–71, 72
Cordelières, F. P., 48, 401t, 407
Cordes, D. B., 424–427, 441, 442
Cornetta, K., 308
Corrie, J. E. T., 92–93, 151, 364–365
Cortes, C., 409
Coskun, H., 189t
Coughlin, P., 149–151
Court, J. B., 419–422, 423–424, 430–435, 431t, 436–437, 460
Cowlishaw, M. F., 176–177
Cowman, A., 141–142
Cowsert, L., 194
Cox, R., 5
Coy, D. L., 92
Craddock, C. P., 166t
Craik, J. S., 151
Cramer, J. F., 258
Crane, J. M., 340–341, 341f, 342, 343–344, 343f, 345, 346f, 347, 348f, 349–350
Cranfield, C. G., 423–424, 442–448, 443t, 449, 453
Cressie, N., 48
Crevenna, A. H., 166t
Crisostomo, A. G., 10–11
Cromme, C., 258
Crouch, R. J., 256–258, 259t, 259–260
Cubitt, A. B., 317–318

Cui, K., 423–424, 449–450
Cui, W., 423–424, 449–450
Cunniffe, S. M. T., 4–5
Cyrne, L., 58–59, 60f, 63–67, 66t, 70–71, 72
Cyr, R. J., 166t
Cytrynbaum, E. N., 149–150
Cywinski, P. J., 423–424, 430–435, 431t, 436–437

D

Dabov, K., 46
Dahan, M., 186–187
Dahms, N. M., 256–258, 263–264
Dai, H. J., 423–424, 442–448, 443t, 449
Dai, H. N., 419–422
Dai, X., 113–114
Dallongeville, S., 189t
Dance, J., 25–26
Dang, B., 5
Daniel, D., 93–94
Danielson, D. C., 277
Danuser, G., 148–149, 150–153, 157–158, 184, 186–187, 189t, 196, 303, 308–309, 373
Dasari, M., 428, 438, 440
Dasgupta, S., 68
Daudin, L., 5
Daugherty, P. S., 318
Davidson, K. G., 340–341
Davidson, M. W., 164–166, 165t, 222, 304–305, 370
Davis, D. M., 224, 231
Davis, E. L., 8, 10–11, 10t, 21
Davis, H. E., 308
Davis, I., 40
Davis, K., 40
Davis, M. M., 224
Davis, S. J., 364–365
Dawson, P. E., 419–422, 424–427, 442–448, 443t, 449
Day, J. P., 283–285
Day, R. N., 110–111, 318–319, 370, 371, 373, 383
de Almeida, R. F. M., 58–60, 60f, 61, 62, 63–67, 64t, 66t, 68, 70–71, 72, 75f, 78
De Angelis, D. A., 129–130, 129t
DeBlasio, S., 164
de Boer, R. J., 192–193, 194, 196, 197
de Chaumont, F., 189t
Deen, P. M., 340–341
De Jonge, C. J., 272–273
de la Fuente, A. J., 419–422, 423–424, 457–458
de Lange, J. H., 113–114
deLara, C. M., 5
de Lartigue, J., 150–151
De La Rue, S. A., 129–130, 129t, 132–133
Delehanty, J. B., 419–422, 424–427, 442–448, 443t, 449

Delest, M., 395t, 411
Demmers, J. A., 8–9
Deniset-Besseau, A., 110
Denk, W., 6–7, 39–40, 110
Dennes, A., 258
Derfus, A. M., 420t, 460
Desai, A., 149–150, 152–153
Desfonds, G., 164–166, 165t
Desteroo, J. M. P., 173–175
Deutsch, U., 340–341
Devaux, P. F., 67–68
de Vries, S., 64t
Devynck, J., 242, 246
Devynck, M.-A., 242
Dewitte, F., 256–258, 259–260, 259t, 262–263
Dhara, K., 419–422, 423–424, 451, 452
Dhonukshe, P. B., 164–166, 173, 174f, 272–273
Diamantopoulos, G. S., 151
Di Benedetto, S., 166t
Dickenson, N. E., 238
Dieringer, J. A., 274–275
Dietrich, C., 90–91
Diez, S., 92–93
Digman, M. A., 112, 119–120, 123
Ding, H., 90, 99–100, 102
DiPilato, L. M., 319, 322
Divecha, N., 168–169
Dix, J. A., 90–91
Doktycz, M. J., 238
Dolbow, J. E., 58–59, 61–62
Domke, K. F., 283–285
Donoso, A., 164–166, 168–169
Doose, S., 84
Dorn, J. F., 184
Douglas, K. T., 424–427, 442–448, 443t, 455, 457
Douglass, A. D., 90–91
Downing, K. H., 92
Doyle, B., 110–111
Draegestein, K., 186–187, 188f
Dragsten, P., 142
Drobizhev, M., 285
Duan, R., 419–422
Dubach, J. M., 427–428, 453
Dubremetz, J. F., 256–258, 259–260, 259t, 262–263
Ducros, M., 111
Duden, R., 265–266
Dudovich, N., 280–281
Dufour, A., 184, 196
Dunne, J. C., 266–267
Dunn, K. W., 294
Dunn, N., 164–166, 168–169
Dunn, R. C., 238
Dunn, T. A., 326–327
Dunsby, C., 69
Dupree, P., 166t
Durr, N. J., 90

Durroux, T., 355–357, 365–366
Dustin, M. L., 224, 231
Dzyubachyk, O., 184, 185–186, 187f, 189t, 192–193, 195, 197

E

Eaton, J. W., 395t
Ebisu, S., 256–258, 259t, 263, 263f, 265, 266f
Eble, J. A., 258
Eccleston, M. E., 113–114
Eckhard, K., 252–253
Edelstein, A., 405–406
Edward, M., 110–111
Egen, J. G., 110
Eggeling, C., 41
Eils, R., 184, 186–187, 189t
Eisenberg, E., 256–258, 259–260, 262, 265–266
Elad, M., 46
Elangovan, M., 110–111
Eliceiri, K. W., 184, 404
Ellenberg, J., 407
Ellisman, M. H., 340–341, 354–355, 357, 370
El-Sayed, I. H., 99
El-Sayed, M. A., 90, 98–99
Elson, D. S., 111–112
Elson, E. L., 58–59, 61–62, 91, 139, 141–142, 173–175, 193–194
Emans, N., 166t
Emiliani, V., 186–187
Enderlein, J., 92–93
Endesfelder, U., 34, 42
Enejder, A. M. K., 90, 280–281
Engel, E., 40–41
Engel, S., 74–76, 75f, 78
Enwright, J. F. III, 383
Epstein, C. J., 340
Epstein, T., 423–424, 442–448, 443t, 449
Erickson, E. S., 238
Erickson, R., 383
Ericson, M. B., 90
Ernst, L., 362
Erogbogbo, F., 102
Ersoy, I., 189t
Essers, J., 187f, 189t
Euskirchen, G., 292
Evanko, D., 40
Evans, C. A., 153
Evans, C. L., 274–276, 277
Ewers, H., 86f, 88, 91, 92–93
Ewing, A. G., 420t

F

Fallier-Becker, P., 340–341
Fan, F. R. F., 238
Fang, N., 88, 89, 93, 94, 95–98, 95f, 97f, 99, 422
Fang, P., 340–341
Fan, J., 90–91

Fan, W., 419–422, 423–424, 428–429, 430–437, 431t, 437f, 438f
Farkas, E. R., 68
Fazekas de St, G. B., 58–59
Feder, T. J., 131f, 139–140
Fedorov, A., 62, 70–71, 72
Fedrizzi, L., 452–453
Fehrentz, J. A., 361
Feigenson, G. W., 61–62, 68
Feldmann, J., 90
Feng, D., 395t
Fenton, R. A., 340–341
Fercher, A. F., 273–274
Ferhatosmanoglu, H., 189t
Fernandes, L. P., 110–111
Fernandez-Suarez, M., 354–355
Ferreira, L. S., 419–422
Fetter, K., 166t
Field, M. J., 164–166, 165t
Fierke, C. A., 423–424, 449–450
Filiz, F., 340
Finney, L. A., 450
Fisahn, J., 419–422, 423–424, 430–435, 431t, 436–437
Fischer, L. H., 423–424, 443t, 449, 457–458
Fisher, D. D., 166t
Fisher, D. S., 151
Florea, B. I., 8–9
Flotte, T., 273–274
Foi, A., 40, 46
Folkard, M., 5
Follen, M., 99
Folmer, V., 58–59
Forkey, J. N., 92–93
Fornasiero, E. F., 228
Förster, T., 110–111, 316–317, 371
Forstner, M. B., 224
Fosbrink, M., 325–326
Fourgeaud, L., 128
Fradkov, A. F., 164–166, 165t
Frame, M. C., 110–111
Francis, L. W., 238
Frank, J., 90–91
Franklin, J., 231
Franzl, T., 90
Frasch, W. D., 93–94
Fraser, S. E., 273–274
Freeman, R., 424–427, 428, 438, 439, 440, 441
French, P. M., 69, 111–112, 224, 231
Fried, E., 58–59, 61–62
Friedl, P., 110, 113
Friedrich, M. W., 326
Frigeri, A., 340
Frigerio, L., 166t
Frigoli, M., 420t, 424–427, 449–450, 451–452
Friman, O., 189t
Friml, J., 164–166, 173, 174f
Fritzsche, W., 423–424, 442–448, 443t, 449, 453

Fromme, R., 93–94
Frommer, W. B., 423–424, 454–455
Fruhwirth, G. O., 110–111
Fuchs, J., 165t
Fu, H., 330
Fujii, N., 355, 358, 364
Fujimura, M., 340
Fujioka, A., 319, 323–324
Fujiwara, T. K., 90–91, 364–365
Fuller, J. E., 419–422
Funatsu, T., 419–422, 428, 457–458
Funston, A., 86f
Furman, C. S., 340–341
Furuike, S., 92–93

G

Gadella, T. W. Jr., 168–169, 272–273
Gaetz, J., 151–153
Gaietta, G., 355, 357
Galbraith, C. G., 40, 222, 223
Galbraith, J. A., 40, 222, 223
Galjart, N., 186–187, 188f
Galle, R. F., 153
Gallicano, G. I., 419–422
Gamsey, S., 424–427, 441, 442
Ganem, N. J., 149–150
Gans, R., 93
Gao, D., 5
Gao, L., 39–40
Gao, X., 325
Garcia-Bereguiain, M. A., 132–133
Gasser, S. M., 189t
Gattas-Asfura, K. M., 424–427, 449–450
Gaus, K., 58–59, 224, 231
Gearheart, L., 93
Geddes, C. D., 420t
Geigenberger, P., 419–422, 423–424, 430–435, 431t, 436–437
Ge, J., 424–427, 441, 442
Gelfand, V. I., 87, 92, 96
Gendreizig, S., 355–357
Genin, G. M., 58–59, 61–62
Gennerich, A., 92, 96
Genovesio, A., 186–187
Georgalis, Y., 90–91
George, N., 355, 357–358, 364–365
Georges, A., 92
Germain, R. N., 110
Getis, A., 231
Ghazani, A. A., 419–422
Ghitani, A., 42
Ghosh, P., 256–258, 263–264
Giasson, S., 419–422, 460
Giepmans, B. N., 354–355, 357, 370
Gierke, S., 305, 308–309
Giglia-Mari, G., 8–9
Gillespie, A., 340

Gill, R., 424–427, 428, 438, 439, 440, 441
Girirajan, T. P. K., 42, 222
Gobin, R., 340–341
Gocayne, J. D., 153
Godinez, W. J., 184, 186–187
Goeppert-Mayer, M., 6
Goff, J., 194
Goldberg, I. G., 184
Goldberg, Y., 127–148
Goldman, R. D., 33, 47, 87, 92
Goldman, Y. E., 92–93
Goldsmith, E. C., 84
Gole, A. M., 84
Golland, P., 189t
Gomez-Godinez, V., 8–9
González-González, I. M., 127–148
Gonzalez, R. C., 44
Gonzhlez-Rodriguez, J., 66t
Goodhead, D. T., 4–5
Gooding, J. J., 231
Gordon, A., 395t
Gorelick-Feldman, D. A., 340–341
Gorka, C., 66t
Gorris, H. H., 423–424, 430–437, 431t
Gorshkova, I., 256–258, 259–260, 259t
Goshima, G., 92, 96, 150–151
Gota, C., 419–422, 428, 457–458
Gouanvé, F., 424–427, 428–429, 449–450, 451–452
Gouget, B., 5
Goulding, S. E., 266–267
Gould, T. J., 224–225
Grafe, A., 419–422, 423–424, 442–448, 443t, 449, 454–455
Granada, B., 373
Grange, R., 89–90
Granger, C. L., 166t
Grant, D., 69
Grassme, H., 58–59
Gratton, E., 84, 112, 119–120, 123, 374–375
Graves, B. J., 383
Gray, V. L., 419–422, 423–424, 430–435, 431t, 436–437, 460
Greene, L. E., 256–258, 259–260, 262, 265–266
Greenway, G. M., 419–422, 423–424, 438, 440
Greenwood, J. S., 164–166, 166t, 173
Gregory, K., 273–274
Greulich, K. O., 7–8
Griesbeck, O., 111, 373
Griffin, B. A., 357
Griffis, E. R., 152
Griffiths, S., 163–183
Grigaravicius, P., 7–8
Grigoriev, I., 186–187, 188f
Grinstein, S., 186–187, 189t
Groc, L., 86f, 87–88, 90–91
Groegel, D. B. M., 423–424, 430–437, 431t
Groen, A. C., 151, 157–158

Gronemeyer, T., 355–357
Gropper, M. A., 340
Gross, D., 452–453
Gross, S. P., 96, 186, 195, 424–427, 449–450
Groves, J. T., 224
Grutter, P., 419–422, 460
Gryczynski, I., 372–373, 420t
Gryczynski, Z., 372–373
Grynkiewicz, G., 316
Guan, X., 424–427, 428–429, 449–450
Guertin, D. A., 189t
Guggino, W. B., 340–341
Gu, H., 423–424, 442–448, 443t, 454–455
Guice, K. B., 419–422, 423–424, 430–435, 431t, 436
Guignet, E. G., 357
Guilford, W. H., 186, 195
Guizetti, J., 39
Gulbins, E., 58–59
Gumbel, M., 188
Gunnarsson, L., 90
Gupta, T., 340–341, 345, 349–350
Gupton, S. L., 151, 298
Gurjar, R., 419–422, 423–424, 428–429, 430–437, 431t, 437f, 438f
Gurskaya, N. G., 164–166, 165t
Gustafsen, C., 258
Gustafsson, M. G. L., 40, 41, 42, 152, 222
Gutierrez-Merino, C., 132–133
Gu, Y., 96–98, 97f, 99
Gyobu, N., 340–341

H

Haataja, M., 90–91
Habuchi, S., 165t
Hadwen, B. J., 33
Haft, C. R., 259t
Haggie, P. M., 342
Hague, C., 364–365
Hah, H., 419–422, 423–424, 428–429, 430–437, 431t, 437f, 438f
Hahnenkamp, A., 258
Ha, J. W., 93–94
Halbhuber, K.-J., 8–9, 238
Hall, E. A. H., 419–422, 424–428, 442–448, 443t, 449, 451–452, 453, 454–455
Hall, E. J., 4–5
Hall, R. A., 364–365
Halo, T. L., 357
Hamachi, I., 357
Hamaguchi, H. O., 283–285
Hama, H., 165t
Hamilton, N. A., 392, 395t, 405f, 407, 409–411
Hamilton, R. S., 49–50
Hammaren, H., 90–91
Hammond, V. J., 419–422, 423–424, 438, 440
Hancock, W. O., 92

Handford, M. G., 166t
Han, K. Y., 41
Hanrahan, J. W., 90–91
Hansen, M. N., 90
Han, W. S., 423–424, 454
Han, Y., 364–365
Harada, Y., 419–422, 428, 457–458
Harbour, M. E., 258, 259t, 266–267, 268
Harder, N., 184
Harder, T., 58–59, 224
Harjes, D. I., 427–428, 453
Harms, G., 110
Harootunian, A. T., 354–355
Harper, I., 141–142
Harper, J. V., 8–9, 10, 21–22
Harries, W. E., 340–341
Harris, F. M., 69
Hartgroves, L. C., 224
Hartland, G. V., 86f
Harttnell, L. M., 259t
Harush-Frenkel, O., 419–422
Harvey, C. D., 324–326
Hasan, M. T., 110
Hasegawa, H., 340
Ha, T., 84, 92–93
Haugland, R. P., 418–419
Hauser, A., 340
Hauser, C. T., 357
Hawes, C., 166t
Haydar, T. F., 419–422
Hayden, J. H., 149–150
Hayes, A. J., 419–422, 423–424, 430–435, 431t, 436–437, 460
Haynes, C. L., 84
Hays, N. M., 369–391
Hazelwood, K. L., 164–166, 165t, 304–305
Heale, J. T., 8–9
Hebert, B., 90–91
Hedde, P. N., 165t
Hediger, F., 189t
Hee, M. R., 273–274
He, H., 87
Heidemann, S. R., 92
Heikal, A. A., 63–67, 64t, 116–118
Heilemann, M., 222
Heilker, R., 165t
Hein, B., 40–41, 222
Heine, M., 86f, 87–88, 90–91
Heintzmann, R., 40, 41
Heinze, T., 423–424, 442–448, 443t, 449, 452–453
Helenius, A., 86f, 88, 91, 92–93
Heller, D. A., 419–422
Hellerer, T., 280–281
Hell, S. W., 40–41, 111–112, 116–118, 222
Helmchen, F., 110
Hemar, A., 128

Henderson, J. N., 164–166, 318
Hengstenberg, A., 239–240, 242
Henkart, P., 142
Henley, J. M., 129–130, 129t, 132–133, 137, 139, 140, 142–143
Henriques, R., 228, 401t
Herault, A., 424–427, 449–450, 451–452
Herbst, K. J., 322, 325–326, 331–332
Hern, J. A., 364–365
Herrero, E., 58–59
Herrmann, A., 61, 64t, 68, 73–76, 75f, 77, 78
Herrmann, H., 189t
Her, S., 428, 438, 440, 441, 442
He, S., 99–100
Hess, H. F., 222
Hess, S. T., 42, 116–118, 222, 224–225
Heveker, N., 364–365
He, W., 90
He, X., 424–427, 449–450
Hexel, C. R., 100–101
He, X. X., 423–424, 442–448, 443t, 449
He, Y., 5, 92–94
Hida, T., 256–258, 259t, 263, 263f, 265, 266f
Higuchi, H., 92
Hildgen, P., 419–422, 460
Hill, D., 164
Hill, M. A., 4–5, 21–22
Hinson, S. R., 340
Hiraki, T., 69
Hiraoka, Y., 46
Hirata, T., 419–422
Hiroaki, Y., 340–341
Hirschberg, K., 265
Hirst, G. J., 6–7, 8–9
Hirst, J., 258, 259t, 266–267, 268
Hlavacka, A., 166t
Hoang, C. P., 130
Hoarau, J., 5
Hobert, E. M., 357
Hobson, M. J., 308
Hodges, R. S., 358
Hodgson, L., 322–323
Hoebe, R. A., 272–273
Hoeijmakers, J. H. J., 8–9
Hoffmann, C., 355, 357
Hoffman, R. M., 112
Hoflack, B., 256–258, 259–260, 259t, 262–263
Hof, M., 113–114, 116–119
Hoi, H., 165t
Ho, J. D., 340–341
Holak, T. A., 166t
Holden, S. J., 42
Holen, T., 340–341
Holschen, J., 450
Holt, R. A., 153
Holtta-Vuori, M., 68
Holzle, A., 165t
Honda, K., 357

Hong, J.-K., 424–427, 455, 457
Hoogstraten, D., 8–9
Hopt, A., 7–8
Hornig, S., 423–424, 442–448, 443t, 449
Hornung, T., 93–94
Hoskins, R. A., 153
Houck, R., 194
Houghtaling, B. R., 151–153
Houtsmuller, A. B., 8–9, 186–187, 188f
Hovius, R., 355, 357–358, 364–365
Hovorka, O., 113–114, 116–119
Howard, J., 92, 96, 149–150
Howell, B. J., 151
Hoyer, M., 419–422, 423–424, 442–448, 443t, 449, 452–453
Hrafnsdottir, S., 67–68
Hsieh, C. L., 89–90
Huang, A. Y., 110
Huang, B., 40, 42, 166–167, 364–365
Huang, C. C., 424–427, 438, 439, 440, 441
Huang, D., 273–274
Huang, F., 259–260, 259t
Huang, L., 424–427, 454
Huang, S. H., 423–424, 443t, 449, 457–458
Huang, T., 100–101
Huang, X. H., 99
Huang, Y., 340–341
Huckabay, H. A., 238
Huff, T. B., 90
Hughes, T. E., 285
Hu, K., 303
Hülskamp, M., 166t
Hu, M., 86f
Hunter, P. R., 166t
Hunt, G., 68
Huppa, J. B., 224
Hu, R., 90, 99–100
Husimi, Y., 424–427, 442–448, 443t, 455, 457
Hwang, I., 164–166, 173, 174f
Hwang, W., 92
Hyman, A. A., 149–150
Hyman, B. T., 110

I

Iaboni, A., 364–365
Ibaraki, K., 129, 137
Ibata, K., 373
Igbavboa, U., 64t
Iglesias, P. A., 196
Iino, R., 87, 90–91
Iizuka, T., 440
Ikeda, H., 256–258, 259t, 263, 263f, 265, 266f
Ikonen, E., 68, 223–224
Inagaki, A., 189t
Irudayaraj, J., 99–100
Isaac, B., 42
Isailovic, D., 89

Ishida, H., 292–293
Ishii, M., 110
Ishimura, Y., 440
Ishitsuka, Y., 84
Ishiwata, S., 92–93
Ishmukhametov, R., 93–94
Isik, S., 245*f*, 250*f*
Itoh, H., 92–93
Ivanchenko, S., 164–166, 165*t*, 166*t*, 222–223
Iwasawa, K., 90–91
Iyer, G., 84

J

Jackson, A. O., 166–167
Jackson, C., 405–406
Jackson, D., 164
Jacobsen, V., 86*f*, 88, 91
Jacobson, K., 84, 87, 90–91, 193–194
Jacquier, V., 364–365
Jaensch, S., 407
Jäger, S., 189*t*
Jain, P. K., 98–99
Jakob, B., 8–9
Jakobsen, I., 423–424, 454–455
Jakobs, S., 40–41, 116–118
James, J. R., 364–365
Jana, N. R., 93
Jane, D., 129–130, 132–133, 139, 140, 142–143
Jang, S. Y., 419–422, 424–427, 455, 456–457
Jaqaman, K., 48, 184, 186–187, 189*t*, 196
Jaque, D., 419–422, 423–424, 457–458
Jares-Erijman, E. A., 370
Jarius, S., 340
Jaskolski, F., 129–130, 132–133, 139, 140, 142–143
Javitch, J. A., 364–365
Jendrossek, V., 58–59
Jenkins, E. L., 129–130, 132–133
Jenne, D., 166*t*
Jenner, J. T., 8, 21
Jenner, T. J., 4–5, 21–22
Jeon, H., 419–422, 424–427, 455, 456–457
Jessup, W., 58–59
Jiang, P., 112
Jia, Y. K., 285–287
Ji, L., 303
Ji, N., 40
Jin, B. J., 345–347, 346*f*
Jin, H., 419–422
Jin, S., 340, 342
Jin, Y. H., 424–427, 442–448, 443*t*, 449
Jockers, R., 364–365
Johnson, F. H., 292
Johnson, J., 189*t*
Johnson, P. F., 383
Johnson, T., 308
Johnsson, K., 355–358, 364–365

Johnsson, N., 357–358
Johnstone, D., 164–166
Jones, A. T., 272–273
Jonesb, A. T., 420*t*
Jones, D. R., 168–169
Jones, J. T., 149–150
Jones, S. A., 42
Jones, T., 395*t*, 408–409, 411
Jones, T. R., 189*t*, 409–410
Joo, C., 84
Josefsen, L. B., 419–422, 423–424, 452–453
Joseph, J., 64*t*
Jose-Yacaman, M., 99
Joshi, P. G., 64*t*, 71
Jovin, T. M., 41, 116–118, 370
Jung, J. H., 423–424, 449–450, 454
Jung, J. M., 91
Jung, J. S., 340–341

K

Kachynski, A. V., 90
Kadakia, M. P., 419–422, 420*t*
Kadhim, M. A., 4–5
Kahya, N., 90–91
Kain, S. R., 130
Kajikawa, E., 90–91
Kamegawa, A., 340–341
Kamentsky, L., 394
Kametaka, S., 256–258, 259–260, 259*t*, 265–267, 267*f*
Kaminski, C. F., 113–114
Kamiyama, D., 386–387
Kanegasaki, S., 440
Kang, B., 99
Kang, D., 423–424, 449–450, 454
Kang, I. H., 189*t*
Kano, H., 283–285
Kao, T. H., 166*t*
Kapoor, T. M., 151–153
Karasawa, S., 165*t*, 318
Karim, S. A., 110–111
Karin, M., 379–380
Karlova, R., 64*t*
Karnik, S. K., 355
Karplus, M., 92
Kasai, R. S., 90–91, 364–365
Kasman, I., 419–422, 427–428, 453–454
Kasper, R., 222
Kawano, K., 361
Kawano, S., 224
Kawata, S., 7–8, 26
Kay, S. M., 44–45
Keating, T. J., 149
Kell, D. B., 189*t*
Kelleher, M., 110–111
Keller, P., 303
Keller, W., 166–167

Kemble, R., 166–167
Kennedy, D. C., 277
Kennedy, S., 189t
Kenworthy, A. K., 265
Keppler, A., 355–357
Keri, G., 110–111
Kervrann, C., 44–45
Kessenbrock, K., 166t
Khaja, S., 428, 438, 440
Khalil, A. S., 92
Khodja, H., 5
Khodjakov, A., 150–151, 157–158
Kiêu, K. K., 48
Kim, B., 428, 438, 439, 440
Kim, G., 419–422, 423–424, 428–429, 430–437, 431t, 437f, 438, 438f, 439f, 440, 442–448, 443t, 449
Kim, H., 92, 96
Kim, I. S., 419–422, 424–427, 455, 456–457
Kim, J. H., 8–9, 419–422, 424–427, 455, 456–457
Kim, K., 419–422, 424–427, 428, 438, 440, 441, 442, 455, 456–457
Kim, M. J., 166–167
Kim, S.-H., 419–422, 424–427, 428, 438, 439, 440, 441, 442, 455, 456
Kim, S. Y., 424–427, 455, 457
King, M., 423–424, 440–441
Kinjo, M., 90–91
Kinosita, K. Jr., 92–93, 151
Kirschner, M., 149–150
Kisurina-Evgenieva, O., 150–151, 157–158
Kiyonaka, S., 357
Klar, T. A., 40–41
Klausing, K., 379–380
Klein, T. M., 67, 166–167
Klimant, I., 419–422, 423–424, 430–437, 431t, 449–450, 457–458
Kline-Smith, S. L., 149–150
Kligauf, J., 259t
Klonis, N., 141–142
Klösgen, R. B., 164
Klotzsch, E., 86f, 88, 91
Klumperman, J., 203f
Klumpp, S., 96
Kneipp, H., 422
Kneipp, J., 422
Kneipp, K., 422
Kner, P., 38, 41, 152
Knobel, S. M., 130
Koay, M. S., 452
Kobilka, B., 364–365
Kochaniak, A. B., 165t
Kochubey, O., 259t
Kocsis, E., 401t
Koebler, D., 194
Koenig, K., 7–8
Kokolic, K., 164–166, 168–169
Kolesnick, R., 58–59
Kolin, D. L., 90–91
Komatsu, H., 424–427, 442–448, 443t, 455, 457
Kombrabail, M., 64t, 71
Koncz, C., 166–167
Kondo, J., 90–91
Kondylis, V., 266–267
Kong, X., 8–9
Konig, I., 74
König, K., 7–9, 238
Koo, L. Y., 110
Koo, S. M., 419–422, 423–424, 428–429, 430–437, 431t, 437f, 438f
Koo, Y.-E. L., 419–422, 420t, 423–424, 428–429, 430–437, 431t, 437f, 438, 438f, 439f, 440–441, 442–448, 443t, 449
Kopelman, R. T., 419–422, 420t, 423–424, 427–429, 430–437, 431t, 437f, 438, 438f, 439f, 440–441, 442–448, 443t, 449–450, 451, 452–455, 459–460
Koppel, D. E., 139, 141–142, 173–175
Kopwitthaya, A., 99, 102
Korlach, J., 61–62
Kornfeld, S., 256–258, 263–264
Korte, T., 64t, 68, 74–76, 77
Kossmann, J. M., 423–424, 454–455
Kost, B., 166t
Kosugi, A., 224
Kosugi, Y., 292–293
Koumoutsakos, P., 48, 186–187, 189t, 193–194, 401t
Koushik, S. V., 370, 371, 379
Kovár, L. L., 113–114, 116–119
Koyama-Honda, I., 364–365
Kozubek, E., 166–167
Kranz, C., 239–240, 242
Krause, M., 165t
Kraynov, V. S., 329
Krebs, N., 74–76, 75f, 78
Kreis, T. E., 151
Krek, W., 189t
Kremer, J. R., 395t, 408
Kremers, G. J., 370
Krezel, A., 357
Krieg, R., 8–9
Krishnamoorthy, G., 64t, 71
Krishna, S., 340
Kristoferitsch, W., 340
Kronebusch, P. J., 149–150
Krooswyk, J. D., 424–427, 442–448, 443t
Kryzer, T. J., 340
Krzic, U., 39–40
Kubicova, L., 419–422, 423–424, 454–455
Kubota, M., 373
Kuenz, B., 340
Kuhl, S., 189t
Kühner, S., 166t
Kukura, P., 88, 92–93

Kumar, N. S., 258
Kumar, P., 189t, 305, 308–309
Kumar, R., 102
Kumar, S., 99, 224, 231
Kural, C., 40, 42, 87, 92, 96
Kusumi, A., 90–91, 364–365
Kuwata, H., 186–187, 189t
Kuzmin, A. N., 90
Kvilekval, K., 395t, 406
Kwon, I. C., 419–422, 424–427, 428, 438, 440, 441, 442, 455, 456–457

L

Labno, A. K., 92
LaFountain, D. J., 149–151
LaFountain, J. R. Jr., 149–151
Lagally, M. G., 5
Lakowicz, J. R., 371, 372–373, 418–419, 420t
Lam, C., 340–341, 345, 349–350
Lampe, M., 186–187
Lamprecht, M. R., 189t
Landis, D. M., 340–341
Landschulz, W. H., 383
Lane, M. D., 383
Langbein, W., 89–90, 280–281, 282–285
Langen, H., 224
Langer, R. S., 419–422
Langford, V. S., 6–7
Lang, M. J., 92
Lang, W., 340
Lanigan, P. M., 69
Lankelma, J., 113–114
Lantoine, F., 242, 246
Larpent, C., 420t, 423–427, 428–429, 442–448, 443t, 449–450, 451–452
Larson, D. R., 47, 229
Larson, T., 90
La Rue, S. A., 129–130, 132–133
Lasne, D., 86f, 87–88
Laurence, T. A., 92–93
Lawler, K., 110–111
Law, W. C., 102
Laxalt, A. M., 168–169
Lazareno, S., 364–365
Leatherbarrow, E. L., 8–9, 10, 21–22
Leaver, C. J., 166t
Leblane, R. M., 424–427, 449–450
Le Borgne, R., 256–258, 259–260, 259t, 262–263
Leduc, C., 91
Lee, C., 374–375
Lee, D., 428, 438, 440
Lee, H. Y., 423–424, 449–450, 454
Lee, K. J., 100–101
Lee, M. D., 340–341, 419–422, 424–427, 455, 456
Lee, R. J., 90
Lee, S. F., 42, 48
Lee, S. S., 419–422, 423–427, 454, 455, 456–457

Lee, S.-Y., 424–427, 455, 457
Lee, Y.-D., 428, 438, 440, 441, 442
Legeais, J. M., 110
Le Harzica, R., 7–8
Lehmann, P., 166–167
Leitz, A. J., 364–365
Lelek, M., 228
Lelimousin, M., 164–166, 165t
Lennon, V. A., 340
Leroi, A. M., 188
Levitt, J. A., 110–111
Levi, V., 84
Levskaya, A., 335
Lew, D., 379–380
Lewis, P. D., 238
Lew, M. D., 47
Liao, J., 90–91
Li, B., 423–424, 430–435, 431t, 441
Li, C., 424–427, 453
Lichtman, J. W., 292–293
Liebert, T., 423–424, 442–448, 443t, 449
Liedl, T., 186–187
Li, F., 189t
Li, J. J., 84, 340
Li, L., 340
Lillemeier, B. F., 224
Lillo, M. P., 66t
Lim, C. J., 325
Lim, C.-K., 428, 438, 440, 441, 442
Lim, D., 452–453
Lim, J. M., 423–424, 449–450, 454
Limoli, C. L., 10
Lince-Faria, M., 149–151, 153–155, 155f
Lin, C. P., 273–274
Lin, C. W., 322–323
Lindahl, E., 90–91
Lindquist, R. A., 189t
Lindwasser, O. W., 222
Lin, G., 189t
Ling, X., 164
Ling, Y. C., 452–453
Lin, J., 224
Linkert, M., 406
Link, S., 90, 93–94
Lin, M. Z., 355
Lin, V. S. Y., 89
Liphardt, J., 98–99
Lipowsky, R., 96
Lippincott-Schwartz, J., 40, 149, 164–166, 222, 256, 265–266, 304–305
Lippitz, M., 87–88, 89–90
Li, P. W., 153
Li, Q., 424–427, 441, 442
Lisenbee, C. S., 355
Litowski, J. R., 358
Littlefield, R., 292
Liu, F., 357–358
Liu, G. J., 7–8

Liu, H., 424–427, 449–450
Liu, J. Z., 5, 151–153, 283–285
Liu, L., 238, 423–424, 430–435, 431t, 441
Liu, L. W., 102
Liu, S. Y., 423–427, 428–429, 442, 449–450, 451, 452, 453
Liu, T., 423–427, 428, 442, 451, 452
Liu, W., 265–266
Liu, Y., 423–424, 442–448, 443t, 449
Livanec, P. W., 238
Li, W., 5
Li, X., 357–358
Li, Y., 424–427, 449–450
Li, Z., 68
Lloyd, D., 419–422, 423–424, 430–435, 431t, 436–437, 460
Lloyd, S. H., 419–422, 423–424, 430–435, 431t, 436–437, 460
Locke, R. J., 5
Lockyer, P. J., 327–328
Lodge, R., 265
Loerke, D., 48, 186–187, 189t
Loewke, K. E., 272–273
Loew, L. M., 273–274
Logan, D. C., 164–166, 166t, 168–169
Löhmannsröben, H. G., 419–422, 423–424, 430–435, 431t, 436–437
Lohse, M. J., 355, 357
Looger, L. L., 164–166, 165t, 304–305
Loog, M., 186, 195
Lopes, S. C. D. N., 58–59
Lorimore, S. A., 4–5
Loschberger, A., 67
Lotan, R., 99
Lotem, H., 275–276
Louit, G., 86
Lounis, B., 86f, 87–88, 90–91
Loura, L. M. S., 59–60, 62, 72
Love, S. A., 84
Lowman, J. E., 100–101
Low, P. S., 90
Lowry, D., 93–94
Lu, D. C., 340
Luider, T. M., 8–9
Luini, A., 202, 203f
Lukyanov, K. A., 164–166, 165t, 304–305, 354–355
Lukyanov, S., 164–166, 165t, 304–305, 354–355
Lu, M., 340–341
Luo, A., 164
Luo, Y. S., 90–91, 93, 94, 95f, 99
Lu, Q., 423–424, 449–450
Lu, S., 325
Lu, X., 423–424, 449–450
Lu, Y., 90
Lyle, K. S., 308–309
Lynch, R., 275–276
Lyn, R. K., 277

M

Ma, B. L., 424–427, 454
MacAulay, N., 340–341
Macdonald, D. A., 4–5
MacDougald, O. A., 383
Machacek, M., 329
Mack, A. F., 340–341
MacKay, J. F., 5
Mackey, M. A., 99
MacKinnon, R., 364–365
Maddox, P. S., 149–151, 152–153, 157–158
Madsen, P., 258
Maestro, L. M., 419–422, 423–424, 457–458
Magee, A. I., 68, 69, 224, 231
Magenau, A., 224, 231
Maiato, H., 149–151, 152–155, 155f, 156
Maier, S. R., 139, 142
Ma, J., 189t
Majoros, I. J., 419–422, 423–424, 455, 456
Majumdar, A., 259t
Makarov, N. S., 285
Makino, H., 139
Makino, R., 440
Malhó, R., 58–59
Malicka, J., 420t
Malik, J., 46
Malinow, R., 139
Malinsky, J., 58–59
Malkusch, S., 34
Mallavarapu, A., 150–151
Malmqvist, K., 5
Maloney, J., 273–274
Malpica, A., 99
Mancin, F., 423–427, 449–450, 452, 454
Manders, E. M., 48, 272–273
Manduchi, R., 46
Manley, G. T., 340
Manley, S., 40
Mano, S., 164
Marc, J., 166t
Marcuz, S., 424–427, 449–450
Marée, A. F. M., 192–193, 194, 196, 197
Marinho, H. S., 58–59, 60f, 63–67, 66t, 70–71, 72
Mari, P. O., 8–9
Markwardt, M. L., 318
Marmodée, B., 419–422, 423–424, 430–435, 431t, 436–437
Marquês, J. T., 72
Marquis, B. J., 84
Marra, P., 202, 203f
Marsden, S. J., 4–5
Martin, B. R., 357
Martinez, J., 361
Martinez, K. L., 355, 357–358, 364–365
Martin-Romero, F. J., 132–133
Martin, S., 129–130, 132–133
Martynov, V. I., 128–129, 129t

Marullo, S., 364–365
Maser, R. S., 5
Mashanov, G. I., 364–365
Masia, F., 89–90
Mason, M. D., 222
Masson, W. K., 5
Masuda, A., 358
Masuhara, H., 86, 419
Ma, T., 340–341
Mateo, C. R., 66t
Mathur, J., 164–167, 166t, 168–169, 173
Mathur, N., 164–166, 166t, 168–169, 173
Matias, A. C., 58–59
Matos, I., 149–151, 153–155, 155f
Matov, A., 151, 189t, 308–309
Matsuda, M., 322–323
Matsuzaki, K., 355, 358, 364
Mattera, R., 256–258, 259–260, 259t, 265–266
Mattes, J., 184
Matthay, M. A., 340
Matthews, D. R., 110–111
Mattheyses, A. L., 303
Mattoussi, H., 419–422, 424–427, 442–448, 443t, 449
Matveeva, E., 420t
Matz, M. V., 354–355
Maurel, D., 355–357, 365–366
Mayo-Martin, B., 129–130, 132–133, 139, 140, 142–143
Mayr, T., 423–424, 430–436, 431t, 449–450, 457–458
McConnell, H. M., 90–91
McCubbin, A. G., 166t
McDonagh, C., 423–424, 452–453
McGhee, E. J., 110–111
McGinty, J., 111–112
McKinley, A. J., 6–7
McKinney, S. A., 92–93, 164–166, 165t, 304–305
McKnight, S. L., 383
McLean, A. J., 364
McNally, J. G., 176
McNeill, J., 424–427, 430–435, 431t, 436
McShane, M. J., 419–422, 423–424, 430–435, 431t, 436
McSweeney, S., 164–166, 165t
Meallet-Renault, R., 424–427, 428–429, 449–450, 451–452
Meas-Yedid, V., 184, 196
Medintz, I. L., 419–422, 424–427, 428–429, 442–448, 443t, 449
Mehta, S. B., 89, 272–273
Mei, B. C., 419–422, 424–427, 442–448, 443t, 449
Meier, J., 128
Meier, R. J., 423–424, 430–437, 431t
Meijering, E., 48, 184, 185–187, 187f, 188f, 189t, 192–193, 195, 197
Meinzer, H. P., 188
Meixner, W., 419

Meldrum, R. A., 6–7, 8–9
Melinger, J. S., 419–422, 424–427, 442–448, 443t, 449
Mellon, P. L., 379–380
Melvin, T., 4–5, 8–9
Menger, F. M., 223–224
Merkx, G., 340–341
Merkx, M., 452
Mettlen, M., 186–187, 189t
Meyer, B. H., 355, 357–358, 364–365
Meyer, T., 149–150
Meyhofer, E., 96
Mhlanga, M. M., 228
Michael, B. D., 5
Michalet, X., 84
Michelmore, R., 166–167
Mie, G., 85
Miercke, L. J., 340–341
Miesenbock, G., 129t
Miettinen, M. S., 90–91
Miki, B., 166–167
Mikoshiba, K., 373
Miller, K. E., 92
Miller, M. T., 419
Miller, T. J., 419–422, 423–424, 427–428, 430–435, 431t, 436–437, 453–455
Millet, C. O., 419–422, 423–424, 430–435, 431t, 436–437, 460
Milligan, G., 355–357, 364–365
Milligan, R. A., 96
Mimura, T., 164
Minneman, K. P., 364–365
Mino, T., 358
Mirkin, M. V., 238
Mironov, A. A., 202, 203f
Mishra, S., 64t
Misselwitz, B., 409–410
Mitchison, T., 149–150
Mitchison, T. J., 149–151, 157–158
Mitsuoka, K., 340–341
Mittag, A., 399–400
Miura, K., 84
Miwa, T., 164
Miyake, K., 224
Miyamoto, D. T., 157–158
Miyata, H., 92–93, 151
Miyawaki, A., 165t, 272–273, 322, 370, 373
Mizuno, H., 165t
Mizutani, K., 92–93
Mochizuki, N., 328
Modesti, M., 8–9
Moeller, H. B., 340–341
Moerner, W. E., 42, 90–91
Moffat, J., 189t
Mogilner, A., 149–150
Mohaghegh, P. S., 164–166, 168–169
Mohamed, M. B., 90
Mohanty, A., 164

Mohanty, S. K., 8–9
Mohr, G. J., 419–422, 423–424, 430–435, 431t, 436–437, 442–448, 443t, 449, 452–453, 454–455
Molloy, J. E., 364–365
Monajembashi, S., 7–8
Mondragon, A., 450
Monson, E., 419, 423–424, 428–429, 451, 452
Montalti, M., 420t, 424–427, 449–450
Monticelli, L., 90–91
Montolli, M., 424–427, 449–450
Moore, J., 406
Mooren, O. L., 238
Moossa, A. R., 112
Moreaux, L., 111
Moreira, I. S., 364–365
Morgan, J. R., 308
Mori, Y., 357
Moro, A. J., 423–424, 430–435, 431t, 436–437
Morotomi-Yano, K., 8–9
Morrison, G. H., 452–453
Mortelmaier, M. A., 224
Mortensen, N. P., 238
Morton, J. P., 110–111
Moshelion, M., 166t
Mosig, A., 189t
Moss, L. G., 78
Motzkus, M., 280
Moutinho-Pereira, S., 153, 154–155
Mravec, J., 164–166, 173, 174f
Mudaliar, D. J., 63–67, 64t
Mukherjee, A., 222
Mukherjee, S., 64t, 68, 71
Müller, B., 186–187
Muller, C., 88, 92–93
Muller, M. J. I., 96
Mulvaney, P., 86f, 90
Munnik, T., 168–169
Munro, I., 69
Murakoshi, H., 90–91
Murase, K., 90–91
Murashige, T., 169
Murphy, C. J., 84, 93, 100–101
Murphy, C. S., 164 166, 165t, 304–305
Murphy, R. F., 392, 395t, 409–410
Murray, J. I., 188, 189t
Murrow, L., 291–313
Murthy, N., 428, 438, 440
Murtola, T., 90–91
Myc, A., 419–422, 423–424, 455, 456
Myers-Payne, S. C., 64t

N

Naccache, R., 419–422, 423–424, 457–458
Nagafuchi, A., 90–91
Nagai, T., 318, 373
Nagaria, P. K., 100–101
Nagasaki, Y., 424–427, 456
Nägerl, U. V., 40–41
Naik, R. R., 419–422, 420t
Nair, G., 424–427, 442–448, 443t
Na, J., 428, 438, 440, 441, 442
Nakada, C., 90–91, 364–365
Nakamura, T., 424–427, 456
Nakano, A., 166t
Nakayama-Ratchford, N., 423–424, 442–448, 443t, 449
Nakazono, M., 173
Nakshatri, H., 99–100
Nalbant, P., 329
Naldini, L., 305
Nallathamby, P. D., 100–101
Nam, H. Y., 419–422, 424–427, 455, 456–457
Nandi, M., 419–422, 423–424, 451, 452
Nan, X. L., 92
Narayanaswamy, A., 189t
Nar, H., 164–166
Nath, S., 189t
Nebel, M., 238
Nebenführ, A., 164
Neefjes, J. J., 132, 141–142
Neely, J. D., 340–341
Neher, E., 7–8
Neil, M. A., 69, 111–112, 224, 231
Nelms, B. E., 5
Nelson, B. K., 164
Nelson, D., 189t
Nelson, G., 189t
Nesterov, A., 259–260, 259t
Neukirchen, D., 166t
Neumann, F. R., 189t
Newberg, J., 409–410
Newman, R. H., 316–319, 322–323
Nguyen, A. W., 318
Nguyen, N. N., 419–422
Nichols, B. J., 265
Nicolson, T. J., 452
Nielsen, E., 166t
Nielsen, L. J., 419–422, 423–424
Niemela, P. S., 90–91
Nienhaus, G. U., 130, 164–166, 165t, 166t, 222–223
Nienhaus, K., 164–166, 165t
Nie, S. M., 99, 419–422
Niessen, W. J., 186–187, 187f, 188f, 189t, 195
Nietzsche, S., 419–422, 423–424, 442–448, 443t, 449, 453, 454–455
Nigg, A. L., 8–9
Nikitin, A. Y., 110
Ni, Q., 319–321, 322–323, 326–327, 370
Nishikawa, K., 340–341
Nishikawa, S., 87, 164
Nishimune, A., 129–130, 132–133, 139, 142
Nishimura, M., 164
Nishimura, S. Y., 90–91

Nishizaka, T., 92–93
Nitin, N., 99
Nitzsche, B., 92–93
Nocera, D. G., 424–427, 442–448, 443t
Noell, S., 340–341
Noji, H., 87, 92–93
Nomura, W., 358
Noshita, N., 340
Novo, C., 86f
Nukina, N., 165t
Nuss, G., 423–424, 430–435, 431t

O

Oberdorff-Maass, S., 355, 357
Octeau, V., 87–88
Odde, D. J., 151
Oddos, S., 224, 231
O'Halloran, T. V., 450
Ohashi, N., 358
Ohba, K., 358
Oh, J. M., 428, 438, 440, 441, 442
Ohsugi, Y., 90–91
Oishi, M., 424–427, 455, 456
Oishi, S., 355, 358, 364
Oiwa, K., 92–93
Ojida, A., 357
Okabe, K., 419–422, 428, 457–458
Okada, S., 440–441
Okada, Y., 132
Okten, Z., 47
Olenych, S., 222
Oliveira, C. L., 258
Oliveira, M. I., 364–365
Oliveira, R. A., 49–50
Olive, P. L., 20
Olivier, N., 110
Olivo-Marin, J. C., 184, 185–187, 189t, 192–193, 195, 196, 197
Oller, H., 364–365
Olsen, L. F., 419–422, 423–424, 428, 438, 439, 440
Olson, M. F., 110–111
Omae, K., 353–370
Omann, G. M., 102
O'Neill, P., 4–5, 8–9, 10, 21–22
Oorschot, V., 203f
Opekarova, M., 58–59
Oprian, D. D., 364–365
Orellana, A., 166t
Oron, D., 89–90, 280–281
Orringer, D. A., 420t, 435–436
Orrit, M., 87–88, 89–90
Orte, A., 424–427, 442–448, 443t
Osgood, C. J., 100–101
Osher, S., 46
Oswald, F., 130, 164–166, 165t, 166t, 222–223
Otsuki, S., 292–293

Ouadahi, K., 420t, 424–427, 449–450, 451–452
Outten, C. E., 450
Owen, D. M., 69, 224, 231
Ow, H. O., 419–422
Oyelere, A. K., 99
Ozalp, V. C., 419–422, 423–424
Ozols, A., 5

P

Padmawar, P., 340
Pakhomov, A. A., 128–129, 129t
Palaniappan, K. P., 189t
Pallon, J., 5
Pansu, R. B., 424–427, 449–450, 451–452
Pantazis, P., 273–274
Panula, P., 68
Papadopoulos, M. C., 340
Parak, W. J., 84, 186–187
Park, C. M., 166–167
Park, E. J., 419–422, 423–424, 454
Parker, A. W., C0005#, 4–5, 8–9, 10–11, 21–22
Park, E. S., 373
Park, H., 419–422, 424–427, 455, 456
Parkinson, M., 258, 259t, 266–267, 268
Park, K., 419–422, 424–427, 455, 456–457
Park, M., 423–424, 449–450
Park, R. W., 419–422, 424–427, 455, 456–457
Park, S. Y., 99
Parmryd, I., 68
Parton, R., 44–45
Parton, R. M., 49–50
Parus, S., 419–422, 423–424, 442–448, 443t, 449, 452–453
Pastor-Soler, N., 340–341
Patel, G., 110–111
Patterson, G. H., 42, 130, 149, 164–166, 165t, 222, 304–305
Pavani, S. R. P., 42
Pavlova, I., 99
Pawley, J., 33, 34–35
Paxon, T. L., 420t
Pebesma, E. J., 48
Pedersen, J. S., 258
Pediani, J. D., 355–357, 364–365
Pedroso, N., 58–59
Peers, C., 419–422, 423–424, 452–453
Pego, R. L., 176
Pemble, H., 291–313
Peng, H.-S., 184, 395t, 407–408, 410–411, 423–427, 442–448, 443t, 449, 457–458
Peng, J. F., 423–424, 442–448, 443t, 449
Peng, X. H., 99
Pereira, A. J., 149–151, 152–155, 155f, 156
Perez, J., 151
Periasamy, A., 73–74, 110–111, 370–371, 373, 374, 383
Perlman, Z. E., 151, 157–158

Perona, P., 46
Perrimon, N., 194–195
Perroy, J., 322–323
Perry, G. L. W., 228
Persengiev, S. P., 8–9
Pertsinidis, A., 47
Pertz, O., 328, 329
Petersen, C. M., 258
Petersen, N. O., 90–91
Peters, P. J., 256–258, 259–260, 262, 265–266
Petrini, J. H. J., 5
Petros, J., 428, 438, 440
Petrov, G. I., 283–285
Petschek, R. G., 92–93
Peuckert, C., 8–9
Pezacki, J. P., 277
Pfeffer, S. R., 259t
Phair, R. D., 265–266
Philbert, M., 428–429
Philbert, M. A., 419–422, 423–424, 427–428, 430–437, 431t, 438, 439f, 440, 442–448, 443t, 449, 452–455, 459
Phillips, G. N. Jr., 78
Pick, H., 355–358, 364–365
Pieper, S., 410–411
Pike, L. J., 58–59, 77
Pinaud, F. F., 84
Pincus, Z., 194–195
Pinedo, H. M., 113–114
Pin, J. P., 355–357, 365–366
Piperakis, S. M., 20
Pishko, M. V., 428, 438, 439, 440
Piston, D. W., 130, 370, 373
Pitta Bauermann, L., 240, 242, 249f
Pittock, S. J., 340
Plamann, K., 110
Planchon, T. A., 39–40
Plazzo, A. P., 64t, 68, 75f, 77, 78
Pliss, A., 90
Pohlmann, R., 258
Poland, S. P., 110–111
Polishchuk, E. V., 202, 203f
Polishchuk, R. S., 202, 203f, 265
Pomorski, A., 357
Pomorski, T., 67–68
Pope, I., 271–292
Pope, S. J. A., 419–422, 423–424, 430–435, 431t, 436–437, 460
Poulsen, A. K., 423–424, 428, 438, 439, 440
Powell, P. R., 420t
Prasad, P. N., 90, 99–100, 102
Prasher, D. C., 292
Preston, G. M., 340–341
Preuss, M., 166t
Prezeau, L., 355–357, 365–366
Prieto, M., 59–60, 62, 70–71, 72, 75f, 78
Prigozhina, N., 298
Prise, K. M., 5

Prodi, L., 420t, 424–427, 449–450
Proppert, S., 67
Prossnitz, E. R., 364–365
Prummer, M., 364–365
Psaltis, D., 89–90
Puertollano, R., 256–258, 259–260, 259t, 262, 265–266
Puhl, H. L. III, 371, 379
Puliafito, C. A., 273–274
Pu, Y., 89–90

Q

Qian, H., 91, 193–194
Qian, L., 194
Qian, W., 99
Qian, X., 423–424, 451–452
Qiao, Y. X., 92–94
Qi, H., 110
Quickenden, T. I., 6–7
Quinlan, M. E., 92–93
Quinn, J. A., 110–111

R

Rabanel, J.-M., 419–422, 460
Rabouille, C., 266–267
Rabut, G., 407
Rademakers, S., 8–9
Radford, D., 164–166, 168–169
Radhamony, R., 164–166, 168–169
Raghuraman, H., 68
Rago, G., 90, 283–285
Ralph, G. S., 129–130, 129t, 132–133
Rampazzo, E., 423–427, 449–450, 452
Rappoport, J. Z., 303
Rash, J. E., 340–341
Rasmussen, S. G., 364–365
Ravichandra, B., 64t, 71
Ray, A., 423–424, 442–448, 443t, 449
Raymond, E. S., 399, 404
Ray, S., 96
Rebane, A., 285
Reck-Peterson, S. L., 92
Redford, G. I., 376
Red, R., 165t
Reese, T. S., 340–341
Reichenzeller, M., 189t
Reijo Pera, R. A., 272–273
Reindl, M., 340
Reinhard, B. M., 98–99
Reiser, J., 308
Reister, E., 90–91
Reits, E. A., 132, 141–142
Remington, S. J., 164–166
Ren, J. C., 87
Renn, A., 88, 92–93
Rentero, C., 224, 231

Repakova, J., 68
Reynaud, E. G., 39–40
Reynolds, A. R., 110–111
Reynolds, P., C0005#, 8–9, 10, 21–22
Richards-Kortum, R., 99
Ridley, A. J., 329
Rieder, C. L., 149–150
Riedl, J., 166t
Riehle, A., 58–59
Riemann, I., 7–8
Rihova, B., 113–114, 116–119
Rijss, J. P., 340–341
Rines, D. R., 47
Ripley, B. D., 231
Ritchie, K., 90–91
Rittscher, J., 184
Rittweger, E., 41
Rivera gil, P., 84
Rives, M. L., 355–357, 365–366
Rizzo, M. A., 318, 373
Roach, E., 164–166, 168–169
Robbins, M. S., 33
Robbins, R. A., 340–341
Roberts, L. M., 166t
Roberts, T. H., 265
Robinson, D. G., 164–166, 173, 174f
Robinson, M. S., 258, 259t, 263–264, 266–267, 268
Rocha-Mendoza, I., 280–281, 282–285
Rocker, C., 165t
Röcker, C., 164–166, 165t, 166t, 222–223
Rodriguez, E. M., 419–422, 423–424, 457–458
Rodriguez, M., 224, 231
Rohr, K., 184, 186–187
Rong, G. X., 98–99
Roots, R., 440–441
Rose, A., 47–48
Rosenzweig, Z., 424–427, 449–450
Rosinski, M., 308
Rossi, A., 340–341, 345–347, 346f, 349–350
Ross, J. L., 92
Ross, W. N., 316
Rossy, J., 224, 231
Rothman, J. E., 129–130, 129t, 256
Rouille, Y., 256–258, 259–260, 259t, 262–263
Roy, I., 90, 99–100, 102
Roy, P., 419–422, 423–424, 451, 452
Roysam, B., 189t
Rozentur, E., 419–422
Rozenzhak, S. M., 419–422, 420t
Rudin, L. I., 46
Ruedas-Rama, M. J., 419–422, 424–428, 442–448, 443t, 449, 451–452, 453, 454–455
Rueden, C., 410–411
Rug, M., 141–142
Ruhnow, F., 92–93
Runions, J., 166t

Rust, M. J., 42, 222
Rutter, G. A., 452
Ryan, T. A., 129–130, 129t

S

Saadoun, S., 340
Sabatini, D. M., 189t
Sabirov, R. Z., 132
Sacan, A., 189t
Saenger, W., 90–91
Sage, D., 189t, 401t
Sahlender, D. A., 258, 259t, 266–267, 268
Saiga, Y., 292
Saitoh, S. I., 224
Saito, K., 90–91
Saito, M., 90–91
Sakai, N., 256–258, 259t, 263, 263f, 265, 266f
Sakakihara, S., 87
Sako, Y., 90–91
Salih, A., 164–166, 165t, 166t, 222–223
Salmon, E. D., 148–153, 154–155, 155f, 157–158
Salmon, T. D., 151
Salmon, W. C., 151, 298
Salonen, E., 68
Samaj, J., 166t
Samelson, L. E., 224
Samhan-Arias, A. K., 132–133
Sammak, P. J., 292
Sammalkorpi, M., 90–91
Sandoghdar, V., 86f, 88, 91, 92–93
Sandstrom, A., 340–341
Sanford, J. C., 166–167
Sankaranarayanan, S., 129–130, 129t
Santella, A., 407
Santi, P. A., 39–40
Sanz-Rodriguez, F., 419–422, 423–424, 457–458
Sapsford, K. E., 428–429
Sarkar, K., 419–422, 423–424, 451, 452
Sartania, N., 364–365
Sasaki, K., 322–323, 419
Sasaki, S., 340–341
Sase, I., 92–93, 151
Sastry, L., 308
Sato, Y., 401t
Sauer, M., 67, 222
Savage, J. R., 4–5
Sawada, N., 258, 259t, 266–267
Saxton, M. J., 50–51, 84, 87, 193–194
Sbalzarini, I. F., 48, 186–187, 189t, 193–194, 401t
Schaefer, L. H., 46
Schanne-Klein, M. C., 110
Scharff-Poulsen, A. M., 423–424, 428, 438, 439, 440, 442–448, 443t, 454–455
Schattat, M., 164
Schaufele, F., 318–319, 383
Schenke-Layland, K., 273–274
Schenkel, M., 164–166

Author Index

Schepartz, A., 357
Scherer, S. E., 153
Schermelleh, L., 39, 40, 42
Schettino, G., 5
Schiffmann, D. A., 406
Schipper, N. W., 113–114
Schleifenbaum, A., 319, 323–324
Schlessinger, J., 139, 141–142, 173–175
Schmälzlin, E., 419–422, 423–424, 430–435, 431t, 436–437
Schmid, B., 401t
Schmid, S. L., 186–187, 189t
Schmidt, C. F., 92
Schmidt, R., 40, 42
Schmitt, F., 164–166, 165t, 166t, 222–223
Schnapp, B. J., 92
Schneider, I., 269
Schnitzer, M. J., 84, 92
Scholey, J. M., 149–151
Scholthof, H. B., 166–167
Scholthof, K. B. G., 166–167
Schönle, A., 116–118
Schroeder, F., 64t, 66t
Schuck, S., 58–59, 78
Schuhmann, W., 238, 239–240, 242, 245f, 249f, 250f, 251f, 252–253
Schuler, J., 90–91
Schulte, A., 238, 239–240, 242, 249f, 251f
Schulz, A., 423–424, 442–448, 443t, 449, 452–453
Schuman, J. S., 273–274
Schuster, T., 424–427, 428–429, 449–450, 451–452
Schüttpelz, M., 222
Schuurhuis, G. J., 113–114
Schwartz, O., 89–90
Schwarz, H., 58–59
Schwarz, J. P., 74, 110–111
Schwille, P., 61–62, 90–91
Scolari, S., C0015#, 74–76, 75f, 78
Sedat, J. W., 38
Seefeldt, B., 222
Segura, J. M., 355, 357–358, 364–365
Seifert, U., 90–91
Selvin, P. R., 42, 47, 87, 92–93, 96
Sengupta, P., 423–424, 442–448, 443t, 449
Seong, B. L., 419–422, 424–427, 455, 456
Seo, S. A., 370, 371, 423–424, 449–450
Serge, A., 128
Serpinskaya, A. S., 87, 92
Shah, N. C., 274–275
Shain, W., 189t
Shaner, N. C., 34, 164–166, 165t, 292, 304–305, 318–319, 373
Shan, X. Y., 90–91
Shao, L., 42
Sharif, W. D., 130
Sharma, R., 419–422

Shaw, M. A., 92–93
Shaw, T. J., 100–101
Shcheglov, A. S., 164–166, 165t
Sheets, E. D., 116–118
Sheetz, M. P., 91, 193–194
Sheng, L., 5
Shen, H., 189t
Shen, J., 424–427, 441, 442
Sheppard, C. J. R., 89
Shiang, Y. C., 424–427, 438, 439, 440, 441
Shields, D., 194
Shi, L. Z., 8–9
Shimizu, H., 165t
Shimizu, M., 292–293
Shimomura, O., 292
Shin, D. M., 99
Shinmi, D., 357
Shi, Z.-Y., 419
Shroff, H., 40, 222, 223
Shubeita, G. T., 186, 195
Shun, T., 194
Sick, T. J., 340
Siegel, A. J., 149–151
Silberberg, Y., 280–281
Silberstein, C., 340–341
Silva, L. C., 70–71, 72
Simmonds, D., 166–167
Simons, K., 58–59, 78, 223–224, 303
Simon, S. M., 303
Simpson, P. J., 4–5
Sims, P. A., 92
Sinclair, A. M., 164–166, 166t, 168–169, 173
Sindelar, C. V., 92
Singaram, B., 424–427, 441, 442
Singaravelu, R., 277
Sinha, M., 64t
Sinka, R., 258, 259t, 266–267, 268
Sisco, P. N., 84
Sixt, M., 166t
Skach, W., 340
Skach, W. R., 341f, 342
Skewis, L. R., 98–99
Skoog, F., 169
Slater, N. K., 113–114
Slattery, J. P., 131f, 139–140
Slaughter, L. S., 93–94
Sligar, S. G., 364–365
Slotkin, J. R., 419–422
Slowing, I. I., 89
Smal, I., 48, 184, 185–187, 188f, 192–193, 195, 197
Smith, A. E., 86f, 88, 91
Smith, B. L., 340–341
Smith, D. K., 90
Smith, K. L., 189t
Smith, M. B., 407
Smith, P. B., 424–427, 442–448, 443t, 449
Smith, R., 422

Smucker, B., 308
Snaebjornsson, M. T., 340–341
Snee, P. T., 424–427, 442–448, 443t
Sokolov, K., 84, 90, 99
Sole, J. G., 419–422, 423–424, 457–458
Soler-Argilaga, C., 66t
Soll, D. R., 189t, 192–193, 194–195
Soloviev, V. Y., 111–112
Somers, R. C., 424–427, 442–448, 443t
Sommi, P., 150–151
Sonek, M., 7–8
Song, A., 419
Song, M., 5
Son, H., 423–424, 454
Sonnichsen, C., 90, 93–94, 98–99
Sorkin, A., 259–260, 259t
Sougrat, R., 222
Souslova, E. A., 165t
Sperling, R. A., 84
Spetzler, D., 93–94
Spielhofer, P., 166t
Spiller, D., 189t
Spindler, K., 164–166, 165t, 166t
Spindler, K.-D., 222–223
Sprague, B. L., 175
Springer, G. H., 373
Srihari, R., 383
Srivastava, M., 90–91
Staehelin, L. A., 340–341
Stalder, R., 151
Staleva, H., 86f
Stanca, S. E., 419–422, 423–424, 430–435, 431t, 436–437, 442–448, 443t, 449, 453, 454–455
Staroverov, D. B., 164–166, 165t
Stauber, C., 379–380
Stavreva, D. A., 176
Stehbens, S., 291–313
Steinbach, P. A., 34, 292, 373
Stellacci, F., 91
Stelzer, E. H. K., 40
Stender, A. S., 88, 95–96, 422
Stenmark, H., 166t
Stepensky, D., 401t
Stephens, D. J., 272–273, 420t
Stephens, J., 8–9
Stephenson, F. A., 90–91
Steup, M., 419–422, 423–424, 430–435, 431t, 436–437
Stevens, D. L., 4–5, 21–22
Stewart, M. H., 419–422, 424–427, 442–448, 443t, 449
Steyer, J. A., 303
Stich, M. I. J., 423–424, 443t, 449, 457–458
Stierhof, Y., 164–166, 173, 174f
Stiles, P. L., 274–275
Stinson, W. G., 273–274
Stöckl, M., 64t, 68, 77

Stöckl, M. T., 61, 73–74
Stoddard, A. K., 423–424, 449–450
Stoller, P., 88
Stolwijk, J. A., 423–427, 430–437, 431t, 442–448, 443t, 449
Stone, J. W., 84
Stone, M. O., 419–422, 420t
Strackea, F., 7–8
Straight, A., 149–151
Strano, M. S., 419–422
Straub, M., 111–112
Strickler, J. H., 6–7, 39–40, 110
Strohalm, J., 113–114, 116–119
Strohalm, M., 113–114, 116–119
Strollo, C., 419–422, 423–424, 428–429, 430–437, 431t, 437f, 438f
Stroud, R. M., 340–341
Stryer, L., 371
Stuurman, N., 150–151, 392, 395t, 405–406
Suarez, F., 419–422, 460
Subra, G., 361
Subramaniam, V., 116–118
Subr, V., 113–114, 116–119
Sugimoto, Y., 355, 358, 364
Su, H., 5
Suhling, K., 110–111
Su, J., 424–427, 449–450
Sullivan, C. J., 238
Sumner, J. P., 423–424, 449–450, 451, 452
Sunahara, R. K., 364–365
Sun, C., 428, 438, 440
Sun, H., 423–424, 442–448, 443t, 449, 454–455
Sun, H. B., 7–8, 26
Sun, I.-C., 424–427, 455, 457
Sun, L.-N., 423–427, 442–448, 443t, 449, 457–458
Sun, W., 88, 89, 93, 94, 95–98, 95f, 97f, 99, 422
Sun, X. M., 423–424, 442–448, 443t, 449
Sun, Y., 5, 370, 371, 407
Suresh, K., 258
Susumu, K., 419–422, 424–427, 442–448, 443t, 449
Suzuki, H., 340–341
Suzuki, K. G., 90–91, 364–365, 424–427, 442–448, 443t, 455, 457
Suzuki, M., 424–427, 442–448, 443t, 455, 457
Svedberg, F., 90
Svoboda, K., 92, 110
Swaminathan, R., 130
Swanson, E. A., 273–274
Swartling, J., 113–114
Swedlow, J. R., 33, 47, 184, 395t, 404, 406
Swoger, J., 39–40
Syed, S., 92–93, 96
Sylvester, A. W., 164

Author Index

T

Tabarin, T., 423–424, 452–453
Tabata, K. V., 87
Tahir, K. B., 111–112
Tajima, M., 341f, 345, 346f, 347, 348f, 349
Takada, K., 7–8, 26
Takahashi, A., 92–93
Takeichi, M., 90–91
Talavera, E. M., 424–427, 442–448, 443t
Tamamura, H., 358
Tamura, A., 424–427, 456
Tamura, M., 90–91
Tanaami, T., 292–293
Tanaka, G., 86
Tanaka, T., 7–8, 26, 358, 440
Tang, B., 424–427, 441, 442
Tang, Q. Q., 383
Taniike, M., 256–258, 259t, 263, 263f, 265, 266f
Tani, K., 340–341
Tanimura, N., 224
Tank, D. W., 110
Tanner, W., 58–59
Tan, W., 423–424, 442–448, 443t, 449
Tassiou, A. M., 20
Taucher-Scholz, G., 8–9
Taylor, S. S., 354–355
Taylor, W. R., 428, 438, 440
Tchebotareva, A. L., 87–88
Tecilla, P., 423–427, 449–450, 452, 454
ten Kate, T. K., 113–114
Teoh, C., 383
Teolato, P., 423–424, 451, 452
Terrillon, S., 355, 357
Thaa, B., 74–76
Thakar, R., 424–427, 442–448, 443t
Thaler, C., 370, 371, 379
Thébaud, A., 184, 196
Theer, P., 110
Theill, L. E., 379–380
Theriot, J. A., 149, 194–195
Thevenaz, P., 401t
Thévenaz, P., 401t
Thirup, S. S., 258
Thoma, C., 189t
Thomann, D., 47
Thomas, T. P., 419–422, 423–424, 455, 456
Thompson, M. A., 42, 47, 48
Thompson, R. E., 47, 229
Thoronton, K., 113–114
Thorsrud, B., 419
Tilley, L., 141–142
Tillo, S. E., 285
Timmers, A. C., 166t
Timpson, P., 110–111
Tinel, N., 355–357, 365–366
Ting, A. Y., 354–355, 370
Tinnefeld, P., 222

Tiret, P., 111
Tirlapur, U. K., 7–9
Tjalkens, R., 419–422, 427–428, 453–454
Tkachenko, E., 329
Tobeña-Santamaria, R., 168–169
Tomasi, C., 46
Tomczak, A., 166–167
Tomita, K., 112
Tomiyama, Y., 256–258, 259t, 263, 263f, 265, 266f
Tomosada, N., 292–293
Tonellato, U., 423–427, 449–450, 452, 454
Tong, L., 90
Tong, Z., 424–427, 449–450
Toomre, D., 40, 303
Toprak, E., 40, 42, 92–93
Townsend, K. M. S., 5
Trammell, S. A., 419–422, 424–427, 442–448, 443t, 449
Travis, K., 99
Trejo, J., 364
Trelease, R. N., 355
Trevin, S., 242, 246
Trewyn, B. G., 89
Trible, R. P., 224
Triller, A., 128
Trinquet, E., 355–357, 365–366
Trobacher, C. P., 164–166, 166t, 173
Tromberg, B. J., 7–8
Tsay, J. M., 84
Tsien, R. Y., 128–129, 184, 292, 317–318, 354–355, 357, 362, 370, 373
Tsuchiya, H., 112
Tsujimoto, G., 355, 358, 364
Tsukita, S., 90–91
Tsunoda, M., 89
Tsutsui, H., 165t
Tsutsumi, H., 358
Tsutsumi, N., 173
Tunnacliffe, A., 419–422, 424–427, 442–448, 443t, 449
Turck, C. W., 340
Turcu, E., 10, 25–26
Turcu, F., 252–253
Turney, S., 91
Tuthill, D. E., 164
Tu, Y., 30–31, 292
Tvaruskó, W., 184
Tyner, K. M., 423–424, 459
Tyrakowski, C. M., 424–427, 442–448, 443t

U

Uberti, M. A., 364–365
Uchida, K., 340
Uchiyama, S., 419–422, 428, 457–458
Uchiyama, Y., 256–258, 259–260, 259t, 262–263, 263f, 265, 266f
Ueda, T., 166t

Uematsu, N., 8–9
Ueno, H., 87
Ulbrich, E., 419–422, 423–424, 428–429, 430–437, 431t, 437f, 438f
Ulbrich, K., 113–114, 116–119
Umemura, Y., 90–91
Uney, J., 129–130, 129t, 132–133
Unser, M., 184, 189t, 401t
Unwin, P. R., 238
Uphoff, S., 42
Upton, K., 149–150
Urizar, E., 364–365
Uronen, R. L., 68
Uwada, T., 86

V

Vachon, J. J., 424–427, 449–450, 451–452
Vale, R. D., 90–91, 92, 96, 150–151
Vallone, B., 165t
Valtorta, F., 228
van Cappellen, W. A., 184, 185–186, 187f, 189t
van de Linde, S., 42, 67, 222
van der Wel, N. N., 256–258, 259–260, 262, 265–266
van Dijk, M. A., 87–88, 89–90
van Dongen, J. T., 419–422, 423–424, 430–435, 431t, 436–437
Van Duyne, R. P., 274–275
van Gent, D. C., 8–9
van Haastert, P. J. M., 189t
van Heijningen, T. H., 113–114
Van Hoek, A. N., 340–341, 341f, 342, 349
van Leeuwen, W., 168–169
van Meer, G., 67–68
Vannier, C., 128
Van Noorden, C. J., 272–273
Van Oijen, A. M., 165t
Van Oven, C. H., 272–273
van Rijnsoever, C., 203f
van Royen, M. E., 186–187, 188f
Van Wilder, V., 166t
Vapnik, V., 409
Varrin-Doyer, M., 340
Vartiainen, E. M., 283–285
Vathy, L. A., 90
Vattulainen, I., 68, 90–91
Veit, M., 74–76, 75f, 78
Velasquez-Castano, J. C., 428, 438, 440
Venetta, B. D., 374
Verbavatz, J. M., 340–341
Verdijk, M. A., 340–341
Verkade, P., 203f
Verkaik, N. S., 8–9
Verkhusha, V. V., 164–166, 165t, 224–225, 304–305
Verkman, A. S., 90–91, 130, 340–341, 341f, 342, 343–344, 343f, 345–347, 346f, 348f, 349–350
Vermeer, J. E., 168–169

Vermeulen, W., 8–9
Verveer, P. J., 39–40, 110–111
Vetrone, F., 419–422, 423–424, 457–458
Vetvicka, D., 113–114, 116–119
Viana, A. S., 72
Vigna, E., 305
Vilardaga, J. P., 355, 357
Vincent, A., 340
Vinkenborg, J. L., 452
Violin, J. D., 326–327
Visser, A. J., 62
Visvardis, E.-E., 20
Vogel, H., 355–358, 364–365
Vogel, S. S., 370, 371, 379
Vogel, V., 88
Voigt, B., 166t
Vojnovic, B., 5, 110–111
Volkmer, A., 277
Volkov, V., 90
von Andrian, U. H., 110
Vonesch, C., 184
Vonesch, J. L., 184
Von, K. G., 58–59
von Kriegsheim, A., 110–111
von Plessen, G., 90
von Vacano, B., 280
Vrljic, M., 90–91

W

Wadsworth, P., 149
Wagenbach, M., 92
Wagenknecht-Wiesner, A., 116–118
Waggoner, A., 362
Waguri, S., 256–258, 259–260, 259t, 262–263, 263f, 265, 266–267, 266f
Wahlby, C., 408–409
Wahl, G. M., 112
Walczak, C. E., 149–150
Walker, S. A., 327–328
Walker, W. F., 48, 186, 195
Wallace, W., 46
Wallrabe, H., 110–111, 370, 371
Walsh, C. T., 357–358
Walter, T., 400, 410
Walvick, R. P., 63–67, 64t
Walz, T., 340–341
Wang, C., 189t
Wang, D., 424–427, 428–429, 442, 449–450, 453
Wang, E., 294
Wang, G., 422
Wang, G. F., 88, 93, 94, 95–98, 95f, 97f, 99
Wang, H., 194
Wang, H. F., 90
Wang, H. N., 86f
Wang, H. Y., 98–99
Wang, J., 165t
Wang, K. M., 423–424, 442–448, 443t, 449
Wang, L., 355

Wang, N., 424–427, 449–450
Wang, S., 423–427, 430–435, 431t, 441, 449–450
Wang, X., 383
Wang, X.-D., 423–424, 430–437, 431t
Wang, X. J., 419–422, 424–427, 442–448, 443t, 449
Wang, Y., 238, 423–424, 430–435, 431t, 441, 442–448, 443t, 449
Wang, Y. L., 149
Wang, Y. Q., 99
Wang, Y. X., 99
Warburton, P., 419–422, 423–424, 452–453
Ward, J. F., 4–5, 10
Ward, R. J., 355–357
Ward, W. W., 292
Warshaw, D. M., 92
Watanabe, H., 340
Watanabe, T. M., 92
Waterman-Storer, C. M., 148–150, 151, 152–153, 292, 298, 303, 304–305
Waterman–Storer, C. M., 148–149, 151–152
Waters, P., 340
Waterston, R. H., 188, 189t
Watson, P., 89–90, 272–273, 283–285, 420t
Watts, P., 419–422, 423–424, 438, 440
Wax, A., 84
Wayne, R. O., 85
Wdliamson, L. S., 66t
Webb, W. W., 6–7, 61–62, 68, 110, 131f, 139–140, 141–142, 173–175, 229
Webster, A., 419–422, 423–424, 438, 440
Wedlich–Soldner, R., 166t
Wegener, J., 423–427, 430–437, 431t, 442–448, 443t, 449
Wehrmann, M., 231
Wei, A., 90
Weidemann, W., 165t
Wei, L.-N., 47
Weinstein, H., 364–365
Weinstein, J., 142
Weiss, A., 224
Weiss, M., 135, 139
Weiss, S., 84, 92–93
Weitsman, G., 110–111
Weller, M., 58–59
Welsher, K., 423–424, 442–448, 443t, 449
Welte, M. A., 96
Werb, Z., 166t
Wessels, D., 189t
Westbrook, T. R., 419–422, 420t
Westerberg, N., 423–424, 449–450
Weterings, E., 8–9
Wharton, C. W., 6–7, 8–9
White, M. R. H., 189t
Whorton, M. R., 364–365
Wichmann, J., 40–41, 222
Wickham, M., 141–142
Wiedenmann, J., 130, 164–166, 165t, 166t, 222–223

Wieland, F. T., 256
Wieschaus, E. F., 96
Wiesner, U., 423–424, 442–448, 443t, 449
Wiesner, U. B., 419–422
Wiles, C., 419–422, 423–424, 438, 440
Wiles, S., 273–274
Wilk, T., 90
Williams, B. D., 386–387
Williams, K. E., 164
Williamson, D. J., 224, 231
Williams, R. M., 110
Willig, K. I., 40–41, 222
Willner, I., 424–427, 428, 438, 439, 440, 441
Wilson, O., 90
Winoto, L., 152
Wiseman, P. W., 90–91
Wittig, B., 422
Wittmann, T., 189t, 292, 298, 301, 304–305, 308–309
Wolburg-Buchholz, K., 340–341
Wolburg, H., 340–341
Wolfbeis, O. S., 423–427, 430–437, 431t, 442–448, 443t, 449, 457–458
Wolf, E., 85
Wolfe, B. L., 364
Wolf, E. D., 166–167
Wolff, M., 165t
Wolf, K., 110
Wollman, R., 149–151, 392
Wolter, S., 67
Wong, C. C., 272–273
Wong, S. T. C., 189t
Woodcroft, B. J., 401t
Woods, L. A., 420t
Woods, R. A., 72–73
Woods, R. E., 44
Wood, W. G., 64t, 66t
Woolley, R., 423–424, 452–453
Wörz, S., 184, 186–187
Wotschadlo, J., 423–424, 442–448, 443t, 449, 452–453
Wouters, F. S., 110–111, 371
Wright, C. J., 238
Wright, E. G., 4–5
Wroblewski, T., 166–167
Wu, C. F., 424–427, 430–435, 431t, 436, 442–448, 443t, 449
Wu, D., 273–274
Wu, J., 423–424, 449–450
Wu, R., 166–167
Wurm, C. A., 42
Wu, S., 424–427, 449–450, 454
Wu, W., 424–427, 441, 442
Wu, X., 256–258
Wu, Y. I., 335
Wyatt, M. D., 100–101
Wydro, M., 166–167

X

Xiao, J., 424–427, 449–450
Xiao, L. H., 92–94
Xia, Y. N., 86f
Xie, G., 364–365
Xie, X. S., 90, 92, 274–276, 277, 285–287
Xiong, Y., 196
Xue, Y., 450
Xu, H., 419–422, 423–424, 428, 430–435, 431t, 436–437, 438, 439f, 440, 441
Xu, K., 424–427, 441, 442
Xu, M., 112
Xu, M. Y., 424–427, 454
Xu, X., 424–427, 449–450
Xu, X. H. N., 100–101
Xu, Y., 423–424, 451–452

Y

Yadavalli, V. K., 428, 438, 439, 440
Yaginuma, K., 379–380
Yajima, J., 92–93
Yakovlev, V. V., 283–285
Yamamoto-Hino, M., 165t
Yamamoto, J., 173
Yamamoto, N., 112, 358
Yamashita, H., 90–91
Yamauchi, K., 112
Yanagida, T., 87
Yan, D., 340
Yang, B., 90–91, 340–341
Yang, F., 78
Yang, G., 150–153, 157–158, 184
Yang, L., 44–45
Yang, M., 112
Yang, X., 5
Yang, Y., 164
Yano, A., 355, 358, 364
Yano, K., 8–9
Yano, Y., 355, 358, 364
Yao, X., 340
Yarar, D., 151
Yarmush, M. L., 308
Yasuda, R., 92–93, 370, 371
Yasumura, T., 340–341
Ye, F. M., 424–427, 442–448, 443t, 449
Yeh, E., 150–151
Yeh, R., 340–341
Yeung, E. S., 88, 89, 92–94, 99
Yildiz, A., 42, 47, 92
Yin, J., 357–358, 424–427, 428–429, 442, 449–450, 453
Yokomori, K., 8–9
Yong, K. T., 90, 99–100, 102
Yoon, J., 423–424, 449–450
York, A. G., 42
York, J., 93–94
Yoshida, M., 92–93
Youn, I., 419–422, 424–427, 455, 456–457
Yu, C. X., 99–100
Yue, Z., 113–114
Yu, H., 194
Yu, J. H., 166t, 423–424, 113t, 449, 457–458
Yu, L., 424–427, 441, 442
Yuseff, M. I., 166t

Z

Zaccheroni, N., 420t, 424–427, 449–450
Zaccolo, M., 322
Zacharias, D. A., 330, 373
Zador, Z., 340
Zadrozny, T., 164
Zamai, M., 112, 119–120, 123
Zamarron, A., 419–422, 423–424, 457–458
Zamvil, S. S., 340
Zanella, M., 84
Zare, R. N., 364–365
Zech, T., 224
Zedayko, T., 423–424, 442–448, 443t, 449
Zeigler, M. B., 64t
Zelenina, M., 340–341
Zelmer, A., 273–274
Zeng, F., 424–427, 449–450, 454
Zhai, Y., 149–150
Zhang, B., 5, 184, 196
Zhang, F., 84
Zhang, H., 340–341
Zhang, J., 272–273, 322–323, 325, 330, 370
Zhang, L., 423–424, 430–435, 431t, 441
Zhang, N., 92
Zhang, W., 224
Zhang, X. H., 5, 102
Zhang, Y., 47, 238
Zhao, J., 424–427, 449–450
Zhao, X., 256–258
Zhao, Y., 90
Zheng, G., 285–287
Zhong, Y., 423–424, 451–452
Zhou, J., 423–424, 451–452
Zhou, S., 424–427, 441, 442
Zhou, T., 424–427, 441, 442
Zhou, Y., 189t
Zhuang, X., 222
Zhu, D., 424–427, 449–450
Zhuo, L., 424–427, 441, 442
Zhu, W., 423–424, 451–452
Zimmer, C., 184, 196, 228
Zimmer, J. P., 424–427, 442–448, 443t
Zimmer, M., 292
Zipfel, W. R., 110
Zou, S. L., 86f
Zugates, G. T., 419–422
Zumbusch, A., 280–281
Zurn, A., 355, 357
Zweifel, D. A., 90

Subject Index

Note: Page numbers followed by *"f"* indicate figures, and *"t"* indicate tables.

A

Acousto-optical tunable filter (AOTF), 298
ACP. *See* Acyl carrier protein
Acyl carrier protein (ACP), 357–358
Agro-infilteration
  materials
    disposals, 167
    plants, 167–168
    reagents, 167
  protocol
    infilteration, 168
    infilteration, before, 168
    postinfilteration, 168
AKAR. *See* A-kinase activity reporter
A-kinase activity reporter (AKAR)
  affinity, PAABD, 330
  cell imaging
    donor and acceptor channels, KIBS, 333
    Hank's balanced salt solution buffer, 333
    imaging, CFP–YFP FRET pair, 333
  cellular expression
    DMEM, 332
    reporters, 332
  comparison, emission ratio change, 330, 331*f*
  controls
    cAMP-induced response, 334
    PKA phosphorylation, 333
  data analysis and FRET quantification, 334
  GTGGSEL link, 330
  microscope setup, 332
  unimolecular *vs.* bimolecular design, 331–332
  variants, CFP and YFP, 330
AOTF. *See* Acousto-optical tunable filter
AQP4. *See* Aquaporin-4
Aquaporin-4 (AQP4)
  CNS and NMO, 340
  diffusion
    amino acid sequence, 343–344, 343*f*
    AQP4-Mz, 345
    NMO autoantibody binding, 345
    primary astrocyte cell cultures, 344
    quantum dot single particle tracking, 343–344, 343*f*
    temperature-dependence data, 344
  GFP-AQP4 chimeras, dynamics and structure
    magnification TIRFM and line scans, 346*f*, 347
    OAP model, 346*f*, 347
    TIRFM, 345–347, 346*f*
  heterotetramers
    implementation, 347, 348*f*
    single-molecule analysis, 347
  imaging
    AQP4-M1 *vs.* AQP4-M23 diffusion, 342
    blue-native gel electrophoresis, cell lysates, 341–342, 341*f*
    FFEM cells transfected, 341–342, 341*f*
    remarkable greater mobility, 341*f*, 342
    SPT, 342
    TIRFM, 341*f*, 342
  Met-1 and Met-23 isoform, 340–341
  monomers, 340–341
  OAPs (*see* Orthogonal arrays of particles)
  photobleaching
    BN-PAGE and freeze-fracture electron microscopy, 349
    confocal fluorescence micrographs, 348*f*, 349
    plasma membrane, 340
  super-resolution imaging, AQP4 OAPs
    PA-GFP, 349–350
    PALM image, 348*f*, 349–350
Atomic force microscopy (AFM), 238

B

Bio-image semantic query user environment (Bisque), 394, 406
Biolistic bombardment, 168
Bisque. *See* Bio-image semantic query user environment

C

Calibration, FLIM
  description, 376–377
  device and materials, 377
  validation
    Coumarin 6 ethanol and HPTS, PB, 377–379, 378*f*
    lifetime value estimation, 377
cAMP-responsive element-binding (CREB) protein, 326

CARS. *See* Coherent anti-stokes Raman
    scattering
CCD cameras. *See* Charge-coupled device cameras
Cell membrane labeling
  chemical dyes
    change, anisotropy, 69
    CHO, 69
    DPH, 69
  fluorescent lipid analogues
    chromophores, 68
    structural basis, 67–68
  fluorescent protein engineering, 69–70
  microscopy, 67
  probes, lipid domains, 67
Cell tracking
  description, 186
  and lineage reconstruction, embryogenesis, 187f
  object and background, 185–186
  tools, 188–191, 189t
CFP. *See* Cyan fluorescence protein
Charge-coupled device (CCD) cameras
  high-resolution cooled, 301–302
  magnification determination, 32
  scientific grade-cooled, 293–294
Chemical specificity, live cell imaging
  cellular structures, 272–273
  experimental setup
    correlative CARS, 285
    dual femtosecond laser sources, 278–280
    dual frequency CARS, 283–285
    picosecond laser sources, 277
    tuning and spectral focusing, 280–283
  maximization, collection efficiency
    CARS, 285–287, 287f
    objective-condenser combinations, 285–287, 286t
    techniques, fluorescence, 285–287
  noninvasive
    CARS, 275–276
    microscopy, 277
    OCT, 273–274
    Raman scattering, 274–275
    SHG, 273–274
Chinese hamster ovary (CHO) cells
  β_2ARs, 358
  host cells, 361
  mGluR-expressing, 365–366
CHO cells. *See* Chinese hamster ovary cells
CLSM. *See* Confocal laser scanning microscope
Coherent anti-stokes Raman scattering (CARS), 90
  correlative, live cell imaging, 285
  dual frequency
    image contrast, 283–285
    spectral lineshape, 283–285
  energy level, 275–276, 275f
  microscopy, 277, 278f
  vibrational resonance, 275–276
Coiled-coil tag-probe labeling

disadvantages, 355
E3-tagged proteins and K probes
  confocal microscopy, 362
  fluorescence intensity, cell membranes, 363, 363f
  stock solutions, 362–363
fluorophore-labeled probe peptides, preparation
  DMF, 361
  phenyl-based columns, 362
  TFA, 361
GFPs, 354–355
helical wheel representation, E3/K3, 358, 359f
living cells, membrane proteins, 354f
methods, 355
oligomerization, receptor
  detection, 365–366, 366f
  glycophorin A G83I mutant, 365–366
  homo-oligomers and hetero-oligomers, 364–365
  multicolor labeling, advantage, 365–366
  protein–protein interactions, 364–365
posttranslational labeling, 354f, 355
principles
  enzymatic reactions, 357–358
  peptide-metal interactions, 357
  protein-ligand interactions, 355–357
receptor internalization
  mechanism, 364
  selective labeling, 364, 365f
tag-fused membrane proteins, design
  EGF receptors, 359–360
  plasmid map and DNA sequence, 360f
  tagged membrane proteins, expression, 360–361
Color recovery after photoconversion (CRAP)
  fluorescence values, 173–175, 176f
  mEosFP probes, 173–175
Comet assay-single-cell gel electrophoresis
  image capture and analysis, 21–22
  immunohistochemical staining method, 20–21
  irradiated cells transfer, microscope slide, 21
Complementary-metal-oxide (CMOS), 32–33
Completely spatially random (CSR), 231–232
Confocal laser scanning microscope (CLSM)
  fluorescence recovery, 304–305
  frame acquisition, 294
  high degree, automation, 170
  postinfilteration, 168
  ROI, 172
  SDC microscopy, 293–294
Correlative light-electron microscopy (CLEM)
  cell identification, EPON blocks, 211–212
  empty slot grid
    CVLEM, 214
    donor slot grid, water, 210f, 215
    and samples processing, 211f, 215
  immunolabeling, NANOGOLD, 207
  living cells and fixation

Subject Index 497

identification methods, 204, 205f
required materials, 203–204
sectioning samples, orientation,
  204, 206f
procedure, 202–203
resolution, 202
sample contrast, locating and embedding
  procedure, 209–211
  required materials, 208
sample orientation and EM sectioning,
  212–214
types, 202–203, 203f
CRAP. See Color recovery after photoconversion
Cuvette lifetimes, *in vivo* applications
and FLIM
  cell-to-cell variability, 74–76
  HA transmembrane domain, 78
  instrument response function, 76
  NBD-labeled lipids, 77
  spatial resolved images, 74
  fluorescence intensity, 70
  lipid domain, 71
  polyene probes, 70–71
  trans-parinaric acid
    global analysis, 73
    light absorption/autofluorescence, 72–73
Cyan fluorescence protein (CFP)
  AKAR2, 330
  laser beam, 264
  and YFP fluorophores, 78, 328
Cyclic voltammetry (CV), 244, 245

D

Deblurring and deconvolution
  microscope image data, 46
  Wiener filter, 46
Differential interference contrast (DIC)
  microscopy
  description, 88
  lateral and axial resolution, 89
  Nomarski type prism, 89
  staining, nanoprobes, 89
Diffusivity, tracking
  analysis, 197
  measures, 193–194, 193t
N-Dimethylformamide (DMF), 361
Diode-pumped solid-state (DPSS) lasers, 298
Dodecenyl succinic anhydride (DDSA), 208
DPSS lasers. See Diode-pumped solid-state lasers
Dulbecco's modified Eagle's medium
  (DMEM), 306
Dynamic light scattering (DLS), 429

E

Electrodeposition paint (EDP)
  EDP-insulated CF, 244
  solution, 244
Electron microscopy (EM)

analysis
  in vivo dynamics and ultrastructure, ER
    carriers, 216, 217f
  multiple labeling, ER-to-Golgi carriers,
    216, 218f
  required materials, 216
and sample orientation, sectioning
  CLEM, 212
  procedure, 212–214
  required materials, 212
Electron multiplication CCD (EMCCD)
  cameras, 301–302
EMCCD cameras. See Electron multiplication
  CCD cameras
Enzymatic reactions, 357–358
Epithelial sheet migration assay
  cell-cell contacts, 308–309
  cell monolayer, 308–309
  cytoskeleton structures, 308–309
  HaCaT keratinocyte, 309
  materials, 309
  preparation, live cell imaging chamber, 310
  sterile tissue culture, 309–310
  Trypsin-EDTA, 309–310

F

FACS. See Fluorescence activated cell sorting
FAQs. See Frequently asked questions
FFEM. See Freeze-fracture electron micrographs
FLIP. See Fluorescence loss in photobleach
Fluorescence activated cell sorting (FACS),
  305, 308
Fluorescence lifetime imaging microscopy
  (FLIM)
  acceptor and donor, 370
  benchmarking, 116
  description, 370
  detection, 370–371
  detector
    microscope design and TCSPC, 114, 115f
    technical parameters, 114–115
  documentation, support and quality,
    373–374
  drug uptake kinetics monitoring
    intensity matching, single-exponential
      sequence, 119–120, 121f
    nuclear DOX accumulation, exponential-fit
      analysis, 121–122
    nuclei migration dynamics and changes,
      119–120, 120f
    phasor analysis, multiple fluorescent species,
      122–123
  FD method, 371
  FRET (*see* Förster resonance energy transfer
    (FRET))
  intensity-based imaging, 386
  measurements
    analysis, FD, 376

Fluorescence lifetime imaging microscopy
  (FLIM) (cont.)
 calibration, 376–379
 FRET-standard proteins, 381–383
 testing, FRET-standard proteins, 379–381
 probe environment, 386–387
 protein-protein interactions, 383–386
 transfection approach, 386
 two-photon system testing, biological setting, 116–118
Fluorescence lifetime microscopy
 DOX, 113–114
 drug uptake kinetics monitoring, FLIM
  nuclear DOX accumulation, exponential-fit analysis, 121–122
  phasor analysis, multiple fluorescent species, 122–123
 FRET, 110–111
 live-cell and intravital studies, 111–112
 MMT cell spheroids preparation, collagen matrix, 112–113
 molecular imaging, 111
 monitoring, DOX uptake
  benchmarking, 116
  detector, 114–116
  microscope setup, 114
  testing, FLIM two-photon system, 116–118
  time-lapse and intensity detection, 118–119
 MPM, 110
Fluorescence lifetime spectroscopy and imaging
 complexity and dynamics
  dynamic composition/critical fluctuations, 61–62
  membrane model systems, 61
 cuvette lifetimes and FLIM, 70–78
 FRET, 61
 labeling cell membranes
  chemical dyes, 68–69
  fluorescent lipid analogues, 67–68
  fluorescent protein engineering, 69–70
 lipid domains, 58–61
 membrane organization, 58–59
 probe selection
  excitation/emission wavelengths, 63, 64t
  global analysis, 60f, 63
  membrane biophysical properties, 63–67, 66t
 time-resolved, 59–60
Fluorescence loss in photobleach (FLIP), 142–145
Fluorescence recovery after photobleach (FRAP)
 bleach mode, 135
 buffer composition, 130–131
 data normalization, 138–139
 equipment
  confocal microscope, 134
  fluorophore-tagged proteins, 133–134
  key parameters, 134
 and FLIP, 142–145

fluorescent probes
 GFP derivative, 128–129, 129t
 quantification, pH-dependence, 129–130
 SEP and eGFP, 129
half-time and mobile fraction
 description, 139
 nonspiny straight dendrites, 140
 recovery kinetics, 139–140
 SEP-GluR, 139
 two-dimensional diffusion model, 139
health and viability, cells, 132–133, 133f
hippocampal cultures, 134
imaging mode
 acquisition photobleaching, 135–136
 measures, 135–136
lateral mobility, 142, 143f
membrane spanning membrane, 128
osmolarity, 132
photobleaching and recovery, 130
protocol, 136–137
raw fluorescence data
 ImageJ plug-ins software, 137
 ImageJ selection tool, 138
 LOCI and ROI, 138
 LSM Toolbox, 138
soma, 140–142
temperature, 132
theory and analysis, 131f, 134
Fluorescence recovery after photobleaching (FRAP)
 exchange kinetics, AP1 and GGAs, 265–266
 fluorescence values, 176f
 mEosFP probes, 173–175
 TGN-endosome transport kinetics, 265
 use, 175
Fluorescence resonance energy transfer (FRET)
 Cy3 to Cy5, receptors, 357–358
 intermolecular, 355
 intramolecular, 357
 RG, TMR, 365–366
Fluorescent proteins (FPs)
 description, 164
 photoactivable and photoconvertible, 164–166
Fluorescent speckle microscopy (FSM)
 actin and tubulin polymers, 149–150
 binomial statistics, 148–149
 CCD and TIRF, 152
 choice, cells, 153
 depolymerization, 150–151
 description, 148
 dynamics, macromolecular structures, 149
 GTP cap, 149–150
 image analysis, 156–158
 kymograph and statistical analysis, 152–153
 labeled and endogenous ratio, 148–149
 live-cell imaging and setup
  description, 153–154
  materials, 154

Subject Index 499

low expression, proteins
  microinjected fluorophore-conjugation, 152
  tiled-array, 151–152
microtubule dynamics, 149–150
potential negative effects, 149
protein expression regulation, 153
slide-and-cluster model, 151
wide-field vs. spinning-disk confocal
  analysis, 154–155, 155f
  SNR and out-of-focus information, 154–155
Förster resonance energy transfer (FRET)
  acceptor and donor, 424–427
  alignment, electromagnetic dipoles, 372–373
  description, 110–111, 371
  efficiency, energy transfer, 371–372, 424–427, 453
  excitation and emission spectra, 372–373, 372f
  and FLIM measurements, 111–112
  and FPs, 373
  sensitized emission, primary acceptor, 450
  spectral overlap, 371–372, 372f
  standard proteins measurement, FLIM
    cell expression, 381–382
    polar plot analysis, 381–382, 381f
  standard proteins testing, FLIM, 379–380
    biological expressions, 379
    cells transfection, electroporation, 379–380
    TRAF domain, 379
FRAP. See Fluorescence recovery after photobleaching
Freeze-fracture electron micrographs (FFEM)
  and BN-PAGE, 341–342
  cell membranes, 340–341
Frequently asked questions (FAQs), 399
FRET. See Fluorescence resonance energy transfer; Förster resonance energy transfer
FRET-based biosensors
  AKAR, 330–334
  cellular and molecular events, 316–317
  classes and units, 317
  description, 316–317
  enzymatic modifications, 322–323
  fluorescent protein pair
    CFPs, 318
    description, 317–318
    donor emission and acceptor excitation, 318
    spectral considerations, 318–319
  GTPases and PKAs, 317
  PKAs, 323–327
  small GTPases
    activation reporters, 328–329
    coordination, migrating cells, 329
    description, 327–328
  unimolecular and bimolecular
    characteristics, 321–322

donor and acceptor, single-chain reporter, 320f, 321
GTPases, 322
sensitivity, 322
unit sensing
  biological monitors, 319
  modular designs, 319, 320f
  PAABD, 319–321
  posttranslational modifications, 319–321
  receiving and sensing segments, 319–321
vital decisions, 316

G

GDP. See Guanosine diphosphate
GEF. See Guanine nucleotide exchange factors
GFPs. See Green fluorescent proteins
Glucose oxidase (GOx), 436–437
Gold NPs (GNPs), 453
GPCRs. See G protein-coupled receptors
G protein-coupled receptors (GPCRs)
  allosteric regulatory mechanisms, 364–365
  HEK 293 cell membranes, 355–357
Green fluorescent proteins (GFPs)
  derivatives, fluorophores, 129t, 130, 317–318
  description, 128–129, 292, 316
  dichroic mirrors, Semrock, 298
  discovery and development, 354–355
Guanine nucleotide exchange factors (GEF), 327–328
Guanosine diphosphate (GDP), 327–328

H

High resolution imaging
  data acquisition and processing
    3D super-resolution microscopy techniques, 42
    ground state depletion (GSD), 41
    improvement, axial resolution, 40
    internal reflection microscopy, 39–40
    localization microscopy techniques, 42
    RESOLFT microscopy, 40–41
    SIM, 41
    super-resolution techniques, 40
  data, analysis
    accurate localization, fluorophores, 47
    distribution and colocalization, 48
    intensity and molecular quantification, 47
    motility statistics, 48–51
    segmentation and particle tracking, 47–48
  deblurring and deconvolution, 46
  denoising methods
    algorithms, 44
    effects, 44–45, 45f
    imaging system, 44
    patch-based algorithm, 44–45
  image restoration, 51
  light microscopy, 30–31

High resolution imaging (cont.)
  physical limitations
    Abbe diffraction limit, 31–32
    fluorophore properties, 34
    photon detection and signal to noise, 32–33
    temporal resolution limits, 34–35
  preparations, fluorescence
    complexity and sophistication, 37–38
    coverslip thickness correction, 38
    intelligent imaging innovations, 38
    microscope setup, 35–37
  temporal resolution, 31

I

Image acquisition
  CCD vs. Nipkow disk, 302, 303f
  EMCCD cameras, 301–302
  images comparison, 301–302, 302f
  PSF, 301
Instrument response function (IRF), 116
Intracellular protein dynamics imaging
  CCD cameras, 293–294
  CLS microscopy, 292–293
  dual disk design, 292–293
  fluorescence saturation, 294
  GFP, 292
  instrument design
    image acquisition, 301–302
    microscope, stage and environmental control, 296–297
    SDC scanner and illumination, 297–301
  PerkinElmer ultraview systems, 294–296
  SDC
    and fast photoactivation, 304–305
    microscope light path, 292–293, 293f
    photograph, microscope system, 294–296, 295f
    and TIRF, 303
  specimen preparation, 305–310
  spinning disk and multipoint confocal systems, 294

L

Laser-induced radiation microbeam technology
  cell culture and NIR multiphoton irradiation
    colocalization, 17–18
    comet assay-single-cell gel electrophoresis, 20–22
    materials and device, 15–16
    plating adherent cells, 16
    preparation, immunofluorescent staining, 18–19
    raster scanning optimization, 16–17, 17f
    real-time visualization, 18, 19f
    visualization, proteins, 18
  confocal imaging, 14–15

construction, light source
  Kerperian telescope, 12–13, 13f
  pulse broadening, 11–12
  SHG, 11–12
development, realtime induction, 25
DNA damage-repair dynamics
  colocalization, 23, 24f
  GFP technology, 22
  real-time studies, 22–23
DNA DSBs determination, 20–22
high-resolution DNA damage
  cellular repair, 8–9
  multielectron ionization, 10
  multiphoton irradiation wavelength, 8–9, 10t
multiphoton absorption
  quantized energy levels, 6–7, 7f
  square dependence, 6
multiphoton scanning system, 13–14
NIR, 5–6
polymerization processes, 26
pulsed laser sources
  biophysical research, 7–8
  high peak intensities, 8
radiation, mammalian cell, 5, 5f
radiobiological experiment, 25–26
spectral detection, 23–24
spectral detection setup, 9f, 10–11
SSB, 4–5
Laser sources
  dual femtosecond
    CARS microscopy, 278–279, 279f
    OPO, 278–279
    vibrational resonances, 278–279
  picosecond, 277
  single femtosecond
    downstream pulse shaping, 280
    WEU, 280
  tuning and spectral focusing
    chirp (spectral focusing), 281–282
    dual frequency CARS, 283–285
    Raman resonance, 282–283
LAT. See Linker for activation of T cells
Lentivirus-mediated stable expression
  lentiviral destination vector
    gateway cloning technology (Invitrogen), 306–307
    Stbl3 cells, 306–307
  materials, 306
  production
    adenovirus and 293FT cells, 307
    geneticin-free DMEM growth, 307
    harvest and pipetting cell, 308
    ViraPower packaging mix, 307
    VSV-G glycoprotein, 307
  reagents and protocol, 305–306
  reagents and protocols, 305–306
  stable cell line, 308

Subject Index

Light microscopy (LM), 202–203
Linker for activation of T cells (LAT)
  description, 224
  typical PALM acquisition, 229
Live-cell imaging. *See also* Chemical specificity, live cell imaging
  drosophila cells
    dGGA localization, Golgi structures, 266–267, 267f
    generation, S2 clones, 267–268
    LERP-positive transport carriers, 268–269
  mammalian cells
    cell culture units, 260, 260f
    dual-color imaging, 263–264
    exchange kinetics, AP1 and GGAs, 265–266
    FRAP analysis, 265
    interaction visualization, 264
    microscope system setup, 261–262
    time-lapse observations, 262, 263f
    transfection, HeLa, 260–261
    transport carriers, photobleaching, 262–263

M

Mannose 6-phosphate receptors (MPRs)
  CIMPR, 256–258
  cycling kinetics, 265, 266f
  transport kinetics, 269
Mean squared displacement (MSD), 341f, 342
  diffusivity, 197
  motion mode, object, 193–194
Methyl nadic anhydride (MNA), 208
Microscopy imaging and analysis, open source software
  automated image classification
    Enhanced CellClassifier and CellProfiler analyst, 409–410
    Micro-Pilot system and Murphy lab, 409–410
    vector machine/neural network, 409
  fluorescent image analysis, stages, 405, 405f
  high-throughput pipelines
    CellProfiler and Blobfinder, 408–409
    large-scale RNA interference, 408
  image capture
    micro-manager/μ manager, 405–406
  image processing and quantification
    FARSIGHT Toolkit and BioImageXD, 407–408
    ImageJ and Fiji, 407
    IMOD, 408
    JACoP, 407
    stand-alone software packages, 407
    V3D and ICY project, 407–408
  open licenses, 404
  storage
    Bisque, 406
    OMERO system, 406
  visualization, data analysis, and modeling
    automation and fluorophore technologies, 410
    Bioconductor and Cell ID, 411
    BioView3D, 410–411
    imaging revolution, 411–412
    OpenOffice and Octave, 411
    Tulip system, 411
    VANO and iCluster system, 410–411
    volume-rendering methods, 410
    Voxx and 3D Slicer, 410–411
Microscopy setup, mEosFP probes
  CLSM, 170
  glass filter cubes, 171
  lens, 170–171
  material and disposals, 171
  multipinhole iris diaphragm, 170
  ROIs, 170
Monomeric green-to-red photoconvertible Eos fluorescent protein (mEosFP)
  CRAP, 173–175
  emission spectra, 164
  fusion proteins expression
    stable transgenic creation, 169
    transient, 166–169
  optical highlighters, 164–166
  organelle fusion, 173
  organelle tracking, 173
  photoconvertible probes, 164–166, 165t
  post acquisition image processing and data creation, 176–177
  probes, visualization
    caveats, 172
    microscopy setup, 170–171
    photoconversion, 171–172
  proteins tracking, 173
  subcellular compartment, 164–166, 166t
Motility statistics and motion models
  dynamic cellular processes, 48–49
  kymograph (wave drawing), 49–50
  mean-squared displacement *vs.* time, 50–51
  particle stats, 49–50, 50f
Mouse mammary tumor (MMT) cell
  DOX uptake, 113–114
  spheroids, collagen matrix
    cell culture and reagents, 112
    generation, multicellular, 112–113
    incorporation, 3D, 113
MSD. *See* Mean squared displacement
Multiphoton laser-scanning microscopy (MPM)
  description, 110
  single-beam, 123
  time-lapse, 118–119

N

Nanoparticle PEBBLE sensors, live cells
  advantages, intracellular sensors construction, 419, 420t

Nanoparticle PEBBLE sensors, live cells (cont.)
  anion sensing, 454–455
  Ca^{2+} sensing, 452–453
  cell, defined, 418
  chemical/physical information, 460
  copper ion sensing
    description, 449–450
    detection, 450
    intracellular application, 450
  designs
    fluorescent reporters, 424–427
    hydrophobic NP, lipophilic components, 427–428
    structure, 422–423
    TYPE 1 PEBBLE, 260f, 423–424
    TYPE 2 PEBBLE, 263f, 424–427
  dissolved oxygen sensing
    confocal image, live C6 glioma cells, 436, 437f
    delivery methods, 436
    fluorescent, 430–435, 431t
    intracellular concentration, 436–437
    PBS, response, 436, 438f
    preparation, hydrophobic matrixes, 435–436
    role, 430
    TYPE 1 single PEBBLEs, 430–435
  enzymatic intracellular processes, sensing
    activities, enzymes, 457
    apoptosis, 456
    phosphorylation, 456–457
  Fe^{3+}, Mg^{2+}, Hg^{2+} and Pb^{2+}, 454
  fluorescent probe molecules, 418–419
  glucose sensing
    FROC, serum and in vivo measurements, 441
    TYPE 2 enzyme, 441
    TYPE 2 ET PEBBLEs, QDs, 442
  hydrogen peroxide sensing
    defined, 438
    H_2O_2 sensing, 439, 440
    induced H_2O_2 selectivity, 438, 439f
    TYPE 2 ET PEBBLEs, 439
  intracellular delivery methods, NPs, 419–422, 421f
  K^+, Na^+ sensing, 453–454
  nanotechnology, 422
  pH sensing
    endocytosis, 449
    fluorescent, 442–448, 443t
    QDs, 448
  physical properties
    electric field sensing, 459
    temperature sensing, 457–458
  preparation/characterization
    physicochemical, 429
    wet chemistry synthetic methods, 428–429
  real-time intracellular measurements, 459–460
  ROS, 440–441
  technique, 419–422

  zinc ion sensing
    designs, 451
    detection range, 452
    TYPE 1 core-shell, 451–452
National Institute of Health (NIH), 228, 233
Near-infrared (NIR)
  laser
    microbeam induction, 8–10, 11–13
    microbeam irradiation, 16
    pulsed, 5–6
  light
    emission, 435
    fluorescence, 422, 457
Neuromyelitis optica (NMO)
  autoantibody, 342
  sclerosis-like disease, 340
NMO. See Neuromyelitis optica
Noise equivalent power (NEP), 33
Nonfluorescent nanoparticle probes, live cells
  absorption-based microscopy
    bright-field, 87
    photothermal effect-based detection, 87–88
  biosensors
    cancer diagnosis and therapy, 99
    multiplexed detection, 99–100
    single particle sensors, 100
  cytotoxicity
    description, 100
    in vivo tracking, 100–101
  description, 84
  DIC microscopy, 88–89
  3D tracking technique, 102
  dynamic biological process, 84
  interferometric detection technique
    focused laser beam, 88
    signal amplification, 88
  multimodality imaging, 102
  NLO microscopy, 89–90
  Raman scattering, 84
  scattering-based microscopy
    advantages, 86
    dark-field, 85–86, 86f
    evanescent wave illumination setup, 87
    metal nanospheres, 85
    microscopic scattering, 85–86
    NIR, 85
    Rayleigh detection, 85
  small nanoparticle probe, 101
  SPT, 90–99
Nonlinear optical (NLO) microscopy
  CARS, 90
  description, 89–90
  weak photoluminescence, nanoparticles, 90

O

OAPs. See Orthogonal arrays of particles
OCT. See Optical coherence tomography

Subject Index

OME. *See* Open microscopy environment
Open microscopy environment (OME), 406
Open source software
  BioImage suite, 412
  commercial image analysis and storage solutions, 392
  common platforms, bio-image analysis
    ImageJ, CellProfiler, and OMERO, 404
    macro user interface plug-in, 400, 401*t*
    plug-ins advantages, 400–404
    Spot Tracker, 404
    syntax and structure, 400
  documentation, support, and quality
    description, 399
    GNU Linux, 399–400
    ImageJ and CellProfiler, 399
    NIH, 399
    poor documentation/support, 399
  flexible, innovative, and verifiable
    advantage and analysis protocol, 393
    black box and levels, 393
    fluorescent image, 392
    high-quality packages and new analysis methods, 393
  high-quality, 392
  microscopy imaging and analysis
    automated image classification, 409–410
    high-throughput pipelines, 408–409
    image capture, 405–406
    image processing and quantification, 407–408
    storage, 406
    visualization, data analysis and modeling, 410–412
  modularity, reuse, and interoperability
    Java programming language and cross-platform, 394
    object-orientated language, 394
    OMERO image database, 394
    project and code selection, 394, 395*t*
    Python and Jython, 394
  OsiriX imaging software, 412
  quantitative and automated analysis methods, 392
  rapid transfer, 412–413
Optical coherence tomography (OCT), 273–274
Optical parametric oscillator (OPO), 278–279
Optical techniques, imaging membrane domains
  cell culture and transfection procedure
    invitrogen NEON, 226
    materials, 225
    PBS, 226
  data analysis
    extraction, quantitative super-resolution data, 228
    Getis and Franklin's local point pattern, 232–233
    numbers extraction, 233–234
    PALM, 228–231
    Ripley's *K*-function, 231–232
  fluorescence microscopy, 222
  LAT, 224
  PALM, 222–223
  photoconvertible fluorescent protein mEos2, 222–223, 223*f*
  synapse formation and data acquisition
    day, imaging, 226
    fixed cell imaging, 227
    live-cell imaging, 227
    TIRF, 227
  T cell, 224–225
Orthogonal arrays of particles (OAPs)
  dynamics and structure, 345–347
  post-golgi assembly, 349
  super-resolution imaging, 349–350

P

PAABD. *See* Phosphoamino acid binding domain
Paraformaldehyde (PFA), 227
Particle tracking
  molecular complexes, 186
  spatiotemporal tracing and graph-based optimization, 186–187
  tools, 189*t*, 192
  vesicle dynamics, 188*f*
PBS. *See* Phosphate buffered saline; Phosphate buffer solution
Peptide-metal interactions
  arsenic and boron, 357
  metal cations, 357
  organic fluorophores, 357
Phosphate buffered saline (PBS), 226, 227
Phosphate buffer solution (PBS)
  hydrodynamic radius, 442
  oxygen concentration, 436–437
Phosphoamino acid binding domain (PAABD)
  AKAR, 330
  bimolecular KARs, 322
  description, 319–321
  phosphotyrosine binding, 324–325
Photoactivated localization microscopy (PALM)
  activated molecules, 223
  data analysis
    data table, 229–230, 230*f*
    live-cell, 230–231
    peak intensity, 229
Photomultiplier tubes (PMTs), 32–33
Photonic explorer for biomedical use with biologically localized embedding (PEBBLE). *See* Nanoparticle PEBBLE sensors, live cells
Point spread functions (PSF)
  circular, 301
  mask and full 2D Gaussian fit, 229
  sensitive camera, 222–223

Polyacrylamide (PAA)
  NP matrixes, enzyme, 428
  PEBBLEs, 419–422
Polydecylmethacrylate (PDMA), 427–428
Polyethyleneimine (PEI)
  cationic polymer, 436
  NP, 456
  PEBBLEs, 456–457
Protein kinases (PKAs)
  description, 323
  design, reporters
    activation, protein pair, 323–324
    Akt-1 activation, 323–324
    indicator, CREB, 326
    KARs, 324–325
    MAPKs, 325
    PAABD and BimKAR, 325–326
  oscillations, 326–327
Protein-ligand interactions, 355–357
Protein-protein interactions, FLIM-FRET
  defined, 383
  dimerization, 384–386
  domain plasmids and transfection, C/EBPα B Zip
    GHFT1 cells, 384
    materials, 383–384
PSF. See Point spread functions

Q

Quantum dots (QDs)
  fluorescence, 424–427, 451–452
  NPs, 422–423
  and pH dye, 427–428
  TYPE 2 ET PEBBLEs, 442

R

Raman scattering
  CARS, 275–276
  frequency of light, 274–275
  SERS, 274–275
Ras binding domain (RBD), 328
RBD. See Ras binding domain
Reactive oxygen species (ROS), 171
  description, 437–438
  hydrogen peroxide, 438
  PEBBLEs, 440–441
Region of interest (ROI)
  CLSM, 172
  organelles, 170
  photoconversion, 172
Resolution, conventional microscope
  Abbe diffraction limit and optical resolution, 31–32
  CCD, 32
  imaging system, 32
  point spread function (PSF), 31–32
  fluorophore properties

localization microscopy techniques, 34
photobleaching and photodamage, 34
STED, 34
photon detection and signal-to-noise
  CMOS, 32–33
  NEP, 33
  PMT, 32–33
  quantum efficiency, 33
  temporal resolution limits, 34–35
Ripley's K-function
  and Getis and Franklin's local point pattern analysis
    calculation, 231
    color map, 233
    CSR, 231–232
    L-function, 231
    L(r) value, molecule calculation, 232
    quantitative data extraction, 232–233
ROI. See Region of interest
ROS. See Reactive oxygen species

S

Scanning electrochemical microscopy (SECM).
  See Shear-Force-Based Constant Distance Scanning Electrochemical Microscopy
Scanning probe microscopy (SPM), 238
SDC microscopy. See Spinning disk confocal microscopy
Second harmonic generation (SHG)
  detect collagen, 273–274
  live cells, chemical specificity, 285–287
  microscopy, 277
SERS. See Surface enhanced Raman spectroscopy
Shear-Force-Based Constant Distance Scanning Electrochemical Microscopy (SF-CD-SECM)
  applications, live cell studies
    cellular respiration, 252–253
    hormones and neurotransmitters, mechanisms, 248–250
    in vitro cell system, 248
    local noradrenalin/adrenalin release measurements, 250–252, 250f
    local NO release measurements, 250–252, 250f
    topology and local oxygen measurements, 252–253, 252f
  execution, cell topography and activity imaging
    feedback distance control, 248
    live cell, 247, 249f
    optical microscope, 247
    resonance frequency, 247
  flexible CF, chemical release measurements
    materials and equipment, 243
    probe preparation, 243–244

Subject Index

flexible nickel phthalocyanine-modified CF, 246
flexible Pt-modified CF, O_2 measurements
   materials and equipment, 244
   probe preparation, 244–246
   probes geometry and surface modification, 242, 243f
   system requirements, 240–242
SHG. *See* Second harmonic generation
Signal particle tracking (SPT)
   human health and bioengineering, 90–91
   membrane diffusion
      dark and bright-field microscopies, 91
      DOPE, 91
      high temporal resolution, 91
      techniques, 91
   membrane process, 90–91
   motor proteins
      linear, *in vitro,*, 406
      xenopus melanophores, 92
   plasmonic ruler, 98–99
   rotation and orientation
      ATP hydrolyzation, 96
      dark-field microscope, 93–94
      description, 92–93
      DIC images, A549 cell, 96–98, 97f
      DIC microscopy, 94
      dynamics and gene delivery vectors, 96–98
      gold nanorod, 2D space, 94, 95f
      molecule fluorescence polarization-based methods, 92–93
      plasmonic gold nanorods, 93
      SPORT, 94
      surface functional groups, 96–98
Signal-to-noise ratio (SNR) tracking
   image quality, 195
   preprocessing, 196
   verification, 196
Single live cell topography and activity imaging
   CD mode scanning, SECM, 239–240
   cellular scanning electrochemical microscopy, 238, 239f
   SF-CD-SECM, live cell studies
      applications, 248–253
      execution instructions, 246–248
      system requirements, 240–242
      tailored vibrating scanning probes, 242–246
Single particle tracking (SPT)
   AQP4-M1 diffuses, 342
   quantum dot, 344
Single strand breaks (SSB), 4–5
SNR. *See* Signal-to-noise ratio
Spinning disk confocal (SDC) microscopy
   and fast photoactivation, 304–305, 304f
   scanner and illumination
      AOTF, 298
      bandpass filters, 299–300

      dichroic mirrors and emission filters, 298, 299t
      DPSS lasers, 298
      Gaussian beam profile, 298
      live cell imaging system, 299–300
      Nipkow disk, 297–298
      shutter control, 300–301, 300f
      streaking pattern, 297–298, 297f
      transistor-transistor logic signal, 300–301
      Yokogawa spinning disk heads, 298
   and TIRF, 303
SPT. *See* Signal particle tracking; Single particle tracking
Stimulated emission depletion (STED)
   GSD, 41
   uses, 40–41
Structured illumination microscopy (SIM), 41
Surface enhanced Raman spectroscopy (SERS), 274–275

T

Tagged membrane proteins expression
   eukaryotic cells, 360–361
   glass bottom dish, 361
   host cells, 361
TFA. *See* Treatment with trifluoroacetic acid
TGN-endosome trafficking, mammalian and drosophila cells
   live-cell imaging
      clathrin coat formation, 269
      dual-color imaging, 263–264
      exchange kinetics, AP1 and GGAs, 265–266
      FRAP analysis, 265
      generation, S2 clones, 267–268
      interaction visualization, 264
      LERP-positive transport carriers, 268–269
      microscope system setup, 261–262
      simple time-lapse observation, 262
      transfection, HeLa cells, 260–261
      transport carriers, photobleaching, 262–263
   lysosomal enzyme receptor protein (LERP), 258
   model, post-Golgi trafficking, 256–258, 257f
   molecular tools
      TGN-endosome transport, 258, 259t
      YFP, 259–260
   MPR, 256–258
Time-correlated single photon counting (TCSPC)
   data, 116–118
   detectors and phasor analysis, 123
   single-point laser-scanning, 111–112
TIRF. *See* Total internal reflection fluorescence

Total internal reflection fluorescence (TIRF)
  localization microscopy, 222
  microscopy (TIRFM)
    cells expressing, 342
    x, y-resolution, 347
    and PALM, 234
    and SDC, 303
Tracking
  analysis
    aggregation, 197
    diffusivity, 197
    velocity, 197
  cell (see Cell tracking)
  computational approach, 187–188
  digital imaging process, 184–185
  imaging
    dimensionality, 195
    frame rate, 196
    quality, 195
  live imaging, dynamic process, 184
  measures
    diffusivity, 193–194, 193t
    morphology, 194–195
    motility, 192–193, 193t
    velocity, 193t, 194
  particle (see Particle tracking)
  percentage, PubMed database, 185f
  preprocessing, 196
  recognition and association, objects, 185
  time-lapse imaging and longitudinal examination, 184
  tool selection, 196
  verification, 196
Trans-Golgi network (TGN).
  See TGN-endosome trafficking, mammalian and drosophila cells
Treatment with trifluoroacetic acid (TFA), 361

U

Uniform resource locator (URL), 406
URL. See Uniform resource locator

V

Velocity, tracking
  analysis, 197
  measures, 193t, 194
Visualization Toolkit (VTK), 410
VTK. See Visualization Toolkit

W

Wavelength extension unit (WEU), 280
WEU. See Wavelength extension unit

Y

Yellow fluorescent protein (YFP)
  and CFP, 78, 373
  FRET pairings, 373
  insertion, 259–260
YFP. See Yellow fluorescent protein

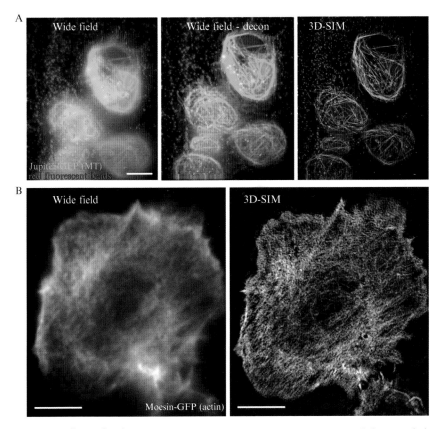

Graeme Ball et al., Figure 2.1 Super-resolution 3D-SIM imaging of the cytoskeleton in living *Drosophila* macrophages captured with an OMX-Blaze microscope (manufactured and distributed by Applied Precision, a GE Healthcare Company). (A) Microtubules (Jupiter GFP) and endocytosed red fluorescent beads. A single time point is shown from a time lapse movie of 11 frames: Conventional wide-field, deconvolved, and SI-reconstructed images derived from the same data set are compared. Each 3D-SIM image requires 240 images/μm of Z depth (15 images/slice, 8 slices/μm) captured at a rate of about 1 s/μm. (B) Actin cytoskeleton in a well-spread living macrophage (Moesin GFP). Wide-field and SI-reconstructed images derived from the same data set are compared as a projected Z series of 0.5 μm from a single time point. Resolution in 3D-SIM was about 120 nm, wide-field 275 nm, and deconvolved wide-field 250 nm.

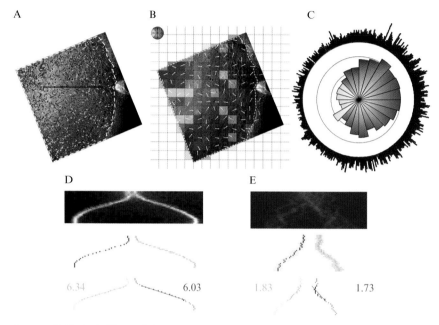

Graeme Ball et al., Figure 2.3 ParticleStats: An open source software package for the analysis of intracellular particle motility. Panels A–C summarize EB1:GFP tracks that mark growing microtubule plus-end directionality in a stage 8 *Drosophila* oocyte. (A) ParticleStats:Directionality track plot, where colors represent directionality as for the wind map. (B) "Wind map" representing sum of track directions in each square, where colors correspond to the four sectors shown top left. (C) "Rose diagram" summarizing overall directionality, where each outer dot represents a track. (D) ParticleStats:Kymograph used to create a kymograph summarizing the movement of eight aligned kinetochore pairs during meiosis in a *Drosophila* embryo. (E) Kymograph for a mutant with poor kinetochore synchrony.

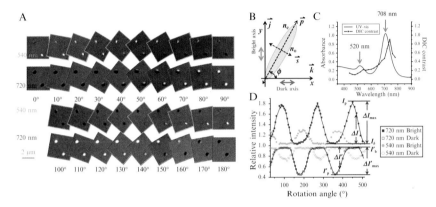

Gufeng Wang and Ning Fang, Figure 4.2 Gold nanorod orientations in 2D space. (A) DIC images of two 25 nm×73 nm gold nanorods at different orientations in 2D space. The same nanorods were illuminated at their transverse (540 nm) or longitudinal plasmonic resonance wavelengths (720 nm). The gold nanorods were positively charged and physically adsorbed on a negatively charged glass slide and submerged in deionized water. The glass slide was fixed on a rotating stage that allows 360° rotation. (B) Definition of the 2D orientation (*azimuth angle* ϕ of a nanorod with respect to the polarization directions of the two illumination beams in a DIC microscope. (C) UV–vis spectrum of the gold nanorod suspension in deionized water (red) and DIC spectrum of an immobilized, randomly oriented gold nanorod (blue). The DIC contrast is defined as the difference between the maximum and the minimum intensities divided by the average local background intensity. The two DIC peaks are red-shifted compared to their plasmon resonance wavelengths. (D) Periodic changes of the bright/dark intensities of a gold nanorod when rotated under a DIC microscope and illuminated at the two DIC peak wavelengths. All intensities are relative to the background level. The periodic patterns at these two illumination wavelengths are shifted by ∼90°, consistent with the relative orientation between the transverse and the longitudinal plasmonic resonance modes. Reprinted with permission from Wang *et al.* (2010b). Copyright 2011 American Chemical Society.

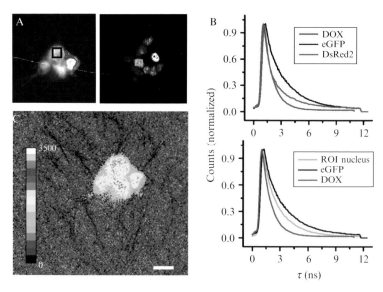

Gert-Jan Bakker et al., Figure 5.2 Fluorescence lifetime analysis of dual-color MMT (MMT-DC) breast cancer cells cultured in a 3D collagen matrix. (A) Summed intensity images of cells before (left) and after (right) 90 min incubation with DOX (5 μg/ml). (B, top panel) Photon-arrival-time histograms of eGFP-H2B in the nucleus and DsRed2 in the cytosol of MMT-DC cells using ROIs depicted in (A). (B, bottom panel) Lifetime histogram from the nucleus of MMT-DC cells after addition of DOX using the ROI in (A), right image. As a reference, a DOX stock solution in PBS (2 mg/ml) was used for the DOX lifetime (blue lines). (C) TCSPC lifetime image from (A) generated by fitting the decay curve at every pixel of the image using a single-exponential function. Lifetimes are presented using false-color coding (ns). (A, C) A single section from a 3D TCSPC datastack. Scale bar: 20 μm.

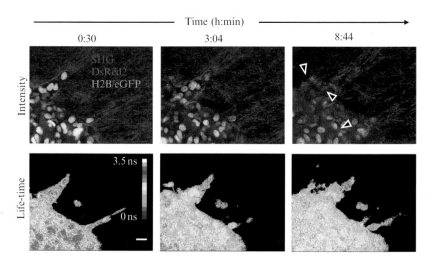

Gert-Jan Bakker et al., Figure 5.3 Simultaneous acquisition of fluorescence intensity (top) and TCSPC lifetime images based on single-exponential fit from a live-cell culture. Time-lapse microscopy of a spheroid of MMT-DC cells invading a 3D collagen matrix before (left panel) and after treatment (middle and right panels) with DOX (see Supplementary Movie S3, http://www.elsevierdirect.com/companions/9780123918574) followed by per-pixel analysis of the spectrally unseparated TCSPC stack. False-color range in bottom panel: 0 to 3.5 ns. Scale bar: 20 μm.

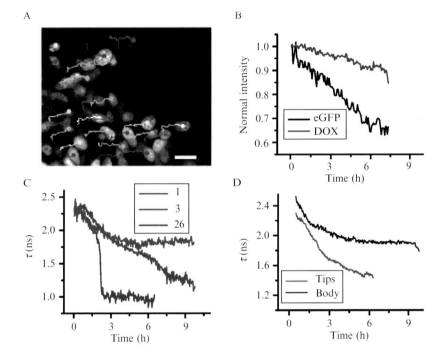

Gert-Jan Bakker et al., Figure 5.4 Migration dynamics of nuclei and changes in lifetime obtained by timelapse imaging. (A) Trajectories of moving nuclei, obtained by manual tracking of the H2B-eGFP intensity from the nucleus (Fiji distribution of ImageJ software; http://fiji.sc/). Scale bar: 20 mm. (B) Bleaching of individual fluorophores, detected as mean green and red intensities of all nuclei as a function of time. Images were obtained after saturation with DOX (26 h after addition to the culture medium). (C) Fast, intermediate, and slow uptake of DOX by individual nuclei. Mean lifetimes from individual nuclei in the spheroid tip (cells 1 and 3 in (A)) and spheroid body (cell 26 in (A)). (D) The average lifetime of all nuclei present in collective invasion zones, compared with nuclei located in the main mass of the spheroid. Images were recorded and analyzed at 5-min intervals.

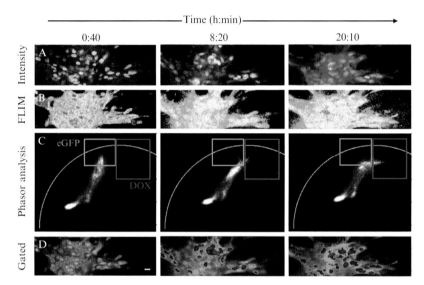

Gert-Jan Bakker et al., Figure 5.5 Matching intensity with single-exponential FLIM sequence with phasor analysis and reverse gating, in order to identify functional subregions. DOX uptake by an MMT-DC cell spheroid, embedded in a 3D collagen matrix, was monitored for 17h (see Supplementary Movie S5, http://www.elsevierdirect.com/companions/9780123918574). (A) Intensity images, eGFP (green), DsRed2 and DOX (red), collagen (SHG; blue). (B) TCSPC image based on single-exponential fit from the spectrally unseparated TCSPC stack (false-color range 0 to 3.5ns). (C) Phasor plot derived from the TCSPC stack (LaVision Phasor analysis software, V4.0.207). Green ROI includes longer lifetimes of eGFP and a DOX degradation product; red ROI includes shorter lifetimes from native DOX. (D) Gated image from the phasor population analysis. Remaining pixels represent the summed TCSPC (intensity) image. Scale bar: 20μm.

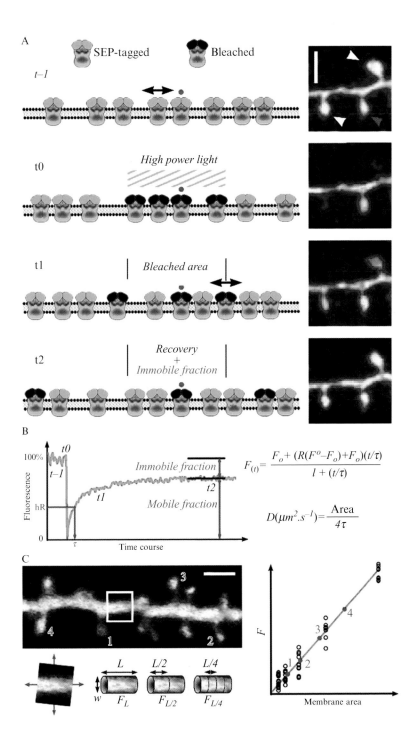

(Continued)

Inmaculada M. González-González et al., Figure 6.1 Fluorescence recovery after photobleaching, theory, and analysis. (A) Schematic representing successive steps of a FRAP experiments using SEP-tagged proteins. A fraction of tagged proteins undergoes motion in the plasma membrane, while another fraction is clustered and thus immobilized (red dot). Left column, simulated FRAP images, t−1 is the prebleach step, t0 is the bleach step (see black bleached proteins), t1 is the initial recovery due to lateral diffusion (black double arrow), t2 corresponds to the steady state late recovery where only immobile proteins (red dot) remains within the bleached area. The right-hand panels show corresponding stages of FRAP in dendritic spines. In the top panel, the white arrows indicate the bleached spines and the red arrow a control spine. The scale bar is 1 μm. (B) A typical FRAP recording and formulas for curve fitting (according to Feder et al. 1996) and diffusion coefficient calculus (see analysis section). (C) Complex membrane area approximation using calibration curves. Dendrites with spines of membrane-anchored eGFP-expressing neuron. A nonspiny region of the shaft is cropped and aligned, and fluorescence is measured for various lengths (L). Using the width (w) of the shaft, the area of a corresponding cylinder is computed. Left, fluorescence is plotted versus calculated area to produce a calibration curve. Membrane area of labeled spines (1–4) can be read on the curve.

A Proteins able to move laterally in the membrane

B Cross-linking with primary antibody

C Cross-linking with secondary antibody

D Antibody Fab fragments do not block lateral diffusion

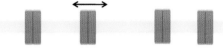
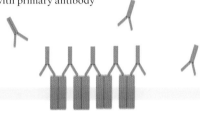
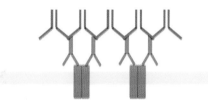

Inmaculada M. González-González *et al.*, Figure 6.4 Antibody cross-linking of surface proteins. (A) Schematic view of the plasma membrane (yellow) with transmembrane proteins (red) able to move lateral in the plane of the membrane. (B) Addition of excess antibody causes cross-linking of proteins due to the divalent binding sites of the antibody, thus restricting lateral motion. (C) Cross-linking can be enhanced by the addition of a secondary antibody. (D) Antibody Fab fragments are monovalent and therefore can cause cross-linking. Therefore, Fab fragments can be used as a control to check that antibody binding itself does not influence lateral movement.

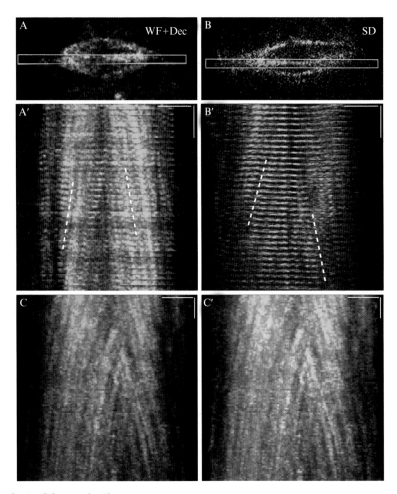

Marin Barisic et al., Figure 7.1 Comparative analysis of FSM by wide-field and spinning-disk confocal. (A–A′ and B–B′) FSM of K-fibers from bioriented chromosomes of *Drosophila* S2 cells stably expressing low levels of GFP-α-tubulin (green) and CID-mCherry (red) obtained, respectively, by wide-field followed by deconvolution (WF+Dec) and spinning-disk confocal (SD). Kymographs show two-dimensional (space×time) representation of selected regions in A and B. Note the poleward flux of tubulin subunits. Horizontal bars: 5μm; vertical bars: 30s. Figure A and A′ were adapted from Matos *et al.* (2009). (C–C′) Comparison between conventional kymograph (C) and chromo-kymograph (C′) of a metaphase to early anaphase spindle in *Drosophila* S2 cells stably expressing low levels of GFP-α-tubulin. Horizontal bars: 5μm; vertical bars: 1min.

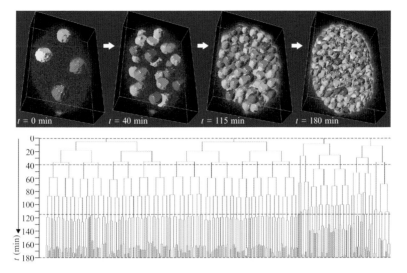

Erik Meijering et al., Figure 9.2 Cell tracking and lineage reconstruction for studying embryogenesis. The top row shows four time points of a 3D+t fluorescence microscopy image data set of a developing *Caenorhabditis elegans* embryo, starting from the 4-cell stage until approximately the 350-cell stage, with the segmentation and tracking results (surface renderings with arbitrary colors) overlaid on the raw image data (volume renderings). In this case, a level-set based model evolution approach was used for segmentation and tracking, modified from Dzyubachyk *et al.* (2010a). The bottom graph shows the lineage tree automatically derived from the tracking results, with the horizontal guidelines (red, dashed) corresponding to the four time points.

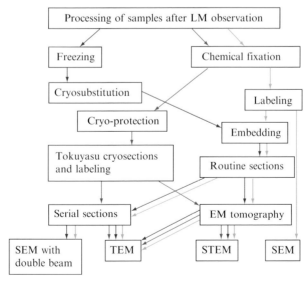

Alexander A. Mironov and Galina V. Beznoussenko, Figure 10.1 Main types of CLEM. Only steps following the LM or fluorescence microscopy (FM) observations are shown. More details see Fig. S1. Green arrows, CLEM based on chemical fixation, immunolabeling, and embedding or scanning EM analysis (Polishchuk *et al.*, 2000). Red arrows, CLEM based on chemical fixation, immunolabeling, and Tokuyasu cryosections (van Rijnsoever *et al.*, 2008). Blue arrows, CLEM based on immunolabeling, quick freezing, cryosubstitution, and epoxy resin embedding (Verkade, 2008).

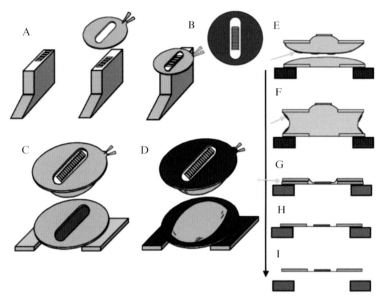

Alexander A. Mironov and Galina V. Beznoussenko, Figure 10.4 Picking up of serial sections with the donor slot grid from the water. (A) Left. Serial sections in the glass bath. Center. Approaching of the donor grid. Right. Orientation of the slot grid related to the position of serial sections within the surface of the bath. (B) Correct positioning of the grid and serial sections. (C) Touching of the water with the empty slot grid leads to the capturing of sections together with the droplet of water by the transfer grid. The acceptor slot grid with carbon-coated formvar film should be placed on the scotch holders (the parallel holders covered with scotch). (D) In order to avoid the hit of dirt on the sections, it is useful to place a very small droplet of distilled water on the acceptor grid that is covered with the formvar film. (E, F). After placement of the transfer grid with sections and droplet of water on the acceptor grid covered by a small droplet of water as well the dirt appeared on the lateral surface of the droplet and after elimination of water with a filter paper sections appeared being attached to the formvar (G). (H) The transfer grid should be eliminated. (I) The acceptor grid should detached from the scotch holder. Green arrows in E–G show dirt that is eliminated from the section during this manipulation being shifted to the lateral surface of grids.

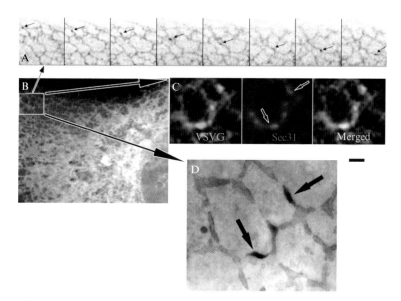

Alexander A. Mironov and Galina V. Beznoussenko, Figure 10.6 *In vivo* dynamics and ultrastructure of individual ER carriers studied using CLEM. Cells transfected with the temperature-sensitive G protein of vesicular stomatitis virus that is tagged with green fluorescence protein were placed at 40 °C for 16h to block folding of this protein and, thus, to prevent its exit from the ER. Then, cells were shifted to the permissive temperature 32 °C and examined under the confocal microscope in 10min after the shift. Video recording was performed during 3min, and then cells were fixed and prepared for immuno EM. The G protein was labeled with antibody against folded protein and then with monovalent Fab fragments of antibody against the primary antibody that were conjugated with peroxidase/DAB. (A) The consecutive frames of the video-film examining the behavior of the ER–to-Golgi carrier (arrows). (B) Cells were fixed. (C) Cells were labeled for VSVG (green) and Sec31 (red). (D) Identification of the very same area in EM images. VSVG was labeled with DAB (thick arrows). Bars: 2 μm (A), 4 μm (B), 1 μm (C), and 0.5 μm (D).

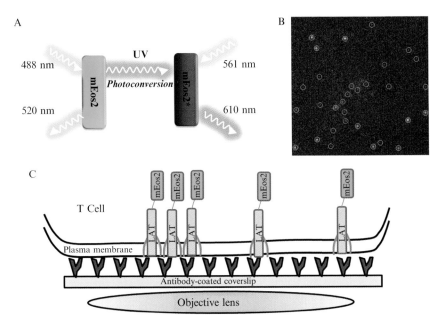

Dylan M. Owen *et al.*, Figure 11.1 (A) The photoconvertible fluorescent protein mEos2 is green in its native form; being excited at 488nm and emitting fluorescence around 520nm. Upon irradiation with UV light at 405nm, the protein converts to a red-fluorescent form. In this state, it is excited at 561nm and emits around 610nm. (B) When imaged under TIRF illumination through a red filter, individual PSFs originating from single mEos2 molecules can be observed and localized. (C) The mEos2 fluorescent protein is fused to LAT and the construct is transfected into Jurkat T cells where it localizes to the plasma membrane. The T cells can then be made to form immunological synapses by placing them on activating antibody-coated glass coverslips.

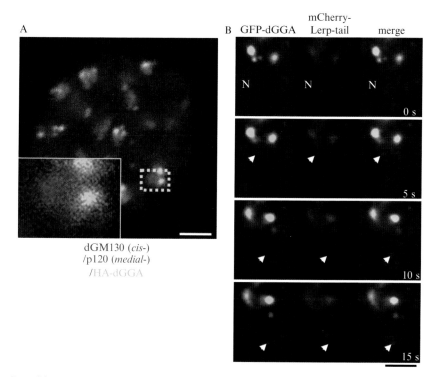

Satoshi Kametaka and Satoshi Waguri, Figure 13.5 (A) dGGA localization in the punctate Golgi structures in the S2 cells. S2 cells transiently expressing HA-dGGA were immunostained with anti-HA (green), anti-p120 Golgi protein (*medial*-Golgi, red), and anti-dGM130 (*cis*-Golgi, blue). (B) Dual-color imaging of GFP-dGGA and mCherry-LERP-tail (Table 13.1). Transport carriers emerging from the Golgi structure (arrowheads) were subjected to the time-lapse imaging with 5 s intervals. N, nucleus. Modified from Kametaka *et al.* (2010), bars: 2 μm.

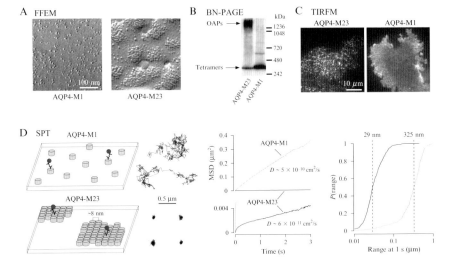

A. S. Verkman et al., Figure 17.1 Approaches to visualize AQP4 and OAPs. (A) Freeze-fracture electron micrographs of the plasma membrane P-face of COS-7 cells expressing the M1 and M23 isoforms of AQP4. (B) AQP4 immunoblot following Blue-native gel electrophoresis of cell lysates from AQP4-expressing COS-7 cells. (C) Total internal reflection fluorescence micrographs of GFP-AQP4 chimeras. (D) (left) Schematic showing the organization of AQP4 tetramers (left) and examples of single particle trajectories of Qdot-labeled AQP4 molecules in the plasma membrane of AQP4-expressing COS-7 cells. Each cylinder represents one AQP4 tetramer in which a subset of AQP4 molecules are labeled with quantum dots (red) for single particle tracking. (Center) Combined mean squared displacement (MSD) versus time plots and averaged diffusion coefficients for AQP4-M1 (gray) and AQP4-M23 (black) in COS-7 cells. (Right) Cumulative probability distribution of range at 1 s (P(range)) deduced from SPT measurements, with dashed lines indicating median range. Adapted from Crane et al. (2008) and Tajima et al. (2010).

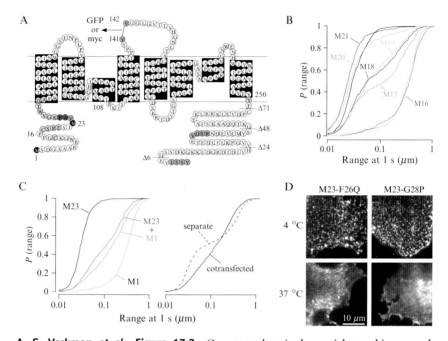

A. S. Verkman et al., Figure 17.2 Quantum dot single particle tracking reveals determinants of AQP4 OAP assembly. (A) AQP4 sequence and topology showing site of GFP or epitope (myc) insertion in the second extracellular loop. Black: Met-1 and Met-23 translation initiation sites; blue: residues where mutations did not affect OAP assembly; red: mutations disrupt OAPs; pink: mutations mildly disrupt OAPs; yellow: mutations reduce plasma membrane expression; green: C-terminal PDZ-binding domains. (B) P(range) for indicated AQP4 truncation mutants. (C.) OAP modulation by coexpression of M1-AQP4 and M23-AQP4. (Left) P(range) for cells transfected with M23 only (black) or M1 only (gray), or cotransfected with M23 (red) and M1 (green) at M23-to-M1 ratios of 1:1. (Right) P(range) comparing M1/M23 cotransfection (solid) versus "separate" (dashed), computed by summing P(range) curves for separate transfections. (D) TIRFM of Alexa-labeled AQP4 in cells expressing M23-F26Q or M23-G28P and fixed at 4 or 37 °C. Adapted from Crane and Verkman (2009a, 2009b) and Crane et al. (2009).

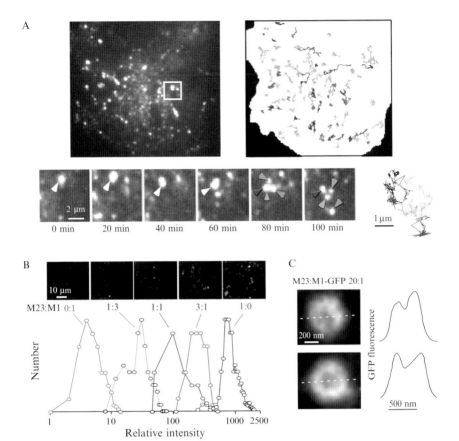

A. S. Verkman et al., Figure 17.3 OAP dynamics and structure revealed by TIRFM of GFP-AQP4 chimeras. (A) TIRFM image (left) showing distinct fluorescent spots in cells expressing M23-AQP4, corresponding to OAPs, with deduced single OAP trajectories over 3h shown at the right. (Bottom) High magnification of boxed region showing spontaneous OAP disruption events; trajectories of original OAP (black) and daughter OAPs (red, green, yellow, blue) shown at the right. (B) U87MG cells were transfected with GFP-M23 and GFP-M1 AQP4 at indicated ratios. Representative TIRF micrographs show fluorescent spot (top). Deduced number histograms of single-spot fluorescence (background-subtracted, area-integrated intensities), proportional to OAP size, shown at the bottom. Unity represents the intensity of monomeric GFP. (C) U87MG cells were transfected with GFP-M23 and (untagged) M1 AQP4 at a ratio of 20:1. TIRFM of two large AQP4 aggregates (left), showing relative concentration of fluorescence at the periphery. Line profiles (dashed white lines at the left) shown at the right. Adapted from Tajima *et al.* (2010) and Jin *et al.* (2011).